AMERICA'S Top Computer and Technical Jobs

Detailed Information on 112 Major Jobs at All Levels of Education and Training

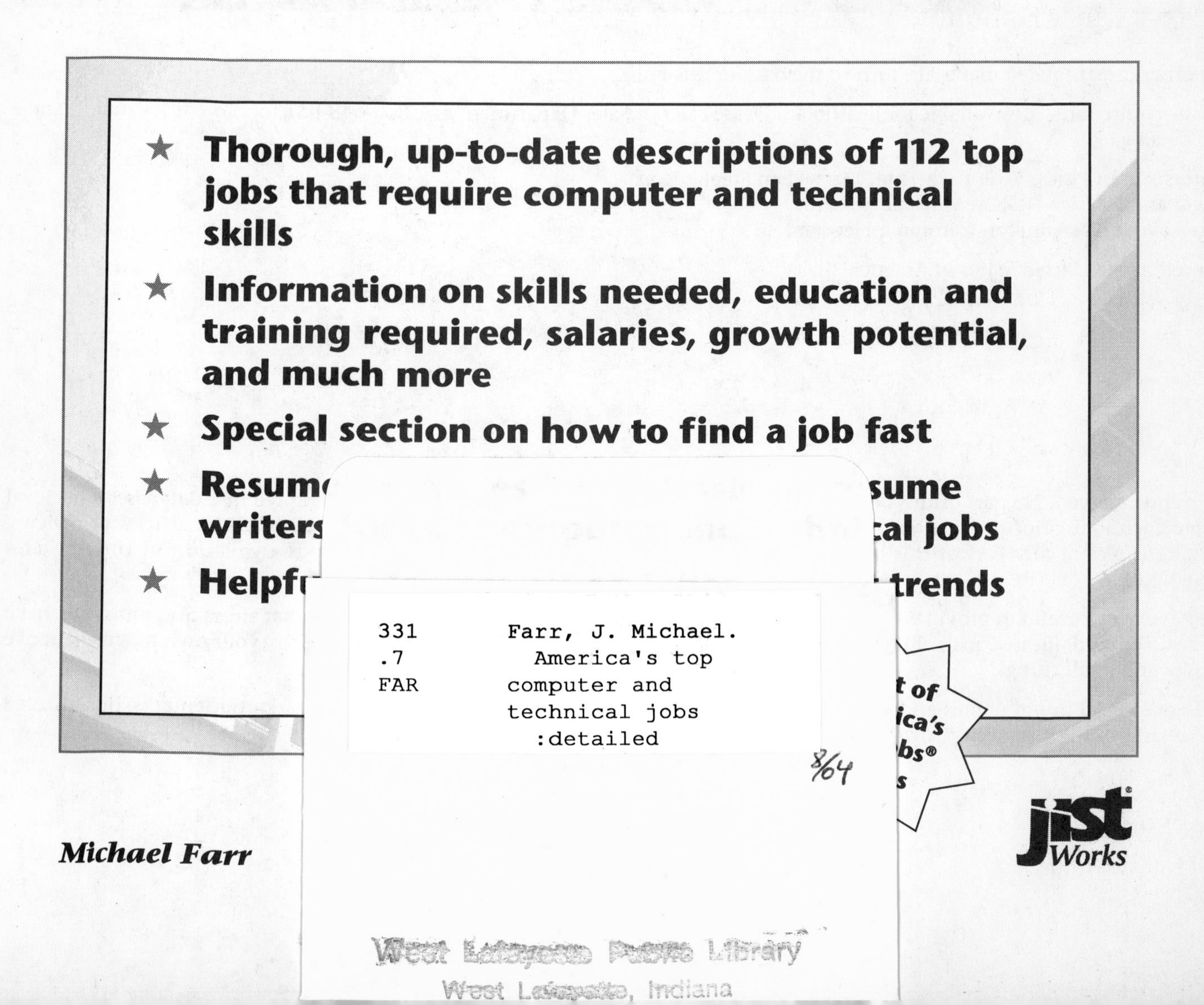

- ★ Thorough, up-to-date descriptions of 112 top jobs that require computer and technical skills
- ★ Information on skills needed, education and training required, salaries, growth potential, and much more
- ★ Special section on how to find a job fast
- ★ Resum[illegible] sume writers [illegible] cal jobs
- ★ Helpfu[illegible] trends

[illegible] t of [illegible] ica's [illegible] bs® [illegible] s

Michael Farr

JIST Works

America's Top Computer and Technical Jobs

Detailed Information on 112 Major Jobs at All Levels of Education and Training

Published by JIST Works, an imprint of JIST Publishing, Inc.
8902 Otis Avenue
Indianapolis, IN 46216-1033
Phone: 800-648-JIST   Fax: 800-JIST-FAX
E-mail: info@jist.com   Web site: www.jist.com

Visit our Web site for more details on JIST, free job search information, book excerpts, and ordering information on our many products!

Some other books by Michael Farr:

The Very Quick Job Search
The Quick Resume & Cover Letter Book
America's Top Resumes for America's Top Jobs
Getting the Job You Really Want
Best Jobs for the 21st Century (LaVerne L. Ludden, coauthor)

Other books in the America's Top Jobs® series:

America's Top 300 Jobs
America's Fastest Growing Jobs
America's Top Jobs for People Without a Four-Year Degree
America's Top Jobs for College Graduates
America's Top Military Careers
Career Guide to America's Top Industries

For other career-related materials, turn to the back of this book.

Quantity discounts are available for JIST books. Please call our Sales Department at 1-800-648-JIST for more information and a free catalog.

Editors: Susan Pines, Veda Dickerson, Mary Ellen Stephenson
Cover and Interior Designer: Aleata Howard
Page Layout Coordinator: Carolyn J. Newland

Printed in the United States of America

05 04 03 02   9 8 7 6 5 4 3 2 1

ISBN 1-56370-883-3

Relax—You Don't Have to Read This Whole Book!

This is a big book, but you don't need to read it all. I've organized it into easy-to-use sections so you can browse just the information you want. To get started, simply scan the table of contents, where you'll find brief explanations of the major sections plus a list of the jobs described in this book. Really, this book is easy to use, and I hope it helps you.

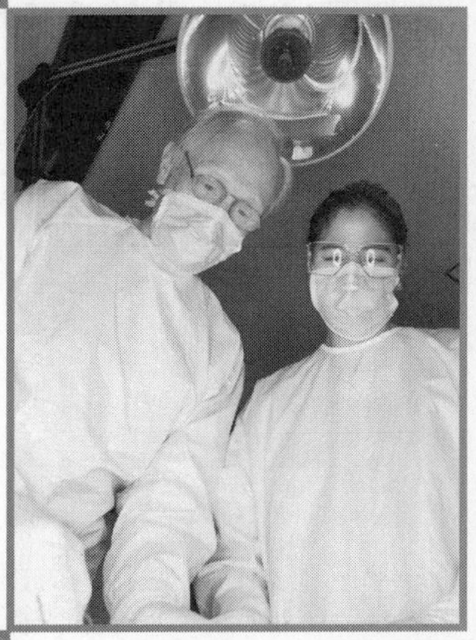

Who Should Use This Book?

This is more than a book of job descriptions. I've spent quite a bit of time thinking about how to make its contents useful for a variety of situations, including

* **Exploring career options.** The job descriptions in Section One give a wealth of information on many of the most desirable jobs in the labor market.
* **Considering more education or training.** The information helps you avoid costly mistakes in choosing a career or deciding on additional training or education—and it increases your chances of planning a bright future.
* **Job seeking.** This book helps you identify new job targets, prepare for interviews, and write targeted resumes. The career planning and job search advice in Section Two has been proven to cut job search time in half!
* **Career planning.** The job descriptions help you explore your options, and Sections Two and Three provide career planning advice and other useful information.

Source of Information

The occupational descriptions in this book come from the good people at the U.S. Department of Labor, as published in the most recent edition of the *Occupational Outlook Handbook*. The *OOH* is the best source of career information available, and the descriptions include the latest data on earnings and other details. Other sections also present solid information from various sources at the U.S. Department of Labor. So, thank you to all the people at the Labor Department who gather, compile, analyze, and make sense of this information. It's good stuff, and I hope you can make good use of it.

Mike Farr

Mike Farr

Table of Contents

Summary of Major Sections

Introduction. The introduction explains each job description element, gives tips on using the book for career exploration and job seeking, and provides other details. *The introduction begins on page 1.*

Section One: Descriptions of 112 Top Computer and Technical Jobs. This section presents thorough descriptions of 112 jobs that require computer skills or other technical training or experience. Education and training requirements for these jobs vary from on-the-job training to a four-year college degree and more. Each description gives information on significant points, nature of the work, working conditions, employment, training, other qualifications, advancement, job outlook, earnings, related occupations, and sources of additional information. The jobs are presented in alphabetical order. The page numbers where specific descriptions begin are listed here in the table of contents. *Section One begins on page 9.*

Section Two: *The Quick Job Search*—Seven Steps to Getting a Good Job in Less Time. This brief but important section offers results-oriented career planning and job search techniques. It includes tips on identifying your key skills, defining your ideal job, using effective job search methods, writing resumes, organizing your time, improving your interviewing skills, and following up on leads. The second part of this section features professionally written and designed resumes for some of America's top computer and technical jobs. *Section Two begins on page 289.*

Section Three: Important Trends in Jobs and Industries. This section includes two well-written articles on labor market trends. The articles are short and worth your time. *Section Three begins on page 341.*

Titles of the articles in Section Three are

* "Tomorrow's Jobs"
* "Employment Trends in Major Industries"

The 112 Jobs Described in Section One

The titles for the 112 jobs described in Section One are listed below, in alphabetical order. The page number where each description begins is also listed. Simply find jobs that interest you, then read those descriptions. An introduction to Section One begins on page 3 and provides additional information on how to interpret the descriptions.

INTRODUCTION

This book is about improving your life, not just about selecting a job. The career you choose will have an enormous impact on how you live your life.

While a huge amount of information is available on occupations, most people don't know where to find accurate, reliable facts to help them make good career decisions—or they don't take the time to look. Important choices such as what to do with your career or whether to get additional training or education deserve your time.

If you are considering more training or education—whether additional coursework, a college degree, or an advanced degree—this book will help with solid information. Training or education beyond high school is now typically required to get better jobs, but many good jobs can be learned through on-the-job experience and other informal or short-term training. The education and training needed for the jobs described in *America's Top Computer and Technical Jobs* vary enormously. This book provides descriptions for major jobs requiring computer and technical skills and gives you the facts you need for exploring your options.

A certain type of work or workplace may interest you as much as a certain type of job. If your interests and values are to work in healthcare, for example, you can do this in a variety of work environments, in a variety of industries, and in a variety of jobs. For this reason, I suggest you begin exploring alternatives by following your interests and finding a career path that allows you to use your talents doing something you enjoy.

Also, remember that money is not everything. The time you spend in career planning can pay off in higher earnings, but being satisfied with your work—and your life—is often more important than how much you earn. This book can help you find the work that suits you best.

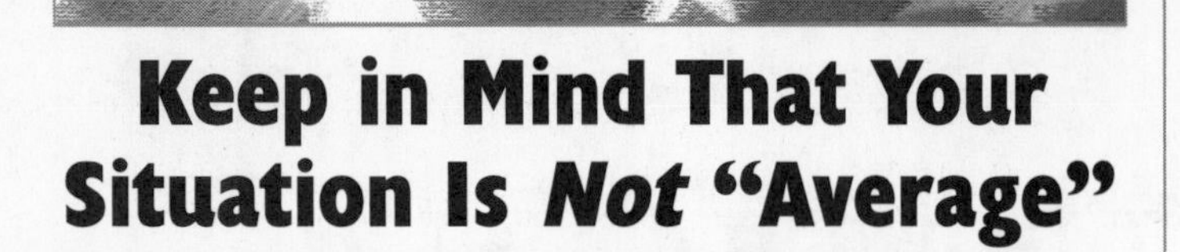

Keep in Mind That Your Situation Is *Not* "Average"

Projected employment growth and earnings trends are quite positive for many occupations and industries. Keep in mind, however, that the averages in this book will not be true for many individuals. Within any field, some people earn more and some less.

My point is this: Your situation is probably not average. Some people do better than others, and some are willing to accept less pay for a more desirable work environment. Earnings vary enormously in different parts of the country, in different occupations, and in different industries. But this book's solid information is a great place to start. Good information will give you a strong foundation for good decisions.

Four Important Labor Market Trends That Will Affect Your Career

Our economy has changed in dramatic ways over the past 10 years, with profound effects on how we work and live. Section Three of this book provides more information on labor market trends but, in case you don't read it, here are four trends that you simply *must* consider.

1. Education and Earnings Are Related

I'm sure you won't be surprised to learn that people with higher levels of education and training have higher average earnings. The data that follows comes from the Department of Labor's Internet site. The site presents the median earnings for people with various levels of education. (The median is the point where half earn more and half earn less.) The site also indicates the average percentage of people at that educational level who are unemployed. Based on this information, I computed the earnings advantage of people at various education levels compared to people with a high school degree.

Earnings for Year-Round, Full-Time Workers Age 25 and Over, by Educational Attainment

Level of Education	Median Annual Earnings	Premium Over High School Grads	Unemployment Rate
Professional degree	$80,230	$51,423 (179%)	1.3%
Doctoral degree	$70,476	$41,669 (145%)	1.4%
Master's degree	$55,302	$26,495 (92%)	1.6%
Bachelor's degree	$46,276	$17,469 (61%)	1.9%
Associate degree	$35,389	$6,582 (23%)	2.5%
Some college, no degree	$32,400	$3,593 (12%)	3.2%
High school graduate	$28,807	——	4.0%
Less than a high school diploma	$21,400	$-7,407 (–25%)	7.1%

Source: Earnings—U.S. Census Bureau, Current Population Survey, PINC-3, updated 12/2001; Unemployment rate—Bureau of Labor Statistics, 1998 data; Earnings—Bureau of the Census, 1997 data.

The earnings gap between a college graduate and someone with a high school education is growing wider and is now $17,469 a year. That's enough to buy a nice car, make a down payment on a house, or even take a month's vacation for two to Europe. As you see, over a lifetime, these additional earnings can make an enormous difference in the college graduate's lifestyle.

And there is more. Jobs that require a four-year college degree are projected to grow about twice as fast as jobs that do not. Research shows that people with higher educational levels are less likely to be unemployed and that they remain unemployed for shorter periods of time. Overall, the data on earnings and other criteria indicate that people with more education and training do better than those with less. There are exceptions, of course, but for most people, more education and training results in higher earnings and lower rates of unemployment.

Many jobs can be obtained without a college degree, but most better-paying jobs require either training beyond high school or substantial work experience.

2. Knowledge of Computer and Other Technologies Is Increasingly Important

Jobs requiring computer and technical skills are projected to be among the fastest growing jobs in America. As you look over the list of jobs in the table of contents for *America's Top Computer and Technical Jobs,* you will notice the enormous variety of technical jobs. The education and training requirements for these jobs range from on-the-job experience to a four-year college degree and more. But even jobs that do not appear to be technical often call for computer literacy or technical skills. Managers, for example, are often expected to understand and use spreadsheet, word-processing, and database software.

In all fields, people without job-related technical and computer skills will have a more difficult time finding good opportunities than people who have these skills. Older workers, by the way, often do not have the computer skills that younger workers do. Employers tend to hire the skills they need, and people without these abilities won't get the best jobs. So consider upgrading your job-related computer and technology skills if you need to—and plan to stay up-to-date on your current and future jobs.

3. Ongoing Education and Training Are Essential

School and work once were separate activities, and most people did not go back to school after they began working. But with rapid changes in technology, most people are now required to learn throughout their work lives. Jobs are constantly upgraded, and today's jobs often cannot be handled by people who have only the knowledge and skills that were adequate for workers a few years ago. To remain competitive, people without technical or computer skills must get them. Those who do not will face increasingly limited job options.

What this means is that you should plan to upgrade your job skills throughout your working life. This may include taking formal courses, reading work-related magazines at home, signing up for on-the-job training, or participating in other forms of education. Upgrading your work-related skills on an ongoing basis is no longer optional for most jobs, and you ignore doing so at your peril.

4. Good Career Planning Has Increased in Importance

Most people spend more time watching TV in a week than they spend on career planning during an entire year. Yet most people will change their jobs many times and make major career changes five to seven times.

While you probably picked up this book for its information on jobs, it also provides a great deal of information on career planning. For example, Section Two gives good career and job search advice, and Section Three has useful information on labor market trends. I urge you to read these and related materials because career-planning and job-seeking skills are the keys to surviving in this new economy.

Information on the Major Sections of This Book

I want this book to be easy to use. In the table of contents, I provide brief comments on each section, and that may be all you need. If not, here are some additional details you may find useful in getting the most out of this book.

Section One: Descriptions of 112 Top Computer and Technical Jobs

Section One is the main part of this book and is probably the reason you picked it up. It contains brief, well-written descriptions for 112 major jobs that require computer and other technical skills. The content for these job descriptions comes from the U.S. Department of Labor and is considered by many people to be the most accurate and up-to-date data available. The jobs are presented in alphabetical order. The table of contents provides a page number that shows where each description begins.

Together, the jobs in Section One provide enormous variety at all levels of earnings and training. One way to explore career options is to go through the table of contents and identify those jobs that seem interesting. If you are interested in medical jobs, for example, you can quickly spot those you want to learn more about. You may also see other jobs that look interesting, and you should consider those as well.

Next, read the descriptions for the jobs that interest you and, based on what you learn, identify those that *most* interest you. These are the jobs you should consider. Sections Two and Three give you additional information on how best to do so.

Each occupational description in this book follows a standard format, making it easier for you to compare jobs. The following overview describes the kinds of information found in each part of a description and offers tips on how to interpret the information.

Job Title

This is the title used for the job in the *Occupational Outlook Handbook,* published by the U.S. Department of Labor.

Numbers

The numbers following the job title refer to closely related job titles from the Occupational Information Network (O*NET). The O*NET was developed by the U.S. Department of Labor to replace the older *Dictionary of Occupational Titles (DOT)*. Like the *DOT* in the past, the O*NET is used by state employment service offices to classify applicants and job openings, and by a variety of career information systems. You can get additional information on the related O*NET titles on the Internet at www.onetcenter.org or at www.careerOINK.com. Reference books that provide O*NET descriptions include the *O*NET Dictionary of Occupational Titles* and the *Enhanced Occupational Outlook Handbook*, both published by JIST Publishing. Your librarian can help you find these books.

Significant Points

The bullet points in this part of a description highlight key characteristics for each job, such as recent trends or education and training requirements.

Nature of the Work

This part of the description discusses what workers typically do in a particular job. Individual job duties may vary by industry or employer. For instance, workers in larger firms tend to be more specialized, whereas those in smaller firms often have a wider variety of duties. Most occupations have several levels of skills and responsibilities through which workers may progress. Beginners may start as trainees performing routine tasks under close supervision. Experienced workers usually undertake more difficult tasks and are expected to perform with less supervision.

In this part of a description, you will also find information about the influence of technological advancements on the way work is done. For example, because of the Internet, reporters are now able to submit stories from remote locations—with just a click of the mouse.

This part also discusses emerging specialties. For instance, webmasters—who are responsible for all the technical aspects involved in operating a website—comprise a specialty within systems analysts, computer scientists, and database administrators.

Working Conditions

This part of the description identifies the typical hours worked, the workplace environment, physical activities, risk of injury, special equipment, and the extent of travel required. For example, stationary engineers and boiler operators are susceptible to injury, while paralegals and legal assistants have high job-related stress. Radiologic technologists and technicians may wear protective clothing or equipment; machinists do physically demanding work; and some engineers travel frequently.

In many occupations, people work regular business hours—40 hours a week, Monday through Friday. In other occupations, they do not. For example, licensed practical nurses often work evenings and weekends. The work setting can range from a hospital, to a mall, to an off-shore oil rig.

Employment

This section reports the number of jobs the occupation recently provided and the key industries where these jobs are found. When significant, the geographic distribution of jobs and the proportion of part-time (less than 35 hours a week) and self-employed workers in the occupation are mentioned. Self-employed workers account for nearly eight percent of the work force but are concentrated in a small number of occupations, such as photographers and landscape architects.

Training, Other Qualifications, and Advancement

After you know what a job is all about, you need to understand how to train for it. This section describes the most significant sources of training, including the training preferred by employers, the typical length of training, and advancement possibilities. Job skills are sometimes acquired through high school, informal on-the-job training, formal training (including apprenticeships), the armed forces, home study, hobbies, or previous work experience. For example, experience is particularly important for prepress technicians and workers. Many professional and technical jobs, on the other hand, require formal postsecondary education—postsecondary vocational or technical training, or college, postgraduate, or professional education.

This section of the job description also mentions desirable skills, aptitudes, and personal characteristics. For some entry-level jobs, personal characteristics are more important than formal training. Employers generally seek people who communicate well, compute accurately, think logically, learn quickly, get along with others, and demonstrate dependability.

Some occupations require certification or licensing to enter the field, to advance, or to practice independently. Certification or licensing generally involves completing courses and passing examinations. More and more occupations require continuing education or skill improvement to keep up with the changing economy or to improve advancement opportunities.

Job Outlook

In planning for the future, you must consider potential job opportunities. This section of a job description indicates what factors will result in growth or decline in the number of jobs. In some cases, the description mentions that an occupation is likely to provide numerous job openings or relatively few openings. Occupations which are large and have high turnover, such as data entry and information processing workers, generally provide the most job openings, reflecting the need to replace workers who transfer to other occupations or who stop working.

Some statements discuss the relationship between the number of jobseekers and the number of job openings. In some occupations, there is a rough balance between jobseekers and openings, whereas other occupations are characterized by shortages or surpluses. Limited training facilities, salary regulations, or undesirable aspects of the work—as in the case of heavy vehicle and mobile equipment service technicians—can cause shortages of entrants. On the other hand, glamorous or potentially high-paying occupations, such as aircraft pilots, generally have surpluses of jobseekers. Variation in job opportunities by industry, size of firm, or geographic location also may be discussed. Even in crowded fields, job openings do exist. Good students and well qualified individuals should not be deterred from undertaking training or seeking entry.

Susceptibility to layoffs due to imports, slowdowns in economic activity, technological advancements, or budget cuts are also addressed in this section. For example, employment of tool and die makers is sensitive to technological advancements, while employment of library technicians is affected by various budget issues.

Key Phrases Used in the Descriptions

This box explains how to interpret the key phrases that describe projected changes in employment. It also explains the terms for the relationship between the number of job openings and the number of job seekers. The descriptions of this relationship in a particular occupation reflect the knowledge and judgment of economists in the Bureau's Office of Occupational Statistics and Employment Projections.

Changing Employment Between 2000 and 2010

If the statement reads:	Employment is projected to:
Grow much faster than average	Increase 36 percent or more
Grow faster than average	Increase 21 to 35 percent
Grow about as fast as average	Increase 10 to 20 percent
Grow more slowly than average	Increase 3 to 9 percent
Little or no change	Increase 0 to 2 percent
Decline	Decrease 1 percent or more

Opportunities and Competition for Jobs

If the statement reads:	Job openings compared to job seekers may be:
Very good to excellent opportunities	More numerous
Good or favorable opportunities	In rough balance
May face keen competition or can expect keen competition	Fewer

Earnings

This section discusses typical earnings and how workers are compensated—annual salaries, hourly wages, commissions, piece rates, tips, or bonuses. Within every occupation, earnings vary by experience, responsibility, performance, tenure, and geographic area. Earnings data from the Bureau of Labor Statistics and, in some cases, from outside sources are included. Data may cover the entire occupation or a specific group within the occupation.

Benefits account for a significant portion of total compensation costs to employers. Benefits such as paid vacation, health insurance, and sick leave may not be mentioned because they are so widespread. Employers may offer other, less-traditional benefits, such as flexible hours and profit-sharing plans to attract and retain highly qualified workers. Less-common benefits also include childcare, tuition for dependents, housing assistance, summers off, and free or discounted merchandise or services.

Related Occupations

Occupations involving similar duties, skills, interests, education, and training are listed here. This allows you to look up these jobs, if they interest you.

Sources of Additional Information

No single publication can completely describe all aspects of an occupation. Thus, this section lists mailing addresses for associations, government agencies, unions, and other organizations that can provide occupational information. In some cases, toll-free phone numbers and Internet addresses also are listed. These are provided for your convenience and do not constitute an endorsement. Free or relatively inexpensive publications offering more information may be mentioned. Some of these are available in libraries, school career centers, guidance offices, or on the Internet.

Section Two: *The Quick Job Search*—Seven Steps to Getting a Good Job in Less Time

For more than 20 years now, I've been interested in helping people find better jobs in less time. If you have ever experienced unemployment, you know it is not pleasant. Unemployment is something most people want to get over quickly—in fact, the quicker the better. Section Two will give you some techniques to help.

I know that most of you who read this book want to improve yourselves. You want to consider career and training options that lead to a better job and life in whatever way you define this—better pay, more flexibility, more-enjoyable or more-meaningful work, proving to your mom that you really can do anything you set your mind to, and so on. That is why I include advice on career planning and job search in the first part of Section Two. It's a short section, but it includes the basics that are most important in planning your career and in reducing the time it takes to get a job. I hope it will make you think about what is important to you in the long run.

The second part of Section Two showcases professionally written resumes for some of America's top computer and technical jobs. Use these as examples when creating your resume.

I know you will resist completing the activities in Section Two, but consider this: It is often not the best person who gets the job, but the best job seeker. People who do their career planning and job search homework often get jobs over those with better credentials, because they have these distinct advantages:

1. **They get more interviews,** including many for jobs that will never be advertised.
2. **They do better in interviews.**

People who understand what they want and what they have to offer employers present their skills more convincingly and are much better at answering problem questions. And, because they have learned more about job search techniques, they are likely to get more interviews with employers who need the skills they have.

Doing better in interviews often makes the difference between getting a job offer or sitting at home. And spending time planning your career can make an enormous difference to your happiness and lifestyle over time. So please consider reading Section Two and completing its activities. I suggest you schedule a time right now to at least read Section Two. An hour or so spent there can help you do just enough better in your career planning, job seeking, and interviewing to make the difference. Go ahead—get out your schedule book and get it over with (nag, nag, nag).

One other thing: If you work through Section Two, and it helps you in some significant way, I'd like to hear from you. Please write or e-mail me via the publisher, whose contact information appears elsewhere in this book.

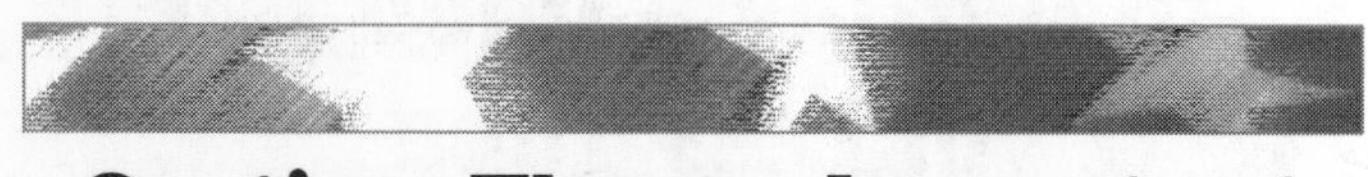

Section Three: Important Trends in Jobs and Industries

This section is made up of two very good articles on labor market trends. These articles come directly from U.S. Department of Labor sources and are interesting, well written, and short. One is on overall trends, with an emphasis on occupational groups; the other is on trends in major industry groups. I know this sounds boring, but both articles are quick reads and will give you a good idea of factors that will impact your career in the years to come.

The first article is titled "Tomorrow's Jobs." It highlights many important trends in employment and includes information on the fastest growing jobs, jobs with high pay at various levels of education, and other details.

The second article is titled "Employment Trends in Major Industries." I included this information because you may find that you can use your skills or training in industries you have not considered. The article provides a good review of major trends with an emphasis on helping you make good employment decisions. This information can help you seek jobs in industries that offer higher pay or that are more likely to interest you. Many people overlook one important fact—the industry you work in is as important as the occupation you choose.

Tips on Using This Book

This book is based on information from government sources and includes the most up-to-date and accurate data available anywhere. The entries are well written and pack a lot of information into short descriptions. *America's Top Computer and Technical Jobs* can be used in many ways. The following discussion provides tips on the four most-frequent uses:

- For people exploring career, education, or training alternatives
- For job seekers
- For employers and business people
- For counselors, instructors, and other career specialists

Tips for People Exploring Career, Education, or Training Alternatives

America's Top Computer and Technical Jobs is an excellent resource for anyone exploring career, education, or training alternatives. Many people take career interest tests to identify career options. This book can be used for the same purpose.

Many people do not have a good idea of what they want to do in their careers. They may be considering additional training or education but may not know what sort they should get. If you are one of these people, *America's Top Computer and Technical Jobs* can help in several ways. Here are a few pointers.

Review the list of jobs. Trust yourself. Research studies indicate that most people have a good sense of their interests. Your interests can be used to guide you to career options you should consider in more detail.

Begin by looking over the occupations listed in the table of contents. Look at all the jobs, because you may identify previously overlooked possibilities. If other people will be using this book, please don't mark in it. Instead, on a separate sheet of paper, list the jobs that interest you. Or make a photocopy of the table of contents and mark the jobs that interest you.

Next, carefully read the descriptions of the jobs that most interest you. A quick review will often eliminate one or more of these jobs based on pay, working conditions, education required, or other considerations. After you have identified the three or four jobs that seem most interesting, research each one more thoroughly before making any important decisions.

Study the jobs and their training and education requirements. Too many people decide to obtain additional training or education without knowing much about the jobs the training will lead to. Reviewing the descriptions in this book is one way to learn more about an occupation before you enroll in an education or training program. If you are currently a student, the job descriptions in this book can also help you decide on a major course of study or learn more about the jobs for which your studies are preparing you.

Do not be too quick to eliminate a job that interests you. If a job requires more education or training than you currently have, you can obtain this training in many ways.

Don't abandon your past experience and education too quickly. If you have significant work experience, training, or education, these should not be abandoned too quickly. Many skills you have learned and used in previous jobs or other settings can apply to related jobs. Many times, after people carefully consider what they want to do, they change careers and find that the skills they have can still be used.

America's Top Computer and Technical Jobs can help you explore career options in several ways. First, carefully review descriptions for jobs you have held in the past. On a separate sheet of paper, list the skills needed in those jobs. Then do the same for jobs that interest you now. By comparing the lists, you will be able to identify skills you used in previous jobs that you could also use in jobs that interest you for the future. These "transferable" skills form the basis for moving to a new career.

You can also identify skills you have developed or used in nonwork activities, such as hobbies, family responsibilities, volunteer work, school, military, and extracurricular interests.

The descriptions can be used even if you want to stay with the same employer. For example, you may identify jobs within your organization that offer more rewarding work, higher pay, or other advantages over your present job. Read the descriptions related to these jobs, and you may be able to transfer into another job rather than leave the organization.

Tips for Job Seekers

You can use the descriptions in this book to give you an edge in finding job openings and in getting job offers—even when you are competing with people who have better credentials. Here are some ways *America's Top Computer and Technical Jobs* can help you in the job search.

Identify related job targets. You may be limiting your job search to a small number of jobs for which you feel qualified, but by doing so you eliminate many jobs you could do and enjoy. Your search for a new job should be broadened to include more possibilities. Go through the entire list of jobs in the table of contents and check any that require skills similar to those you have. Look at all the jobs, since doing so sometimes helps you identify targets you would otherwise overlook.

Many people are not aware of the many specialized jobs related to their training or experience. The descriptions in *America's Top Computer and Technical Jobs* are for major job titles, but a variety of more specialized jobs may require similar skills. Reference books that list more specialized job titles include the *Enhanced Occupational Outlook Handbook* and the *O*NET Dictionary of Occupational Titles*. Both are published by JIST.

The descriptions can also point out jobs that interest you but that have higher responsibility or compensation levels. While you may not consider yourself qualified for such jobs now, you should think about seeking jobs that are above your previous levels but within your ability to handle.

Prepare for interviews. This book's job descriptions are an essential source of information to help you prepare for interviews. If you carefully review the description of a job before an interview, you will be much better prepared to emphasize your key skills. You should also review descriptions for past jobs and identify skills needed in the new job.

Negotiate pay. The job descriptions in this book will help you know what pay range to expect. Note that local pay and other details can differ substantially from the national averages in the descriptions.

Tips for Employers and Business People

If you are an employer, a human resource professional, or other business person, you can use this book's information to write job descriptions, study pay ranges, and set criteria for new employees. The information can also help you conduct more-effective interviews by providing a list of key skills needed by new hires.

Tips for Counselors, Instructors, and Other Career Specialists

Counselors, instructors, and other career specialists will find suggestions in the previous tips for using this book to help people explore career options or find jobs. My best suggestion to professionals is to get this book off the shelf and into the hands of the people who need it. Leave it on a table or desk and show people how the information can help them. Wear this book out—its real value is as a tool used often and well.

Additional Information About the Projections

Readers interested in more information about projections and details on the labor force, economic growth, industry and occupational employment, or the methods and assumptions used in this book's projections should consult the November 2001 edition of the *Monthly Labor Review,* published by the Bureau of Labor Statistics. It provides information on the limitations inherent in economic and employment projections.

For more information about employment change, job openings, earnings, unemployment rates, and training requirements by occupation, consult *Occupational Projections and Training Data,* published by the Bureau of Labor Statistics.

For occupational information from an industry perspective, including some occupations and career paths that *America's Top Computer and Technical Jobs* does not cover, consult another BLS publication, the *Career Guide to Industries*. This book is also available from JIST, under the title *Career Guide to America's Top Industries*.

Section One

Descriptions of 112 Top Computer and Technical Jobs

This is the book's major section. It contains descriptions for 112 major occupations that require computer and technical skills. The jobs are arranged in alphabetical order. Refer to the table of contents for a list of the jobs and the page numbers where their descriptions begin.

The table of contents can also help you identify jobs you want to explore. If you are interested in medical jobs, for example, you can go through the list and quickly find those you want to learn more about. Also, you may spot other jobs that might be interesting, and you should consider those as well. Read the descriptions for any jobs that sound interesting.

While the descriptions in this section are easy to understand, the introduction to this book provides additional information for interpreting them. When reading the descriptions, keep in mind that they present information that is the average for the country. Conditions in your area and with specific employers may be quite different.

Also, you may come across jobs that sound interesting but require additional training or education. Don't eliminate them too soon. There are many ways to obtain training and education, and most people change jobs and careers many times. You probably have more skills than you realize that can transfer to new jobs, so consider taking some chances. Get out of your rut. Do what it takes to fulfill your dreams. Be creative. You often have more opportunities than barriers, but you have to go out and find the opportunities.

Actors, Producers, and Directors

O*NET 27-2011.00, 27-2012.01, 27-2012.02, 27-2012.03, 27-2012.04, 27-2012.05

Significant Points

- Actors endure long periods of unemployment, intense competition for roles, and frequent rejections in auditions.
- Formal training through a university or acting conservatory is typical; however, many actors, producers, and directors find work based on experience and talent alone.
- Because earnings for actors are erratic, many supplement their incomes by holding jobs in other fields.

Nature of the Work

Actors, producers, and directors express ideas and create images in theater, film, radio, television, and other performing arts media. They interpret a writer's script to entertain, inform, or instruct an audience. Although the most famous actors, producers, and directors work in film, network television, or theater in New York or Los Angeles, far more work in local or regional television studios, theaters, or film production companies engaged in advertising, public relations, or independent, small-scale movie productions.

Actors perform in stage, radio, television, video, or motion picture productions. They also work in cabarets, nightclubs, theme parks, and commercials, and in "industrial" films produced for training and educational purposes. Most actors struggle to find steady work; only a few ever achieve recognition as stars. Some well-known, experienced performers may be cast in supporting roles. Others work as "extras," with no lines to deliver, or make brief, cameo appearances speaking only one or two lines. Some actors also teach in high school or university drama departments, acting conservatories, or public programs.

Producers are entrepreneurs, overseeing the business and financial decisions of a production. They select scripts and approve development of script ideas, arrange financing, and determine the size and cost of stage, radio, television, video, or motion picture productions. Producers hire directors, principal cast members, and key production staff members. They also negotiate contracts with artistic and design personnel in accordance with collective bargaining agreements and guarantee payment of salaries, rent, and other expenses. Producers coordinate the activities of writers, directors, managers, and agents to ensure that each project stays on schedule and within budget.

Directors are responsible for the creative decisions of a production. They interpret scripts, express concepts to set and costume designers, audition and select cast members, conduct rehearsals, and direct the work of cast and crew. Directors approve the design elements of a production, including sets, costumes, choreography, and music.

Working Conditions

Actors, producers, and directors work under constant pressure. To succeed, they need patience and commitment to their craft. Actors strive to deliver flawless performances, often while working in undesirable and unpleasant conditions. Producers and directors experience stress from the need to adhere to budgets, union work rules, and production schedules; organize rehearsals; and meet with designers, financial backers, and production executives.

Acting assignments typically are short term (ranging from 1 day to a few months), which means that there often are long periods of unemployment between jobs. The uncertain nature of the work results in unpredictable earnings and intense competition for even the lowest paid jobs. Often, actors, producers, and directors must hold other jobs to sustain a living.

When performing, actors typically work long, irregular hours. For example, stage actors may perform one show at night while rehearsing another during the day. They also might travel with a show when it tours the country. Movie actors may work on location, sometimes under adverse weather conditions, and may spend considerable time in their trailers or dressing rooms waiting to perform their scenes. Actors who perform in television often appear on camera with little preparation time because scripts tend to be revised frequently or written moments before taping.

Evening and weekend work is a regular part of a stage actor's life. On weekends, more than one performance may be held per day. Actors and directors working on movies or television programs, especially those who shoot on location, may work in the early morning or late evening hours to do nighttime filming or to tape scenes inside public facilities outside of normal business hours.

Actors should be in good physical condition and have the necessary stamina and coordination to move about theater stages and large movie and television studio lots. They also need to maneuver about complex technical sets while staying in character and projecting their voices audibly. Actors must be fit to endure heat from stage or studio lights and the weight of heavy costumes. Producers and directors should anticipate such hazards and ensure the safety of actors by conducting extra rehearsals on the set so that actors can learn the layout of set pieces and props, allowing time for warm-ups and stretching exercises to guard against physical and vocal injuries, and providing adequate breaks to prevent heat exhaustion and dehydration.

Employment

In 2000, actors, producers, and directors held about 158,000 jobs, primarily in motion pictures, theater, television, and radio. Because many others were between jobs, the total number of actors, producers, and directors available for work was higher. Employment in the theater is cyclical—higher in the fall and spring seasons—and concentrated in New York and other major cities with large commercial houses for musicals and touring productions. Also, many cities support established professional regional theaters that operate on a seasonal or year-round basis.

In summer, stock companies in suburban and resort areas also provide employment opportunities. Actors, producers, and directors may find work on cruise lines and in theme parks. Many smaller nonprofit professional companies, such as repertory companies, dinner theaters, and theaters affiliated with drama schools, acting conservatories, and universities provide employment opportunities for local amateur talent and professional entertainers. Auditions typically are held in New York for many productions across the country and for shows that go on the road.

Employment in motion pictures and films for television is centered in New York and in Hollywood. However, small studios are located throughout the country. Many films are shot on location and may employ local professional and nonprofessional actors. In television, opportunities are concentrated in the network centers of New York and Los Angeles, but cable television services and local television stations around the country also employ many actors, producers, and directors.

Training, Other Qualifications, and Advancement

Persons who become actors, producers, and directors follow many paths. Employers generally look for people with the creative instincts, innate talent, and intellectual capacity to perform. Actors should possess a passion for performing and enjoy entertaining others. Most aspiring actors participate in high school and college plays, work in college radio stations, or perform with local community theater groups. Local and regional theater experience and work in summer stock, on cruise lines, or in theme parks help many young actors hone their skills and earn qualifying credits towards membership in one of the actors' unions. Union membership and work experience in smaller communities may lead to work in larger cities, notably New York or Los Angeles. In television and film, actors and directors typically start in smaller television markets or with independent movie production companies, then work their way up to larger media markets and major studio productions. Intense competition, however, ensures that only a few actors reach star billing.

Formal dramatic training, either through an acting conservatory or a university program, generally is necessary; however, some people successfully enter the field without it. Most people studying for a bachelor's degree take courses in radio and television broadcasting, communications, film, theater, drama, or dramatic literature. Many continue their academic training and receive a Master of Fine Arts (MFA) degree. Advanced curriculums may include courses in stage speech and movement, directing, playwriting, and design, as well as intensive acting workshops.

Actors at all experience levels may pursue workshop training through acting conservatories or by being mentored by a drama coach. Actors also research roles so that they can grasp concepts quickly during rehearsals and understand the story's setting and background. Sometimes actors learn a foreign language or train with a dialect coach to develop an accent to make their characters more realistic.

Actors need talent, creative ability, and training that will enable them to portray different characters. Because competition for parts is fierce, versatility and a wide range of related performance skills, such as singing, dancing, skating, juggling, and miming are especially useful in lifting actors above the average and getting them noticed by producers and directors. Actors must have poise, stage presence, the capability to affect an audience, and the ability to follow direction. Modeling experience also may be helpful. Physical appearance often is a deciding factor in being selected for particular roles.

Many professional actors rely on agents or managers to find work, negotiate contracts, and plan their careers. Agents generally earn a percentage of the pay specified in an actor's contract. Other actors rely solely on attending open auditions for parts. Trade publications list the time, date, and location of these auditions.

To become a movie extra, one must usually be listed by a casting agency, such as Central Casting, a no-fee agency that supplies extras to the major movie studios in Hollywood. Applicants are accepted only when the number of persons of a particular type on the list—for example, athletic young women, old men, or small children—falls below the foreseeable need. In recent years, only a very small proportion of applicants has succeeded in being listed.

There are no specific training requirements for producers. They come from many different backgrounds. Talent, experience, and business acumen are very important determinants of success for producers. Actors, writers, film editors, and business managers commonly enter the field. Also, many people who start out as actors move into directing, while some directors might try their hand at acting. Producers often start in a theatrical management office, working for a press agent, managing director, or business manager. Some start in a performing arts union or service organization. Others work behind the scenes with successful directors, serve on boards of directors, or promote their own projects. No formal training exists for producers; however, a growing number of colleges and universities now offer degree programs in arts management and managing nonprofits.

As the reputations and box-office draw of actors, producers, and directors grow, they might work on bigger budget productions, on network or syndicated broadcasts, or in more prestigious theaters. Actors may advance to lead roles and receive star billing. A few actors move into acting-related jobs, such as drama coaches or directors of stage, television, radio, or motion picture productions. Some teach drama privately or in colleges and universities.

Job Outlook

Employment of actors, producers, and directors is expected to grow faster than the average for all occupations through 2010. Although a growing number of people will aspire to enter these professions, many will leave the field early because the work, when it is available, is hard, the hours are long, and the pay is low. Despite faster-than-average employment growth, competition for jobs will be stiff, in part because of the large number of highly trained and talented actors auditioning for roles. Only performers with the most stamina and talent will regularly find employment.

Expanding cable and satellite television operations, increasing production and distribution of major studio and independent films, and continued growth and development of interactive media, such as direct-for-web movies and videos, should increase demand for actors, producers, and directors. A strong Broadway and off-Broadway community and vibrant regional theater network are expected to offer many job opportunities.

Earnings

Median annual earnings of actors were $25,920 in 2000. The middle 50 percent earned between $16,950 and $59,769. The lowest 10 percent earned less than $12,700, and the highest 10 percent earned more than $93,620. Median annual earnings in the industries employing the largest numbers of actors were as follows:

Motion picture production and services	$54,440
Producers, orchestras, and entertainers	28,310
Miscellaneous amusement and recreation services	13,500

Minimum salaries, hours of work, and other conditions of employment are covered in collective bargaining agreements between show producers and the unions representing workers. Actors' Equity Association (Equity) represents stage actors; Screen Actors Guild (SAG) covers actors in motion pictures, including television, commercials, and films; and the American Federation of Television and Radio Artists (AFTRA) represents television and radio studio performers. While these unions generally determine minimum salaries, any actor or director may negotiate for a salary higher than the minimum.

On July 1, 2001, the members of SAG and AFTRA negotiated a new joint contract covering all unionized employment. Under the contract, motion picture and television actors with speaking parts earned a minimum daily rate of $636 or $2,206 for a 5-day week. Actors also receive contributions to their health and pension plans and additional compensation for reruns and foreign telecasts of the productions in which they appear.

According to Equity, the minimum weekly salary for actors in Broadway productions as of June 25, 2001 was $1,252. Actors in off-Broadway theaters received minimums ranging from $440 to $551 a week as of October 30, 2000, depending on the seating capacity of the theater. Regional theaters that operate under an Equity agreement pay actors $500 to $728 per week. For touring productions, actors receive an additional $106 per day for living expenses ($112 per day in larger, higher-cost cities). According to Equity, fewer than 15 percent of its dues-paying members actually worked during any given week during 2000. Median earnings for those able to find employment in 2000 were less than $10,000.

Some well-known actors—stars—earn well above the minimum; their salaries are many times the figures cited, creating the false impression that all actors are highly paid. For example, of the nearly 100,000 SAG members, only about 50 might be considered stars. The average income that SAG members earn from acting, less than $5,000 a year, is low because employment is erratic. Therefore, most actors must supplement their incomes by holding jobs in other fields.

Many actors who work more than a set number of weeks per year are covered by a union health, welfare, and pension fund, which includes hospitalization insurance and to which employers contribute. Under some employment conditions, Equity and AFTRA members receive paid vacations and sick leave.

Median annual earnings of producers and directors were $41,030 in 2000. The middle 50 percent earned between $29,000 and $60,330. The lowest 10 percent earned less than $21,050, and the highest 10 percent earned more than $87,770. Median annual earnings in the industries employing the largest numbers of producers and directors were as follows:

Motion picture production and services	$50,280
Producers, orchestras, and entertainers	38,820
Radio and television broadcasting	34,630

Many stage directors belong to the Society of Stage Directors and Choreographers (SSDC), and film and television directors belong to the Directors Guild of America (DGA). Earnings of stage directors vary greatly. According to the SSDC, summer theaters offer compensation, including "royalties" (based on the number of performances), usually ranging from $2,500 to $8,000 for a 3- to 4-week run. Directing a production at a dinner theater usually will pay less than directing one at a summer theater, but has more potential for income from royalties. Regional theaters may hire directors for longer periods, increasing compensation accordingly. The highest paid directors work on Broadway and commonly earn $50,000 per show. However, they also receive payment in the form of royalties—a negotiated percentage of gross box office receipts—that can exceed their contract fee for long-running box office successes.

Producers seldom get a set fee; instead, they get a percentage of a show's earnings or ticket sales.

Related Occupations

People who work in performing arts occupations that may require acting skills include announcers; dancers and choreographers; and musicians, singers, and related workers. Others working in theater-related occupations are hairdressers, hairstylists, and cosmetologists; fashion designers; set and exhibit designers; sound engineering technicians; and writers and authors.

Sources of Additional Information

For general information about theater arts and a list of accredited college-level programs, contact:

- National Association of Schools of Theater, 11250 Roger Bacon Dr., Suite 21, Reston, VA 20190. Internet: http://www.arts-accredit.org/nast/default.htm

For general information on actors, producers, and directors, contact:

- Actors Equity Association, 165 West 46th St., New York, NY 10036. Internet: http://www.actorsequity.org
- Screen Actors Guild, 5757 Wilshire Blvd., Los Angeles, CA 90036-3600. Internet: http://www.sag.org
- American Federation of Television and Radio Artists—Screen Actors Guild, 4340 East-West Hwy., Suite 204, Bethesda, MD 20814-4411

Actuaries

O*NET 15-2011.00

Significant Points

- A strong background in mathematics is essential.
- About 7 out of 10 actuaries are employed in the insurance industry.
- This small occupation generates relatively few job openings; the fastest employment growth is expected in the computer and data processing services, health services, and management and actuarial consulting industries.

Nature of the Work

Actuaries are essential employees because they determine future risk, make price decisions, and formulate investment strategies. Some actuaries also design insurance, financial, and pension plans and ensure that these plans are maintained on a sound financial basis. Most actuaries specialize in life and health or property and casualty insurance; others work primarily in finance or employee benefits. Some use a broad knowledge of business and mathematics in investment, risk classification, or pension planning.

Regardless of specialty, actuaries assemble and analyze data to estimate probabilities of an event taking place, such as death, sickness, injury, disability, or property loss. They also address financial questions, including those involving the level of pension contributions required to produce a certain retirement income level or how a company should invest resources to maximize return on investment in light of potential risk. Moreover, actuaries may help determine company policy and sometimes explain complex technical matters to company executives, government officials, shareholders, policyholders, or the public in general. They may testify before public agencies on proposed legislation affecting their businesses or explain changes in contract provisions to customers. They also may help companies develop plans to enter new lines of business or new geographic markets with existing lines of business by forecasting demand in competitive settings.

Most actuaries are employed in the insurance industry, in which they estimate the amount a company will pay in claims. For example, property/casualty actuaries calculate the expected amount of claims resulting from automobile accidents, which varies depending on the insured person's age, sex, driving history, type of car, and other factors. Actuaries ensure that the price, or premium, charged for such insurance will enable the company to cover claims and other expenses. This premium must be profitable and yet competitive with other insurance companies.

Actuaries employed in other industries perform several different functions. The small but growing group of actuaries in the financial services industry, for example, manages credit and helps price corporate security offerings. Because banks now offer their customers investment products such as annuities and asset management services, actuaries increasingly help financial institutions manage the substantial risks associated with these products. Actuaries employed as pension actuaries enrolled under the provisions of the Employee Retirement Income Security Act (ERISA) of 1974 evaluate pension plans covered by that act and report on their financial soundness to plan members, sponsors, and federal regulators. Actuaries working in government help manage social programs such as Social Security and Medicare.

In addition to salaried actuaries, numerous consulting actuaries provide advice to clients on a contract basis. Their clients include insurance companies, corporations, health maintenance organizations, healthcare providers, government agencies, and attorneys. The duties of most consulting actuaries are similar to those of other actuaries. For example, some design pension plans through calculating the future value of current deductions from earnings and by determining the amount of employer contributions. Others provide advice to healthcare plans or financial services firms. Consultants sometimes testify in court regarding the value of potential lifetime earnings of a person who is disabled or killed in an accident, the current value of future pension benefits in divorce cases, or other complex calculations. Many consulting actuaries work in reinsurance, in which one insurance company arranges to share a large prospective liability policy with another insurance company in exchange for a percentage of the premium.

Working Conditions

Actuaries have desk jobs, and their offices usually are comfortable and pleasant. They often work at least 40 hours a week. Some actuaries, particularly consulting actuaries, may travel to meet with clients. Consulting actuaries also may experience more erratic employment and be expected to work more than 40 hours per week.

Employment

Actuaries held about 14,000 jobs in 2000. Over seven-tenths of the actuaries who were wage and salary workers were employed in the insurance industry. Some had jobs in life and health insurance companies, while property and casualty insurance companies, pension funds, or insurance agents and brokers employed others. Most of the remaining actuaries worked for firms providing a variety of corporate services, especially management and public relations, or for actuarial consulting services. A relatively small number of actuaries was employed by security and commodity brokers or by government agencies. Some developed computer software for actuarial calculations.

Training, Other Qualifications, and Advancement

As with many business positions, actuaries need a strong background in mathematics and general business. Applicants for beginning actuarial jobs usually have a bachelor's degree in mathematics, actuarial science, statistics, or a business-related discipline, such as economics, finance, or accounting. About 100 colleges and universities offer an actuarial science program, and most colleges and universities offer a degree in mathematics or statistics. Some companies hire applicants without specifying a major, provided that the applicant has a working knowledge of mathematics, including calculus, probability, and statistics, and has demonstrated this ability by passing one or two actuarial exams required for professional designation. Courses in economics, accounting, finance, and insurance also are useful. Companies increasingly prefer well-rounded individuals who, in addition to a strong technical background, have some training in liberal arts and business, and possess strong communication skills.

In addition to knowledge of mathematics, computer skills are becoming increasingly important. Actuaries should be able to develop and use spreadsheets and databases, as well as standard statistical analysis software. Knowledge of computer programming languages, such as Visual Basic, also is useful.

Two professional societies sponsor programs leading to full professional status in their specialty. The first, the Society of Actuaries (SOA), administers a series of actuarial examinations for the life and health insurance, pension, and finance and investment fields. The Casualty Actuarial Society (CAS), on the other hand, gives a series of examinations for the property and casualty field, which includes fire, accident, medical malpractice; workers compensation; and personal injury liability.

The first four exams of the SOA and CAS examination series are jointly sponsored by the two societies and cover the same material. For this reason, students do not need to commit themselves to a specialty until they have taken the initial examinations. These test an individual's competence in probability, calculus, statistics, and other branches of mathematics. The first few examinations help students evaluate their potential as actuaries. Many prospective actuaries begin taking the exams in college with the help of self-study guides and courses. Those who pass one or more examinations have better opportunities for employment at higher starting salaries than those who do not.

Actuaries are encouraged to complete the entire series of examinations as soon as possible, advancing first to the associate level, and then to the fellowship level. Advanced casualty topics include investment and assets, dynamic financial analysis, and valuation of insurance topics. Candidates in the SOA examination series must choose a specialty—group and health benefits, individual life and annuities, pensions, investments, or finance. Examinations are given twice a year, in the spring and the fall. Although many companies allot time to their employees for study, extensive home study is required to pass the examinations, and many actuaries study for months to prepare for each examination. It is likewise common for employers to pay the hundreds of dollars for fees and study materials. Most reach the associate level within 4 to 6 years and the fellowship level a few years later.

Specific requirements apply for pension actuaries, who verify the financial status of defined benefit pension plans to the federal government. These actuaries must be enrolled by the Joint Board for the Enrollment of Actuaries. To qualify for enrollment, applicants must meet certain experience and examination requirements, as stipulated by the Joint Board.

To perform their duties effectively, actuaries must keep up with current economic and social trends and legislation, as well as with developments in health, business, finance, and economics that could affect insurance or investment practices. Good communication and interpersonal skills also are important, particularly for prospective consulting actuaries.

Beginning actuaries often rotate among different jobs in an organization to learn various actuarial operations and phases of insurance work, such as marketing, underwriting, and product development. At first, they prepare data for actuarial projects or perform other simple tasks. As they gain experience, actuaries may supervise clerks, prepare correspondence, draft reports, and conduct research. They may move from one company to another early in their careers as they move up to higher positions.

Advancement depends largely on job performance and the number of actuarial examinations passed. Actuaries with a broad knowledge of the insurance, pension, investment, or employee benefits fields can advance to administrative and executive positions in their companies. Actuaries with supervisory ability may advance to management positions in other areas, such as underwriting, accounting, data processing, marketing, or advertising. Some actuaries assume college and university faculty positions.

Job Outlook

This small occupation generates relatively few job openings from employment growth and the need to replace those who leave the occupation each year. The fastest employment growth is expected in the computer and data processing services, health services, and management and actuarial consulting industries. Employment of actuaries is expected to grow more slowly than the average for all occupations through 2010, as projected job growth in these industries is offset by a slowdown in actuarial employment growth in insurance industries, which traditionally employ the majority of actuaries.

New employment opportunities should become available in health services, in medical and health insurance industries, and in government—in healthcare and Social Security. Changes in managed healthcare and the desire to contain healthcare costs will continue to provide opportunities for actuaries. Some actuaries also are evaluating the risks associated with controversial medical issues, such as genetic testing or the impact of diseases such as AIDS. Others in this field are involved in drafting healthcare legislation. As healthcare issues and Social Security reform continue to receive growing attention, opportunities for actuaries should increase.

Actuaries will continue to be needed to evaluate risks associated with catastrophes, such as earthquakes, tornadoes, hurricanes, floods, and other natural disasters. Growing areas in property and casualty insurance are environmental and international risk management. Actuaries evaluate risks such as the likelihood of a toxic waste spill, or the costs and benefits of implementing pollution control equipment in a factory. As economic globalization continues and companies expand their operations abroad, they increasingly rely on actuaries to evaluate the risk of setting up a new factory or acquiring a foreign subsidiary.

The banking and securities and commodities industries also should provide additional jobs for actuaries. As financial services continue to consolidate and insurance firms, banks, and securities firms enter one another's markets, new opportunities will emerge. Actuaries will be needed to analyze the risks associated with entering a new market, such as launching a new service or merging with an already established company.

At the same time, changes in consumer preferences for retirement investment plans will adversely affect employment in the life insurance and pension funds industries. The overall decline in the life insurance industry, reflecting fewer life insurance policies sold in favor of investments earning higher returns, will continue to affect the need for actuaries. Similarly, more people are choosing to invest in defined contribution plans, which are less complicated to analyze and, therefore, require fewer actuaries than defined pension systems. Actuaries in the pension funds industry are more likely to be involved in financial planning—helping people manage their retirement money.

Layoffs in the insurance and financial industries due to downsizing and mergers also affect employment. Many of the actuaries released from insurance firms are choosing to establish consulting practices. Jobs should be available for actuaries working in consulting as firms who do not employ their own actuarial staff continue to hire consulting actuaries to analyze various risks.

Earnings

Median annual earnings of actuaries were $66,590 in 2000. The middle 50 percent earned between $47,260 and $93,140. The lowest 10 percent had earnings of less than $37,130, while the top 10 percent earned over $127,360. The average salary for actuaries employed by the federal government was $78,120 in 2001.

According to the National Association of Colleges and Employers, annual starting salaries for bachelor's degree graduates in mathematics/actuarial science averaged $45,753 in 2001.

Insurance companies and consulting firms give merit increases to actuaries as they gain experience and pass examinations. Some companies also offer cash bonuses for each professional designation achieved. A 2001 salary survey of insurance and financial services companies, conducted by the Life Office Management Association, Inc., indicated that the average base salary for an entry-level actuary with the largest U.S. companies was $44,546. Associate actuaries with the largest U.S. companies, who direct and provide leadership in the design, pricing, and implementation of insurance products, received an average salary of $91,544. Actuaries at the highest technical level without managerial responsibilities in the same size companies earned an average of $108,777.

Related Occupations

Actuaries need a strong background in mathematics, statistics, and related fields. Other workers whose jobs involve related skills include accountants and auditors, budget analysts, economists and market and survey researchers, financial analysts and personal financial advisors, insurance underwriters, mathematicians, and statisticians.

Sources of Additional Information

Career information on actuaries specializing in pensions is available from:

- American Society of Pension Actuaries, 4245 N. Fairfax Dr., Suite 750, Arlington, VA 22203. Internet: http://www.aspa.org

For information about actuarial careers in life and health insurance, employee benefits and pensions, and finance and investments, contact:

- Society of Actuaries (SOA), 475 N. Martingale Rd., Suite 800, Schaumburg, IL 60173-2226. Internet: http://www.soa.org

For information about actuarial careers in property and casualty insurance, contact:

- Casualty Actuarial Society (CAS), 1100 N. Glebe Rd., Suite 600, Arlington, VA 22201. Internet: http://www.casact.org

The SOA and CAS jointly sponsor a Web site for those interested in pursuing an actuarial career at:

- Internet: http://www.BeAnActuary.org

For general facts about actuarial careers, contact:

- American Academy of Actuaries, 1100 17th St. N.W., 7th Floor, Washington, DC 20036. Internet: http://www.actuary.org/index.htm

Aerospace Engineers

O*NET 17-2011.00

Significant Points

- Overall job opportunities in engineering are expected to be good.
- A bachelor's degree is required for most entry-level jobs.
- Starting salaries are significantly higher than those of college graduates in other fields.
- Continuing education is critical to keep abreast of the latest technology.

Nature of the Work

Aerospace engineers are responsible for developing extraordinary machines, from airplanes that weigh over a half a million pounds to spacecraft that travel over 17,000 miles an hour. They design, develop, and test aircraft, spacecraft, and missiles and supervise the manufacturing of these products. Aerospace engineers who work with aircraft are considered *aeronautical engineers*, and those working specifically with spacecraft are considered *astronautical engineers*.

Aerospace engineers develop new technologies for use in aviation, defense systems, and space exploration, often specializing in areas such as structural design, guidance, navigation and control, instrumentation and communication, or production methods. They often use Computer Aided Design (CAD), robotics, and lasers and advanced electronic optics to assist them. They also may specialize in a particular type of aerospace product, such as commercial transports, military fighter jets, helicopters, spacecraft, or missiles and rockets. Aerospace engineers may be experts in aerodynamics, thermodynamics, celestial mechanics, propulsion, acoustics, or guidance and control systems.

Aerospace engineers typically are employed within the aerospace industry, although their skills are becoming increasingly valuable in other fields. For example, aerospace engineers in the motor vehicles manufacturing industry design vehicles that have lower air resistance, increasing the fuel efficiency of vehicles.

Working Conditions

Most engineers work in office buildings, laboratories, or industrial plants. Others may spend time outdoors at construction sites, mines, and oil and gas exploration and production sites, where they monitor or direct operations or solve onsite problems. Some engineers travel extensively to plants or work sites.

Many engineers work a standard 40-hour week. At times, deadlines or design standards may bring extra pressure to a job. When this happens, engineers may work longer hours and experience considerable stress.

Employment

Aerospace engineers held about 50,000 jobs in 2000. Almost one-half worked in the aircraft and parts and guided missile and space vehicle manufacturing industries. Federal government agencies, primarily the Department of Defense and the National Aeronautics and Space Administration, provided almost 15 percent of jobs. Engineering and architectural services, research and testing services, and search and navigation equipment firms accounted for most of the remaining jobs.

Training, Other Qualifications, and Advancement

A bachelor's degree in engineering is required for almost all entry-level engineering jobs. College graduates with a degree in a physical science or mathematics occasionally may qualify for some engineering jobs, especially in specialties in high demand. Most engineering degrees are granted in electrical, electronics, mechanical, or civil engineering. However, engineers trained in one branch may work in related branches. For example, many aerospace engineers have training in mechanical engineering. This flexibility allows employers to meet staffing needs in new technologies and specialties in which engineers are in short supply. It also allows engineers to shift to fields with better employment prospects or to those that more closely match their interests.

Most engineering programs involve a concentration of study in an engineering specialty, along with courses in both mathematics and science. Most programs include a design course, sometimes accompanied by a computer or laboratory class or both.

In addition to the standard engineering degree, many colleges offer 2- or 4-year degree programs in engineering technology. These programs, which usually include various hands-on laboratory classes that focus on current issues, prepare students for practical design and production work, rather than for jobs which require more theoretical and scientific knowledge. Graduates of 4-year technology programs may get jobs similar to those obtained by graduates with a bachelor's degree in engineering. Engineering technology graduates, however, are not qualified to register as professional engineers under the same terms as graduates with degrees in engineering. Some employers regard technology program graduates as having skills between those of a technician and an engineer.

Graduate training is essential for engineering faculty positions and many research and development programs, but is not required for the majority of entry-level engineering jobs. Many engineers obtain graduate degrees in engineering or business administration to learn new technology and broaden their education. Many high-level executives in government and industry began their careers as engineers.

About 330 colleges and universities offer bachelor's degree programs in engineering that are accredited by the Accreditation Board for Engineering and Technology (ABET), and about 250 colleges offer accredited bachelor's degree programs in engineering technology. ABET accreditation is based on an examination of an engineering program's student achievement, program improvement, faculty, curricular content, facilities, and institutional commitment. Although most institutions offer programs in the major branches of engineering, only a few offer programs in the smaller specialties. Also, programs of the same title may vary in content. For example, some programs emphasize industrial practices, preparing students for a job in industry, whereas others are more theoretical and are designed to prepare students for graduate work. Therefore, students should investigate curricula and check accreditations carefully before selecting a college.

Admissions requirements for undergraduate engineering schools include a solid background in mathematics (algebra, geometry, trigonometry, and calculus) and sciences (biology, chemistry, and physics), and courses in English, social studies, humanities, and computers. Bachelor's degree programs in engineering typically are designed to last 4 years, but many students find that it takes between 4 and 5 years to complete their studies. In a typical 4-year college curriculum, the first 2 years are spent studying mathematics, basic sciences, introductory engineering, humanities, and social sciences. In the last 2 years, most courses are in engineering, usually with a concentration in one branch. Some programs offer a general engineering curriculum; students then specialize in graduate school or on the job.

Some engineering schools and 2-year colleges have agreements whereby the 2-year college provides the initial engineering education, and the engineering school automatically admits students for their last 2 years. In addition, a few engineering schools have arrangements whereby a student spends 3 years in a liberal arts college studying pre-engineering subjects and 2 years in an engineering school studying core subjects, and then receives a bachelor's degree from each school. Some colleges and universities offer 5-year master's degree programs. Some 5- or even 6-year cooperative plans combine classroom study and practical work, permitting students to gain valuable experience and finance part of their education. All 50 states and the District of Columbia usually require licensure for engineers who offer their services directly to the public. Engineers who are licensed are called Professional Engineers (PE). This licensure generally requires a degree from an ABET-accredited engineering program, 4 years of relevant work experience, and successful completion of a state examination. Recent graduates can start the licensing process by taking the examination in two stages. The initial Fundamentals of Engineering (FE) examination can be taken upon graduation. Engineers who pass this examination commonly are called Engineers in Training (EIT) or Engineer Interns (EI). The EIT certification usually is valid for 10 years. After acquiring suitable work experience, EITs can take the second examination, the Principles and Practice of Engineering Exam. Several states have imposed mandatory continuing education requirements for relicensure. Most states recognize licensure from other states.

Engineers should be creative, inquisitive, analytical, and detail oriented. They should be able to work as part of a team and to communicate well, both orally and in writing. Communication abilities are becoming more important because much of their work is becoming more diversified, meaning that engineers interact with specialists in a wide range of fields outside engineering.

Beginning engineering graduates usually work under the supervision of experienced engineers and, in large companies, also may receive formal classroom or seminar-type training. As new engineers gain knowledge and experience, they are assigned more difficult projects with greater independence to develop designs, solve problems, and make decisions. Engineers may advance to become technical specialists or to supervise a staff or team of engineers and technicians. Some may eventually become engineering managers or enter other managerial or sales jobs.

Job Outlook

Employment of aerospace engineers is expected to grow about as fast as the average for all occupations through 2010. The decline in defense department expenditures for military aircraft, missiles, and other aerospace systems has restricted defense-related employment opportunities in recent years. However, an expected increase in defense spending in these areas may result in increased employment of aerospace engineers in defense-related areas during the 2000–2010 period. Demand should increase for aerospace engineers to design and produce civilian aircraft, due to the need to accommodate increasing passenger traffic and to replace much of the present fleet with quieter and more fuel-efficient aircraft. Additional opportunities for aerospace engineers will be created with aircraft manufacturers to search for ways to use existing technology for new purposes. Some employment opportunities also will occur in industries not typically associated with aerospace, such as motor vehicles. Most job openings, however, will result from the need to replace aerospace engineers who transfer to other occupations or leave the labor force.

Earnings

Median annual earnings of aerospace engineers were $67,930 in 2000. The middle 50 percent earned between $56,410 and $82,570. The lowest 10 percent earned less than $47,700, and the highest 10 percent earned more than $94,310. Median annual earnings in the industries employing the largest numbers of aerospace engineers in 2000 were:

Federal government	$74,170
Search and navigation equipment	71,020
Aircraft and parts	68,230
Guided missiles, space vehicles, and parts	65,830

According to a 2001 salary survey by the National Association of Colleges and Employers, bachelor's degree candidates in aerospace engineering received starting offers averaging $46,918 a year, master's degree candidates were offered $59,955, and PhD candidates were offered $64,167.

Related Occupations

Aerospace engineers apply the principles of physical science and mathematics in their work. Other workers who use scientific and mathematical principles include architects, except landscape and naval; engineering and natural sciences managers; computer and information systems managers; mathematicians; drafters; engineering technicians; sales engineers; science technicians; and physical and life scientists, including agricultural and food scientists, biological and medical scientists, conservation scientists and foresters, atmospheric scientists, chemists and materials scientists, environmental scientists and geoscientists, and physicists and astronomers.

Sources of Additional Information

For further information about aerospace engineers, contact:

- Aerospace Industries Association, 1250 Eye St. NW, Washington, DC 20005. Internet: http://www.aia-aerospace.org
- American Institute of Aeronautics and Astronautics, Inc., Suite 500, 1801 Alexander Bell Dr., Reston, VA 20191-4344. Internet: http://www.aiaa.org

High school students interested in obtaining a full package of guidance materials and information (product number SP-01) on a variety of engineering disciplines should contact the Junior Engineering Technical Society by sending $3.50 to:

- JETS-Guidance, 1420 King St., Suite 405, Alexandria, VA 22314-2794. Internet: http://www.jets.org

High school students interested in obtaining information on ABET-accredited engineering programs should contact:

- The Accreditation Board for Engineering and Technology, Inc., 111 Market Place, Suite 1050, Baltimore, MD 21202-4012. Internet: http://www.abet.org

Non-licensed engineers and college students interested in obtaining information on Professional Engineer licensure should contact:

- The National Society of Professional Engineers, 1420 King St., Alexandria, VA 22314-2794. Internet: http://www.nspe.org
- National Council of Examiners for Engineers and Surveying, P.O. Box 1686, Clemson, SC 29633-1686. Internet: http://www.ncees.org

Information on general engineering education and career resources is available from:

- American Society for Engineering Education, 1818 N St. N.W., Suite 600, Washington, DC 20036-2479. Internet: http://www.asee.org

Information on obtaining an engineering position with the federal government is available from the Office of Personnel Management (OPM) through a telephone-based system. Consult your telephone directory under U.S. Government for a local number or call (912) 757-3000; Federal Relay Service: (800) 877-8339. The first number is not toll free, and charges may result. Information also is available from the OPM Internet site: http://www.usajobs.opm.gov

Non-high school students and those wanting more detailed information should contact societies representing the individual branches of engineering. Each can provide information about careers in the particular branch. The individual statements elsewhere in this book also provide other information in detail on agricultural; biomedical; chemical; civil; computer hardware; electrical and electronics; environmental; industrial, including health and safety; materials; mechanical; mining and geological, including mining safety; nuclear; and petroleum engineering.

Agricultural and Food Scientists

O*NET 19-1011.00, 19-1012.00, 19-1013.01, 19-1013.02

Significant Points

- A large proportion, about 41 percent, of salaried agricultural and food scientists works for federal, state, and local governments.
- A bachelor's degree in agricultural science is sufficient for some jobs in applied research; a master's or doctoral degree is required for basic research.

Nature of the Work

The work of agricultural and food scientists plays an important part in maintaining the nation's food supply by ensuring agricultural productivity and the safety of the food supply. Agricultural scientists study farm crops and animals, and develop ways of improving their quantity and quality. They look for ways to improve crop yield with less labor, control pests and weeds more safely and effectively, and conserve soil and water. They research methods of converting raw agricultural commodities into attractive and healthy food products for consumers.

Agricultural science is closely related to biological science, and agricultural scientists use the principles of biology, chemistry, physics, mathematics, and other sciences to solve problems in agriculture. They often work with biological scientists on basic biological research and on applying to agriculture the advances in knowledge brought about by biotechnology.

Many agricultural scientists work in basic or applied research and development. Others manage or administer research and development programs, or manage marketing or production operations in companies that produce food products or agricultural chemicals, supplies, and machinery. Some agricultural scientists are consultants to business firms, private clients, or government.

Depending on the agricultural or food scientist's area of specialization, the nature of the work performed varies.

Food science. Food scientists and technologists usually work in the food processing industry, universities, or the federal government, and help meet consumer demand for food products that are healthful, safe, palatable, and convenient. To do this, they use their knowledge of chemistry, microbiology, and other sciences to develop new or better ways of preserving, processing, packaging, storing, and delivering foods. Some food scientists engage in basic research, discovering new food sources; analyzing food content to determine levels of vitamins, fat, sugar, or protein; or searching for substitutes for harmful or undesirable additives, such as nitrites. They also develop ways to process, preserve, package, or store food according to industry and government regulations. Others enforce government regulations, inspecting food processing areas and ensuring that sanitation, safety, quality, and waste management standards are met. Food technologists generally work in product development, applying the findings from food science research to the selection, preservation, processing, packaging, distribution, and use of safe, nutritious, and wholesome food.

Plant science. Agronomy, crop science, entomology, and plant breeding are included in plant science. Scientists in these disciplines study plants and their growth in soils, helping producers of food, feed, and fiber crops to continue to feed a growing population while conserving natural resources and maintaining the environment. Agronomists and crop scientists not only help increase productivity, but also study ways to improve the nutritional value of crops and the quality of seed. Some crop scientists study the breeding, physiology, and management of crops and use genetic engineering to develop crops resistant to pests and drought. Entomologists conduct research to develop new technologies to control or eliminate pests in infested areas and to prevent the spread of harmful pests to new areas, as well as technologies that are compatible with the environment. They also conduct research or engage in oversight activities aimed at halting the spread of insect-borne disease.

Soil science. Soil scientists study the chemical, physical, biological, and mineralogical composition of soils as they relate to plant or crop growth. They also study the responses of various soil types to fertilizers, tillage practices, and crop rotation. Many soil scientists who work for the federal government conduct soil surveys, classifying and mapping soils. They provide information and recommendations to farmers and other landowners regarding the best use of land, plant growth, and methods to avoid or correct problems such as erosion. They may also consult with engineers and other technical personnel working on construction projects about the effects of, and solutions to, soil problems. Because soil science is closely related to environmental science, persons trained in soil science also apply their knowledge to ensure environmental quality and effective land use.

Animal science. Animal scientists work to develop better, more efficient ways of producing and processing meat, poultry, eggs, and milk. Dairy scientists, poultry scientists, animal breeders, and other related scientists study the genetics, nutrition, reproduction, growth, and development of domestic farm animals. Some animal scientists inspect and grade livestock food products, purchase livestock, or work in technical sales or marketing. As extension agents or consultants, animal scientists advise agricultural producers on how to upgrade animal housing facilities properly, lower mortality rates, handle waste matter, or increase production of animal products, such as milk or eggs.

Working Conditions

Agricultural scientists involved in management or basic research tend to work regular hours in offices and laboratories. The work environment for those engaged in applied research or product development varies, depending on the discipline of agricultural science and on the type of employer. For example, food scientists in private industry may work in test kitchens while investigating new processing techniques. Animal scientists working for federal, state, or university research stations may spend part of their time at dairies, farrowing houses, feedlots, or farm animal facilities or outdoors conducting research associated with livestock. Soil and crop scientists also spend time outdoors conducting research on farms and agricultural research stations. Entomologists work in laboratories, insectories, or agricultural research stations, and may also spend time outdoors studying or collecting insects in their natural habitat.

Employment

Agricultural and food scientists held about 17,000 jobs in 2000. In addition, several thousand persons held agricultural science faculty positions in colleges and universities.

About 41 percent of all nonfaculty salaried agricultural and food scientists work for federal, state, or local governments. Nearly 2 out of 3 worked for the federal government in 2000, mostly in the Department of Agriculture. In addition, large numbers worked for state governments at state agricultural colleges or agricultural research stations. Some worked for agricultural service companies; others worked for commercial research and development laboratories, seed companies, pharmaceutical companies, wholesale distributors, and food products companies. About 4,000 agricultural scientists were self-employed in 2000, mainly as consultants.

Training, Other Qualifications, and Advancement

Training requirements for agricultural scientists depend on their specialty and on the type of work they perform. A bachelor's degree in agricultural science is sufficient for some jobs in applied research or for assisting in basic research, but a master's or doctoral degree is required for basic research. A PhD in agricultural science usually is needed for college teaching and for advancement to administrative research positions. Degrees in related sciences such as biology, chemistry, or physics or in related engineering specialties also may qualify persons for some agricultural science jobs.

All states have a land-grant college that offers agricultural science degrees. Many other colleges and universities also offer agricultural science degrees or some agricultural science courses. However, not every school offers all specialties. A typical undergraduate agricultural science curriculum includes communications, economics, business, and physical and life sciences courses, in addition to a wide variety of technical agricultural science courses. For prospective animal scientists, these technical agricultural science courses might include animal breeding, reproductive physiology, nutrition, and meats and muscle biology.

Students preparing as food scientists take courses such as food chemistry, food analysis, food microbiology, food engineering, and food processing operations. Those preparing as crop or soil scientists take courses in plant pathology, soil chemistry, entomology, plant physiology, and biochemistry, among others. Advanced degree programs include classroom and fieldwork, laboratory research, and a thesis or dissertation based on independent research.

Agricultural and food scientists should be able to work independently or as part of a team and be able to communicate clearly and concisely, both orally and in writing. Most of these scientists also need an understanding of basic business principles, and the ability to apply basic statistical techniques. Employers increasingly prefer job applicants who are able to apply computer skills to determine solutions to problems, to collect and analyze data, and for the control of processes.

The American Society of Agronomy offers certification programs in crops, agronomy, crop advising, soils, horticulture, plant pathology, and weed science. To become certified, applicants must pass designated examinations and meet certain standards with respect to education and professional work experience.

Agricultural scientists who have advanced degrees usually begin in research or teaching. With experience, they may advance to jobs such as supervisors of research programs or managers of other agriculture-related activities.

Job Outlook

Employment of agricultural scientists is expected to grow more slowly than the average for all occupations through 2010. Additionally, the need to replace agricultural and food scientists who retire or otherwise leave the occupation permanently will account for many more job openings than will projected growth, particularly in academia.

Past agricultural research has resulted in the development of higher yielding crops, crops with better resistance to pests and plant pathogens, and chemically based fertilizers and pesticides. Further research is necessary as insects and diseases continue to adapt to pesticides, and as soil fertility and water quality continue to need improvement. Agricultural scientists are using new avenues of research in biotechnology to develop plants and food crops that require less fertilizer, fewer pesticides and herbicides, and even less water for growth. Agricultural scientists will be needed to balance increased agricultural output with protection and preservation of soil, water, and ecosystems. They will increasingly encourage the practice of "sustainable agriculture" by developing and implementing plans to manage pests, crops, soil fertility and erosion, and animal waste in ways that reduce the use of harmful chemicals and do little damage to the natural environment. Also, an expanding population and an increasing public focus on diet, health, and food safety will result in job opportunities for food scientists and technologists.

Graduates with advanced degrees will be in the best position to enter jobs as agricultural scientists. Bachelor's degree holders can work in some applied research and product development positions, but usually only in certain subfields, such as food science and technology. Also, the federal government hires bachelor's degree holders to work as soil scientists. Despite the more limited opportunities for those with only a bachelor's degree to obtain jobs as agricultural scientists, a bachelor's degree in agricultural science is useful for managerial jobs in businesses that deal with ranchers and farmers. These businesses include feed, fertilizer, seed, and farm equipment manufacturers; retailers or wholesalers; and farm credit institutions. Four-year degrees also may help persons enter occupations such as farmer, or farm or ranch manager; cooperative extension service agent; agricultural products inspector; or purchasing or sales agent for agricultural commodity or farm supply companies.

Earnings

Median annual earnings of agricultural and food scientists were $52,160 in 2000. The middle 50 percent earned between $40,720 and $66,370. The lowest 10 percent earned less than $31,910, and the highest 10 percent earned more than $83,740.

Average federal salaries for employees in nonsupervisory, supervisory and managerial positions in certain agricultural science specialties in 2001 were as follows:

Animal science	$76,582
Entomology	70,133
Agronomy	62,311
Horticulture	59,472
Soil science	58,878

According to the National Association of Colleges and Employers, beginning salary offers in 2001 for graduates with a bachelor's degree in animal science averaged $28,031 a year.

Related Occupations

The work of agricultural scientists is closely related to that of biologists and other natural scientists, such as chemists, conservation scientists, and foresters. It is also related to managers of agricultural production, such as farmers, ranchers, agricultural managers. Certain specialties of agricultural science also are related to other occupations. For example, the work of animal scientists is related to that of veterinarians and horticulturists perform duties similar to those of landscape architects.

Sources of Additional Information

Information on careers in agricultural science is available from:

- American Society of Agronomy, Crop Science Society of America, Soil Science Society of America, 677 S. Segoe Rd., Madison, WI 53711-1086
- Food and Agricultural Careers for Tomorrow, Purdue University, 1140 Agricultural Administration Bldg., West Lafayette, IN 47907-1140

For information on careers in food technology, write to:

- Institute of Food Technologists, Suite 300, 221 N. LaSalle St., Chicago IL 60601-1291

Information on acquiring a job as an agricultural scientist with the federal government is available from the Office of Personnel Management through a telephone-based system. Consult your telephone directory under U.S. Government for a local number or call (912) 757-3000; Federal Relay Service: (800) 877-8339. The first number is not toll free, and charges may result. Information also is available from the Internet site: http://www.usajobs.opm.gov

Agricultural Engineers

O*NET 17-2021.00

Significant Points

- Overall job opportunities in engineering are expected to be good.
- A bachelor's degree is required for most entry-level jobs.
- Starting salaries are significantly higher than those of college graduates in other fields.
- Continuing education is critical to keep abreast of the latest technology.

Nature of the Work

Agricultural engineers apply knowledge of engineering technology and biological science to agriculture. They design agricultural machinery and equipment and agricultural structures. They develop ways to conserve soil and water and to improve the processing of agricultural products. Agricultural engineers work in research and development, production, sales, or management.

Working Conditions

Most engineers work in office buildings, laboratories, or industrial plants. Others may spend time outdoors at construction sites, mines, and oil and gas exploration and production sites, where they monitor or direct operations or solve onsite problems. Some engineers travel extensively to plants or work sites.

Many engineers work a standard 40-hour week. At times, deadlines or design standards may bring extra pressure to a job. When this happens, engineers may work longer hours and experience considerable stress.

Employment

More than one third of the 2,400 agricultural engineers employed in 2000 worked for engineering and management services, supplying consultant services to farmers and farm-related industries. Others worked in a wide variety of industries, including crops and livestock as well as manufacturing and government.

Training, Other Qualifications, and Advancement

A bachelor's degree in engineering is required for almost all entry-level engineering jobs. College graduates with a degree in a physical science or mathematics occasionally may qualify for some engineering jobs, especially in specialties in high demand. Most engineering degrees are granted in electrical, electronics, mechanical, or civil engineering. However, engineers trained in one branch may work in related branches. This flexibility allows employers to meet staffing needs in new technologies and specialties in which engineers are in short supply. It also allows engineers to shift to fields with better employment prospects or to those that more closely match their interests.

Most engineering programs involve a concentration of study in an engineering specialty, along with courses in both mathematics and science. Most programs include a design course, sometimes accompanied by a computer or laboratory class or both.

In addition to the standard engineering degree, many colleges offer 2- or 4-year degree programs in engineering technology. These programs, which usually include various hands-on laboratory classes that focus on current issues, prepare students for practical design and production work, rather than for jobs which require more theoretical and scientific knowledge. Graduates of 4-year technology programs may get jobs similar to those obtained by graduates with a bachelor's degree in engineering. Engineering technology graduates, however, are not qualified to register as professional engineers under the same terms as graduates with degrees in engineering. Some employers regard technology program graduates as having skills between those of a technician and an engineer.

Graduate training is essential for engineering faculty positions and many research and development programs, but is not required for the majority of entry-level engineering jobs. Many engineers obtain graduate degrees in engineering or business administration to learn new technology and broaden their education. Many high-level executives in government and industry began their careers as engineers.

About 330 colleges and universities offer bachelor's degree programs in engineering that are accredited by the Accreditation Board for Engineering and Technology (ABET), and about 250 colleges offer accredited bachelor's degree programs in engineering technology. ABET accreditation is based on an examination of an engineering program's student achievement, program improvement, faculty, curricular content, facilities, and institutional commitment. Although most institutions offer programs in the major branches of engineering, only a few offer programs in the smaller specialties. Also, programs of the same title may vary in content. For example, some programs emphasize industrial practices, preparing students for a job in industry, whereas others are more theoretical and are designed to prepare students for graduate work. Therefore, students should investigate curricula and check accreditations carefully before selecting a college.

Admissions requirements for undergraduate engineering schools include a solid background in mathematics (algebra, geometry, trigonometry, and calculus) and sciences (biology, chemistry, and physics), and courses in English, social studies, humanities, and computers. Bachelor's degree programs in engineering typically are designed to last 4 years, but many students find that it takes between 4 and 5 years to complete their studies. In a typical 4-year college curriculum, the first 2 years are spent studying mathematics, basic sciences, introductory engineering, humanities, and social sciences. In the last 2 years, most courses are in engineering, usually with a concentration in one branch. Some programs offer a general engineering curriculum; students then specialize in graduate school or on the job.

Some engineering schools and 2-year colleges have agreements whereby the 2-year college provides the initial engineering education, and the engineering school automatically admits students for their last 2 years. In addition, a few engineering schools have arrangements whereby a student spends 3 years in a liberal arts college studying pre-engineering subjects and 2 years in an engineering school studying core subjects, and then receives a bachelor's degree from each school. Some colleges and universities offer 5-year master's degree programs. Some 5- or even 6-year cooperative plans combine classroom study and practical work, permitting students to gain valuable experience and finance part of their education. All 50 states and the District of Columbia usually require licensure for engineers who offer their services directly to the public. Engineers who are licensed are called Professional Engineers (PE). This licensure generally requires a degree from an ABET-accredited engineering program, 4 years of relevant work experience, and successful completion of a state examination. Recent graduates can start the licensing process by taking the examination in two stages. The initial Fundamentals of Engineering (FE) examination can be taken upon graduation. Engineers who pass this examination commonly are called Engineers in Training (EIT) or Engineer Interns (EI). The EIT certification usually is valid for 10 years. After acquiring suitable work experience, EITs can take the second examination, the Principles and Practice of Engineering Exam. Several states have imposed mandatory continuing education requirements for relicensure. Most states recognize licensure from other states.

Engineers should be creative, inquisitive, analytical, and detail oriented. They should be able to work as part of a team and to communicate well, both orally and in writing. Communication abilities are becoming more important because much of their work is becoming more diversified, meaning that engineers interact with specialists in a wide range of fields outside engineering.

Beginning engineering graduates usually work under the supervision of experienced engineers and, in large companies, also may receive formal classroom or seminar-type training. As new engineers gain knowledge and experience, they are assigned more difficult projects with greater independence to develop designs, solve problems, and make decisions. Engineers may advance to become technical specialists or to supervise a staff or team of engineers and technicians. Some may eventually become engineering managers or enter other managerial or sales jobs.

Job Outlook

Employment of agricultural engineers is expected to increase about as fast as the average for all occupations through 2010. Increasing demand for agricultural products, continued efforts for more efficient agricultural production, and increasing emphasis on the conservation of resources should result in job opportunities for agricultural engineers. However, most openings will be created by the need to replace agricultural engineers who transfer to other occupations or leave the labor force.

Earnings

Median annual earnings of agricultural engineers were $55,850 in 2000. The middle 50 percent earned between $44,220 and $71,460. The lowest 10 percent earned less than $33,660, and the highest 10 percent earned more than $91,600.

According to a 2001 salary survey by the National Association of Colleges and Employers, bachelor's degree candidates in agricultural engineering received starting offers averaging $46,065 a year, and master's degree candidates, on average, were offered $49,808.

Related Occupations

Agricultural engineers apply the principles of physical science and mathematics in their work. Other workers who use scientific and mathematical principles include architects, except landscape and naval; engineering and natural sciences managers; computer and information systems managers; mathematicians; drafters; engineering technicians; sales engineers; science technicians; and physical and life scientists, including agricultural and food scientists, biological and medical scientists, conservation scientists and foresters, atmospheric scientists, chemists and materials scientists, environmental scientists and geoscientists, and physicists and astronomers.

Sources of Additional Information

General information about agricultural engineers can be obtained from:

- American Society of Agricultural Engineers, 2950 Niles Rd., St. Joseph, MI 49085-9659. Internet: http://www.asae.org

High school students interested in obtaining a full package of guidance materials and information (product number SP-01) on a variety of engineering disciplines should contact the Junior Engineering Technical Society by sending $3.50 to:

- JETS-Guidance, 1420 King St., Suite 405, Alexandria, VA 22314-2794. Internet: http://www.jets.org

High school students interested in obtaining information on ABET-accredited engineering programs should contact:

- The Accreditation Board for Engineering and Technology, Inc., 111 Market Place, Suite 1050, Baltimore, MD 21202-4012. Internet: http://www.abet.org

Non-licensed engineers and college students interested in obtaining information on Professional Engineer licensure should contact:

- The National Society of Professional Engineers, 1420 King St., Alexandria, VA 22314-2794. Internet: http://www.nspe.org
- National Council of Examiners for Engineers and Surveying, P.O. Box 1686, Clemson, SC 29633-1686. Internet: http://www.ncees.org

Information on general engineering education and career resources is available from:

- American Society for Engineering Education, 1818 N St. N.W., Suite 600, Washington, DC 20036-2479. Internet: http://www.asee.org

Information on obtaining an engineering position with the federal government is available from the Office of Personnel Management (OPM) through a telephone-based system. Consult your telephone directory under U.S. Government for a local number or call (912) 757-3000; Federal Relay Service: (800) 877-8339. The first number is not toll free, and charges may result. Information also is available from the OPM Internet site: http://www.usajobs.opm.gov

Non-high school students and those wanting more detailed information should contact societies representing the individual branches of engineering. Each can provide information about careers in the particular branch. The individual statements elsewhere in this book also provide other information in detail on aerospace; biomedical; chemical; civil; computer hardware; electrical and electronics; environmental; industrial, including health and safety; materials; mechanical; mining and geological, including mining safety; nuclear; and petroleum engineering.

Aircraft and Avionics Equipment Mechanics and Service Technicians

O*NET 49-2091.00, 49-3011.01, 49-3011.02, 49-3011.03

Significant Points

- The majority of these workers learn their job in 1 of about 200 trade schools certified by the Federal Aviation Administration.
- Opportunities should be favorable, but keen competition is likely for the best paying airline jobs.

Nature of the Work

To keep aircraft in peak operating condition, aircraft and avionics equipment mechanics and service technicians perform scheduled maintenance, make repairs, and complete inspections required by the Federal Aviation Administration (FAA).

Many aircraft mechanics, also called airframe, powerplant, and avionics aviation maintenance technicians, specialize in preventive maintenance. They inspect engines, landing gear, instruments, pressurized sections, accessories—brakes, valves, pumps, and air-conditioning systems, for example—and other parts of the aircraft, and do the necessary maintenance and replacement of parts. Inspections take place following a schedule based on the number of hours the aircraft has flown, calendar days since the last inspection, cycles of operation, or a combination of these factors. Large, sophisticated planes are equipped with aircraft monitoring systems, consisting of electronic boxes and consoles that monitor the aircraft's basic operations and provide valuable diagnostic information to the mechanic. To examine an engine, aircraft mechanics work through specially designed openings while standing on ladders or scaffolds, or use hoists or lifts to remove the entire engine from the craft. After taking an engine apart, mechanics use precision instruments to measure parts for wear and use X-ray and magnetic inspection equipment to check for invisible cracks. Worn or defective parts are repaired or replaced. Mechanics may also repair sheet metal or composite surfaces, measure the tension of control cables, and check for corrosion, distortion, and cracks in the fuselage, wings, and tail. After completing all repairs, they must test the equipment to ensure that it works properly.

Mechanics specializing in repairwork rely on the pilot's description of a problem to find and fix faulty equipment. For example, during a preflight check, a pilot may discover that the aircraft's fuel gauge does not work. To solve the problem, mechanics may troubleshoot the electrical system, using electrical test equipment to make sure that no wires are broken or shorted out, and replace any defective electrical or electronic components. Mechanics work as fast as safety permits so that the aircraft can be put back into service quickly.

Some mechanics work on one or many different types of aircraft, such as jets, propeller-driven airplanes, and helicopters. Others specialize in one section of a particular type of aircraft, such as the engine, hydraulics, or electrical system. *Powerplant mechanics* are authorized to work on engines and do limited work on propellers. *Airframe mechanics* are authorized to work on any part of the aircraft except the instruments, powerplants, and propellers. *Combination airframe-and-powerplant mechanics*—called A & P mechanics—work on all parts of the plane, except instruments. The majority of mechanics working on civilian aircraft today are A & P mechanics. In small, independent repairshops, mechanics usually inspect and repair many different types of aircraft.

Avionics systems are now an integral part of aircraft design and have vastly increased aircraft capability. *Avionics technicians* repair and maintain components used for aircraft navigation and radio communications, weather radar systems, and other instruments and computers that control flight, engine, and other primary functions. These duties may require additional licenses, such as a radiotelephone license issued by the U.S. Federal Communications Commission (FCC). Because of technological advances, an increasing amount of time is spent repairing electronic systems, such as computerized controls. Technicians also may be required to analyze and develop solutions to complex electronic problems.

Working Conditions

Mechanics usually work in hangars or in other indoor areas, although they can work outdoors—sometimes in unpleasant weather—when hangars are full or when repairs must be made quickly. Mechanics often work under time pressure to maintain flight schedules or, in general aviation, to keep from inconveniencing customers. At the same time, mechanics have a tremendous responsibility to maintain safety standards, and this can cause the job to be stressful.

Frequently, mechanics must lift or pull objects weighing as much as 70 pounds. They often stand, lie, or kneel in awkward positions and occasionally must work in precarious positions on scaffolds or ladders. Noise and vibration are common when engines are being tested, so ear protection is necessary. Aircraft mechanics usually work 40 hours a week on 8-hour shifts around the clock. Overtime work is frequent.

Employment

Aircraft mechanics and service technicians held about 173,000 jobs in 2000; fewer than 10 percent were avionic technicians. About two-thirds of all salaried mechanics worked for airlines or airports and flying fields, about 12 percent worked for the federal government, and about 9 percent worked for aircraft assembly firms. Most of the rest were general aviation mechanics, the majority of whom worked for independent repairshops or for companies that operate their own planes to transport executives and cargo. Few mechanics were self-employed.

Most airline mechanics work at major airports near large cities. Civilian mechanics employed by the armed forces work at military installations. Large proportions of mechanics who work for aircraft assembly firms are located in California or in Washington state. Others work for the FAA, many at the facilities in Oklahoma City, Atlantic City, Wichita, or Washington, DC. Mechanics for independent repairshops work at airports in every part of the country.

Training, Other Qualifications, and Advancement

The majority of mechanics who work on civilian aircraft are certified by the FAA as "airframe mechanic," "powerplant mechanic," or "avionics repair specialist." Mechanics who also have an inspector's authorization can certify work completed by other mechanics and perform required inspections. Uncertified mechanics are supervised by those with certificates.

The FAA requires at least 18 months of work experience for an airframe, powerplant, or avionics repairer's certificate. For a combined A & P certificate, at least 30 months of experience working with both engines and airframes is required. Completion of a program at an FAA-certified mechanic school can substitute for the work experience requirement. Applicants for all certificates also must pass written and oral tests and demonstrate that they can do the work authorized by the certificate. To obtain an inspector's authorization, a mechanic must have held an A & P certificate for at least 3 years. Most airlines require that mechanics have a high school diploma and an A & P certificate.

Although a few people become mechanics through on-the-job training, most learn their job in 1 of about 200 trade schools certified by the FAA. About one-third of these schools award 2- and 4-year degrees in avionics, aviation technology, or aviation maintenance management.

FAA standards established by law require that certified mechanic schools offer students a minimum of 1,900 actual class hours. Courses in these trade schools normally last from 24 to 30 months and provide training with the tools and equipment used on the job. Aircraft trade schools are placing more emphasis on technologies such as turbine engines, composite materials—including graphite, fiberglass, and boron—and aviation electronics, which are increasingly being used in the construction of new aircraft. Less emphasis is being placed on old technologies, such as woodworking and welding. Additionally, employers prefer mechanics who can perform a variety of tasks.

Some aircraft mechanics in the armed forces acquire enough general experience to satisfy the work experience requirements for the FAA certificate. With additional study, they may pass the certifying exam. In general, however, jobs in the military services are too specialized to provide the broad experience required by the FAA. Most armed forces mechanics have to complete the entire training program at a trade school, although a few receive some credit for the material they learned in the service. In any case, military experience is a great advantage when seeking employment; employers consider trade school graduates who have this experience to be the most desirable applicants.

Courses in mathematics, physics, chemistry, electronics, computer science, and mechanical drawing are helpful, because they demonstrate many of the principles involved in the operation of aircraft, and knowledge of these principles is often necessary to make repairs. Courses that develop writing skills also are important because mechanics often are required to submit reports.

FAA regulations require current experience to keep the A & P certificate valid. Applicants must have at least 1,000 hours of work experience in the previous 24 months or take a refresher course. As new and more complex aircraft are designed, more employers are requiring mechanics to take ongoing training to update their skills. Recent technological advances in aircraft maintenance necessitate a strong background in electronics—both for acquiring and retaining jobs in this field. FAA certification standards also make ongoing training mandatory. Every 24 months, mechanics are required to take at least 16 hours of training to keep their certificate. Many mechanics take courses offered by manufacturers or employers, usually through outside contractors.

Aircraft mechanics must do careful and thorough work that requires a high degree of mechanical aptitude. Employers seek applicants who are self-motivated, hard-working, enthusiastic, and able to diagnose and solve complex mechanical problems. Agility is important for the reaching and climbing necessary to do the job. Because they may work on the tops of wings and fuselages on large jet planes, aircraft mechanics must not be afraid of heights.

As aircraft mechanics gain experience, they may advance to lead mechanic (or crew chief), inspector, lead inspector, or shop supervisor positions. Opportunities are best for those who have an aircraft inspector's authorization. In the airlines, where promotion often is determined by examination, supervisors sometimes advance to executive positions. Those with broad experience in maintenance and overhaul might become inspectors with the FAA. With additional business and management training, some open their own aircraft maintenance facilities. Mechanics learn many different skills in their training that can be applied to other jobs, and some transfer to other skilled repairer occupations or electronics technician jobs.

Job Outlook

The outlook for aircraft and avionics equipment mechanics and service technicians should be favorable over the next 10 years. The likelihood of fewer entrants from the military and a large number of retirements, point to good employment conditions for students just beginning training.

Job opportunities are likely to be the best at small commuter and regional airlines, at FAA repair stations, and in general aviation. Wages in these companies tend to be relatively low, so there are fewer applicants for these jobs than for those with the major airlines. Also, some jobs will become available as experienced mechanics leave for higher paying jobs with airlines or transfer to another occupation. At the same time, aircraft are becoming increasingly sophisticated in general aviation and in regional carriers, boosting the demand for qualified mechanics. Mechanics will face competition for jobs with large airlines because the high wages and travel benefits that these jobs offer attract more qualified applicants than there are openings. Prospects will be best for applicants with significant experience. Mechanics who keep abreast of technological advances in electronics, composite materials, and other areas will be in greatest demand. The number of job openings for aircraft mechanics in the federal government should decline as the size of the U.S. Armed Forces is reduced.

Employment of aircraft mechanics is expected to increase about as fast as the average for all occupations through the year 2010. A growing population and rising incomes are expected to stimulate the demand for airline transportation, and the number of aircraft is expected to grow. However, employment growth will be somewhat restricted as consolidation within the air carrier industry continues and as productivity increases due to greater use of automated inventory control and modular systems, which speeds repairs and parts replacement.

Most job openings for aircraft mechanics through the year 2010 will stem from replacement needs. Each year, as mechanics transfer to other occupations or retire, several thousand job openings will arise. Aircraft mechanics have a comparatively strong attachment to the occupation, reflecting their significant investment in training and a love for aviation. However, because aircraft mechanics' skills are transferable to other occupations, some mechanics leave for work in related fields.

During recessions, declines in air travel force airlines to curtail the number of flights, which results in less aircraft maintenance and, consequently, layoffs for aircraft mechanics.

Earnings

Median hourly earnings of aircraft mechanics and service technicians were about $19.50 in 2000. The middle 50 percent earned between $15.65 and $23.65. The lowest 10 percent earned less than $12.06, and the highest 10 percent earned more than $26.97. Median hourly earnings in the industries employing the largest numbers of aircraft mechanics and service technicians in 2000 are shown in the following table:

Air transportation, scheduled	$21.57
Aircraft and parts	19.77
Air transportation, nonscheduled	19.16
Federal government	19.11
Airports, flying fields, and services	16.26

Median hourly earnings of avionics technicians were about $19.86 in 2000. The middle 50 percent earned between $16.31 and $24.01. The lowest 10 percent earned less than $13.22, and the highest 10 percent earned more than $27.02.

Mechanics who work on jets for the major airlines generally earn more than those working on other aircraft. Airline mechanics and their immediate families receive reduced-fare transportation on their own and most other airlines.

Almost one-half of all aircraft mechanics, including those employed by some major airlines, are covered by union agreements. The principal unions are the International Association of Machinists and Aerospace Workers and the Transport Workers Union of America. Some mechanics are represented by the International Brotherhood of Teamsters.

Related Occupations

Workers in some other occupations that involve similar mechanical and electrical work are electricians, electrical and electronics installers and repairers, and elevator installers and repairers.

Sources of Additional Information

Information about jobs with a particular airline can be obtained by writing to the personnel manager of the company.

For general information about aircraft and avionics equipment mechanics and service technicians, write to:

- Professional Aviation Maintenance Association, 1707 H St. N.W., Suite 700, Washington, DC 20006

For information on jobs in a particular area, contact employers at local airports or local offices of the state employment service.

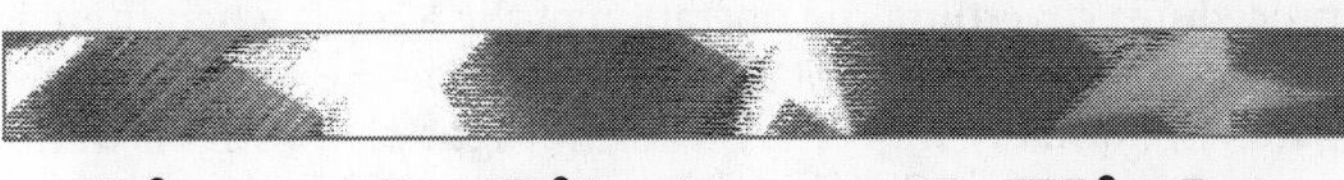

Aircraft Pilots and Flight Engineers

O*NET 53-2011.00, 53-2012.00

Significant Points

- Strong competition is expected for jobs because aircraft pilots have very high earnings, especially those employed by national airlines.
- Pilots usually start with smaller commuter and regional airlines to acquire the experience needed to qualify for higher paying jobs with national airlines.
- Most pilots traditionally have learned to fly in the military, but growing numbers have college degrees with flight training from civilian flying schools that are certified by the Federal Aviation Administration (FAA).

Nature of the Work

Pilots are highly trained professionals who fly airplanes and helicopters to carry out a wide variety of tasks. Four out of five are *airline pilots, copilots,* and *flight engineers* who transport passengers and cargo. Others are *commercial pilots* involved in more unusual tasks, such as dusting crops, spreading seed for reforestation, testing aircraft, flying passengers and cargo to areas not service by regular airlines, directing firefighting efforts, tracking criminals, monitoring traffic, and rescuing and evacuating injured persons.

Except on small aircraft, two pilots usually make up the cockpit crew. Generally, the most experienced pilot, the *captain,* is in command and supervises all other crew members. The pilot and copilot share flying and other duties, such as communicating with air traffic controllers and monitoring the instruments. Some large aircraft have a third pilot—the *flight engineer*—who assists the other pilots by monitoring and operating many of the instruments and systems, making minor in-flight repairs, and watching for other aircraft. New technology can perform many flight tasks, however, and virtually all-new aircraft now fly with only two pilots, who rely more heavily on computerized controls. As older, less technologically sophisticated aircraft continue to be retired from airline fleets, the number of flight engineer jobs will decrease.

Before departure, pilots plan their flights carefully. They thoroughly check their aircraft to make sure that the engines, controls, instruments, and other systems are functioning properly. They also make sure that baggage or cargo has been loaded correctly. They confer with flight dispatchers and aviation weather forecasters to find out about weather conditions en route and at their destination. Based on this information, they choose a route, altitude, and speed that will provide the fastest, safest, and smoothest flight. When flying under instrument flight rules—procedures governing the operation of the aircraft when there is poor visibility—the pilot in command, or the company dispatcher, normally files an instrument flight plan with air traffic control so that the flight can be coordinated with other air traffic.

Takeoff and landing are the most difficult parts of the flight, and require close coordination between the pilot and first officer. For example, as the plane accelerates for takeoff, the pilot concentrates on the runway while the first officer scans the instrument panel. To calculate the speed they must attain to become airborne, pilots consider the altitude of the airport, outside temperature, weight of the plane, and speed and direction of the wind. The moment the plane reaches takeoff speed, the first officer informs the pilot, who then pulls back on the controls to raise the nose of the plane.

Unless the weather is bad, the actual flight is relatively easy. Airplane pilots, with the assistance of autopilot and the flight management computer, steer the plane along their planned route and are monitored by the air traffic control stations they pass along the way. They regularly scan the instrument panel to check their fuel supply, the condition of their engines, and the air-conditioning, hydraulic, and other systems. Pilots may request a change in altitude or route if circumstances dictate. For example, if the ride is rougher than expected, they may ask air traffic control if pilots flying at other altitudes have reported better conditions. If so, they may request an altitude change. This procedure also may be used to find a stronger tailwind or a weaker headwind to save fuel and increase speed.

In contrast, helicopters are used for short trips at relatively low altitude, so pilots must be constantly on the lookout for trees, bridges, power lines, transmission towers, and other dangerous obstacles. Regardless of the type of aircraft, all pilots must monitor warning devices designed to help detect sudden shifts in wind conditions that can cause crashes.

Pilots must rely completely on their instruments when visibility is poor. On the basis of altimeter readings, they know how high above ground they are and whether they can fly safely over mountains and other obstacles. Special navigation radios give pilots precise information that, with the help of special maps, tells them their exact position. Other very sophisticated equipment provides directions to a point just above the end of a runway and enables pilots to land completely "blind." Once on the ground, pilots must complete records on their flight for their organization and the FAA report.

The number of nonflying duties that pilots have depends on the employment setting. Airline pilots have the services of large support staffs, and consequently, perform few nonflying duties. Pilots employed by other organizations such as charter operators or businesses have many other duties. They may load the aircraft, handle all passenger luggage to ensure a balanced load, and supervise refueling; other nonflying responsibilities include keeping records, scheduling flights, arranging for major maintenance, and performing minor aircraft maintenance and repairwork.

Some pilots are instructors. They teach their students the principles of flight in ground-school classes and demonstrate how to operate aircraft in dual-controlled planes and helicopters. A few specially trained pilots are "examiners" or "check pilots." They periodically fly with other pilots or pilot's license applicants to make sure that they are proficient.

Working Conditions

By law, airline pilots cannot fly more than 100 hours a month or more than 1,000 hours a year. Most airline pilots fly an average of 75 hours a month and work an additional 75 hours a month performing nonflying duties. About one-fourth of all pilots work more than 40 hours a week. Most spend a considerable amount of time away from home because the majority of flights involve overnight layovers. When pilots are away from home, the airlines provide hotel accommodations, transportation between the hotel and airport, and an allowance for meals and other expenses. Airlines operate flights at all hours of the day and night, so work schedules often are irregular. Flight assignments are based on seniority.

Those pilots not employed by the airlines often have irregular schedules as well; they may fly 30 hours one month and 90 hours the next. Because these pilots frequently have many nonflying responsibilities, they have much less free time than do airline pilots. Except for business pilots, most do not remain away from home overnight. They may work odd hours. Flight instructors may have irregular and seasonal work schedules, depending on their students' available time and the weather. Instructors frequently work at night or on weekends.

Airline pilots, especially those on international routes, often suffer jet lag—fatigue caused by many hours of flying through different time zones. To guard against excessive pilot fatigue that could result in unsafe flying conditions, the FAA requires airlines to allow pilots at least 8 hours of uninterrupted rest in the 24 hours before finishing their flight duty. The work of test pilots, who check the flight performance of new and experimental planes, may be dangerous. Pilots who are crop dusters may be exposed to toxic chemicals and seldom have the benefit of a regular landing strip. Helicopter pilots involved in police work may be subject to personal injury.

Although flying does not involve much physical effort, the mental stress of being responsible for a safe flight, no matter what the weather, can be tiring. Pilots must be alert and quick to react if something goes wrong, particularly during takeoff and landing.

Employment

Civilian aircraft pilots and flight engineers commercial held about 117,000 jobs in 2000. About 84 percent worked as airline pilots, copilots, and flight engineers. The remainder were commercial pilots who worked as flight instructors at local airports or for large businesses that fly company cargo and executives in their own airplanes or helicopters. Some commercial pilots flew small planes for air taxi companies, usually to or from lightly traveled airports not served by major airlines. Others worked for a variety of businesses, performing tasks such as crop dusting, inspecting pipelines, or conducting sightseeing trips. Federal, state, and local governments also employed pilots. A few pilots were self-employed.

The employment of airplane pilots is not distributed like the population. Pilots are more concentrated in California, New York, Illinois, Washington, Michigan, Georgia, New Jersey, Florida, the District of Columbia, and Texas, which have a high amount of flying activity relative to their population.

Training, Other Qualifications, and Advancement

All pilots who are paid to transport passengers or cargo must have a commercial pilot's license with an instrument rating issued by the FAA. Helicopter pilots must hold a commercial pilot's certificate with a helicopter rating. To qualify for these licenses, applicants must be at least 18 years old and have at least 250 hours of flight experience. The experience required can be reduced through participation in certain flight school curricula approved by the FAA. Applicants also must pass a strict physical examination to make sure that they are in good health and have 20/20 vision with or without glasses, good hearing, and no physical handicaps that could impair their performance. They must pass a written test that includes questions on the principles of safe flight, navigation techniques, and FAA regulations and must demonstrate their flying ability to FAA or designated examiners.

To fly in periods of low visibility, pilots must be rated by the FAA to fly by instruments. Pilots may qualify for this rating by having 105 hours of flight experience, including 40 hours of experience in flying by instruments; they also must pass a written examination on procedures and FAA regulations covering instrument flying and demonstrate to an examiner their ability to fly by instruments.

Airline pilots must fulfill additional requirements. Pilots must have an airline transport pilot's license. Applicants for this license must be at least 23 years old and have a minimum of 1,500 hours of flying experience, including night and instrument flying, and must pass FAA written and flight examinations. Usually, they also have one or more advanced ratings, such as multi-engine aircraft or aircraft type ratings dependent upon the requirements of their particular flying jobs. Because pilots must be able to make quick decisions and accurate judgments under pressure, many airline companies reject applicants who do not pass required psychological and aptitude tests. All licenses are valid so long as a pilot can pass the periodic physical examinations and tests of flying skills required by federal government and company regulations. Depending on their physical condition, a pilot license may have a Class I, II, and III Medical certificate. A Class I Medical Certificate requires the highest standards for vision, hearing, equilibrium, and general physical condition. Requirements for a Class II Medical Certificate are less rigid, but still require a high degree of physical health and an excellent medical history. A Class III Medical Certificate has the least stringent physical requirements. All three classes of medical certificates allow the pilot to wear glasses provided the correction is within the prescribed limits of vision.

The armed forces have always been an important source of trained pilots for civilian jobs. Military pilots gain valuable experience on jet aircraft and helicopters, and persons with this experience usually are preferred for civilian pilot jobs. This primarily reflects the extensive flying time military pilots receive. Persons without armed forces training may become pilots by attending flight schools. The FAA has certified about 600 civilian flying schools, including some colleges and universities that offer degree credit for pilot training. Over the projection period, federal budget reductions are expected to reduce military pilot training. As a result, FAA-certified schools will train a larger share of pilots than in the past. Prospective pilots also may learn to fly by taking lessons from individual FAA-certified flight instructors.

Although some small airlines will hire high school graduates, most airlines require at least 2 years of college and prefer to hire college graduates. In fact, most entrants to this occupation have a college degree. Because the number of college educated applicants continues to increase, many employers are making a college degree an educational requirement.

Depending on the type of aircraft, new airline pilots start as first officers or flight engineers. Although some airlines favor applicants who already have a flight engineer's license, they may provide flight engineer training for those who have only the commercial license. Many pilots begin with smaller regional or commuter airlines, where they obtain experience flying passengers on scheduled flights into busy airports in all weather conditions. These jobs often lead to higher paying jobs with bigger, national airlines.

Initial training for airline pilots includes a week of company indoctrination, 3 to 6 weeks of ground school and simulator training, and 25 hours of initial operating experience, including a check-ride with an FAA aviation safety inspector. Once trained and "on the line," pilots are required to attend recurrent training and simulator checks twice a year throughout their career.

Organizations other than airlines usually require less flying experience. However, a commercial pilot's license is a minimum requirement, and employers prefer applicants who have experience in the type of craft they will be flying. New employees usually start as first officers, or fly less sophisticated equipment. Test pilots often are required to have an engineering degree.

Advancement for all pilots usually is limited to other flying jobs. Many pilots start as flight instructors, building up their flying hours while they earn money teaching. As they become more experienced, these pilots occasionally fly charter planes or perhaps get jobs with small air transportation firms, such as air taxi companies. Some advance to business flying jobs. A small number get flight engineer jobs with the airlines.

In the airlines, advancement usually depends on seniority provisions of union contracts. After 1 to 5 years, flight engineers advance according to seniority to first officer and, after 5 to 15 years, to captain. Seniority also determines which pilots get the more desirable routes. In a nonairline job, a first officer may advance to pilot and, in large companies, to chief pilot or director of aviation in charge of aircraft scheduling, maintenance, and flight procedures.

Job Outlook

Pilots are expected to face strong competition for jobs through the year 2010. Many qualified persons seek jobs in this occupation because it offers very high earnings, glamour, prestige, and free or low-cost travel benefits. As time passes, some pilots will fail to maintain their qualifications, and the number of applicants competing for each opening should decline. Factors affecting demand, however, are not expected to ease that competition.

Relatively few jobs will be created from rising demand for pilots, even though employment is expected to increase about as fast as the average for all occupations through 2010. Expected growth in domestic and international airline passenger and cargo traffic will create a need for more airliners, pilots, and flight instructors. However, computerized flight management systems on new aircraft will continue to eliminate the need for flight engineers on those planes, thus restricting the growth of pilot employment. In addition, the trend toward using larger planes in the airline industry will increase pilot productivity. Future business travel could also be adversely affected by the growing use of teleconferencing, facsimile mail, and electronic communications—such as e-mail—as well as by the elimination of middle management positions in corporate downsizing. Employment of business pilots is expected to grow more slowly than in the past as more businesses opt to fly with regional and smaller airlines serving their area rather than to buy and operate their own aircraft. The number of job openings resulting from the need to replace pilots who retire or leave the occupation traditionally has been very low. Aircraft pilots usually have a strong attachment to their occupation because it requires a substantial investment in specialized training that is not transferable to other fields, and it commonly offers very high earnings. However, many of the pilots who were hired in the late 1960s are approaching the age for mandatory retirement and, thus, several thousand job openings are expected to be generated each year.

Pilots who have logged the greatest number of flying hours in the more sophisticated equipment typically have the best prospects. For this reason, military pilots often have an advantage over other applicants. Job seekers with the most FAA licenses also will have a competitive advantage. Opportunities for pilots in the regional commuter airlines and international service are expected to be more favorable, as these segments are expected to grow faster than other segments of the industry.

Employment of pilots is sensitive to cyclical swings in the economy. During recessions, when a decline in the demand for air travel forces airlines to curtail the number of flights, airlines may temporarily furlough some pilots. Commercial and corporate flying, flight instruction, and testing of new aircraft also decline during recessions, adversely affecting the employment of pilots in those areas.

Earnings

Earnings of aircraft pilots and flight engineers vary greatly depending whether they work as airline or commercial pilots. Earnings of airline pilots are among the highest in the nation, and depend on factors such as the type, size, and maximum speed of the plane and the number of hours and miles flown. For example, pilots who fly jet aircraft usually earn higher salaries than do pilots who fly turboprops. Airline pilots and flight engineers may earn extra pay for night and international flights. In 2000, median annual earnings of airline pilots, copilots, and flight engineers were $110,940. The lowest 10 percent earned less than $36,110. Over 25 percent earned more than $145,000.

Median annual earnings of commercial pilots were $43,300 in 2000. The middle 50 percent earned between $31,500 and $61,230. The lowest 10 percent earned less than $24,290, and the highest 10 percent earned more than $92,000.

Airline pilots usually are eligible for life and health insurance plans financed by the airlines. They also receive retirement benefits and, if they fail the FAA physical examination at some point in their careers, they get disability payments. In addition, pilots receive an expense allowance, or "per diem," for every hour they are away from home. Per diem can represent up to $500 each month in addition to their salary. Some airlines also provide allowances to pilots for purchasing and cleaning their uniforms. As an additional benefit, pilots and their immediate families usually are entitled to free or reduced fare transportation on their own and other airlines.

More than one-half of all aircraft pilots are members of unions. Most of the pilots who fly for the major airlines are members of the Airline Pilots Association, International, but those employed by one major airline are members of the Allied Pilots Association. Some flight engineers are members of the Flight Engineers' International Association.

Related Occupations

Although they are not in the cockpit, air traffic controllers and airfield operation specialists also play an important role in making sure flights are safe and on schedule, and participate in many of the decisions that pilots must make.

Sources of Additional Information

Information about job opportunities, salaries for a particular airline, and qualifications required may be obtained by writing to the personnel manager of the airline.

For information on airline pilots, contact:

- Airline Pilots Association, 1625 Massachusetts Ave. N.W., Washington, DC 20036
- Air Transport Association of America, Inc., 1301 Pennsylvania Ave. N.W., Suite 1100, Washington, DC 20004

For information on helicopter pilots, contact:

- Helicopter Association International, 1619 Duke St., Alexandria, VA 22314

For a copy of the List of Certificated Pilot Schools, write to:

- Superintendent of Documents, U.S. Government Printing Office, Washington, DC 20402. There is a charge for this publication.

For information about job opportunities in companies other than airlines, consult the classified section of aviation trade magazines and apply to companies that operate aircraft at local airports.

Air Traffic Controllers

O*NET 53-2021.00

Significant Points

- Nearly all air traffic controllers are employed and trained by the federal government.
- Keen competition is expected for the few job openings in this occupation.
- Aircraft controllers earn relatively high pay and have good benefits.

Nature of the Work

The air traffic control system is a vast network of people and equipment that ensures the safe operation of commercial and private aircraft. Air traffic controllers coordinate the movement of air traffic to make certain that planes stay a safe distance apart. Their immediate concern is safety, but controllers also must direct planes efficiently to minimize delays. Some regulate airport traffic; others regulate flights between airports.

Although *airport tower* or *terminal controllers* watch over all planes traveling through the airport's airspace, their main responsibility is to organize the flow of aircraft in and out of the airport. Relying on radar and visual observation, they closely monitor each plane to ensure a safe distance between all aircraft and to guide pilots between the hangar or ramp and the end of the airport's airspace. In addition, controllers keep pilots informed about changes in weather conditions such as wind shear—a sudden change in the velocity or direction of the wind that can cause the pilot to lose control of the aircraft.

During arrival or departure, several controllers direct each plane. As a plane approaches an airport, the pilot radios ahead to inform the terminal of its presence. The controller in the radar room, just beneath the control tower, has a copy of the plane's flight plan and already has observed the plane on radar. If the path is clear, the controller directs the pilot to a runway; if the airport is busy, the plane is fitted into a traffic pattern with other aircraft waiting to land. As the plane nears the runway, the pilot is asked to contact the tower. There, another controller, who also is watching the plane on radar, monitors the aircraft the last mile or so to the runway, delaying any departures that would interfere with the plane's landing. Once the plane has landed, a ground controller in the tower directs it along the taxiways to its assigned gate. The ground controller usually works entirely by sight, but may use radar if visibility is very poor.

The procedure is reversed for departures. The ground controller directs the plane to the proper runway. The local controller then informs the pilot about conditions at the airport, such as weather, speed and direction of wind, and visibility. The local controller also issues runway clearance for the pilot to take off. Once in the air, the plane is guided out of the airport's airspace by the departure controller.

After each plane departs, airport tower controllers notify *en route controllers* who will take charge next. There are 21 en route control centers located around the country, each employing 300 to 700 controllers, with more than 150 on duty during peak hours at the busier facilities. Airplanes usually fly along designated routes; each center is assigned a certain airspace containing many different routes. En route controllers work in teams of up to three members, depending on how heavy traffic is; each team is responsible for a section of the center's airspace. A team, for example, might be responsible for all planes that are between 30 to 100 miles north of an airport and flying at an altitude between 6,000 and 18,000 feet.

To prepare for planes about to enter the team's airspace, the radar associate controller organizes flight plans coming off a printer. If two planes are scheduled to enter the team's airspace at nearly the same time, location, and altitude, this controller may arrange with the preceding control unit for one plane to change its flight path. The previous unit may have been another team at the same or an adjacent center, or a departure controller at a neighboring terminal. As a plane approaches a team's airspace, the radar controller accepts responsibility for the plane from the previous controlling unit. The controller also delegates responsibility for the plane to the next controlling unit when the plane leaves the team's airspace.

The radar controller, who is the senior team member, observes the planes in the team's airspace on radar and communicates with the pilots when necessary. Radar controllers warn pilots about nearby planes, bad weather conditions, and other potential hazards. Two planes on a collision course will be directed around each other. If a pilot wants to change altitude in search of better flying conditions, the controller will check to determine that no other planes will be along the proposed path. As the flight progresses, the team responsible for the aircraft notifies the next team in charge. Through team coordination, the plane arrives safely at its destination.

Both airport tower and en route controllers usually control several planes at a time; often, they have to make quick decisions about completely different activities. For example, a controller might direct a plane on its landing approach and at the same time provide pilots entering the airport's airspace with information about conditions at the airport. While instructing these pilots, the controller also would observe other planes in the vicinity, such as those in a holding pattern waiting for permission to land, to ensure that they remain well separated.

In addition to airport towers and en route centers, air traffic controllers also work in flight service stations operated at more than 100 locations. These *flight service specialists* provide pilots with information on the station's particular area, including terrain, preflight and in-flight weather information, suggested routes, and other information important to the safety of a flight. Flight service station specialists help pilots in emergency situations and initiate and coordinate searches for missing or overdue aircraft. However, they are not involved in actively managing air traffic.

Some air traffic controllers work at the Federal Aviation Administration's (FAA) Air Traffic Control Systems Command Center in Herndon, Virginia, where they oversee the entire system. They look for situations that will create bottlenecks or other problems in the system, then respond with a management plan for traffic into and out of the troubled sector. The objective is to keep traffic levels in the trouble spots manageable for the controllers working at en route centers.

Currently, the FAA is in the midst of developing and implementing a new automated air traffic control system that will allow controllers to more efficiently deal with the demands of increased air traffic. For example, some traditional air traffic controller tasks—like determining how far apart planes should be kept—will be done by computer. Present separation standards call for a 2,000-foot vertical spacing between two aircraft operating above 29,000 feet and flying the same ground track. With the aid of new technologies, the FAA will be able to reduce this vertical separation standard to 1,000 feet. Improved communication between computers on airplanes and those on the ground also is making the controller's job a little easier.

At present controllers sit at consoles with green-glowing screens that display radar images generated by a computer. In the future, controllers will work at a modern workstation computer that depicts air routes in full-color on a 20- by 20-inch screen. The controllers will select radio channels simply by touching on-screen buttons instead of turning dials or switching switches. The new technology will also enable controllers to zoom in on selected corners of the air space that is their responsibility and get better images of moving traffic than is possible with today's machines. The new automated air traffic control system is expected to become operational in several phases over the next 8 years.

The FAA is also considering implementing a system called "free flight" which would give pilots much more freedom in operating their aircraft. The change will require new concepts of shared responsibility between controllers and pilots. Air traffic controllers will still be central to the safe operation of the system, but their responsibilities will eventually shift from controlling to monitoring flights. At present, controllers assign routes, altitudes, and speeds. Under the new system, airlines and pilots would choose them. Controllers would intervene only to ensure that aircraft remained at safe distances from one another, to prevent congestion in terminal areas and entry into closed airspace, or to otherwise ensure safety. Today's practices often result in planes zigzagging from point to point along corridors rather than flying from city to city in a straight line. This results in lost time and fuel. However, it may be several years before a free flight system is implemented, despite its potential advantages. For the system to work, new equipment must be added for pilots and controllers, and new procedures developed to accommodate both the tightly controlled and flexible aspects of free flight. Budget constraints within the federal government may delay or slow implementation.

Working Conditions

Controllers work a basic 40-hour week; however, they may work additional hours for which they receive overtime pay or equal time off. Because most control towers and centers operate 24 hours a day, 7 days a week, controllers rotate night and weekend shifts.

During busy times, controllers must work rapidly and efficiently. This requires total concentration to keep track of several planes at the same time and make certain all pilots receive correct instructions. The mental stress of being responsible for the safety of several aircraft and their passengers can be exhausting for some persons.

Employment

Air traffic controllers held about 27,000 jobs in 2000. They were employed by the federal government at airports—in towers and flight service stations—and in en route traffic control centers. The overwhelming majority worked for the FAA. Some professional controllers conduct research at the FAA's national experimental center near Atlantic City, NJ. Others serve as instructors at the FAA Academy in Oklahoma City, OK. A small number of civilian controllers worked for the U.S. Department of Defense. In addition to controllers employed by the federal government, some worked for private air traffic control companies providing service to non-FAA towers.

Training, Other Qualifications, and Advancement

Air traffic controller trainees are selected through the competitive federal Civil Service system. Applicants must pass a written test that measures their ability to learn the controller's duties. Applicants with experience as a pilot, navigator, or military controller can improve their rating by scoring well on the occupational knowledge portion of the examination. Abstract reasoning and three-dimensional spatial visualization are among the aptitudes the exam measures. In addition, applicants usually must have 3 years of general work experience or 4 years of college, or a combination of both. Applicants also must survive a week of screening at the FAA Academy in Oklahoma City, which includes aptitude tests using computer simulators and physical and psychological examinations. Successful applicants receive drug-screening tests. For airport tower and en route center positions, applicants must be less than 31 years old. Those 31 years old and over are eligible for positions at flight service stations.

Controllers must be articulate, because pilots must be given directions quickly and clearly. Intelligence and a good memory also are important because controllers constantly receive information that they must immediately grasp, interpret, and remember. Decisiveness also is required because controllers often have to make quick decisions. The ability to concentrate is crucial because controllers must make these decisions in the midst of noise and other distractions.

Trainees learn their jobs through a combination of formal and on-the-job training. They receive 7 months of intensive training at the FAA academy, where they learn the fundamentals of the airway system, FAA regulations, controller equipment, aircraft performance characteristics, as well as more specialized tasks. To receive a job offer, trainees must successfully complete the training and pass a series of examinations, including a controller skills test that measures speed and accuracy in recognizing and correctly solving air traffic control problems. The test requires judgments on spatial relationships and requires application of the rules and procedures contained in the Air Traffic Control Handbook. Based on aptitude and test scores, trainees are selected to work at either an en route center or a tower.

After graduation, it takes several years of progressively more responsible work experience, interspersed with considerable classroom instruction and independent study, to become a fully qualified controller. This training includes instruction in the operation of the new, more automated air traffic control system—including the automated Microwave Landing System that enables pilots to receive instructions over automated data links that is being installed in control sites across the country.

Controllers who fail to complete either the academy or the on-the-job portion of the training are usually dismissed. Controllers must pass a physical examination each year and a job performance examination twice each year. Failure to become certified in any position at a facility within a specified time also may result in dismissal. Controllers also are subject to drug screening as a condition of continuing employment.

At airports, new controllers begin by supplying pilots with basic flight data and airport information. They then advance to ground controller, then local controller, departure controller, and finally, arrival controller. At an en route traffic control center, new controllers first deliver printed flight plans to teams, gradually advancing to radar associate controller and then radar controller.

Controllers can transfer to jobs at different locations or advance to supervisory positions, including management or staff jobs in air traffic control and top administrative jobs in the FAA. However, there are only limited opportunities for a controller to switch from a position in an en route center to a tower.

Job Outlook

Extremely keen competition is expected for air traffic controller jobs because the occupation attracts many more qualified applicants than the small number of job openings that result mostly from replacement needs. Replacement needs are very low because of the relatively high pay, liberal retirement benefits, and controllers' very strong attachment to the occupation. A new FAA hiring policy, allowing eligible retired military air traffic controllers to apply for FAA positions, will make competition even keener.

Employment of air traffic controllers is expected to grow more slowly than average through the year 2010. Employment growth is not expected to keep pace with growth in the number of aircraft flying because of the implementation of a new air traffic control system over the next several years. This computerized system will assist the controller by automatically making many of the routine decisions. Automation will allow controllers to handle more traffic, thus increasing their productivity.

Air traffic controllers who continue to meet the proficiency and medical requirements enjoy more job security than most workers do. The demand for air travel and the workloads of air traffic controllers decline during recessions, but controllers seldom are laid off.

Earnings

Median annual earnings of air traffic controllers in 2000 were $82,520. The middle 50 percent earned between $62,250 and $101,570. The lowest 10 percent earned less than $44,760, and the highest 10 percent earned more than $111,150.

The average annual salary, excluding overtime earnings, for air traffic controllers in the federal government—which employs 89 percent of the total—in nonsupervisory, supervisory, and managerial positions was $53,313 in 2001. Both the worker's job responsibilities and the complexity of the particular facility determine a controller's pay. For example, controllers who work at the FAA's busiest air traffic control facilities earn higher pay.

Depending on length of service, air traffic controllers receive 13 to 26 days of paid vacation and 13 days of paid sick leave each year, life insurance, and health benefits. In addition, controllers can retire at an earlier age and with fewer years of service than other federal employees can. Air traffic controllers are eligible to retire at age 50 with 20 years of service as an active air traffic controller or after 25 years of active service at any age. There is a mandatory retirement age of 56 for controllers who manage air traffic.

Related Occupations

Airfield operations specialists also are involved in the direction and control of traffic in air transportation.

Sources of Additional Information

Information on acquiring a job as an air traffic controller with the federal government may be obtained from the Office of Personnel Management through a telephone-based system. Consult your telephone directory under U.S. Government for a local number, or call (912) 757-3000; Federal Relay Service: (800) 877-8339. That number is not toll free and charges may result. Information also is available on the Internet: http://www.usajobs.opm.gov

Architects, Except Landscape and Naval

O*NET 17-1011.00

Significant Points

- More than 28 percent were self-employed—about four times the proportion for all professional and related occupations.
- Licensing requirements include a professional degree in architecture, a period of practical training and the passing of all divisions of the Architect Registration Examination.

- Architecture graduates may face competition, especially for jobs in the most prestigious firms; experience from working in a firm during school and knowledge of computer-aided design and drafting technology are advantages.

Nature of the Work

People need places in which to live, work, play, learn, worship, meet, govern, shop, and eat. These places may be private or public; indoors or out; rooms, buildings, or complexes, and together comprise neighborhoods, towns, suburbs and cities. *Architects*—licensed professionals trained in the art and science of building design—transform these needs into concepts and then develop the concepts into building images and plans that can be constructed by others.

Architects design the overall aesthetic and functional look of buildings and other structures. The design of a building involves far more than its appearance. Buildings also must be functional, safe, and economical, and must suit the needs of the people who use them. Architects take all these things into consideration when they design buildings and other structures.

Architects provide professional services to individuals and organizations planning a construction project. They may be involved in all phases of development, from the initial discussion with the client through the entire construction process. Their duties require specific skills—designing, engineering, managing, supervising, and communicating with clients and builders.

The architect and client discuss the objectives, requirements, and budget of a project. In some cases, architects provide various predesign services—conducting feasibility and environmental impact studies, selecting a site, or specifying the requirements the design must meet. For example, they may determine space requirements by researching the number and type of potential users of a building. The architect then prepares drawings and a report presenting ideas for the client to review.

After the initial proposals are discussed and accepted, architects develop final construction plans. These plans show the building's appearance and details for its construction. Accompanying these are drawings of the structural system; air-conditioning, heating, and ventilating systems; electrical systems; plumbing; and possibly site and landscape plans. They also specify the building materials and, in some cases, the interior furnishings. In developing designs, architects follow building codes, zoning laws, fire regulations, and other ordinances, such as those requiring easy access by disabled persons. Throughout the planning stage, they make necessary changes. Although they have traditionally used pencil and paper to produce design and construction drawings, architects are increasingly turning to computer-aided design and drafting (CADD) technology for these important tasks.

Architects may also assist the client in obtaining construction bids, selecting a contractor, and negotiating the construction contract. As construction proceeds, they may visit the building site to ensure the contractor is following the design, adhering to the schedule, using the specified materials, and meeting quality work standards. The job is not complete until all construction is finished, required tests are made, and construction costs are paid. Sometimes, architects also provide postconstruction services, such as facilities management. They advise on energy efficiency measures, evaluate how well the building design adapts to the needs of occupants, and make necessary improvements.

Architects design a wide variety of buildings, such as office and apartment buildings, schools, churches, factories, hospitals, houses, and airport terminals. They also design complexes such as urban centers, college campuses, industrial parks, and entire communities. They also may advise on the selection of building sites, prepare cost analysis and land-use studies, and do long-range planning for land development.

Architects sometimes specialize in one phase of work. Some specialize in the design of one type of building—for example, hospitals, schools, or housing. Others focus on planning and predesign services or construction management, and do minimal design work. They often work with engineers, urban planners, interior designers, landscape architects, and other professionals. In fact, architects spend a great deal of their time in coordinating information from, and the work of, others engaged in the same project. Consequently, architects—particularly at larger firms—are now using the Internet to update designs and communicate changes for the sake of speed and cost savings.

During the required training period leading up to licensing as architects, entry-level workers are called interns. This training period, which generally lasts 3 years, gives them practical work experience which aids interns in preparing for the Architect Registration Examination (ARE). Typical duties may include preparing construction drawings on CADD, building models, or assisting in the design of one part of a project.

Working Conditions

Architects usually work in a comfortable environment. Most of their time is spent in offices consulting with clients, developing reports and drawings, and working with other architects and engineers. However, they often visit construction sites to review the progress of projects.

Architects may occasionally be under stress, working nights and weekends to meet deadlines. In 2000, almost half of all architects worked more than 40 hours a week, in contrast to about 1 in 4 workers in all occupations combined.

Employment

Architects held about 102,000 jobs in 2000. The majority of jobs were in architectural firms—most of which employ fewer than 5 workers. A few worked for general building contractors, and for government agencies responsible for housing, planning, or community development, such as the U.S. Departments of Defense and Interior, and the General Services Administration. Nearly 3 in 10 architects were self-employed.

Training, Other Qualifications, and Advancement

All states and the District of Columbia require individuals to be licensed (registered) before they may call themselves architects or contract to provide architectural services. Nevertheless, many architecture school graduates work in the field while they are in the process of becoming licensed. However, a licensed architect is required to take legal responsibility for all work. Licensing requirements include a professional degree in architecture, a period of practical training or internship, and passage of all divisions of the ARE.

In most states, the professional degree in architecture must be from one of the 111 schools of architecture with degree programs accredited by the National Architectural Accrediting Board (NAAB). However, state architectural registration boards set their own standards, so graduation from a non-NAAB–accredited program may meet the educational requirement for licensing in a few states. Three types of professional degrees in architecture are available through colleges and universities. The majority of all architectural degrees are from 5-year Bachelor of Architecture programs, intended for students entering from high school or with no previous architectural training. In addition, a number of schools offer a 2-year Master of Architecture program for students with a preprofessional undergraduate degree in architecture or a related area, or a 3- or 4-year Master of Architecture program for students with a degree in another discipline.

The choice of degree type depends upon each individual's preference and educational background. Prospective architecture students should consider the available options before committing to a program. For example, although the 5-year Bachelor of Architecture program offers the fastest route to the professional degree, courses are specialized and, if the student does not complete the program, transferring to a nonarchitectural program may be difficult. A typical program includes courses in architectural history and theory, building design, structures, technology, construction methods, professional practice, math, physical sciences, and liberal arts. Central to most architectural programs is the design studio, where students put into practice the skills and concepts learned in the classroom. During the final semester of many programs, students devote their studio time to creating an architectural project from beginning to end, culminating in a 3-dimensional model of their design.

Many schools of architecture also offer post-professional degrees for those who already have a bachelor's or master's degree in architecture or other areas. Although graduate education beyond the professional degree is not required for practicing architects, it may be for research, teaching, and certain specialties.

Architects must be able to visually communicate their ideas to clients. Artistic and drawing ability is very helpful in doing this, but not essential. More important are a visual orientation and the ability to conceptualize and understand spatial relationships. Good communication skills, the ability to work independently or as part of a team, and creativity are important qualities for anyone interested in becoming an architect. Computer literacy also is required as most firms use computers for writing specifications, 2- and 3-dimensional drafting, and financial management. Knowledge of computer-aided design and drafting (CADD) is helpful and will become essential as architectural firms continue to adopt this technology. Recently, the profession recognized National CAD Standards (NCS); architecture students who master NCS may have an advantage in the job market.

All state architectural registration boards require a training period before candidates may sit for the ARE and become licensed. Most states have adopted the training standards established by the Intern Development Program, a program of the American Institute of Architects and the National Council of Architectural Registration Boards (NCARB). These standards stipulate broad and diversified training under the supervision of a licensed architect over a 3-year period. New graduates usually begin as intern-architects in architectural firms, where they assist in preparing architectural documents or drawings. They also may do research on building codes and materials, or write specifications for building materials, installation criteria, the quality of finishes, and other related details. Graduates with degrees in architecture also enter related fields such as graphic, interior, or industrial design; urban planning; real estate development; civil engineering; or construction management. After completing the on-the-job training period, interns are eligible to sit for the ARE. The examination tests candidates for their knowledge, skills, and ability to provide the various services required in the design and construction of buildings. Candidates who pass the ARE and meet all standards established by their state board are licensed to practice in that state.

After becoming licensed and gaining experience, architects take on increasingly responsible duties, eventually managing entire projects. In large firms, architects may advance to supervisory or managerial positions. Some architects become partners in established firms; others set up their own practice.

Several states require continuing education to maintain a license, and many more states are expected to adopt mandatory continuing education. Requirements vary by state, but usually involve the completion of a certain number of credits every year or two through seminars, workshops, formal university classes, conferences, self-study courses, or other sources. A growing number of architects voluntarily seek certification by NCARB, which can facilitate their getting licensed to practice in additional states. Certification is awarded after independent verification of the applying architect's educational transcripts, employment record, and professional references. It is the primary requirement for reciprocity of licensing among state boards that are NCARB members.

Job Outlook

Prospective architects may face competition for entry-level positions, especially if the number of architectural degrees awarded remains at current levels or increases. Employment of architects is projected to grow about as fast as the average for all occupations through 2010 and additional job openings will stem from the need to replace architects who retire or leave the labor force for other reasons. However, many individuals are attracted to this occupation, and the number of applicants often exceeds the number of available jobs, especially in the most prestigious firms. Prospective architects who gain career-related experience in an architectural firm while in school and who know CADD technology (especially that which conforms to the new national standards) will have a distinct advantage in obtaining an intern-architect position after graduation.

Employment of architects is strongly tied to the level of local construction, particularly nonresidential structures such as office buildings, shopping centers, schools, and healthcare facilities. After a boom in nonresidential construction during the 1980s, building slowed significantly during the first half of the 1990s. This trend is expected to continue because of slower labor force growth and increases in telecommuting and flexiplace work. However, as the stock of buildings ages, demand for remodeling and repair work should grow considerably. The needed renovation and rehabilitation of old buildings, particularly in urban areas where space for new buildings is becoming limited, is expected to provide many job opportunities for architects. In addition, demographic trends and changes in healthcare delivery are influencing the demand for certain institutional structures, and should also provide more jobs for architects in the future. For example, increases in the school-age population have resulted in new school construction. Additions to existing schools (especially colleges and universities), as well as overall modernization, will continue to add to demand for architects through 2010. Growth is expected in the number of adult care centers, assisted-living facilities, and community health clinics, all of which are preferable, less costly alternatives to hospitals and nursing homes.

Because construction—particularly office and retail—is sensitive to cyclical changes in the economy, architects will face particularly strong competition for jobs or clients during recessions, and layoffs may occur. Those involved in the design of institutional buildings such as schools, hospitals, nursing homes, and correctional facilities will be less affected by fluctuations in the economy.

Even in times of overall good job opportunities, however, there may be areas of the country with poor opportunities. Architects who are licensed to practice in one state must meet the licensing requirements of other states before practicing elsewhere. Obtaining licensure in other states, after initially receiving licensure in one state, is known as "reciprocity", and is much easier if an architect has received certification from the National Council of Architectural Registration Boards.

Earnings

Median annual earnings of architects were $52,510 in 2000. The middle 50 percent earned between $41,060 and $67,720. The lowest 10 percent earned less than $32,540 and the highest 10 percent earned more than $85,670.

Earnings of partners in established architectural firms may fluctuate because of changing business conditions. Some architects may have difficulty establishing their own practices and may go through a period when their expenses are greater than their income, requiring substantial financial resources.

Related Occupations

Architects design buildings and related structures. Construction managers, like architects, are also engaged in the planning and coordinating of activities concerned with the construction and maintenance of buildings and facilities. Others who engage in similar work are landscape architects, civil engineers, urban and regional planners, and designers, including interior designers, commercial and industrial designers, and graphic designers.

Sources of Additional Information

Information about education and careers in architecture can be obtained from:

- Practice Management Professional Interest Area, The American Institute of Architects, 1735 New York Ave. N.W., Washington, DC 20006
- Intern Development Program, National Council of Architectural Registration Boards, Suite 1100K, 1801 K Street N.W., Washington, DC 20006-1310. Internet: http://www.ncarb.org
- Consortium for Design and Construction Careers, P.O. Box 1515, Oak Park, IL 60304-1515. Internet: http://www.archcareers.net

Archivists, Curators, and Museum Technicians

O*NET 25-4011.00, 25-4012.00, 25-4013.00

Significant Points

- Employment usually requires graduate education and related work experience.
- Keen competition is expected because qualified applicants outnumber the most desirable job openings.

Nature of the Work

Archivists, curators, and museum technicians search for, acquire, appraise, analyze, describe, arrange, catalogue, restore, preserve, exhibit, maintain, and store valuable items that can be used by researchers or for exhibitions, publications, broadcasting, and other educational programs. Depending on the occupation, these items include historical documents, audiovisual materials, institutional records, works of art, coins, stamps, minerals, clothing, maps, living and preserved plants and animals, buildings, computer records, or historic sites.

Archivists and curators plan and oversee the arrangement, cataloguing, and exhibition of collections and, along with technicians and conservators, maintain collections. Archivists and curators may coordinate educational and public outreach programs, such as tours, workshops, lectures, and classes, and may work with the boards of institutions to administer plans and policies. They also may research topics or items relevant to their collections. Although some duties of archivists and curators are similar, the types of items they deal with differ. Curators usually handle objects found in cultural, biological, or historical collections, such as sculptures, textiles, and paintings, while archivists mainly handle valuable records, documents, or objects that are retained because they originally accompanied and relate specifically to the document.

Archivists determine what portion of the vast amount of records maintained by various organizations, such as government agencies, corporations, or educational institutions, or by families and individuals, should be made part of permanent historical holdings, and which of these records should be put on exhibit. They maintain records in their original arrangement according to the creator's organizational scheme, and describe records to facilitate retrieval. Records may be saved on any medium, including paper, film, videotape, audiotape, electronic disk, or computer. They also may be copied onto some other format to protect the original, and to make them more accessible to researchers who use the records. As computers and various storage media evolve, archivists must keep abreast of technological advances in electronic information storage.

Archives may be part of a library, museum, or historical society, or may exist as a distinct unit within an organization or company. Archivists consider any medium containing recorded information as documents, including letters, books, and other paper documents, photographs, blueprints, audiovisual materials, and computer records. Any document that reflects organizational transactions, hierarchy, or procedures can be considered a record. Archivists often specialize in an area of history or technology so they can better determine what records in that area qualify for retention and should become part of the archives. Archivists also may work with specialized forms of records, such as manuscripts, electronic records, photographs, cartographic records, motion pictures, and sound recordings.

Computers are increasingly used to generate and maintain archival records. Professional standards for use of computers in handling archival records are still evolving. However, computers are expected to transform many aspects of archival collections as computer capabilities, including multimedia and Worldwide-Web use, expand and allow more records to be stored and exhibited electronically.

Curators oversee collections in museums, zoos, aquariums, botanical gardens, nature centers, and historic sites. They acquire items through purchases, gifts, field exploration, intermuseum exchanges, or, in the case of some plants and animals, reproduction. Curators also plan and prepare exhibits. In natural history museums, curators collect and observe specimens in their natural habitat. Their work involves describing and classifying species, while specially trained collection managers and technicians provide hands-on care of natural history collections. Most curators use computer databases to catalogue and organize their collections. Many also use the Internet to make information available to other curators and the public. Increasingly, curators are expected to participate in grantwriting and fundraising to support their projects.

Most curators specialize in a field, such as botany, art, paleontology, or history. Those working in large institutions may be highly specialized. A large natural history museum, for example, would employ specialists in birds, fishes, insects, and mollusks. Some curators maintain the collection, others do research, and others perform administrative tasks. Registrars, for example, keep track of and move objects in the collection. In small institutions, with only one or a few curators, one curator may be responsible for multiple tasks, from maintaining collections to directing the affairs of museums.

Conservators manage, care for, preserve, treat, and document works of art, artifacts, and specimens. This may require substantial historical, scientific, and archaeological research. They use X rays, chemical testing, microscopes, special lights, and other laboratory equipment and techniques to examine objects and determine their condition, the need for treatment or restoration, and the appropriate method for preservation. They then document their findings and treat items to minimize deterioration or restore items to their original state. Conservators usually specialize in a particular material or group of objects, such as documents and books, paintings, decorative arts, textiles, metals, or architectural material.

Museum technicians assist curators by performing various preparatory and maintenance tasks on museum items. Some museum technicians also may assist curators with research. Archives technicians help archivists organize, maintain, and provide access to historical documentary materials.

Museum directors formulate policies, plan budgets, and raise funds for their museums. They coordinate activities of their staff to establish and maintain collections. As their role has evolved, museum directors increasingly need business backgrounds in addition to an understanding of the subject matter of their collections.

Working Conditions

The working conditions of archivists and curators vary. Some spend most of their time working with the public, providing reference assistance and educational services. Others perform research or process records, which often means working alone or in offices with only a few people. Those who restore and install exhibits or work with bulky, heavy record containers may climb, stretch, or lift. Those in zoos, botanical gardens, and other outdoor museums or historic sites frequently walk great distances.

Curators who work in large institutions may travel extensively to evaluate potential additions to the collection, organize exhibitions, and conduct research in their area of expertise. However, travel is rare for curators employed in small institutions.

Employment

Archivists, curators, and museum technicians held about 21,000 jobs in 2000. About 34 percent were employed in museums, botanical gardens, and zoos, and 18 percent worked in educational services, mainly in college and university libraries. Nearly one-third worked in federal, state, and local government. Most federal archivists work for the National Archives and Records Administration; others manage military archives in the U.S. Department of Defense. Most federal government curators work at the Smithsonian Institution, in the military museums of the Department of Defense, and in archaeological and other museums managed by the U.S. Department of Interior. All state governments have archival or historical records sections employing archivists. State and local governments have numerous historical museums, parks, libraries, and zoos employing curators.

Some large corporations have archives or records centers, employing archivists to manage the growing volume of records created or maintained as required by law or necessary to the firms' operations. Religious and fraternal organizations, professional associations, conservation organizations, major private collectors, and research firms also employ archivists and curators.

Conservators may work under contract to treat particular items, rather than as regular employees of a museum or other institution. These conservators may work on their own as private contractors, or as an employee of a conservation laboratory or regional conservation center that contracts their services to museums.

Training, Other Qualifications, and Advancement

Employment as an archivist, conservator, or curator usually requires graduate education and related work experience. Many archivists and curators work in archives or museums while completing their formal education, to gain the "hands-on" experience that many employers seek when hiring.

Employers usually look for archivists with undergraduate and graduate degrees in history or library science, with courses in archival science. Some positions may require knowledge of the discipline related to the collection, such as business or medicine. An increasing number of archivists have a double master's degree in history and library science. There are currently no programs offering bachelor's or master's degrees in archival science. However, approximately 65 colleges and universities offer courses or practical training in archival science as part of history, library science, or another discipline. The Academy of Certified Archivists offers voluntary certification for archivists. Certification requires the applicant to have experience in the field and to pass an examination offered by the academy.

Archivists need research and analytical ability to understand the content of documents and the context in which they were created, and to decipher deteriorated or poor quality printed matter, handwritten manuscripts, or photographs and films. A background in preservation management is often required of archivists because they are responsible for taking proper care of their records. Archivists also must be able to organize large amounts of information and write clear instructions for its retrieval and use. In addition, computer skills and the ability to work with electronic records and databases are increasingly important.

Many archives are very small, including one-person shops, with limited promotion opportunities. Archivists typically advance by transferring to a larger unit with supervisory positions. A doctorate in history, library science, or a related field may be needed for some advanced positions, such as director of a state archive.

For employment as a curator, most museums require a master's degree in an appropriate discipline of the museum's specialty—art, history, or archaeology—or museum studies. Many employers prefer a doctoral degree, particularly for curators in natural history or science museums. Earning two graduate degrees—in museum studies (museology) and a specialized subject—gives a candidate a distinct advantage in this competitive job market. In small museums, curatorial positions may be available to individuals with a bachelor's degree. For some positions, an internship of full-time museum work supplemented by courses in museum practices is needed.

Curatorial positions often require knowledge in a number of fields. For historic and artistic conservation, courses in chemistry, physics, and art are desirable. Since curators—particularly those in small museums—may have administrative and managerial responsibilities, courses in business administration, public relations, marketing, and fundraising also are recommended. Similar to archivists, curators need computer skills and the ability to work with electronic databases. Curators also need to be familiar with digital imaging, scanning technology, and copyright infringement, since many are responsible for posting information on the Internet.

Curators must be flexible because of their wide variety of duties. They need to design and present exhibits and, in small museums, manual dexterity to build exhibits or restore objects. Leadership ability and business skills are important for museum directors, while marketing skills are valuable for increasing museum attendance and fundraising.

In large museums, curators may advance through several levels of responsibility, eventually to museum director. Curators in smaller museums often advance to larger ones. Individual research and publications are important for advancement in larger institutions.

When hiring conservators, employers look for a master's degree in conservation, or in a closely related field, and substantial experience. There are only a few graduate programs in museum conservation techniques in the United States. Competition for entry to these programs is keen; to qualify, a student must have a background in chemistry, archaeology or studio art, and art history, as well as work experience. For some programs, knowledge of a foreign language is also helpful. Conservation apprenticeships or internships as an undergraduate can also enhance one's admission prospects. Graduate programs last 2 to 4 years; the latter years include internship training. A few individuals enter conservation through apprenticeships with museums, nonprofit organizations, and conservators in private practice. Apprenticeships should be supplemented with courses in chemistry, studio art, and history. Apprenticeship training, although accepted, usually is a more difficult route into the conservation profession.

Museum technicians usually need a bachelor's degree in an appropriate discipline of the museum's specialty, museum studies training, or previous museum work experience, particularly in exhibit design. Similarly, archives technicians usually need a bachelor's degree in library science or history, or relevant work experience. Technician positions often serve as a stepping stone for individuals interested in archival and curatorial work. With the exception of small museums, a master's degree is needed for advancement.

Relatively few schools grant a bachelor's degree in museum studies. More common are undergraduate minors or tracks of study that are part of an undergraduate degree in a related field, such as art history, history, or archaeology. Students interested in further study may obtain a master's degree in museum studies. Colleges and universities throughout the country offer master's degrees in museum studies. However, many employers feel that, while museum studies are helpful, a thorough knowledge of the museum's specialty and museum work experience are more important.

Continuing education, which enables archivists, curators, and museum technicians to keep up with developments in the field, is available through meetings, conferences, and workshops sponsored by archival, historical, and museum associations. Some larger organizations, such as the National Archives, offer such training in-house.

Job Outlook

Competition for jobs as archivists, curators, and museum technicians is expected to be keen as qualified applicants outnumber job openings. Graduates with highly specialized training, such as master's degrees in both library science and history, with a concentration in archives or records management, and extensive computer skills should have the best opportunities for jobs as archivists. A curator job is attractive to many people, and many applicants have the necessary training and subject knowledge; but there are only a few openings. Consequently, candidates may have to work part-time, as an intern, or even as a volunteer assistant curator or research associate after completing their formal education. Substantial work experience in collection management, exhibit design, or restoration, as well as database management skills, will be necessary for permanent status. Job opportunities for curators should be best in art and history museums, since these are the largest employers in the museum industry.

The job outlook for conservators may be more favorable, particularly for graduates of conservation programs. However, competition is stiff for the limited number of openings in these programs, and applicants need a technical background. Students who qualify and successfully complete the program, have knowledge of a foreign language, and are willing to relocate, will have an advantage over less qualified candidates.

Employment of archivists, curators, and museum technicians is expected to increase about as fast as the average for all occupations through 2010. Jobs are expected to grow as public and private organizations emphasize establishing archives and organizing records and information, and as public interest in science, art, history, and technology increases. Although overall museum attendance is increasing, public interest in smaller, specialized museums with unique collections is expected to increase faster. However, museums and other cultural institutions are often subject to funding cuts during recessions or periods of budget tightening, reducing demand for archivists and curators. Although the rate of turnover among archivists and curators is relatively low, the need to replace workers who leave the occupation or stop working will create some additional job openings.

Earnings

Median annual earnings of archivists, curators, and museum technicians in 2000 were $33,080. The middle 50 percent earned between $24,740 and $45,490. The lowest 10 percent earned less than $19,200, and the highest 10 percent earned more than $61,490.

Median annual earnings of archivists, curators, and museum technicians in 2000 were $31,460 in museums and art galleries.

Earnings of archivists and curators vary considerably by type and size of employer, and often by specialty. Average salaries in the federal government, for example, are usually higher than those in religious organizations. Salaries of curators in large, well-funded museums can be several times higher than those in small ones.

The average annual salary for archivists in the federal government in nonsupervisory, supervisory, and managerial positions was $63,299 in 2001; museum curators, $64,616; museum specialists and technicians, $44,711; and archives technicians, $33,934.

Related Occupations

The skills that archivists, curators, and museum technicians use in preserving, organizing, and displaying objects or information of historical interest are shared by artists and related workers; librarians; and anthropologists and archeologists, historians, and other social scientists.

Sources of Additional Information

For information on archivists and on schools offering courses in archival studies, contact:

- Society of American Archivists, 527 South Wells St., 5th floor, Chicago, IL 60607-3922. Internet: http://www.archivists.org

For general information about careers as a curator and schools offering courses in museum studies, contact:

- American Association of Museums, 1575 I St. NW, Suite 400, Washington, DC 20005. Internet: http://www.aam-us.org

For information about conservation and preservation careers and education programs, contact:

- American Institute for Conservation of Historic and Artistic Works, 1717 K St. N.W., Suite 301, Washington, DC 20006. Internet: http://palimpsest.stanford.edu/aic

Atmospheric Scientists

O*NET 19-2021.00

Significant Points

- The federal government employs more than 4 out of 10 atmospheric scientists and is their largest employer.
- A bachelor's degree in meteorology, or in a closely related field with courses in meteorology, is the minimum educational requirement; a master's degree is necessary for some positions, and a PhD is required for most research positions.

- Applicants may face competition for jobs if the number of degrees awarded in atmospheric science and meteorology remain near current levels.

Nature of the Work

Atmospheric science is the study of the atmosphere—the blanket of air covering the earth. *Atmospheric scientists*, commonly called *meteorologists*, study the atmosphere's physical characteristics, motions, and processes, and the way it affects the rest of our environment. The best known application of this knowledge is in forecasting the weather. However, weather information and meteorological research are also applied in air-pollution control, agriculture, air and sea transportation, defense, and the study of trends in earth's climate such as global warming, droughts, or ozone depletion.

Atmospheric scientists who forecast the weather, known professionally as *operational meteorologists*, form the largest group of specialists. They study information on air pressure, temperature, humidity, and wind velocity; and apply physical and mathematical relationships to make short- and long-range weather forecasts. Their data come from weather satellites, weather radars, and sensors and observers in many parts of the world. Meteorologists use sophisticated computer models of the world's atmosphere to make long-term, short-term, and local-area forecasts. These forecasts inform not only the general public, but also those who need accurate weather information for both economic and safety reasons, as in the shipping, air transportation, agriculture, fishing, and utilities industries.

The use of weather balloons, launched a few times a day to measure wind, temperature, and humidity in the upper atmosphere, is currently supplemented by sophisticated atmospheric monitoring equipment that transmits data as frequently as every few minutes. Doppler radar, for example, can detect airflow patterns in violent storm systems—allowing forecasters to better predict tornadoes and other hazardous winds, as well as to monitor the storm's direction and intensity. Combined radar and satellite observations allow meteorologists to predict flash floods.

Some atmospheric scientists work in research. *Physical meteorologists*, for example, study the atmosphere's chemical and physical properties; the transmission of light, sound, and radio waves; and the transfer of energy in the atmosphere. They also study factors affecting the formation of clouds, rain, snow, and other weather phenomena, such as severe storms. *Synoptic meteorologists* develop new tools for weather forecasting using computers and sophisticated mathematical models. *Climatologists* collect, analyze, and interpret past records of wind, rainfall, sunshine, and temperature in specific areas or regions. Their studies are used to design buildings, plan heating and cooling systems, and aid in effective land use and agricultural production. Other research meteorologists examine the most effective ways to control or diminish air pollution.

Working Conditions

Most weather stations operate around the clock 7 days a week. Jobs in such facilities usually involve night, weekend, and holiday work, often with rotating shifts. During weather emergencies, such as hurricanes, operational meteorologists may work overtime. Operational meteorologists are also often under pressure to meet forecast deadlines. Weather stations are found all over—at airports, in or near cities, and in isolated and remote areas. Some atmospheric scientists also spend time observing weather conditions and collecting data from aircraft. Weather forecasters who work for radio or television stations broadcast their reports from station studios, and may work evenings and weekends. Meteorologists in smaller weather offices often work alone; in larger ones, they work as part of a team. Meteorologists not involved in forecasting tasks work regular hours, usually in offices. Those who work for private consulting firms or for companies analyzing and monitoring emissions to improve air quality usually work with other scientists or engineers.

Employment

Atmospheric scientists held about 6,900 jobs in 2000. The federal government is the largest single employer of civilian meteorologists, employing about 3,000. The National Oceanic and Atmospheric Administration (NOAA) employed most federal meteorologists in the National Weather Service stations throughout the nation; the remainder of NOAA's meteorologists worked mainly in research and development or management. The Department of Defense employed several hundred civilian meteorologists. Others worked for research and testing services, private weather consulting services, radio and television broadcasting, air carriers, and computer and data processing services.

Although several hundred people teach atmospheric science and related courses in college and university departments of meteorology or atmospheric science, physics, earth science, and geophysics, these individuals are classified as college or university faculty, rather than atmospheric scientists.

In addition to civilian meteorologists, hundreds of armed forces members are involved in forecasting and other meteorological work.

Training, Other Qualifications, and Advancement

A bachelor's degree in meteorology or atmospheric science, or in a closely related field with courses in meteorology, usually is the minimum educational requirement for an entry-level position as an atmospheric scientist.

The preferred educational requirement for entry-level meteorologists in the federal government is a bachelor's degree, not necessarily in meteorology. The degree should have at least 24 semester hours of meteorology courses, including 6 hours in the analysis and prediction of weather systems and 2 hours of remote sensing of the atmosphere or instrumentation. Other required courses include differential and integral calculus, differential equations, 6 hours of college physics, and at least 9 hours of courses appropriate for a physical science major—such as statistics, computer science, chemistry, physical oceanography, or physical climatology. Sometimes, a combination of experience and education may be substituted for a degree.

Although positions in operational meteorology are available for those with only a bachelor's degree, obtaining a master's degree enhances employment opportunities and advancement potential. A master's degree usually is necessary for conducting applied research and development, and a PhD is required for most basic research positions. Students planning on a career in research and development need not necessarily major in atmospheric science or meteorology as an undergraduate. In fact, a bachelor's degree in mathematics, physics, or engineering provides excellent preparation for graduate study in atmospheric science.

Because atmospheric science is a small field, relatively few colleges and universities offer degrees in meteorology or atmospheric science, although many departments of physics, earth science, geography, and geophysics offer atmospheric science and related courses. Prospective students should make certain that courses required by the National Weather Service and other employers are offered at the college they are considering. Computer science courses, additional meteorology courses, a strong background in mathematics and physics, and good communication skills are important to prospective employers. Many programs combine the study of meteorology with another field, such as agriculture, oceanography, engineering, or physics. For example, hydrometeorology is the blending of hydrology (the science of earth's water) and meteorology, and is the field concerned with the effect of precipitation on the hydrologic cycle and the environment. Students who wish to become broadcast meteorologists for radio or television stations should develop excellent communication skills through courses in speech, journalism, and related fields. Those interested in air quality work should take courses in chemistry and supplement their technical training with coursework in policy or government affairs.

Beginning atmospheric scientists often do routine data collection, computation, or analysis, and some basic forecasting. Entry-level operational meteorologists in the federal government usually are placed in intern positions for training and experience. During this period, they learn about the weather service's forecasting equipment and procedures, and rotate to different offices to learn about various weather systems. After completing the training period, they are assigned a permanent duty station. Experienced meteorologists may advance to supervisory or administrative jobs, or may handle more complex forecasting jobs. After several years of experience, some meteorologists establish their own weather consulting services.

The American Meteorological Society offers professional certification of consulting meteorologists, administered by a Board of Certified Consulting Meteorologists. Applicants must meet formal education requirements (although not necessarily have a college degree), pass an examination to demonstrate thorough meteorological knowledge, have a minimum of 5 years of experience or a combination of experience plus an advanced degree, and provide character references from fellow professionals.

Job Outlook

Employment of atmospheric scientists is projected to increase about as fast as the average for all occupations through 2010, but prospective atmospheric scientists may face competition if the number of degrees awarded in atmospheric science and meteorology remain near current levels. The National Weather Service (NWS) has completed an extensive modernization of its weather forecasting equipment and finished all hiring of meteorologists needed to staff the upgraded stations. The NWS has no plans to increase the number of weather stations or the number of meteorologists in existing stations for many years. Employment of meteorologists in other federal agencies is expected to decline slightly as efforts to reduce the federal government workforce continue.

On the other hand, job opportunities for atmospheric scientists in private industry are expected to be better than in the federal government over the 2000–2010 period. As research leads to continuing improvements in weather forecasting, demand should grow for private weather consulting firms to provide more detailed information than has formerly been available, especially to weather-sensitive industries. Farmers, commodity investors, radio and television stations, and utilities, transportation, and construction firms can greatly benefit from additional weather information more closely targeted to their needs than the general information provided by the National Weather Service. Additionally, research on seasonal and other long-range forecasting is yielding positive results, which should spur demand for more atmospheric scientists to interpret these forecasts and advise weather-sensitive industries. However, because many customers for private weather services are in industries sensitive to fluctuations in the economy, the sales and growth of private weather services depend on the health of the economy.

There will continue to be demand for atmospheric scientists to analyze and monitor the dispersion of pollutants into the air to ensure compliance with federal environmental regulations outlined in the Clean Air Act of 1990, but employment increases are expected to be small.

Earnings

Median annual earnings of atmospheric scientists in 2000 were $58,510. The middle 50 percent earned between $39,780 and $72,740. The lowest 10 percent earned less than $29,880, and the highest 10 percent earned more than $89,060.

The average salary for meteorologists in nonsupervisory, supervisory, and managerial positions employed by the federal government was about $68,100 in 2001. Meteorologists in the federal government with a bachelor's degree and no experience received a starting salary of $24,245 or $29,440, depending on their college grades. Those with a master's degree could start at $29,440 or $36,606; those with the PhD, at $47,039 or $59,661. Beginning salaries for all degree levels are slightly higher in selected areas of the country where the prevailing local pay level is higher.

Related Occupations

Workers in other occupations concerned with the physical environment include environmental scientists and geoscientists, physicists and astronomers, mathematicians, and civil, chemical, and environmental engineers.

Sources of Additional Information

Information about careers in meteorology is available from:

- American Meteorological Society, 45 Beacon St., Boston, MA 02108. Internet: http://www.ametsoc.org/AMS

Information on obtaining a meteorologist position with the federal government is available from the Office of Personnel Management through a telephone-based system. Consult your telephone directory under U.S. Government for a local number or call (912) 757-3000; Federal Relay Service: (800) 877-8339. The first number is not toll free, and charges may result. Information also is available from the Internet site: http://www.usajobs.opm.gov

Automotive Service Technicians and Mechanics

O*NET 49-3023.01, 49-3023.02

Significant Points

- Formal automotive technician training is the best preparation for these challenging technology-based jobs.
- Opportunities should be very good for automotive service technicians and mechanics with good diagnostic and problem-solving skills and knowledge of electronics and mathematics.
- Automotive service technicians and mechanics must continually adapt to changing technology and repair techniques as vehicle components and systems become increasingly sophisticated.

Nature of the Work

Anyone whose car or light truck has broken down knows the importance of the jobs of automotive service technicians and mechanics. The ability to diagnose the source of a problem quickly and accurately—a most valuable skill—requires good reasoning ability and a thorough knowledge of automobiles. Many technicians consider diagnosing hard-to-find troubles one of their most challenging and satisfying duties.

The work of automotive service technicians and mechanics has evolved from simply mechanical to high technology. Today integrated electronic systems and complex computers run vehicles and measure their performance while on the road. Automotive service technicians have developed into diagnostic, high-tech problem solvers. Technicians must have an increasingly broad base of knowledge about how vehicles' complex components work and interact, as well as the ability to work with electronic diagnostic equipment and computer-based technical reference materials.

Automotive service technicians and mechanics use these high-tech skills to inspect, maintain, and repair automobiles and light trucks with gasoline engines. The increasing sophistication of automotive technology now relies on workers who can use computerized shop equipment and work with electronic components, while maintaining their skills with traditional handtools. Because of these changes in the occupation, workers are increasingly called "automotive service technicians," and the title "mechanic" is being used less and less frequently. Diesel service technicians and mechanics work on diesel-powered trucks, buses, and equipment. Motorcycle mechanics repair and service motorcycles, motor scooters, mopeds, and, occasionally, small all-terrain vehicles.

When mechanical or electrical troubles occur, technicians first get a description of the symptoms from the owner or, if they work in a large shop, the repair service estimator who wrote the repair order. To locate the problem, technicians use a diagnostic approach. First, they test to see if components and systems are proper and secure, and then isolate those components or systems that could not logically be the cause of the problem. For example, if an air conditioner malfunctions, the technician's diagnostic approach can pinpoint a problem as simple as a low coolant level or as complex as a bad drive-train connection that has shorted out the air conditioner. Technicians may have to test drive the vehicle or use a variety of testing equipment, such as onboard and hand-held diagnostic computers or compression gauges, to identify the source of the problem. These tests may indicate whether a component is salvageable or if a new one is required to get the vehicle back in working order.

During routine service inspections, technicians test and lubricate engines and other major components. In some cases, the technician may repair or replace worn parts before they cause breakdowns that could damage critical components of the vehicle. Technicians usually follow a checklist to ensure that they examine every critical part. Belts, hoses, plugs, brake and fuel systems, and other potentially troublesome items are among those closely watched.

Service technicians use a variety of tools in their work. They use power tools, such as pneumatic wrenches to remove bolts quickly, machine tools like lathes and grinding machines to rebuild brakes, welding and flame-cutting equipment to remove and repair exhaust systems, and jacks and hoists to lift cars and engines. They also use common handtools like screwdrivers, pliers, and wrenches to work on small parts and in hard-to-reach places.

In modern repair shops, service technicians compare the readouts from diagnostic testing devices to the benchmarked standards given by the manufacturer of the components being tested. Deviations outside of acceptable levels are an indication to the technician that further attention to an area is necessary. The testing devices diagnose problems and make precision adjustments with precise calculations downloaded from large computerized databases. The computerized systems provide automatic updates to technical manuals and unlimited access to manufacturers' service information, technical service bulletins, and other information databases, which allow technicians to keep current on trouble spots and to learn new procedures.

Automotive service technicians in large shops have increasingly become specialized. For example, *transmission technicians and rebuilders* work on gear trains, couplings, hydraulic pumps, and other parts of transmissions. Extensive knowledge of computer controls, diagnosis of electrical and hydraulic problems, and other specialized skills are needed to work on these complex components, which employ some of the most sophisticated technology used in vehicles. *Tune-up technicians* adjust the ignition timing and valves, and adjust or replace spark plugs and other parts to ensure efficient engine performance. They often use electronic test equipment to isolate and adjust malfunctions in fuel, ignition, and emissions control systems.

Automotive air-conditioning repairers install and repair air conditioners and service components, such as compressors, condensers, and controls. These workers require special training in federal and state regulations governing the handling and disposal of refrigerants. *Front-end mechanics* align and balance wheels and repair steering mechanisms and suspension systems. They frequently use special alignment equipment and wheel-balancing machines. *Brake repairers* adjust brakes, replace brake linings and pads, and make other repairs on brake systems. Some technicians and mechanics specialize in both brake and front-end work.

Working Conditions

Almost half of automotive service technicians work a standard 40-hour week, but over 30 percent work more than 40 hours a week. Many of those working extended hours are self-employed technicians. To satisfy customer service needs, some service shops offer evening and weekend service. Generally, service technicians work indoors in well-ventilated and lighted repair shops. However, some shops are drafty and noisy. Although they fix some problems with simple computerized adjustments, technicians frequently work with dirty and greasy parts, and in awkward positions. They often lift heavy parts and tools. Minor cuts, burns, and bruises are common, but technicians usually avoid serious accidents when the shop is kept clean and orderly and safety practices are observed.

Employment

Automotive service technicians and mechanics held about 840,000 jobs in 2000. The majority worked for retail and wholesale automotive dealers, independent automotive repair shops, or automotive service facilities at department, automotive, and home supply stores. Others found employment in gasoline service stations; taxicab and automobile leasing companies; federal, state, and local governments; and other organizations. About 18 percent of service technicians were self-employed.

Training, Other Qualifications, and Advancement

Automotive technology is rapidly increasing in sophistication, and most training authorities strongly recommend that persons seeking automotive service technician and mechanic jobs complete a formal training program in high school or in a postsecondary vocational school. However, some service technicians still learn the trade solely by assisting and learning from experienced workers.

Many high schools, community colleges, and public and private vocational and technical schools offer automotive service technician training programs. The traditional postsecondary programs usually provide a thorough career preparation that expands upon the student's high school repair experience.

Postsecondary automotive technician training programs vary greatly in format, but normally provide intensive career preparation through a combination of classroom instruction and hands-on practice. Some trade and technical school programs provide concentrated training for 6 months to a year, depending on how many hours the student attends each week. Community college programs normally spread the training over 2 years; supplement the automotive training with instruction in English, basic mathematics, computers, and other subjects; and award an associate degree or certificate. Some students earn repair certificates and opt to leave the program to begin their career before graduation. Recently, some programs have added to their curriculums training on employability skills such as customer service and stress management. Employers find that these skills help technicians handle the additional responsibilities of dealing with the customers and parts vendors.

High school programs, while an asset, vary greatly in quality. The better programs, such as the Automotive Youth Education Service (AYES), with 150 participating schools and more than 300 participating dealers, conclude with the students receiving their technician's certification and high school diploma. Other programs offer only an introduction to automotive technology and service for the future consumer or hobbyist. Still others aim to equip graduates with enough skills to get a job as a mechanic's helper or trainee mechanic.

The various automobile manufacturers and their participating dealers sponsor 2-year associate degree programs at postsecondary schools across the nation. The Accrediting Commission of Career Schools and Colleges of Technology (ACCSCT) currently certifies a number of automotive and diesel technology schools. Schools update their curriculums frequently to reflect changing technology and equipment. Students in these programs typically spend alternate 6- to 12-week periods attending classes full time and working full time in the service departments of sponsoring dealers. At these dealerships, students get practical experience while assigned to an experienced worker who provides hands-on instruction and time-saving tips.

The National Automotive Technicians Education Foundation (NATEF), an affiliate of the National Institute for Automotive Service Excellence (ASE), establishes the standards by which training facilities become certified. Once the training facility achieves these minimal standards, NATEF recommends the facility to ASE for certification. The ASE certification is a nationally recognized standard for programs offered by high schools, postsecondary trade schools, technical institutes, and community colleges that train automobile service technicians, collision repair and refinish technicians, engine machinists, and medium/heavy truck technicians. Automotive manufacturers provide ASE certified instruction, service equipment, and current model cars on which students practice new skills and learn the latest automotive technology. While ASE certification is voluntary, it does signify that the program meets uniform standards for instructional facilities, equipment, staff credentials, and curriculum. To ensure that programs keep up with ever-changing technology, repair techniques, and ASE standards, the certified programs are subjected to periodic compliance reviews and mandatory recertification. NATEF program experts also review and update program standards to match the level of training and skill-level achievement necessary for success in the occupation. In mid-2000, 1,491 high school and postsecondary automotive service technician training programs had been certified by ASE. Of these, 1,200 trained automobile service technicians, 224 instructed collision specialists, and 62 trained diesel and medium/heavy truck specialists.

For trainee automotive service technician jobs, employers look for people with strong communication and analytical skills. Technicians need good reading, mathematics, and computer skills to study technical manuals and to keep abreast of new technology and learn new service and repair procedures and specifications. Trainees also must possess mechanical aptitude and knowledge of how automobiles work. Most employers regard the successful completion of a vocational training program in automotive service technology as the best preparation for trainee positions. Experience working on motor vehicles in the armed forces or as a hobby also is valuable. Because of the complexity of new vehicles, a growing number of employers require completion of high school and additional postsecondary training. Courses in automotive repair, electronics, physics, chemistry, English, computers, and mathematics provide a good educational background for a career as a service technician.

There are more computers aboard a car today than aboard the first spacecraft. A new car has from 10 to 15 onboard computers, operating everything from the engine to the radio. Some of the more advanced vehicles have global positioning systems, Internet access, and other high-tech features integrated into the functions of the vehicle. Therefore, knowledge of electronics and computers has grown increasingly important for service technicians. Engine controls and dashboard instruments were among the first components to use electronics, but now, everything from brakes to transmissions and air-conditioning systems to steering systems is run primarily by computers and electronic components. In the past, a specialist usually handled any problems involving electrical systems or electronics. Now that electronics are so common, it is essential for service technicians to be familiar with at least the basic principles of electronics. Electrical components or a series of related components account for nearly all malfunctions in modern vehicles.

In addition to electronics and computers, automotive service technicians will have to learn and understand the science behind the alternate fuel vehicles that have begun to enter the market. The fuel for these vehicles will come from the dehydrogenization of water, electric fuel cells, natural gas, solar power, and other nonpetroleum-based sources. Some vehicles will even capture the energy from brakes and use it as fuel. As vehicles with these new technologies become more common, technicians will need additional training to learn the science and engineering that makes them possible.

Beginners usually start as trainee technicians, mechanics' helpers, lubrication workers, or gasoline service station attendants, and gradually acquire and practice their skills by working with experienced mechanics and technicians. With a few months' experience, beginners perform many routine service tasks and make simple repairs. It usually takes 2 to 5 years of experience to become a journey-level service technician, who is expected to quickly perform the more difficult types of routine service and repairs. However, some graduates of postsecondary automotive training programs are often able to earn promotion to the journey level after only a few months on the job. An additional 1- to 2-years' experience familiarizes mechanics and technicians with all types of repairs. Difficult specialties, such as transmission repair, require another year or two of training and experience. In contrast, brake specialists may learn their jobs in considerably less time because they do not need a complete knowledge of automotive repair.

In the past, many persons became automotive service technicians through 3- or 4-year formal apprenticeship programs. However, apprenticeships have become rare, as formal vocational training programs in automotive service technology have become more common.

At work, the most important possessions of technicians and mechanics are their handtools. Technicians and mechanics usually provide their own tools, and many experienced workers have thousands of dollars invested in them. Employers typically furnish expensive power tools, engine analyzers, and other diagnostic equipment, but technicians accumulate handtools with experience. Some formal training programs have alliances with tool manufacturers that help entry-level technicians accumulate tools during their training period.

Employers increasingly send experienced automotive service technicians to manufacturer training centers to learn to repair new models or to receive special training in the repair of components, such as electronic fuel injection or air-conditioners. Motor vehicle dealers also may send promising beginners to manufacturer-sponsored mechanic training programs. Employers typically furnish this additional training to maintain or upgrade employee skills and increase their value to the dealership. Factory representatives also visit many shops to conduct short training sessions.

Voluntary certification by Automotive Service Excellence (ASE) has become a standard credential for automotive service technicians. Certification is available in 1 or more of 8 different service areas, such as electrical systems, engine repair, brake systems, suspension and steering, and heating and air conditioning. For certification in each area, technicians must have at least 2 years of experience and pass a written examination. Completion of an automotive training program in high school, vocational or trade school, or community or junior college may be substituted for 1 year of experience. In some cases, graduates of ASE-certified programs achieve certification in up to three specialties. For certification as a master automotive mechanic, technicians must be certified in all eight areas. Mechanics and technicians must retake each examination at least every 5 years to maintain their certifications.

Experienced technicians who have leadership ability sometimes advance to shop supervisor or service manager. Those who work well with customers may become automotive repair service estimators. Some with sufficient funds open independent repair shops.

Job Outlook

Job opportunities in this occupation are expected to be very good for persons who complete automotive training programs in high school, vocational and technical schools, or community colleges. Persons with good diagnostic and problem-solving skills, and whose training includes basic electronics skills, should have the best opportunities. For well-prepared people with a technical background, automotive service technician careers offer an excellent opportunity for good pay and the satisfaction of highly skilled work with vehicles incorporating the latest in high technology. However, persons without formal automotive training are likely to face competition for entry-level jobs.

Employment of automotive service technicians and mechanics is expected to increase about as fast as the average through the year 2010. The growing complexity of automotive technology necessitates service by skilled workers, contributing to the growth in demand for highly trained mechanics and technicians. Employment growth will continue to be concentrated in motor vehicle dealerships and independent automotive repair shops. Many new jobs will also be created in small retail operations that offer after-warranty repairs, such as oil changes, brake repair, air conditioner service, and other minor repairs generally taking less than 4 hours to complete. Fewer national department store chains will provide auto repair services in large shops. Employment of automotive service technicians and mechanics in gasoline service stations will continue to decline, as fewer stations offer repair services.

In addition to job openings due to growth, a substantial number of openings will be created by the need to replace experienced technicians who transfer to other occupations, retire, or stop working for other reasons. Most persons who enter the occupation can expect steady work, because changes in general economic conditions and developments in other industries have little effect on the automotive repair business.

Earnings

Median hourly earnings of automotive service technicians and mechanics, including commission, were $13.70 in 2000. The middle 50 percent earned between $9.86 and $18.67 an hour. The lowest 10 percent earned less than $7.59, and the highest 10 percent earned more than $23.67 an hour. Median annual earnings in the industries employing the largest numbers of service technicians in 2000 were as follows:

Local government	$16.90
New and used car dealers	16.87
Auto and home supply stores	12.35
Automotive repair shops	12.15
Gasoline service stations	11.86

Many experienced technicians employed by automotive dealers and independent repair shops receive a commission related to the labor cost charged to the customer. Under this method, weekly earnings depend on the amount of work completed. Employers frequently guarantee commissioned mechanics and technicians a minimum weekly salary. Many master technicians earn from $70,000 to $100,000 annually.

Some automotive service technicians are members of labor unions such as the International Association of Machinists and Aerospace Workers; the International Union, United Automobile, Aerospace and Agricultural Implement Workers of America; the Sheet Metal Workers' International Association; and the International Brotherhood of Teamsters.

Related Occupations

Other workers who repair and service motor vehicles include automotive body and related repairers, diesel service technicians and mechanics, and small engine mechanics.

Sources of Additional Information

For more details about work opportunities, contact local automotive dealers and repair shops or local offices of the state employment service. The state employment service also may have information about training programs.

A list of certified automotive technician training programs can be obtained from:

- National Automotive Technicians Education Foundation, 101 Blue Seal Dr. S.E., Suite 101, Leesburg, VA 20175. Internet: http://www.natef.org

For a directory of accredited private trade and technical schools that offer programs in automotive technician training, contact:

- Accrediting Commission of Career Schools and Colleges of Technology, 2101 Wilson Blvd., Suite 302, Arlington, VA 22201 Internet: http://www.accsct.org

For a list of public automotive technician training programs, contact:

- SkillsUSA-VICA, P.O. Box 3000, 1401 James Monroe Hwy., Leesburg, VA 22075. Internet: http://www.skillsusa.org

Information on automobile manufacturer-sponsored programs in automotive service technology can be obtained from:

- Automotive Youth Educational Systems (AYES), 2701 Troy Center Dr., Suite 450, Troy, MI 48084. Internet: http://www.ayes.org

Information on how to become a certified automotive service technician is available from:

- ASE, 101 Blue Seal Dr. S.E., Suite 101, Leesburg, VA 20175. Internet: http://www.asecert.org

For general information about the work of automotive service technicians and mechanics, contact:

- National Automobile Dealers Association, 8400 Westpark Dr., McLean, VA 22102. Internet: http://www.nada.org

Biological and Medical Scientists

O*NET 19-1021.01, 19-1021.02, 19-1022.00, 19-1023.00, 19-1029.99, 19-1041.00, 19-1042.00, 19-1099.99

Significant Points

- A PhD degree usually is required for independent research, but a master's degree is sufficient for some jobs in applied research or product development; a bachelor's degree is adequate for some nonresearch jobs.
- Medical scientist jobs require a PhD degree in a biological science, but some jobs need a medical degree.
- Doctoral degree holders face considerable competition for independent research positions; holders of bachelor's or master's degrees in biological science can expect better opportunities in nonresearch positions.

Nature of the Work

Biological and medical scientists study living organisms and their relationship to their environment. They research problems dealing with life processes. Most specialize in some area of biology such as zoology (the study of animals) or microbiology (the study of microscopic organisms).

Many biological scientists and virtually all medical scientists work in research and development. Some conduct basic research to advance knowledge of living organisms, including viruses, bacteria, and other infectious agents. Past research has resulted in the development of vaccines, medicines, and treatments for cancer and other diseases. Basic biological and medical research continues to provide the building blocks necessary to develop solutions to human health problems, and to preserve and repair the natural environment. Biological and medical scientists mostly work independently in private industry, university, or government laboratories, often exploring new areas of research or expanding on specialized research started in graduate school. Those who are not wage and salary workers in private industry typically submit grant proposals to obtain funding for their projects. Colleges and universities, private industry, and federal government agencies, such as the National Institutes of Health and the National Science Foundation, contribute to the support of scientists whose research proposals are determined to be financially feasible and have the potential to advance new ideas or processes.

Biological and medical scientists who work in applied research or product development use knowledge provided by basic research to develop new drugs and medical treatments, increase crop yields, and protect and clean up the environment. They usually have less autonomy than basic researchers to choose the emphasis of their research, relying instead on market-driven directions based on the firm's products and goals. Biological and medical scientists doing applied research and product development in private industry may be required to express their research plans or results to nonscientists who are in a position to veto or approve their ideas, and they must understand the business impact of their work. Scientists increasingly are working as part of teams, interacting with engineers, scientists of other disciplines, business managers, and technicians. Some biological and medical scientists also work with customers or suppliers, and manage budgets.

Those who conduct research usually work in laboratories and use electron microscopes, computers, thermal cyclers, or a wide variety of other equipment. Some conduct experiments using laboratory animals or greenhouse plants. This is particularly true of botanists, physiologists, and zoologists. For some biological scientists, a good deal of research is performed outside of laboratories. For example, a botanist may do research in tropical rain forests to see what plants grow there, or an ecologist may study how a forest area recovers after a fire.

Some biological and medical scientists work in managerial or administrative positions, usually after spending some time doing research and learning about the firm, agency, or project. They may plan and administer programs for testing foods and drugs, for example, or direct activities at zoos or botanical gardens. Some work as consultants to business firms or to government, while others test and inspect foods, drugs, and other products.

In the 1980s, swift advances in basic biological knowledge related to genetics and molecules spurred growth in the field of biotechnology. Biological and medical scientists using this technology manipulate the genetic material of animals or plants, attempting to make organisms more productive or resistant to disease. Research using biotechnology techniques, such as recombining DNA, has led to the discovery of important drugs, including human insulin and growth hormone. Many other substances not previously available in large quantities are starting to be produced by biotechnological means; some may be useful in treating cancer and other diseases. Today, many of these scientists are involved in biotechnology, including those who work on the Human Genome project, isolating, identifying, and sequencing human genes and then determining their functionality. This work continues to lead to the discovery of the genes associated with specific diseases and inherited traits, such as certain types of cancer or obesity. These advances in biotechnology have opened up research opportunities in almost all areas of biology, including commercial applications in agriculture, environmental remediation, and the food and chemical industries.

Most biological scientists who come under the category of *biologist* are further classified by the type of organism they study or by the specific activity they perform, although recent advances in the understanding of basic life processes at the molecular and cellular levels have blurred some traditional classifications.

Aquatic biologists study plants and animals living in water. *Marine biologists* study saltwater organisms, and *limnologists* study fresh water organisms. Marine biologists are sometimes mistakenly called oceanographers, but oceanography is the study of the physical characteristics of oceans and the ocean floor. *Biochemists* study the chemical composition of living things. They analyze the complex chemical combinations and reactions involved in metabolism, reproduction, growth, and heredity. Biochemists and molecular biologists do most of their work in biotechnology, which involves understanding the complex chemistry of life.

Botanists study plants and their environment. Some study all aspects of plant life; others specialize in areas such as identification and classification of plants, the structure and function of plant parts, the biochemistry of plant processes, the causes and cures of plant diseases, and the geological record of plants.

Microbiologists investigate the growth and characteristics of microscopic organisms such as bacteria, algae, or fungi. *Medical microbiologists* study the relationship between organisms and disease or the effect of antibiotics on microorganisms. Other microbiologists specialize in environmental, food, agricultural, or industrial microbiology, virology (the study of viruses), or immunology (the study of mechanisms that fight infections). Many microbiologists use biotechnology to advance knowledge of cell reproduction and human disease.

Physiologists study life functions of plants and animals, both in the whole organism and at the cellular or molecular level, under normal and abnormal conditions. Physiologists often specialize in functions such as growth, reproduction, photosynthesis, respiration, or movement, or in the physiology of a certain area or system of the organism.

Biophysicists study the application of principles of physics, such as electrical and mechanical energy and related phenomena, to living cells and organisms.

Zoologists and wildlife biologists study animals and wildlife—their origin, behavior, diseases, and life processes. Some experiment with live animals in controlled or natural surroundings while others dissect dead animals to study their structure. They may also collect and analyze biological data to determine the environmental effects of current and potential use of land and water areas. Zoologists usually are identified by the animal group studied—ornithologists (birds), mammalogists (mammals), herpetologists (reptiles), and ichthyologists (fish).

Ecologists study the relationships among organisms and between organisms and their environments and the effects of influences such as population size, pollutants, rainfall, temperature, and altitude. Utilizing knowledge of various scientific disciplines, they may collect, study, and report data on air, food, soil, and water.

Soil scientists study soil characteristics, map soil types, and investigate responses of soil to determine its capabilities and productivity.

Biological scientists who do biomedical research are usually called *medical scientists*. Medical scientists work on basic research into normal biological systems to understand the causes of and to discover treatment for disease and other health problems. Medical scientists try to identify changes in a cell, chromosome, or even gene that signal the development of medical problems, such as different types of cancer. After identifying structures of or changes in organisms that provide clues to health problems, medical scientists work on the treatment of problems. For example, a medical scientist involved in cancer research may formulate a combination of drugs that will lessen the effects of the disease. Medical scientists with a medical degree can administer these drugs to patients in clinical trials, monitor their reactions, and observe the results. (Medical scientists without a medical degree normally collaborate with a medical doctor who deals directly with patients.) The medical scientist will return to the laboratory to examine the results and, if necessary, adjust the dosage levels to reduce negative side effects or to try to induce even better results. In addition to using basic research to develop treatments for health problems, medical scientists attempt to discover ways to prevent health problems from developing, such as affirming the link between smoking and increased risk of lung cancer, or between alcoholism and liver disease.

Working Conditions

Biological and medical scientists usually work regular hours in offices or laboratories and usually are not exposed to unsafe or unhealthy conditions. Those who work with dangerous organisms or toxic substances in the laboratory must follow strict safety procedures to avoid contamination. Medical scientists also spend time working in clinics and hospitals administering drugs and treatments to patients in clinical trials. Many biological scientists such as botanists, ecologists, and zoologists take field trips that involve strenuous physical activity and primitive living conditions.

Some biological and medical scientists depend on grant money to support their research. They may be under pressure to meet deadlines and to conform to rigid grant-writing specifications when preparing proposals to seek new or extended funding.

Employment

Biological and medical scientists held about 138,000 jobs in 2000; about half were biological scientists. Federal, state, and local governments employ 4 in 10 biological scientists. Federal biological scientists worked mainly in the U.S. Departments of Agriculture, Interior, and Defense, and in the National Institutes of Health. Most of the rest worked in the drug industry, which includes pharmaceutical and biotechnology establishments, hospitals, or research and testing laboratories. About 1 in 8 medical scientists worked in government, with most of the remainder found in research and testing laboratories, educational institutions, the drug industry, and hospitals.

In addition, many biological and medical scientists held biology faculty positions in colleges and universities.

Training, Other Qualifications, and Advancement

For biological scientists, the PhD degree usually is necessary for independent research and for advancement to administrative positions. A master's degree is sufficient for some jobs in applied research or product development and for jobs in management, inspection, sales, and service. The bachelor's degree is adequate for some nonresearch jobs. For example, some graduates with a bachelor's degree start as biological scientists in testing and inspection, or get jobs related to biological science, such as technical sales or service representatives. In some cases, graduates with a bachelor's degree are able to work in a laboratory environment on their own projects, but this is unusual. Some may work as research assistants. Others become biological technicians, medical laboratory technologists or, with courses in education, high school biology teachers. Many with a bachelor's degree in biology enter medical, dental, veterinary, or other health profession schools.

In addition to required courses in chemistry and biology, undergraduate biological science majors usually study allied disciplines such as mathematics, physics, and computer science. Computer courses are essential, as employers increasingly prefer job applicants who are able to apply computer skills to modeling and simulation tasks and to operate computerized laboratory equipment. Those interested in studying the environment also should take courses in environmental studies and become familiar with current legislation and regulations.

Most colleges and universities offer bachelor's degrees in biological science and many offer advanced degrees. Curriculums for advanced degrees often emphasize a subfield such as microbiology or botany, but not all universities offer all curriculums. Advanced degree programs include classroom and fieldwork, laboratory research, and a thesis or dissertation. Biological scientists who have advanced degrees often take temporary postdoctoral research positions that provide specialized research experience. In private industry, some may become managers or administrators within the field of biology; others leave biology for nontechnical managerial, administrative, or sales jobs.

Biological scientists should be able to work independently or as part of a team and be able to communicate clearly and concisely, both orally and in writing. Those in private industry, especially those who aspire to management or administrative positions, should possess strong business and communication skills and be familiar with regulatory issues and marketing and management techniques. Those doing field research in remote areas must have physical stamina.

The PhD degree in a biological science is the minimum education required for prospective medical scientists because the work of medical scientists is almost entirely research oriented. A PhD degree qualifies one to do research on basic life processes or on particular medical problems or diseases, and to analyze and interpret the results of experiments on patients. Medical scientists who administer drug or gene therapy to human patients, or who otherwise interact medically with patients—such as drawing blood, excising tissue, or performing other invasive procedures—must have a medical degree. It is particularly helpful for medical scientists to earn both PhD and medical degrees.

In addition to formal education, medical scientists usually spend several years in a postdoctoral position before they apply for permanent jobs. Postdoctoral work provides valuable laboratory experience, including experience in specific processes and techniques, such as gene splicing, which are transferable to other research projects. In some institutions, the postdoctoral position can lead to a permanent position.

Job Outlook

Despite prospects of faster-than-average job growth for biological and medical scientists over the 2000–2010 period, doctoral degree holders can expect to face considerable competition for basic research positions. The federal government funds much basic research and development, including many areas of medical research. Recent budget tightening has led to smaller increases in federal basic research and development expenditures, further limiting the dollar amount of each grant, although the number of grants awarded to researchers remains fairly constant. At the same time, the number of newly trained scientists has continued to increase at a steady rate, so both new and established scientists have experienced greater difficulty winning and renewing research grants. If the number of advanced degrees awarded continues to grow unabated, this competitive scenario is likely to persist. Additionally, applied research positions in private industry may become more difficult to obtain if more scientists seek jobs in private industry than have done so in the past due to the competitive job market for college and university faculty.

Opportunities for those with a bachelor's or master's degree in biological science are expected to be better. The number of science-related jobs in sales, marketing, and research management, for which non-PhDs usually qualify, are expected to be more plentiful than independent research positions. Non-PhDs also may fill positions as science or engineering technicians or health technologists and technicians. Some become high school biology teachers, while those with a doctorate in biological science may become college and university faculty.

Biological and medical scientists enjoyed very rapid gains in employment between the mid-1980s and mid-1990s, in part reflecting increased staffing requirements in new biotechnology companies. Employment growth should slow somewhat as increases in the number of new biotechnology firms slow and existing firms merge or are absorbed into larger ones. However, much of the basic biological research done in recent years has resulted in new knowledge, including the isolation and identification of new genes. Biological and medical scientists will be needed to take this knowledge to the next stage, which is the understanding of how certain genes function within an entire organism, so that gene therapies can be developed to treat diseases. Even pharmaceutical and other firms not solely engaged in biotechnology are expected to increasingly use biotechnology techniques, spurring employment increases for biological and medical scientists. In addition, efforts to discover new and improved ways to clean up and preserve the environment will continue to add to growth. More biological scientists will be needed to determine the environmental impact of industry and government actions and to prevent or correct environmental problems. Expected expansion in research related to health issues such as AIDS, cancer, and Alzheimer's disease also should result in employment growth.

Biological and medical scientists are less likely to lose their jobs during recessions than are those in many other occupations because many are employed on long-term research projects. However, a recession could further influence the amount of money allocated to new research and development efforts, particularly in areas of risky or innovative research. A recession could also limit the possibility of extension or renewal of existing projects.

Earnings

Median annual earnings of biological scientists were $49,239 in 2000. Median annual earnings of medical scientists were $57,196 in 2000, with epidemiologists earning $48,390 and medical scientists, except epidemiologists, earning $57,810. Median annual earnings of medical scientists were $54,260 in research and testing laboratories and $41,010 in hospitals in 1999.

According to the National Association of Colleges and Employers, beginning salary offers in 2000 averaged $29,235 a year for bachelor's degree recipients in biological science, $35,667 for master's degree recipients, and $42,744 for doctoral degree recipients.

In the federal government in 2001, general biological scientists in nonsupervisory, supervisory, and managerial positions earned an average salary of $61,236; microbiologists, $67,835; ecologists, $61,936; physiologists, $78,366; and geneticists, $72,510.

Related Occupations

Many other occupations deal with living organisms and require a level of training similar to that of biological and medical scientists. These include agricultural and food scientists, and conservation scientists and foresters, as well as health occupations such as physicians and surgeons, dentists, and veterinarians.

Sources of Additional Information

For information on careers in the biological sciences, contact:

- American Institute of Biological Sciences, Suite 200, 1444 I St. N.W., Washington, DC 20005. Internet: http://www.aibs.org

For information on careers in physiology, contact:

- American Physiological Society, Education Office, 9650 Rockville Pike, Bethesda, MD 20814. Internet: http://www.the-aps.org

For information on careers in biochemistry or biological sciences, contact:

- Federation of American Societies for Experimental Biology, 9650 Rockville Pike, Bethesda, MD 20814. Internet: http://www.faseb.org

For a brochure entitled *Is a Career in the Pharmaceutical Sciences Right for Me?* contact:

- American Association of Pharmaceutical Scientists (AAPS), 2107 Wilson Blvd., Suite #700, Arlington, VA 22201. Internet: http://www.aaps.org/sciaffairs/careerinps.htm

For information on careers in microbiology, contact:

- American Society for Microbiology, Office of Education and Training—Career Information, 1325 Massachusetts Ave. NW, Washington, DC 20005. Internet: http://www.asmusa.org

Information on obtaining a biological or medical scientist position with the federal government is available from the Office of Personnel Management (OPM) through a telephone-based system. Consult your telephone directory under U.S. Government for a local number or call (912) 757-3000; Federal Relay Service: (800) 877-8339. The first number is not toll free, and charges may result. Information also is available from the OPM Internet site: http://www.usajobs.opm.gov

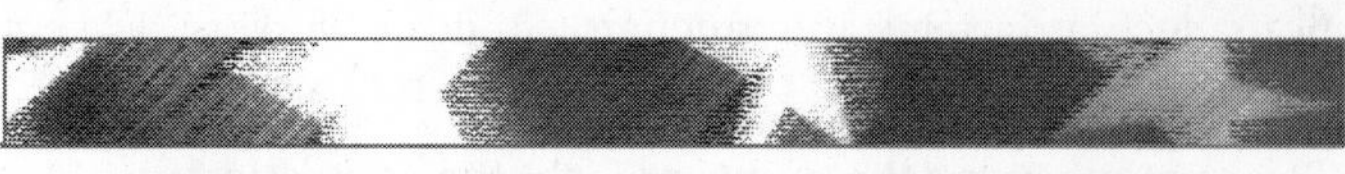

Biomedical Engineers

O*NET 17-2031.00

Significant Points

- Overall job opportunities in engineering are expected to be good.
- A bachelor's degree is required for most entry-level jobs.
- Starting salaries are significantly higher than those of college graduates in other fields.
- Continuing education is critical to keep abreast of the latest technology.

Nature of the Work

By combining biology and medicine with engineering, biomedical engineers develop devices and procedures that solve medical and health-related problems. Many do research, along with life scientists, chemists, and medical scientists, on the engineering aspects of the biological systems of humans and animals. Biomedical engineers also design devices used in various medical procedures, such as the computers used to analyze blood or the laser systems used in corrective eye surgery. They develop artificial organs, imaging systems such as ultrasound, and devices for automating insulin injections or controlling body functions. Most engineers in this specialty require a sound background in one of the more basic engineering specialties, such as mechanical or electronics engineering, in addition to specialized biomedical training. Some specialties within biomedical engineering include biomaterials, biomechanics, medical imaging, rehabilitation, and orthopedic engineering.

Working Conditions

Most engineers work in office buildings, laboratories, or industrial plants. Others may spend time outdoors at construction sites, mines, and oil and gas exploration and production sites, where they monitor or direct operations or solve onsite problems. Some engineers travel extensively to plants or work sites.

Many engineers work a standard 40-hour week. At times, deadlines or design standards may bring extra pressure to a job. When this happens, engineers may work longer hours and experience considerable stress.

Employment

This section reports the number of jobs the occupation provided in 2000 and the key industries where these jobs are found. When significant, the geographic distribution of jobs and the proportion of part-time (less than 35 hours a week) and self-employed workers in the occupation are mentioned. Self-employed workers accounted for nearly eight percent of the work force in 2000; however, they were concentrated in a small number of occupations, such as farmers and ranchers, childcare workers, lawyers, health practitioners, and the construction trades.

This section reports the number of jobs the occupation provided in 2000 and the key industries where these jobs are found. When significant, the geographic distribution of jobs and the proportion of part-time (less than 35 hours a week) and self-employed workers in the occupation are mentioned. Self-employed workers accounted for nearly eight percent of the work force in 2000; however, they were concentrated in a small number of occupations, such as farmers and ranchers, childcare workers, lawyers, health practitioners, and the construction trades.

Training, Other Qualifications, and Advancement

A bachelor's degree in engineering is required for almost all entry-level engineering jobs. College graduates with a degree in a physical science or mathematics occasionally may qualify for some engineering jobs, especially in specialties in high demand. Most engineering degrees are granted in electrical, electronics, mechanical, or civil engineering. However, engineers trained in one branch may work in related branches. For example, many aerospace engineers have training in mechanical engineering. This flexibility allows employers to meet staffing needs in new technologies and specialties in which engineers are in short supply. It also allows engineers to shift to fields with better employment prospects or to those that more closely match their interests.

Most engineering programs involve a concentration of study in an engineering specialty, along with courses in both mathematics and science. Most programs include a design course, sometimes accompanied by a computer or laboratory class or both.

In addition to the standard engineering degree, many colleges offer 2- or 4-year degree programs in engineering technology. These programs, which usually include various hands-on laboratory classes that focus on current issues, prepare students for practical design and production work, rather than for jobs which require more theoretical and scientific knowledge. Graduates of 4-year technology programs may get jobs similar to those obtained by graduates with a bachelor's degree in engineering. Engineering technology graduates, however, are not qualified to register as professional engineers under the same terms as graduates with degrees in engineering. Some employers regard technology program graduates as having skills between those of a technician and an engineer.

Graduate training is essential for engineering faculty positions and many research and development programs, but is not required for the majority of entry-level engineering jobs. Many engineers obtain graduate degrees in engineering or business administration to learn new technology and broaden their education. Many high-level executives in government and industry began their careers as engineers.

About 330 colleges and universities offer bachelor's degree programs in engineering that are accredited by the Accreditation Board for Engineering and Technology (ABET), and about 250 colleges offer accredited bachelor's degree programs in engineering technology. ABET accreditation is based on an examination of an engineering program's student achievement, program improvement, faculty, curricular content, facilities, and institutional commitment. Although most institutions offer programs in the major branches of engineering, only a few offer programs in the smaller specialties. Also, programs of the same title may vary in content. For example, some programs emphasize industrial practices, preparing students for a job in industry, whereas others are more theoretical and are designed to prepare students for graduate work. Therefore, students should investigate curricula and check accreditations carefully before selecting a college.

Admissions requirements for undergraduate engineering schools include a solid background in mathematics (algebra, geometry, trigonometry, and calculus) and sciences (biology, chemistry, and physics), and courses in English, social studies, humanities, and computers. Bachelor's degree programs in engineering typically are designed to last 4 years, but many students find that it takes between 4 and 5 years to complete their studies. In a typical 4-year college curriculum, the first 2 years are spent studying mathematics, basic sciences, introductory engineering, humanities, and social sciences. In the last 2 years, most courses are in engineering, usually with a concentration in one branch. Some programs offer a general engineering curriculum; students then specialize in graduate school or on the job.

Some engineering schools and 2-year colleges have agreements whereby the 2-year college provides the initial engineering education, and the engineering school automatically admits students for their last 2 years. In addition, a few engineering schools have arrangements whereby a student spends 3 years in a liberal arts college studying pre-engineering subjects and 2 years in an engineering school studying core subjects, and then receives a bachelor's degree from each school. Some colleges and universities offer 5-year master's degree programs. Some 5- or even 6-year cooperative plans combine classroom study and practical work, permitting students to gain valuable experience and finance part of their education. All 50 states and the District of Columbia usually require licensure for engineers who offer their services directly to the public. Engineers who are licensed are called Professional Engineers (PE). This licensure generally requires a degree from an ABET-accredited engineering program, 4 years of relevant work experience, and successful completion of a state examination. Recent graduates can start the licensing process by taking the examination in two stages. The initial Fundamentals of Engineering (FE) examination can be taken upon graduation. Engineers who pass this examination commonly are called Engineers in Training (EIT) or Engineer Interns (EI). The EIT certification usually is valid for 10 years. After acquiring suitable work experience, EITs can take the second examination, the Principles and Practice of Engineering Exam. Several states have imposed mandatory continuing education requirements for relicensure. Most states recognize licensure from other states.

Engineers should be creative, inquisitive, analytical, and detail oriented. They should be able to work as part of a team and to communicate well, both orally and in writing. Communication abilities are becoming more important because much of their work is becoming more diversified, meaning that engineers interact with specialists in a wide range of fields outside engineering.

Beginning engineering graduates usually work under the supervision of experienced engineers and, in large companies, also may receive formal classroom or seminar-type training. As new engineers gain knowledge and experience, they are assigned more difficult projects with greater independence to develop designs, solve problems, and make decisions. Engineers may advance to become technical specialists or to supervise a staff or team of engineers and technicians. Some may eventually become engineering managers or enter other managerial or sales jobs.

Job Outlook

Employment of biomedical engineers is expected to increase faster than the average for all occupations through 2010. The aging population and the focus on health issues will increase the demand for better medical devices and systems designed by biomedical engineers. For example, computer-assisted surgery and cellular and tissue engineering are being more heavily researched and are developing rapidly. In addition, the rehabilitation and orthopedic engineering specialties are growing quickly, increasing the need for more biomedical engineers. Along with the demand for more sophisticated medical equipment and procedures is an increased concern for cost efficiency and effectiveness that also will increase the need for biomedical engineers.

Earnings

Median annual earnings of biomedical engineers were $57,480 in 2000. The middle 50 percent earned between $45,760 and $74,120. The lowest 10 percent earned less than $36,860 and the highest 10 percent earned more than $90,530.

According to a 2001 salary survey by the National Association of Colleges and Employers, bachelor's degree candidates in biomedical engineering received starting offers averaging $47,850 a year and master's degree candidates, on average, were offered $62,600.

Related Occupations

Biomedical engineers apply the principles of physical science and mathematics in their work. Other workers who use scientific and mathematical principles include architects, except landscape and naval; engineering and natural sciences managers; computer and information systems managers; mathematicians; drafters; engineering technicians; sales engineers; science technicians; and physical and life scientists, including agricultural and food scientists, biological and medical scientists, conservation scientists and foresters, atmospheric scientists, chemists and materials scientists, environmental scientists and geoscientists, and physicists and astronomers.

Sources of Additional Information

For further information about biomedical engineers, contact:

- Biomedical Engineering Society, 8401 Corporate Dr., Suite 110, Landover, MD 20785-2224. Internet: http://mecca.org/BME/BMES/society/index.htm

High school students interested in obtaining a full package of guidance materials and information (product number SP-01) on a variety of engineering disciplines should contact the Junior Engineering Technical Society by sending $3.50 to:

- JETS-Guidance, 1420 King St., Suite 405, Alexandria, VA 22314-2794. Internet: http://www.jets.org

High school students interested in obtaining information on ABET-accredited engineering programs should contact:

- The Accreditation Board for Engineering and Technology, Inc., 111 Market Place, Suite 1050, Baltimore, MD 21202-4012. Internet: http://www.abet.org

Non-licensed engineers and college students interested in obtaining information on Professional Engineer licensure should contact:

- The National Society of Professional Engineers, 1420 King St., Alexandria, VA 22314-2794. Internet: http://www.nspe.org
- National Council of Examiners for Engineers and Surveying, P.O. Box 1686, Clemson, SC 29633-1686. Internet: http://www.ncees.org

Information on general engineering education and career resources is available from:

- American Society for Engineering Education, 1818 N St. N.W., Suite 600, Washington, DC 20036-2479. Internet: http://www.asee.org

Information on obtaining an engineering position with the federal government is available from the Office of Personnel Management (OPM) through a telephone-based system. Consult your telephone directory under U.S. Government for a local number or call (912) 757-3000; Federal Relay Service: (800) 877-8339. The first number is not toll free, and charges may result. Information also is available from the OPM Internet site: http://www.usajobs.opm.gov

Non-high school students and those wanting more detailed information should contact societies representing the individual branches of engineering. Each can provide information about careers in the particular branch. The individual statements elsewhere in this book also provide other information in detail on aerospace; agricultural; chemical; civil; computer hardware; electrical and electronics; environmental; industrial, including health and safety; materials; mechanical; mining and geological, including mining safety; nuclear; and petroleum engineering.

Broadcast and Sound Engineering Technicians and Radio Operators

O*NET 27-4011.00, 27-4012.00, 27-4013.00, 27-4014.00

Significant Points

- Job applicants will face strong competition for the better paying jobs at radio and television stations serving large cities.
- Television stations employ, on average, many more technicians than do radio stations.
- Evening, weekend, and holiday work is common.

Nature of the Work

Broadcast and sound engineering technicians install, test, repair, set up, and operate the electronic equipment used to record and transmit radio and television programs, cable programs, and motion pictures. They work with television cameras, microphones, tape recorders, lighting, sound effects, transmitters, antennas, and other equipment. Some broadcast and sound engineering technicians produce movie soundtracks in motion picture production studios, control the sound of live events, such as concerts, or record music in a recording studio.

In the control room of a radio or television-broadcasting studio, these technicians operate equipment that regulates the signal strength, clarity, and range of sounds and colors of recordings or broadcasts. They also operate control panels to select the source of the material. Technicians may switch from one camera or studio to another, from film to live programming, or from network to local programming. By means of hand signals and, in television, telephone headsets, they give technical directions to other studio personnel.

Audio and video equipment operators operate specialized electronic equipment to record stage productions, live programs or events, and studio recordings. They edit and reproduce tapes for compact discs, records and cassettes, for radio and television broadcasting and for motion picture productions. The duties of audio and video equipment operators can be divided into two categories: technical and production activities used in the production of sound and picture images for film or videotape from set design to camera operation and post production activities where raw images are transformed to a final print or tape.

Radio operators mainly receive and transmit communications using a variety of tools. They are also responsible for repairing equipment using such devices as electronic testing equipment, hand tools, and power tools. These help to maintain communication systems in an operative condition.

Broadcast and sound engineering technicians and radio operators perform a variety of duties in small stations. In large stations and at the networks, technicians are more specialized, although job assignments may change from day to day. The terms "operator," "engineer," and "technician" often are used interchangeably to describe these jobs. *Transmitter operators* monitor and log outgoing signals and operate transmitters. *Maintenance technicians* set up, adjust, service, and repair electronic broadcasting equipment. *Audio control* engineers regulate volume and sound quality of television broadcasts, while *video control* engineers regulate their fidelity, brightness, and contrast. *Recording* engineers operate and maintain video and sound recording equipment. They may operate equipment designed to produce special effects, such as the illusions of a bolt of lightning or a police siren. *Sound mixers* or *re-recording mixers* produce the sound track of a movie, television, or radio program. After filming or recording, they may use a process called dubbing to insert sounds. *Field technicians* set up and operate broadcasting portable field transmission equipment outside the studio. Television news coverage requires so much electronic equipment, and the technology is changing so rapidly, that many stations assign technicians exclusively to news.

Chief engineers, transmission engineers, and *broadcast field supervisors* supervise the technicians who operate and maintain broadcasting equipment.

Working Conditions

Broadcast, sound engineering, audio and video equipment technicians, and radio operators generally work indoors in pleasant surroundings. However, those who broadcast news and other programs from locations outside the studio may work outdoors in all types of weather. Technicians doing maintenance may climb poles or antenna towers, while those setting up equipment do heavy lifting.

Technicians in large stations and the networks usually work a 40-hour week under great pressure to meet broadcast deadlines, but may occasionally work overtime. Technicians in small stations routinely work more than 40 hours a week. Evening, weekend, and holiday work is usual, because most stations are on the air 18 to 24 hours a day, 7 days a week.

Those who work on motion pictures may be on a tight schedule to finish according to contract agreements.

Employment

Broadcast and sound engineering technicians and radio operators held about 87, 000 jobs in 2000. Their employment was distributed among the following detailed occupations:

Audio and video equipment technicians	37,000
Broadcast technicians	36,000
Sound engineering technicians	11,000
Radio operators	2,900

About 1 out of 3 worked in radio and television broadcasting. Almost 15 percent worked in the motion picture industry. About 4 percent worked for cable and other pay-television services. A few were self-employed. Television stations employ, on average, many more technicians than do radio stations. Some technicians are employed in other industries, producing employee communications, sales, and training programs. Technician jobs in television are located in virtually all cities, whereas jobs in radio are also found in many small towns. The highest paying and most specialized jobs are concentrated in New York City, Los Angeles, Chicago, and Washington, DC—the originating centers for most network programs. Motion picture production jobs are concentrated in Los Angeles and New York City.

Training, Other Qualifications, and Advancement

The best way to prepare for a broadcast and sound engineering technician job is to obtain technical school, community college, or college training in broadcast technology or in engineering or electronics. This is particularly true for those who hope to advance to supervisory positions or jobs at large stations or the networks. In the motion picture industry people are hired as apprentice editorial assistants and work their way up to more skilled jobs. Employers in the motion picture industry usually hire experienced freelance technicians on a picture-by-picture basis. Reputation and determination are important in getting jobs.

Beginners learn skills on the job from experienced technicians and supervisors. They often begin their careers in small stations and, once experienced, move on to larger ones. Large stations usually only hire technicians with experience. Many employers pay tuition and expenses for courses or seminars to help technicians keep abreast of developments in the field.

Audio and video equipment technicians generally need a high school diploma. Many recent entrants have a community college degree or various other forms of post-secondary degrees, although that is not always a requirement. They may substitute on-the-job training for formal education requirements. Experience in a recording studio, as an assistant is a great way of getting experience and knowledge simultaneously.

Radio operators do not usually require any formal training. This is an entry-level position that is generally suited with on-the-job training.

The Federal Communications Commission no longer requires the licensing of broadcast technicians, as the Telecommunications Act of 1996 eliminated this licensing requirement. Certification by the Society of Broadcast Engineers is a mark of competence and experience. The certificate is issued to experienced technicians who pass an examination. By offering the Radio Operator and the Television Operator levels of certification, the Society of Broadcast Engineers has filled the void left by the elimination of the FCC license.

Prospective technicians should take high school courses in math, physics, and electronics. Building electronic equipment from hobby kits and operating a "ham," or amateur radio, are good experience, as is work in college radio and television stations.

Broadcast and sound engineering technicians and radio operators must have manual dexterity and an aptitude for working with electrical, electronic, and mechanical systems and equipment.

Experienced technicians can become supervisory technicians or chief engineers. A college degree in engineering is needed to become chief engineer at a large TV station.

Job Outlook

People seeking entry-level jobs as technicians in the field of radio and television broadcasting are expected to face strong competition in major metropolitan areas, where pay generally is higher and the number of qualified job seekers exceed the number of openings. There, stations seek highly experienced personnel. Prospects for entry-level positions generally are better in small cities and towns for beginners with appropriate training.

The overall employment of broadcast and sound engineering technicians and radio operators is expected to grow about as fast as the average for all occupations through the year 2010. An increase in the number of programming hours should require additional technicians. However, employment growth in radio and television broadcasting may be tempered somewhat because of slow growth in the number of new radio and television stations and laborsaving technical advances, such as computer-controlled programming and remote control of transmitters. Technicians who know how to install transmitters will be in demand as television stations replace existing analog transmitters with digital transmitters. Stations will begin broadcasting in both analog and digital formats, eventually switching entirely to digital.

Employment of broadcast and sound engineering technicians is expected to grow about as fast as average through 2010. The advancements in technology will enhance the capabilities of technicians to help produce a higher quality of programming for radio and television. Employment of audio and video equipment technicians also is expected to grow about as fast as average through 2010. Not only will these workers have to set up audio and video equipment, but it will be necessary for them to maintain and repair this machinery. Employment of radio operators, on the other hand, will grow more slowly than other areas in this field of work. Automation will negatively impact these workers as many stations now operate transmitters and control programming remotely.

Employment of broadcast and sound engineering technicians and radio operators in the cable industry should grow rapidly because of new products coming to market. Such products include cable modems, which deliver high-speed Internet access to personal computers, and digital set-top boxes, which transmit better sound and pictures, allowing cable operators to offer many more channels than in the past. These new products should cause traditional cable subscribers to sign up for additional services.

Employment in the motion picture industry also will grow fast. However, job prospects are expected to remain competitive, because of the large number of people attracted to this relatively small field.

Numerous job openings also will result from the need to replace experienced technicians who leave the occupations. Many leave these occupations for electronic jobs in other areas, such as computer technology or commercial and industrial repair.

Earnings

Television stations usually pay higher salaries than radio stations; commercial broadcasting usually pays more than public broadcasting; and stations in large markets pay more than those in small ones.

Median annual earnings of broadcast technicians in 2000 were $26,950. The middle 50 percent earned between $18,060 and $44,410. The lowest 10 percent earned less than $13,860, and the highest 10 percent earned more than $63,340.

Median annual earnings of sound engineering technicians in 2000 were $39,480. The middle 50 percent earned between $24,730 and $73,720. The lowest 10 percent earned less than $17,560, and the highest 10 percent earned more than $119,400.

Median annual earnings of audio and video equipment technicians in 2000 were $30,310. The middle 50 percent earned between $21,980 and $44,970. The lowest 10 percent earned less than $16,630, and the highest 10 percent earned more than $68,720.

Median annual earnings of radio operators in 2000 were $29,260. The middle 50 percent earned between $23,090 and $39,830. The lowest 10 percent earned less than $17,570, and the highest 10 percent earned more than $54,590.

Related Occupations

Broadcast and sound engineering technicians and radio operators need the electronics training and hand coordination necessary to operate technical equipment, and they generally complete specialized postsecondary programs. Similar occupations include engineering technicians, science technicians, health technologists and technicians, electrical and electronics installers and repairers, and communications equipment operators.

Sources of Additional Information

For information on careers for broadcast and sound engineering technicians and radio operators, write to:

- National Association of Broadcasters, 1771 N St. NW, Washington, DC 20036. Internet: http://www.nab.org

For information on certification, contact:

- Society of Broadcast Engineers, 9247 North Meridian St., Suite 305, Indianapolis, IN 46260. Internet: http://www.sbe.org

For information on careers in the motion picture and television industry, contact:

- Society of Motion Picture and Television Engineers (SMPTE), 595 West Hartsdale Ave., White Plains, NY 10607. Internet: http://www.smpte.org

Budget Analysts

O*NET 13-2031.00

Significant Points

- Two out of five budget analysts work in federal, state, and local governments.
- A bachelor's degree generally is the minimum educational requirement; however, some employers require a master's degree.
- Competition for jobs should remain keen due to the substantial number of qualified applicants; those with a master's degree should have the best prospects.

Nature of the Work

Deciding how to efficiently distribute limited financial resources is an important challenge in all organizations. In most large and complex organizations, this task would be nearly impossible were it not for budget analysts. These professionals play the primary role in the development, analysis, and execution of budgets, which are used to allocate current resources and estimate future financial requirements. Without effective analysis of and feedback about budgetary problems, many private and public organizations could become bankrupt.

Budget analysts can be found in private industry, nonprofit organizations, and the public sector. In private sector firms, a budget analyst examines, analyzes, and seeks new ways to improve efficiency and increase profits. Although analysts working in nonprofit and governmental organizations usually are not concerned with profits, they still try to find the most efficient distribution of funds and other resources among various departments and programs.

Budget analysts have many responsibilities in these organizations, but their primary task is providing advice and technical assistance in the preparation of annual budgets. At the beginning of each budget cycle, managers and department heads submit proposed operational and financial plans to budget analysts for review. These plans outline expected programs, including proposed monetary increases and new initiatives, estimated costs and expenses, and capital expenditures needed to finance these programs.

Analysts examine the budget estimates or proposals for completeness, accuracy, and conformance with established procedures, regulations, and organizational objectives. Sometimes, they employ cost-benefit analysis to review financial requests, assess program trade-offs, and explore alternative funding methods. They also examine past and current budgets and research economic and financial developments that affect the organization's spending. This process enables analysts to evaluate proposals in terms of the organization's priorities and financial resources.

After this initial review process, budget analysts consolidate the individual departmental budgets into operating and capital budget summaries. These summaries contain comments and supporting statements that support or argue against funding requests. Budget summaries then are submitted to senior management or, as is often the case in local and state governments, to appointed or elected officials. Budget analysts then help the chief operating officer, agency head, or other top managers analyze the proposed plan and devise possible alternatives if the projected results are unsatisfactory. The final decision to approve the budget, however, usually is made by the organization head in a private firm or by elected officials in government, such as the state legislative body.

Throughout the remainder of the year, analysts periodically monitor the budget by reviewing reports and accounting records to determine if allocated funds have been spent as specified. If deviations appear between the approved budget and actual performance, budget analysts may write a report explaining the causes of the variations along with recommendations for new or revised budget procedures. In order to avoid or alleviate deficits, they may recommend program cuts or reallocation of excess funds. They also inform program managers and others within their organization of the status and availability of funds in different budget accounts. Before any changes are made to an existing program or a new one is implemented, a budget analyst assesses its efficiency and effectiveness. Analysts also may be involved in long-range planning activities such as projecting future budget needs.

The budget analyst's role has broadened as limited funding has led to downsizing and restructuring throughout private industry and government. Not only do they develop guidelines and policies governing the formulation and maintenance of the budget, but they also measure organizational performance, assess the effect of various programs and policies on the budget, and help draft budget-related legislation. In addition, budget analysts sometimes conduct training sessions for company or government agency personnel regarding new budget procedures.

Working Conditions

Budget analysts usually work in a comfortable office setting. Long hours are common among these workers, especially during the initial development and mid-year and final reviews of budgets. The pressure of deadlines and tight work schedules during these periods can be extremely stressful, and analysts usually are required to work more than the routine 40-hour week.

Budget analysts spend the majority of their time working independently, compiling and analyzing data, and preparing budget proposals. Nevertheless, their schedule sometimes is interrupted by special budget requests, meetings, and training sessions. Some budget analysts travel to obtain budget details and explanations of various programs from coworkers, or to personally observe funding allocation.

Employment

Budget analysts held about 70,000 jobs throughout private industry and government in 2000. Federal, state, and local governments are major employers, accounting for two-fifths of all budget analyst jobs. The U.S. Department of Defense employed 7 of every 10 budget analysts working for the federal government. Other major employers include schools, hospitals, and banks.

Training, Other Qualifications, and Advancement

Private firms and government agencies generally require candidates for budget analyst positions to have at least a bachelor's degree. Within the federal government, a bachelor's degree in any field is sufficient for an entry-level budget analyst position. State and local governments have varying requirements, but a bachelor's degree in one of many areas—accounting, finance, business or public administration, economics, political science, statistics, or a social science such as sociology—may qualify one for entry into the occupation. Sometimes, a field closely related to the employing industry or organization, such as engineering, may be preferred. An increasing number of state governments and other employers require a candidate to possess a master's degree to ensure adequate analytical and communication skills. Some firms prefer candidates with backgrounds in business because business courses emphasize quantitative and analytical skills. Occasionally, budget and financial experience can be substituted for formal education.

Because developing a budget involves manipulating numbers and requires strong analytical skills, courses in statistics or accounting are helpful, regardless of the prospective budget analyst's major field of study. Financial analysis is automated in almost every organization and, therefore, familiarity with word processing and the financial software packages used in budget analysis often is required. Software packages commonly used by budget analysts include electronic spreadsheet, database, and graphics software. Employers usually prefer job candidates who already possess these computer skills.

In addition to analytical and computer skills, those seeking a career as a budget analyst also must be able to work under strict time constraints. Strong oral and written communication skills are essential for analysts because they must prepare, present, and defend budget proposals to decision-makers.

Entry-level budget analysts may receive some formal training when they begin their jobs, but most employers feel that the best training is obtained by working through one complete budget cycle. During the cycle, which typically is one year, analysts become familiar with the various steps involved in the budgeting process. The federal government, on the other hand, offers extensive on-the-job and classroom training for entry-level trainees. In addition to on-the-job training, budget analysts are encouraged to participate in the various classes offered throughout their careers.

Some budget analysts employed in the federal, state, or local level may choose to receive the Certified Government Financial Manager (CGFM) designation granted by the Association of Government Accountants. Other government financial officers also may receive this designation. Candidates must have a minimum of a bachelor's degree, 24 hours of study in financial management, and 2 years' experience in government, plus pass a series of 3 exams. The exams cover topics in governmental environment; governmental accounting, financial reporting, and budgeting; and financial management and control.

Budget analysts start their careers with limited responsibilities. In the federal government, for example, beginning budget analysts compare projected costs with prior expenditures, consolidate and enter data prepared by others, and assist higher-grade analysts by doing research. As analysts progress, they begin to develop and formulate budget estimates and justification statements, perform in-depth analyses of budget requests, write statements supporting funding requests, advise program managers and others on the status and availability of funds in different budget activities, and present and defend budget proposals to senior managers.

Beginning analysts usually work under close supervision. Capable entry-level analysts can be promoted into intermediate-level positions within 1 to 2 years, and then into senior positions within a few more years. Progressing to a higher level means added budgetary responsibility and can lead to a supervisory role. Because of the importance and high visibility of their jobs, senior budget analysts are prime candidates for promotion to management positions in various parts of the organization.

Job Outlook

Competition for budget analyst jobs should remain keen due to the substantial number of qualified applicants. Candidates with a master's degree should have the best job opportunities. Familiarity with computer financial software packages also should enhance a job seeker's employment prospects.

Employment of budget analysts is expected to grow about as fast as the average for all occupations through 2010. Employment growth will be driven by the continuing demand for sound financial analysis in both public and private sector organizations. In addition to employment growth, many job openings will result from the need to replace experienced budget analysts who transfer to other occupations or leave the labor force.

The expanding use of computer applications in budget analysis increases worker productivity by enabling analysts to process more data in less time. However, because analysts now have a greater supply of data available to them, their jobs are becoming more complicated. In addition, as businesses become increasingly complex and specialization within organizations becomes more common, planning and financial control increasingly demand attention. These factors should offset any adverse effects of computer usage on employment of budget analysts.

In coming years, companies will continue to rely heavily on budget analysts to examine, analyze, and develop budgets. Because the financial analysis performed by budget analysts is an important function in every large organization, the employment of budget analysts has remained relatively unaffected by downsizing in the nation's workplaces. Because financial and budget reports must be completed during periods of economic growth and slowdowns, budget analysts usually are less subject to layoffs during economic downturns than many other workers are.

Earnings

Salaries of budget analysts vary widely by experience, education, and employer. Median annual earnings of budget analysts in 2000 were $48,370. The middle 50 percent earned between $38,400 and $61,030. The lowest 10 percent earned less than $31,260, and the highest 10 percent earned more than $74,030.

According to a 2001 survey conducted by Robert Half International, a staffing services firm specializing in accounting and finance, starting salaries of budget and other financial analysts in small firms ranged from $29,750 to $35,500; in large organizations compensation ranged from $33,250 to $40,000. In small firms, analysts with 1 to 3 years of experience earned from $34,500 to $42,750; in large companies they made from $40,250 to $51,000. Senior analysts in small firms earned from $37,750 to $55,750; in large firms they made from $51,500 to $64,000. Earnings of managers in this field ranged from $40,500 to $62,750 a year in small firms, while managers in large organizations earned between $63,500 and $81,250.

In the federal government, budget analysts usually started as trainees earning $21,947 or $27,185 a year in 2001. Candidates with a master's degree might begin at $33,254. Beginning salaries were slightly higher in selected areas where the prevailing local pay level was higher. Budget analysts employed by the federal government in nonsupervisory, supervisory, and managerial positions in 2001 received an average annual salary of $56,710.

Related Occupations

Budget analysts review, analyze, and interpret financial data; make recommendations for the future; and assist in the implementation of new ideas and financial strategies. Workers who use these skills in other occupations include accountants and auditors, cost estimators, economists and market and survey researchers, financial analysts and personal financial advisors, financial managers, and loan counselors and officers.

Sources of Additional Information

Information about career opportunities as a budget analyst may be available from your state or local employment service.

Information on careers in government financial management and the CGFM designation may be obtained from:

- Association of Government Accountants, 2208 Mount Vernon Ave., Alexandria, VA 22301. Internet: http://www.agacgfm.org

Information on careers in budget analysis at the state government level can be obtained from:

- National Association of State Budget Officers, Hall of the States Building, Suite 642, 444 North Capitol St. NW, Washington, DC 20001. Internet: http://www.nasbo.org

Information on obtaining a budget analyst position with the federal government is available from the Office of Personnel Management (OPM) through a telephone-based system. Consult your telephone directory under U.S. Government for a local number or call (912) 757-3000; Federal Relay Service: (800) 877-8339. The first number is not toll free, and charges may result. Information also is available from the OPM Internet site: http://www.usajobs.opm.gov

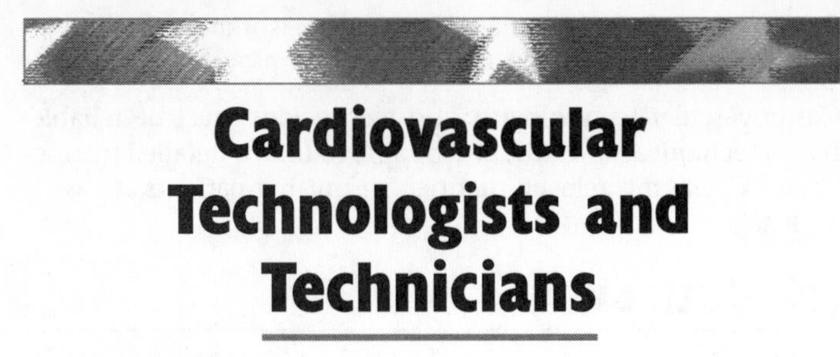

Cardiovascular Technologists and Technicians

O*NET 29-2031.00

Significant Points

- Employment will grow faster than the average, but the number of job openings created will be low, because the occupation is small.
- Job prospects will be good due to an aging population and increased need for vascular technology and sonography as an alternative for more costly and invasive heart surgery.
- About 7 out of 10 jobs are in hospitals, in both inpatient and outpatient settings.

Nature of the Work

Cardiovascular technologists and technicians assist physicians in diagnosing and treating cardiac (heart) and peripheral vascular (blood vessel) ailments. Cardiovascular technologists may specialize in three areas of practice: Invasive cardiology, echocardiography, and vascular technology. Cardiovascular technicians who specialize in electrocardiograms (EKGs), stress testing, and Holter monitors are known as *cardiographic* or *EKG technicians.*

Cardiovascular technologists specializing in invasive procedures are called *cardiology technologists*. They assist physicians with cardiac catheterization procedures in which a small tube, or catheter, is wound through a patient's blood vessel from a spot on the patient's leg into the heart. The procedure can determine if a blockage exists in the blood vessels that supply the heart muscle and help diagnose other problems. Part of the procedure may involve balloon angioplasty, which can be used to treat blockages of blood vessels or heart valves, without the need for heart surgery. Cardiology technologists assist physicians as they insert a catheter with a balloon on the end to the point of the obstruction.

Technologists prepare patients for cardiac catheterization and balloon angioplasty by first positioning them on an examining table and then shaving, cleaning, and administering anesthesia to the top of the patient's leg near the groin. During the procedures, they monitor patients' blood pressure and heart rate using EKG equipment and notify the physician if something appears wrong. Technologists also may prepare and monitor patients during open-heart surgery and the implantation of pacemakers.

Cardiovascular technologists who specialize in echocardiography or vascular technology often run noninvasive tests using ultrasound instrumentation, such as doppler ultrasound. Tests are called "noninvasive" if they do not require the insertion of probes or other instruments into the patient's body. The ultrasound instrumentation transmits high frequency sound waves into areas of the patient's body and then processes reflected echoes of the sound waves to form an image. Technologists view the ultrasound image on a screen that may be recorded on videotape or photographed for interpretation and diagnosis by a physician. While performing the scan, technologists check the image on the screen for subtle differences between healthy and diseased areas, decide which images to include, and judge if the images are satisfactory for diagnostic purposes. They also explain the procedure to patients, record additional medical history, select appropriate equipment settings, and change the patient's position as necessary.

Those who assist physicians in the diagnosis of disorders affecting circulation are known as *vascular technologists* or *vascular sonographers*. They perform a medical history and evaluate pulses by listening to the sounds of the arteries for abnormalities. Then they perform a noninvasive procedure using ultrasound instrumentation to record vascular information, such as vascular blood flow, blood pressure, limb volume changes, oxygen saturation, cerebral circulation, peripheral circulation, and abdominal circulation. Many of these tests are performed during or immediately after surgery.

Technologists who use ultrasound to examine the heart chambers, valves, and vessels are referred to as *cardiac sonographers*, or *echocardiographers*. They use ultrasound instrumentation to create images called echocardiograms. This may be done while the patient is either resting or physically active. Technologists may administer medication to a physically active patient to assess their heart function. Cardiac sonographers may also assist physicians who perform transesophageal echocardiography, which involves placing a tube in the patient's esophagus to obtain ultrasound images.

Cardiovascular technicians who obtain EKGs are known as *electrocardiograph* (or *EKG) technicians*. To take a basic EKG, which traces electrical impulses transmitted by the heart, technicians attach electrodes to the patient's chest, arms, and legs, and then manipulate switches on an EKG machine to obtain a reading. The physician makes a printout for interpretation. This test is done before most kinds of surgery and as part of a routine physical examination, especially for persons who have reached middle age or have a history of cardiovascular problems.

EKG technicians with advanced training perform Holter monitor and stress testing. For Holter monitoring, technicians place electrodes on the patient's chest and attach a portable EKG monitor to the patient's belt. Following 24 or more hours of normal activity for the patient, the technician removes a tape from the monitor and places it in a scanner. After checking the quality of the recorded impulses on an electronic screen, the technician usually prints the information from the tape so that a physician can interpret it later. Physicians use the output from the scanner to diagnose heart ailments, such as heart rhythm abnormalities or problems with pacemakers.

For a treadmill stress test, EKG technicians document the patient's medical history, explain the procedure, connect the patient to an EKG monitor, and obtain a baseline reading and resting blood pressure. Next, they monitor the heart's performance while the patient is walking on a treadmill, gradually increasing the treadmill's speed to observe the effect of increased exertion. Like vascular technologists and cardiac sonographers, cardiographic technicians who perform EKG, Holter monitor, and stress tests are known as "noninvasive" technicians.

Some cardiovascular technologists and technicians schedule appointments, type doctor interpretations, maintain patient files, and care for equipment.

Working Conditions

Technologists and technicians generally work a 5-day, 40-hour week that may include weekends. Those in catheterization labs tend to work longer hours and may work evenings. They also may be on call during the night and on weekends.

Cardiovascular technologists and technicians spend a lot of time walking and standing. Those who work in catheterization labs may face stressful working conditions, because they are in close contact with patients with serious heart ailments. Some patients, for example, may encounter complications from time to time that have life or death implications.

Employment

Cardiovascular technologists and technicians held about 39,000 jobs in 2000. Most worked in hospital cardiology departments, whereas some worked in offices of cardiologists or other physicians, cardiac rehabilitation centers, or ambulatory surgery centers.

Training, Other Qualifications, and Advancement

Although a few cardiovascular technologists, vascular technologists, and cardiac sonographers are currently trained on the job, most receive training in 2- to 4-year programs. Cardiovascular technologists, vascular technologists, and cardiac sonographers normally complete a 2-year junior or community college program. One year is dedicated to core courses followed by a year of specialized instruction in either invasive, noninvasive cardiovascular, or noninvasive vascular technology. Those who are qualified in a related allied health profession only need to complete the year of specialized instruction.

Graduates from the 23 programs accredited by the Joint Review Committee on Education in Cardiovascular Technology are eligible to obtain professional certification through Cardiovascular Credentialing International in cardiac catheterization, echocardiography, vascular ultrasound, and cardiographic techniques. Cardiac sonographers and vascular technologists may also obtain certification with the American Registry of Diagnostic Medical Sonographers.

For basic EKGs, Holter monitoring, and stress testing, 1-year certificate programs exist; but most EKG technicians are still trained on the job by an EKG supervisor or a cardiologist. On-the-job training usually lasts about 8 to 16 weeks. Most employers prefer to train people already in the healthcare field—nursing aides, for example. Some EKG technicians are students enrolled in 2-year programs to become technologists, working part-time to gain experience and make contact with employers.

Cardiovascular technologists and technicians must be reliable, have mechanical aptitude, and be able to follow detailed instructions. A pleasant, relaxed manner for putting patients at ease is an asset.

Job Outlook

Employment of cardiovascular technologists and technicians is expected to grow faster than the average for all occupations through the year 2010. Growth will occur as the population ages, because older people have a higher incidence of heart problems. Employment of *vascular technologists* and *echocardiographers* will grow as advances in vascular technology and sonography reduce the need for more costly and invasive procedures. Employment of *EKG technicians* is expected to decline, as hospitals train nursing aides and others to perform basic EKG procedures. Individuals trained in Holter monitoring and stress testing are expected to have more favorable job prospects than those who can only perform a basic EKG.

Some job openings for cardiovascular technologists and technicians will arise from replacement needs, as individuals transfer to other jobs or leave the labor force. Relatively few job openings, due to both growth and replacement needs are expected, however, because the occupation is small.

Earnings

Median annual earnings of cardiovascular technologists and technicians were $33,350 in 2000. The middle 50 percent earned between $24,590 and $43,450. The lowest 10 percent earned less than $19,540, and the highest 10 percent earned more than $52,930. Median annual earnings of cardiovascular technologists and technicians in 2000 were $33,100 in offices and clinics of medical doctors and $32,860 in hospitals.

Related Occupations

Cardiovascular technologists and technicians operate sophisticated equipment that helps physicians and other health practitioners diagnose and treat patients. So do diagnostic medical sonographers, nuclear medicine technologists, radiation therapists, radiologic technologists and technicians, and respiratory therapists.

Sources of Additional Information

For general information about a career in cardiovascular technology, contact:

- Alliance of Cardiovascular Professionals, 4456 Corporation Ln., Suite 165, Virginia Beach, VA 23462. Internet: http://www.acp-online.org/index.html

For a list of accredited programs in cardiovascular technology, contact:

- Joint Review Committee on Education in Cardiovascular Technology, 3525 Ellicott Mills Dr., Suite N, Ellicott City, MD 21043-4547

For information on vascular technology, contact:

- The Society of Vascular Technology, 4601 Presidents Dr., Suite 260, Lanham, MD 20706-4365. Internet: http://www.svtnet.org

For information on echocardiography, contact:

- American Society of Echocardiography, 4101 Lake Boone Trail, Suite 201, Raleigh, NC 27607. Internet: http://www.asecho.org

For information regarding registration and certification, contact:

- Cardiovascular Credentialing International, 4456 Corporation Ln., Suite 110, Virginia Beach, VA 23462. Internet: http://www.cci-online.org
- American Registry of Diagnostic Medical Sonographers, 600 Jefferson Plaza, Suite 360, Rockville, MD 20852-1150. Internet: http://www.ardms.org

Chemical Engineers

O*NET 17-2041.00

Significant Points

- Overall job opportunities in engineering are expected to be good.
- A bachelor's degree is required for most entry-level jobs.
- Starting salaries are significantly higher than those of college graduates in other fields.
- Continuing education is critical to keep abreast of the latest technology.

Nature of the Work

Chemical engineers apply the principles of chemistry and engineering to solve problems involving the production or use of chemicals, building a bridge between science and manufacturing. They design equipment and develop processes for large-scale chemical manufacturing, plan and test methods of manufacturing the products and treating the by-products, and supervise production. Chemical engineers also work in a variety of manufacturing industries other than chemical manufacturing, such as those producing electronics, photographic equipment, clothing, and pulp and paper. They also work in the healthcare, biotechnology, and business services industries.

The knowledge and duties of chemical engineers overlap many fields. Chemical engineers apply principles of chemistry, physics, mathematics, and mechanical and electrical engineering. They frequently specialize in a particular operation such as oxidation or polymerization. Others specialize in a particular area, such as pollution control or the production of specific products such as fertilizers and pesticides, automotive plastics, or chlorine bleach. They must be aware of all aspects of chemicals manufacturing and how it affects the environment, the safety of workers, and customers. Because chemical engineers use computer technology to optimize all phases of research and production, they need to understand how to apply computer skills to process analysis, automated control systems, and statistical quality control.

Working Conditions

Most engineers work in office buildings, laboratories, or industrial plants. Others may spend time outdoors at construction sites, mines, and oil and gas exploration and production sites, where they monitor or direct operations or solve onsite problems. Some engineers travel extensively to plants or work sites.

Many engineers work a standard 40-hour week. At times, deadlines or design standards may bring extra pressure to a job. When this happens, engineers may work longer hours and experience considerable stress.

Employment

Chemical engineers held about 33,000 jobs in 2000. Manufacturing industries employed 73 percent of all chemical engineers, primarily in the chemicals, electronics, petroleum refining, paper, and related industries. Most others worked for engineering services, research and testing services, or consulting firms that design chemical plants. Some also worked on a contract basis for government agencies or as independent consultants.

Training, Other Qualifications, and Advancement

A bachelor's degree in engineering is required for almost all entry-level engineering jobs. College graduates with a degree in a physical science or mathematics occasionally may qualify for some engineering jobs, especially in specialties in high demand. Most engineering degrees are granted in electrical, electronics, mechanical, or civil engineering. However, engineers trained in one branch may work in related branches. This flexibility allows employers to meet staffing needs in new technologies and specialties in which engineers are in short supply. It also allows engineers to shift to fields with better employment prospects or to those that more closely match their interests.

Most engineering programs involve a concentration of study in an engineering specialty, along with courses in both mathematics and science. Most programs include a design course, sometimes accompanied by a computer or laboratory class or both.

In addition to the standard engineering degree, many colleges offer 2- or 4-year degree programs in engineering technology. These programs, which usually include various hands-on laboratory classes that focus on current issues, prepare students for practical design and production work, rather than for jobs which require more theoretical and scientific knowledge. Graduates of 4-year technology programs may get jobs similar to those obtained by graduates with a bachelor's degree in engineering. Engineering technology graduates, however, are not qualified to register as professional engineers under the same terms as graduates with degrees in engineering. Some employers regard technology program graduates as having skills between those of a technician and an engineer.

Graduate training is essential for engineering faculty positions and many research and development programs, but is not required for the majority of entry-level engineering jobs. Many engineers obtain graduate degrees in engineering or business administration to learn new technology and broaden their education. Many high-level executives in government and industry began their careers as engineers.

About 330 colleges and universities offer bachelor's degree programs in engineering that are accredited by the Accreditation Board for Engineering and Technology (ABET), and about 250 colleges offer accredited bachelor's degree programs in engineering technology. ABET accreditation is based on an examination of an engineering program's student achievement, program improvement, faculty, curricular content, facilities, and institutional commitment. Although most institutions offer programs in the major branches of engineering, only a few offer programs in the smaller specialties. Also, programs of the same title may vary in content. For example, some programs emphasize industrial practices, preparing students for a job in industry, whereas others are more theoretical and are designed to prepare students for graduate work. Therefore, students should investigate curricula and check accreditations carefully before selecting a college.

Admissions requirements for undergraduate engineering schools include a solid background in mathematics (algebra, geometry, trigonometry, and calculus) and sciences (biology, chemistry, and physics), and courses in English, social studies, humanities, and computers. Bachelor's degree programs in engineering typically are designed to last 4 years, but many students find that it takes between 4 and 5 years to complete their studies. In a typical 4-year college curriculum, the first 2 years are spent studying mathematics, basic sciences, introductory engineering, humanities, and social sciences. In the last 2 years, most courses are in engineering, usually with a concentration in one branch. Some programs offer a general engineering curriculum; students then specialize in graduate school or on the job.

Some engineering schools and 2-year colleges have agreements whereby the 2-year college provides the initial engineering education, and the engineering school automatically admits students for their last 2 years. In addition, a few engineering schools have arrangements whereby a student spends 3 years in a liberal arts college studying pre-engineering subjects and 2 years in an engineering school studying core subjects, and then receives a bachelor's degree from each school. Some colleges and universities offer 5-year master's degree programs. Some 5- or even 6-year cooperative plans combine classroom study and practical work, permitting students to gain valuable experience and finance part of their education. All 50 states and the District of Columbia usually require licensure for engineers who offer their services directly to the public. Engineers who are licensed are called Professional Engineers (PE). This licensure generally requires a degree from an ABET-accredited engineering program, 4 years of relevant work experience, and successful completion of a state examination. Recent graduates can start the licensing process by taking the examination in two stages. The initial Fundamentals of Engineering (FE) examination can be taken upon graduation. Engineers who pass this examination commonly are called Engineers in Training (EIT) or Engineer Interns (EI). The EIT certification usually is valid for 10 years. After acquiring suitable work experience, EITs can take the second examination, the Principles and Practice of Engineering Exam. Several states have imposed mandatory continuing education requirements for relicensure. Most states recognize licensure from other states. Many chemical engineers are licensed as PEs.

Engineers should be creative, inquisitive, analytical, and detail oriented. They should be able to work as part of a team and to communicate well, both orally and in writing. Communication abilities are becoming more important because much of their work is becoming more diversified, meaning that engineers interact with specialists in a wide range of fields outside engineering.

Beginning engineering graduates usually work under the supervision of experienced engineers and, in large companies, also may receive formal classroom or seminar-type training. As new engineers gain knowledge and experience, they are assigned more difficult projects with greater independence to develop designs, solve problems, and make decisions. Engineers may advance to become technical specialists or to supervise a staff or team of engineers and technicians. Some may eventually become engineering managers or enter other managerial or sales jobs.

Job Outlook

Chemical engineering graduates may face competition for jobs as the number of openings in traditional fields is projected to be lower than the number of graduates. Employment of chemical engineers is projected to grow more slowly than the average for all occupations through 2010. Although overall employment in the chemical manufacturing industry is expected to decline, chemical companies will continue to research and develop new chemicals and more efficient processes to increase output of existing chemicals, resulting in some new jobs for chemical engineers. Among manufacturing industries, specialty chemicals, plastics materials, pharmaceuticals, biotechnology, and electronics may provide the best opportunities. Much of the projected growth in employment of chemical engineers, however, will be in nonmanufacturing industries, especially services industries such as research and testing services.

Earnings

Median annual earnings of chemical engineers were $65,960 in 2000. The middle 50 percent earned between $53,440 and $80,840. The lowest 10 percent earned less than $45,200, and the highest 10 percent earned more than $93,430.

According to a 2001 salary survey by the National Association of Colleges and Employers, bachelor's degree candidates in chemical engineering received starting offers averaging $51,073 a year, master's degree candidates averaged $57,221, and PhD candidates averaged $75,521.

Related Occupations

Chemical engineers apply the principles of physical science and mathematics in their work. Other workers who use scientific and mathematical principles include architects, except landscape and naval; engineering and natural sciences managers; computer and information systems managers; mathematicians; drafters; engineering technicians; sales engineers; science technicians; and physical and life scientists, including agricultural and food scientists, biological and medical scientists, conservation scientists and foresters, atmospheric scientists, chemists and materials scientists, environmental scientists and geoscientists, and physicists and astronomers.

Sources of Additional Information

Further information about chemical engineers is available from:

- American Institute of Chemical Engineers, Three Park Ave., New York, NY 10016-5901. Internet: http://www.aiche.org
- American Chemical Society, Department of Career Services, 1155 16th St. N.W., Washington, DC 20036. Internet: http://www.acs.org

High school students interested in obtaining a full package of guidance materials and information (product number SP-01) on a variety of engineering disciplines should contact the Junior Engineering Technical Society by sending $3.50 to:

- JETS-Guidance, 1420 King St., Suite 405, Alexandria, VA 22314-2794. Internet: http://www.jets.org

High school students interested in obtaining information on ABET-accredited engineering programs should contact:

- The Accreditation Board for Engineering and Technology, Inc., 111 Market Place, Suite 1050, Baltimore, MD 21202-4012. Internet: http://www.abet.org

Non-licensed engineers and college students interested in obtaining information on Professional Engineer licensure should contact:

- The National Society of Professional Engineers, 1420 King St., Alexandria, VA 22314-2794. Internet: http://www.nspe.org
- National Council of Examiners for Engineers and Surveying, P.O. Box 1686, Clemson, SC 29633-1686. Internet: http://www.ncees.org

Information on general engineering education and career resources is available from:

- American Society for Engineering Education, 1818 N St. N.W., Suite 600, Washington, DC 20036-2479. Internet: http://www.asee.org

Information on obtaining an engineering position with the federal government is available from the Office of Personnel Management (OPM) through a telephone-based system. Consult your telephone directory under U.S. Government for a local number or call (912) 757-3000; Federal Relay Service: (800) 877-8339. The first number is not toll free, and charges may result. Information also is available from the OPM Internet site: http://www.usajobs.opm.gov

Non-high school students and those wanting more detailed information should contact societies representing the individual branches of engineering. Each can provide information about careers in the particular branch. The individual statements elsewhere in this book also provide other information in detail on aerospace; agricultural; biomedical; civil; computer hardware; electrical and electronics; environmental; industrial, including health and safety; materials; mechanical; mining and geological, including mining safety; nuclear; and petroleum engineering.

Chemists and Materials Scientists

O*NET 19-2031.00, 19-2032.00

Significant Points

- A bachelor's degree in chemistry or a related discipline is the minimum educational requirement; however, many research jobs require a PhD.
- Job growth will be concentrated in pharmaceutical companies and in research and testing services firms.
- Strong demand will exist for those with a master's or PhD degree.

Nature of the Work

Everything in the environment, whether naturally occurring or of human design, is composed of chemicals. Chemists and materials scientists search for and use new knowledge about chemicals. Chemical research has led to the discovery and development of new and improved synthetic fibers, paints, adhesives, drugs, cosmetics, electronic components, lubricants, and thousands of other products. Chemists and materials scientists also develop processes that save energy and reduce pollution, such as improved oil refining and petrochemical processing methods. Research on the chemistry of living things spurs advances in medicine, agriculture, food processing, and other fields.

Materials scientists research and study the structures and chemical properties of various materials to develop new products or enhance existing ones. They also determine ways to strengthen or combine materials or develop new materials for use in a variety of products. Materials science encompasses the natural and synthetic materials used in a wide range of products and structures, from airplanes, cars, and bridges to clothing and household goods. Companies whose products are made of metals, ceramics, and rubber employ most material scientists. Other applications of this field include studies of superconducting materials, graphite materials, integrated-circuit chips, and fuel cells. Materials scientists, applying chemistry and physics, study all aspects of these materials. Chemistry plays an increasingly dominant role in materials science, because it provides information about the structure and composition of materials.

Many chemists and materials scientists work in research and development (R&D). In basic research, they investigate properties, composition, and structure of matter and the laws that govern the combination of elements and reactions of substances. In applied R&D, they create new products and processes or improve existing ones, often using knowledge gained from basic research. For example, synthetic rubber and plastics resulted from research on small molecules uniting to form large ones, a process called polymerization. R&D chemists and material scientists use computers and a wide variety of sophisticated laboratory instrumentation for modeling and simulation in their work.

The use of computers to analyze complex data has had the dramatic impact of allowing chemists and materials scientists to practice combinatorial chemistry. This technique makes and tests large quantities of chemical compounds simultaneously in order to find compounds with desired properties. As an integral part of drug and materials discovery, combinatorial chemistry speeds up material designing and research and development, permitting useful compounds to be developed more quickly and inexpensively than was formerly possible. Combinatorial chemistry has allowed chemists to produce thousands of compounds each year and to assist in the completion of sequencing human genes.

Chemists also work in production and quality control in chemical manufacturing plants. They prepare instructions for plant workers that specify ingredients, mixing times, and temperatures for each stage in the process. They also monitor automated processes to ensure proper product yield, and test samples of raw materials or finished products to ensure that they meet industry and government standards, including the regulations governing pollution. Chemists report and document test results and analyze those results in hopes of further improving existing theories or developing new test methods.

Chemists often specialize in a subfield. *Analytical chemists* determine the structure, composition, and nature of substances by examining and identifying the various elements or compounds that make up a substance. They are absolutely crucial to the pharmaceutical industry because pharmaceutical companies need to know the identity of compounds that they hope to turn into drugs. Furthermore, they study the relations and interactions of the parts of compounds and develop analytical techniques. They also identify the presence and concentration of chemical pollutants in air, water, and soil. *Organic chemists* study the chemistry of the vast number of carbon compounds that make up all living things. Organic chemists who synthesize elements or simple compounds to create new compounds or substances that have different properties and applications have developed many commercial products, such as drugs, plastics, and elastomers (elastic substances similar to rubber). *Inorganic chemists* study compounds consisting mainly of elements other than carbon, such as those in electronic components. *Physical and theoretical chemists* study the physical characteristics of atoms and molecules and the theoretical properties of matter, and investigate how chemical reactions work. Their research may result in new and better energy sources. *Macromolecular chemists* study the behavior of atoms and molecules. *Medicinal chemists* study the structural properties of compounds intended for applications to human medicine. *Materials chemists* study and develop new materials to improve existing products or make new ones. In fact, virtually all chemists are involved in this quest in one way or another. Developments in the field of chemistry that involve life sciences will expand, resulting in more interaction between biologists and chemists.

Materials scientists also may specialize in specific areas such as ceramics or metals.

Working Conditions

Chemists and materials scientists usually work regular hours in offices and laboratories. Research and development chemists and materials scientists spend much time in laboratories, but also work in offices when they do theoretical research or plan, record, and report on their lab research. Although some laboratories are small, others are large enough to incorporate prototype chemical manufacturing facilities as well as advanced equipment for chemists. In addition to working in a laboratory, materials scientists also work with engineers and processing specialists in industrial manufacturing facilities. After a material is sold, materials scientists often help customers tailor the material to suit their needs. Chemists do some of their work in a chemical plant or outdoors—while gathering water samples to test for pollutants, for example. Some chemists are exposed to health or safety hazards when handling certain chemicals, but there is little risk if proper procedures are followed.

Employment

Chemists and materials scientists held about 92,000 jobs in 2000. Over half of all chemists are employed in manufacturing firms—mostly in the chemical manufacturing industry, which includes firms that produce plastics and synthetic materials, drugs, soaps and cleaners, paints, industrial organic chemicals, and other miscellaneous chemical products. Chemists also work for state and local governments and for federal agencies. The U.S. Department of Health and Human Services (which includes the Food and Drug Administration, the National Institutes of Health, and the Center for Disease Control) is the major federal employer of chemists. The Departments of Defense and Agriculture and the Environmental Protection Agency also employ chemists. Other chemists work for research, development, and testing services. In addition, thousands of persons with a background in chemistry and materials science hold teaching positions in high schools and in colleges and universities.

Chemists and materials scientists are employed in all parts of the country, but they are mainly concentrated in large industrial areas.

Training, Other Qualifications, and Advancement

A bachelor's degree in chemistry or a related discipline is usually the minimum educational requirement for entry-level chemist jobs. However, many research jobs require a PhD. While some materials scientists hold a degree in materials science, a bachelor's degree in chemistry, physics, or electric engineering also is accepted. For research and development jobs, a PhD in materials science or a related science is often required.

Many colleges and universities offer a bachelor's degree program in chemistry; about 620 are approved by the American Chemical Society (ACS). The number of colleges that offer a degree program in materials science is small, but gradually increasing. Several hundred colleges and universities also offer advanced degree programs in chemistry; around 320 master's programs, and about 190 doctoral programs are ACS-approved.

Students planning careers as chemists and materials scientists should take courses in science and mathematics, and should like working with their hands building scientific apparatus and performing laboratory experiments and computer modeling. Perseverance, curiosity, and the ability to concentrate on detail and to work independently are essential. Interaction among specialists in this field is increasing, especially for chemists in drug development. One type of chemist often relies on the findings of another type of chemist. For example, an organic chemist must understand findings on the identity of compounds prepared by an analytical chemist.

In addition to required courses in analytical, inorganic, organic, and physical chemistry, undergraduate chemistry majors usually study biological sciences, mathematics, and physics. Those interested in the environmental field should also take courses in environmental studies and become familiar with current legislation and regulations. Computer courses are essential, as employers increasingly prefer job applicants who are able to apply computer skills to modeling and simulation tasks and operate computerized laboratory equipment. This is increasingly important as combinatorial chemistry techniques are more widely applied. Scientists with outdated skills or who are unfamiliar with combinatorial chemistry are often retrained by companies inhouse.

Because research and development chemists and materials scientists are increasingly expected to work on interdisciplinary teams, some understanding of other disciplines, including business and marketing or economics, is desirable, along with leadership ability and good oral and written communication skills. Experience, either in academic laboratories or through internships, fellowships, or co-op programs in industry, also is useful. Some employers of research chemists, particularly in the pharmaceutical industry, prefer to hire individuals with several years of postdoctoral experience.

Graduate students typically specialize in a subfield of chemistry, such as analytical chemistry or polymer chemistry, depending on their interests and the kind of work they wish to do. For example, those interested in doing drug research in the pharmaceutical industry usually develop a strong background in synthetic organic chemistry. However, students normally need not specialize at the undergraduate level. In fact, undergraduates who are broadly trained have more flexibility when job hunting or changing jobs than if they narrowly define their interests. Most employers provide new graduates additional training or education.

In government or industry, beginning chemists with a bachelor's degree work in quality control, perform analytical testing, or assist senior chemists in research and development laboratories. Many employers prefer chemists and material scientists with a PhD or at least a master's degree to lead basic and applied research. Nonetheless, relevant work experience is an asset. Chemists who hold a PhD and have previous industrial experience may be particularly attractive to employers because such people are more likely to understand the complex regulations that apply to the pharmaceutical industry. Within materials science, a broad background in various sciences is preferred. This broad base may be obtained through degrees in physics, engineering, or chemistry. While many companies prefer hiring PhDs, many materials scientists have bachelor's and master's degrees. Additionally, both chemists and materials scientists need the ability to apply basic statistical techniques.

Job Outlook

Employment of chemists is expected to grow about as fast as the average for all occupations through 2010. Job growth will be concentrated in drug manufacturing and in research, development, and testing services firms. The chemical industry, the major employer of chemists, should face continued demand for goods such as new and better pharmaceuticals and personal care products, as well as for more specialty chemicals designed to address specific problems or applications. To meet these demands, chemical firms will continue to devote money to research and development—through in-house teams or outside contractors—spurring employment growth of chemists. Strong demand is expected for chemists with a master's or PhD degree.

Within the chemical industry, job opportunities are expected to be most plentiful in pharmaceutical and biotechnology firms. Biotechnological research, including studies of human genes, continues to offer possibilities for the development of new drugs and products to combat illnesses and diseases which have previously been unresponsive to treatments derived by traditional chemical processes. Stronger competition among drug companies and an aging population are contributing to the need for innovative and improved drugs discovered through scientific research. Chemical firms that develop and manufacture personal products such as toiletries and cosmetics must continually innovate and develop new and better products to remain competitive. Additionally, as the population grows and becomes better informed, the demand for different or improved grooming products—including vegetable-based products, products with milder formulas, treatments for aging skin, and products that have been developed using more benign chemical processes than in the past—will remain strong, spurring the need for chemists.

In most of the remaining segments of the chemical industry, employment growth is expected to decline as companies downsize and turn to outside contractors to provide specialized services. As a result, research and testing firms will experience healthy growth. To control costs, some chemical companies, including drug manufacturers, are increasingly turning to these firms to perform specialized research and other work formerly done by in-house chemists. Despite downsizing, some job openings will result from the need to replace chemists who retire or otherwise leave the labor force. Quality control will continue to be an important issue in the chemical and other industries that use chemicals in their manufacturing processes. Chemists also will be needed to develop and improve the technologies and processes used to produce chemicals for all purposes, and to monitor and measure air and water pollutants to ensure compliance with local, state, and federal environmental regulations.

Environmental research will offer many new opportunities for chemists and materials scientists. To satisfy public concerns and to comply with government regulations, the chemical industry will continue to invest billions of dollars each year for technology that reduces pollution and cleans up existing waste sites. Chemists also are needed to find ways to use less energy and to discover new sources of energy.

During periods of economic recession, layoffs of chemists may occur—especially in the industrial chemicals industry. This industry provides many of the raw materials to the auto manufacturing and construction industries, both of which are vulnerable to temporary slowdowns during recessions.

Earnings

Median annual earnings of chemists in 2000 were $50,080. The middle 50 percent earned between $37,480 and $68,240. The lowest 10 percent earned less than $29,620, and the highest 10 percent earned more than $88,030. Median annual earnings in the industries employing the largest numbers of chemists in 2000 were:

Federal government	$65,950
Drugs	50,820
Research and testing services	41,820

A survey by the American Chemical Society reports that the median salary of all their members with a bachelor's degree was $55,000 a year in 2000; with a master's degree, $65,000; and with a PhD, $82,200. Median salaries were highest for those working in private industry; those in academia earned the least. According to an ACS survey of recent graduates, inexperienced chemistry graduates with a bachelor's degree earned a median starting salary of $33,500 in 2000; with a master's degree, $44,100; and with a PhD, $64,500. Among bachelor's degree graduates, those who had completed internships or had other work experience while in school commanded the highest starting salaries.

In 2001, chemists in nonsupervisory, supervisory, and managerial positions in the federal government averaged $70,435 a year.

Related Occupations

The research and analysis conducted by chemists and materials scientists is closely related to work done by agricultural and food scientists, biological and medical scientists, chemical engineers, materials engineers, physicists, and science technicians.

Sources of Additional Information

General information on career opportunities and earnings for chemists is available from:

- American Chemical Society, Education Division, 1155 16th St. N.W., Washington, DC 20036. Internet: http://www.acs.org

Information on obtaining a position as a chemist with the federal government is available from the Office of Personnel Management (OPM) through a telephone-based system. Consult your telephone directory under U.S. Government for a local number or call (912) 757-3000; Federal Relay Service: (800) 877-8339. The first number is not toll free, and charges may result. Information also is available from the OPM Internet site: http://www.usajobs.opm.gov

For general information on materials science, contact:

- Materials Research Society (MRS), 506 Keystone Dr., Warrendale, PA 15086-7573. Internet: http://www.mrs.org

Civil Engineers

O*NET 17-2051.00

Significant Points

- Overall job opportunities in engineering are expected to be good.
- A bachelor's degree is required for most entry-level jobs.
- Starting salaries are significantly higher than those of college graduates in other fields.
- Continuing education is critical to keep abreast of the latest technology.

Nature of the Work

Civil engineers design and supervise the construction of roads, buildings, airports, tunnels, dams, bridges, and water supply and sewage systems. Civil engineering, considered one of the oldest engineering disciplines, encompasses many specialties. The major specialties within civil engineering are structural, water resources, environmental, construction, transportation, and geotechnical engineering.

Many civil engineers hold supervisory or administrative positions, from supervisor of a construction site to city engineer. Others may work in design, construction, research, and teaching.

Working Conditions

Most engineers work in office buildings, laboratories, or industrial plants. Others may spend time outdoors at construction sites, mines, and oil and gas exploration and production sites, where they monitor or direct operations or solve onsite problems. Some engineers travel extensively to plants or work sites.

Many engineers work a standard 40-hour week. At times, deadlines or design standards may bring extra pressure to a job. When this happens, engineers may work longer hours and experience considerable stress.

Employment

Civil engineers held about 232,000 jobs in 2000. A little over half were employed by firms providing engineering consulting services, primarily developing designs for new construction projects. Almost one third of the jobs were in federal, state, and local government agencies. The construction and manufacturing industries accounted for most of the remaining employment. About 12,000 civil engineers were self-employed, many as consultants.

Civil engineers usually work near major industrial and commercial centers, often at construction sites. Some projects are situated in remote areas or in foreign countries. In some jobs, civil engineers move from place to place to work on different projects.

Training, Other Qualifications, and Advancement

A bachelor's degree in engineering is required for almost all entry-level engineering jobs. College graduates with a degree in a physical science or mathematics occasionally may qualify for some engineering jobs, especially in specialties in high demand. Most engineering degrees are granted in electrical, electronics, mechanical, or civil engineering. However, engineers trained in one branch may work in related branches. This flexibility allows employers to meet staffing needs in new technologies and specialties in which engineers are in short supply. It also allows engineers to shift to fields with better employment prospects or to those that more closely match their interests.

Most engineering programs involve a concentration of study in an engineering specialty, along with courses in both mathematics and science. Most programs include a design course, sometimes accompanied by a computer or laboratory class or both.

In addition to the standard engineering degree, many colleges offer 2- or 4-year degree programs in engineering technology. These programs, which usually include various hands-on laboratory classes that focus on current issues, prepare students for practical design and production work, rather than for jobs which require more theoretical and scientific knowledge. Graduates of 4-year technology programs may get jobs similar to those obtained by graduates with a bachelor's degree in engineering. Engineering technology graduates, however, are not qualified to register as professional engineers under the same terms as graduates with degrees in engineering. Some employers regard technology program graduates as having skills between those of a technician and an engineer.

Graduate training is essential for engineering faculty positions and many research and development programs, but is not required for the majority of entry-level engineering jobs. Many engineers obtain graduate degrees in engineering or business administration to learn new technology and broaden their education. Many high-level executives in government and industry began their careers as engineers.

About 330 colleges and universities offer bachelor's degree programs in engineering that are accredited by the Accreditation Board for Engineering and Technology (ABET), and about 250 colleges offer accredited bachelor's degree programs in engineering technology. ABET accreditation is based on an examination of an engineering program's student achievement, program improvement, faculty, curricular content, facilities, and institutional commitment. Although most institutions offer programs in the major branches of engineering, only a few offer programs in the smaller specialties. Also, programs of the same title may vary in content. For example, some programs emphasize industrial practices, preparing students for a job in industry, whereas others are more theoretical and are designed to prepare students for graduate work. Therefore, students should investigate curricula and check accreditations carefully before selecting a college.

Admissions requirements for undergraduate engineering schools include a solid background in mathematics (algebra, geometry, trigonometry, and calculus) and sciences (biology, chemistry, and physics), and courses in English, social studies, humanities, and computers. Bachelor's degree programs in engineering typically are designed to last 4 years, but many students find that it takes between 4 and 5 years to complete their studies. In a typical 4-year college curriculum, the first 2 years are spent studying mathematics, basic sciences, introductory engineering, humanities, and social sciences. In the last 2 years, most courses are in engineering, usually with a concentration in one branch. Some programs offer a general engineering curriculum; students then specialize in graduate school or on the job.

Some engineering schools and 2-year colleges have agreements whereby the 2-year college provides the initial engineering education, and the engineering school automatically admits students for their last 2 years. In addition, a few engineering schools have arrangements whereby a student spends 3 years in a liberal arts college studying pre-engineering subjects and 2 years in an engineering school studying core subjects, and then receives a bachelor's degree from each school. Some colleges and universities offer 5-year master's degree programs. Some 5- or even 6-year cooperative plans combine classroom study and practical work, permitting students to gain valuable experience and finance part of their education. All 50 states and the District of Columbia usually require licensure for engineers who offer their services directly to the public. Engineers who are licensed are called Professional Engineers (PE). This licensure generally requires a degree from an ABET-accredited engineering program, 4 years of relevant work experience, and successful completion of a state examination. Recent graduates can start the licensing process by taking the examination in two stages. The initial Fundamentals of Engineering (FE) examination can be taken upon graduation. Engineers who pass this examination commonly are called Engineers in Training (EIT) or Engineer Interns (EI). The EIT certification usually is valid for 10 years. After acquiring suitable work experience, EITs can take the second examination, the Principles and Practice of Engineering Exam. Several states have imposed mandatory continuing education requirements for relicensure. Most states recognize licensure from other states. Many civil engineers are licensed as PEs.

Engineers should be creative, inquisitive, analytical, and detail oriented. They should be able to work as part of a team and to communicate well, both orally and in writing. Communication abilities are becoming more important because much of their work is becoming more diversified, meaning that engineers interact with specialists in a wide range of fields outside engineering.

Beginning engineering graduates usually work under the supervision of experienced engineers and, in large companies, also may receive formal classroom or seminar-type training. As new engineers gain knowledge and experience, they are assigned more difficult projects with greater independence to develop designs, solve problems, and make decisions. Engineers may advance to become technical specialists or to supervise a staff or team of engineers and technicians. Some may eventually become engineering managers or enter other managerial or sales jobs.

Job Outlook

Employment of civil engineers is expected to increase about as fast as the average for all occupations through 2010. Spurred by general population growth and an expanding economy, more civil engineers will be needed to design and construct higher capacity transportation, water supply, pollution control systems, and large buildings and building complexes. They also will be needed to repair or replace existing roads, bridges, and other public structures. There may be additional opportunities within noncivil engineering firms, such as management consulting or computer services firms. In addition to job growth, openings will result from the need to replace civil engineers that transfer to other occupations or leave the labor force.

Because construction and related industries—including those providing design services—employ many civil engineers, employment opportunities will vary by geographic area and may decrease during economic slowdowns, when construction often is curtailed.

Earnings

Median annual earnings of civil engineers were $55,740 in 2000. The middle 50 percent earned between $45,150 and $69,470. The lowest 10 percent earned less than $37,430, and the highest 10 percent earned more than $86,000. Median annual earnings in the industries employing the largest numbers of civil engineers in 2000 were:

Federal government	$63,530
Heavy construction, except highway	62,010
Local government	56,830
State government	54,630
Engineering and architectural services	54,550

According to a 2001 salary survey by the National Association of Colleges and Employers, bachelor's degree candidates in civil engineering received starting offers averaging $40,616 a year, master's degree candidates received an average offer of $44,080, and PhD candidates were offered $62,280 as an initial salary.

Related Occupations

Engineers apply the principles of physical science and mathematics in their work. Other workers who use scientific and mathematical principles include architects, except landscape and naval; engineering and natural sciences managers; computer and information systems managers; mathematicians; drafters; engineering technicians; sales engineers; science technicians; and physical and life scientists, including agricultural and food scientists, biological and medical scientists, conservation scientists and foresters, atmospheric scientists, chemists and materials scientists, environmental scientists and geoscientists, and physicists and astronomers.

Sources of Additional Information

Further information about civil engineers can be obtained from:

- American Society of Civil Engineers, 1801 Alexander Bell Dr., Reston, VA 20191-4400. Internet: http://www.asce.org

High school students interested in obtaining a full package of guidance materials and information (product number SP-01) on a variety of engineering disciplines should contact the Junior Engineering Technical Society by sending $3.50 to:

- JETS-Guidance, 1420 King St., Suite 405, Alexandria, VA 22314-2794. Internet: http://www.jets.org

High school students interested in obtaining information on ABET-accredited engineering programs should contact:

- The Accreditation Board for Engineering and Technology, Inc., 111 Market Place, Suite 1050, Baltimore, MD 21202-4012. Internet: http://www.abet.org

Non-licensed engineers and college students interested in obtaining information on Professional Engineer licensure should contact:

- The National Society of Professional Engineers, 1420 King St., Alexandria, VA 22314-2794. Internet: http://www.nspe.org
- National Council of Examiners for Engineers and Surveying, P.O. Box 1686, Clemson, SC 29633-1686. Internet: http://www.ncees.org

Information on general engineering education and career resources is available from:

- American Society for Engineering Education, 1818 N St. N.W., Suite 600, Washington, DC 20036-2479. Internet: http://www.asee.org

Information on obtaining an engineering position with the federal government is available from the Office of Personnel Management (OPM) through a telephone-based system. Consult your telephone directory under U.S. Government for a local number or call (912) 757-3000; Federal Relay Service: (800) 877-8339. The first number is not toll free, and charges may result. Information also is available from the OPM Internet site: http://www.usajobs.opm.gov

Non-high school students and those wanting more detailed information should contact societies representing the individual branches of engineering. Each can provide information about careers in the particular branch. The individual statements elsewhere in this book also provide other information in detail on aerospace; agricultural; biomedical; chemical; computer hardware; electrical and electronics; environmental; industrial, including health and safety; materials; mechanical; mining and geological, including mining safety; nuclear; and petroleum engineering.

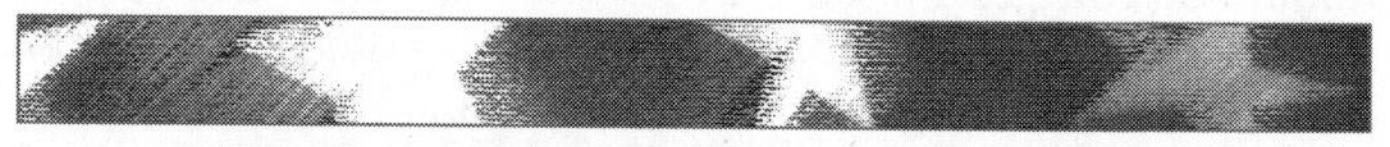

Clinical Laboratory Technologists and Technicians

O*NET 29-2011.00, 29-2012.00

Significant Points

- Clinical laboratory technologists usually have a bachelor's degree with a major in medical technology or in one of the life sciences; clinical laboratory technicians need either an associate degree or a certificate.

- Employment is expected to grow as fast as average as the volume of laboratory tests increases with population growth and the development of new types of tests.

Nature of the Work

Clinical laboratory testing plays a crucial role in the detection, diagnosis, and treatment of disease. Clinical laboratory technologists, also referred to as *clinical laboratory scientists* or *medical technologists*, and clinical laboratory technicians, also known as *medical technicians* or *medical laboratory technicians*, perform most of these tests.

Clinical laboratory personnel examine and analyze body fluids, tissues, and cells. They look for bacteria, parasites, and other microorganisms; analyze the chemical content of fluids; match blood for transfusions; and test for drug levels in the blood to show how a patient is responding to treatment. These technologists also prepare specimens for examination, count cells, and look for abnormal cells. They use automated equipment and instruments capable of performing a number of tests simultaneously, as well as microscopes, cell counters, and other sophisticated laboratory equipment. Then, they analyze the results and relay them to physicians. With increasing automation and the use of computer technology, the work of technologists and technicians has become less hands-on and more analytical.

The complexity of tests performed, the level of judgment needed, and the amount of responsibility workers assume depend largely on the amount of education and experience they have.

Medical and clinical laboratory technologists generally have a bachelor's degree in medical technology or in one of the life sciences, or they have a combination of formal training and work experience. They perform complex chemical, biological, hematological, immunologic, microscopic, and bacteriological tests. Technologists microscopically examine blood, tissue, and other body substances. They make cultures of body fluid and tissue samples, to determine the presence of bacteria, fungi, parasites, or other microorganisms. They analyze samples for chemical content or reaction and determine blood glucose and cholesterol levels. They also type and cross match blood samples for transfusions.

Medical and clinical laboratory technologists evaluate test results, develop and modify procedures, and establish and monitor programs, to ensure the accuracy of tests. Some medical and clinical laboratory technologists supervise medical and clinical laboratory technicians.

Technologists in small laboratories perform many types of tests, whereas those in large laboratories generally specialize. Technologists who prepare specimens and analyze the chemical and hormonal contents of body fluids are *clinical chemistry technologists*. Those who examine and identify bacteria and other microorganisms are *microbiology technologists*. *Blood bank technologists*, or *immunohematology technologists*, collect, type, and prepare blood and its components for transfusions. *Immunology technologists* examine elements and responses of the human immune system to foreign bodies. *Cytotechnologists* prepare slides of body cells and microscopically examine these cells for abnormalities that may signal the beginning of a cancerous growth. *Molecular biology technologists* perform complex genetic testing on cell samples.

Medical and clinical laboratory technicians perform less complex tests and laboratory procedures than technologists. Technicians may prepare specimens and operate automated analyzers, for example, or they may perform manual tests following detailed instructions. Like technologists, they may work in several areas of the clinical laboratory or specialize in just one. *Histology technicians* cut and stain tissue specimens for microscopic examination by pathologists, and *phlebotomists* collect blood samples. They usually work under the supervision of medical and clinical laboratory technologists or laboratory managers.

Working Conditions

Hours and other working conditions of clinical laboratory technologists and technicians vary, according to the size and type of employment setting. In large hospitals or in independent laboratories that operate continuously, personnel usually work the day, evening, or night shift and may work weekends and holidays. Laboratory personnel in small facilities may work on rotating shifts, rather than on a regular shift. In some facilities, laboratory personnel are on call several nights a week or on weekends, in case of an emergency.

Clinical laboratory personnel are trained to work with infectious specimens. When proper methods of infection control and sterilization are followed, few hazards exist. Protective masks, gloves, and goggles are often necessary to ensure the safety of laboratory personnel.

Laboratories usually are well lighted and clean; however, specimens, solutions, and reagents used in the laboratory sometimes produce fumes. Laboratory workers may spend a great deal of time on their feet.

Employment

Clinical laboratory technologists and technicians held about 295,000 jobs in 2000. About half worked in hospitals. Most of the remaining jobs were found in medical laboratories or offices and clinics of physicians. A small number were in blood banks, research and testing laboratories, and in the federal government at U.S. Department of Veterans Affairs hospitals and U.S. Public Health Service facilities.

Training, Other Qualifications, and Advancement

The usual requirement for an entry-level position as a medical or clinical laboratory technologist is a bachelor's degree with a major in medical technology or in one of the life sciences. Universities and hospitals offer medical technology programs. It also is possible to qualify through a combination of education, on-the-job, and specialized training.

Bachelor's degree programs in medical technology include courses in chemistry, biological sciences, microbiology, mathematics, statistics, and specialized courses devoted to knowledge and skills used in the clinical laboratory. Many programs also offer or require courses in management, business, and computer applications. The Clinical Laboratory Improvement Act (CLIA) requires technologists who perform certain highly complex tests to have at least an associate degree.

Medical and clinical laboratory technicians generally have either an associate degree from a community or junior college or a certificate from a hospital, vocational or technical school, or from one of the U.S. Armed Forces. A few technicians learn their skills on the job.

The National Accrediting Agency for Clinical Laboratory Sciences (NAACLS) fully accredits 503 programs for medical and clinical laboratory technologists, medical and clinical laboratory technicians, histologic technologists and technicians, and pathologists' assistants. NAACLS also approves 70 programs in phlebotomy, cytogenetic technology, molecular biology, and clinical assisting. Other nationally recognized accrediting agencies include the Commission on Accreditation of Allied Health Education Programs (CAAHEP) and the Accrediting Bureau of Health Education Schools (ABHES).

Some states require laboratory personnel to be licensed or registered. Information on licensure is available from state departments of health or boards of occupational licensing. Certification is a voluntary process by which a nongovernmental organization, such as a professional society or certifying agency, grants recognition to an individual whose professional competence meets prescribed standards. Widely accepted by employers in the health industry, certification is a prerequisite for most jobs and often is necessary for advancement. Agencies certifying medical and clinical laboratory technologists and technicians include the Board of Registry of the American Society for Clinical Pathology, the American Medical Technologists, the National Credentialing Agency for Laboratory Personnel, and the Board of Registry of the American Association of Bioanalysts. These agencies have different requirements for certification and different organizational sponsors.

Clinical laboratory personnel need good analytical judgment and the ability to work under pressure. Close attention to detail is essential, because small differences or changes in test substances or numerical readouts can be crucial for patient care. Manual dexterity and normal color vision are highly desirable. With the widespread use of automated laboratory equipment, computer skills are important. In addition, technologists in particular are expected to be good at problem solving.

Technologists may advance to supervisory positions in laboratory work or become chief medical or clinical laboratory technologists or laboratory managers in hospitals. Manufacturers of home diagnostic testing kits and laboratory equipment and supplies seek experienced technologists to work in product development, marketing, and sales. Graduate education in medical technology, one of the biological sciences, chemistry, management, or education usually speeds advancement. A doctorate is needed to become a laboratory director. However, federal regulation allows directors of moderate complexity laboratories to have either a master's degree or a bachelor's degree combined with the appropriate amount of training and experience. Technicians can become technologists through additional education and experience.

Job Outlook

Employment of clinical laboratory workers is expected to grow about as fast as the average for all occupations through the year 2010, as the volume of laboratory tests increases with population growth and the development of new types of tests.

Technological advances will continue to have two opposing effects on employment through 2010. New, increasingly powerful diagnostic tests will encourage additional testing and spur employment. On the other hand, research and development efforts targeted at simplifying routine testing procedures may enhance the ability of nonlaboratory personnel, physicians and patients, in particular, to perform tests now done in laboratories.

Although significant, growth will not be the only source of opportunities. As in most occupations, many openings will result from the need to replace workers who transfer to other occupations, retire, or stop working for some other reason.

Earnings

Median annual earnings of medical and clinical laboratory technologists were $40,510 in 2000. The middle 50 percent earned between $34,220 and $47,460. The lowest 10 percent earned less than $29,240, and the highest 10 percent earned more than $55,560. Median annual earnings in the industries employing the largest numbers of medical and clinical laboratory technologists in 2000 were as follows:

Hospitals	$40,840
Medical and dental laboratories	39,780
Offices and clinics of medical doctors	38,850

Median annual earnings of medical and clinical laboratory technicians were $27,540 in 2000. The middle 50 percent earned between $22,260 and $34,320. The lowest 10 percent earned less than $18,550, and the highest 10 percent earned more than $42,370. Median annual earnings in the industries employing the largest numbers of medical and clinical laboratory technicians in 2000 were as follows:

Hospitals	$28,860
Colleges and universities	27,810
Offices and clinics of medical doctors	27,180
Medical and dental laboratories	25,250
Health and allied health services, not elsewhere classified	24,370

Related Occupations

Clinical laboratory technologists and technicians analyze body fluids, tissue, and other substances using a variety of tests. Similar or related procedures are performed by chemists and material scientists, science technicians, and veterinary technologists, technicians, and assistants.

Sources of Additional Information

For a list of accredited and approved educational programs for clinical laboratory personnel, contact:

- National Accrediting Agency for Clinical Laboratory Sciences, 8410 W. Bryn Mawr Ave., Suite 670, Chicago, IL 60631. Internet: http://www.naacls.org

Information on certification is available from:

- American Association of Bioanalysts, 917 Locust St., Suite 1100, St. Louis, MO 63101. Internet: http://www.aab.org
- American Medical Technologists, 710 Higgins Rd., Park Ridge, IL 60068
- American Society for Clinical Pathology, Board of Registry, 2100 West Harrison St., Chicago, IL 60612. Internet: http://www.ascp.org/bor
- National Credentialing Agency for Laboratory Personnel, P.O. Box 15945-289, Lenexa, KS 66285-5935. Internet: http://www.nca-info.org

Additional career information is available from:

- American Association of Blood Banks, 8101 Glenbrook Rd., Bethesda, MD 20814-2749. Internet: http://www.aabb.org
- American Society for Clinical Laboratory Science, 7910 Woodmont Ave., Suite 530, Bethesda, MD 20814. Internet: http://www.ascls.org
- American Society for Clinical Pathology, 2100 West Harrison St., Chicago, IL 60612. Internet: http://www.ascp.org

Communications Equipment Operators

O*NET 43-2011.00, 43-2021.01, 43-2021.02, 43-2099.99

Significant Points

- Switchboard operators constitute 3 out of 4 of these workers.
- Workers train on the job.
- Employment is expected to decline due to new labor-saving communications technologies and consolidation of jobs.

Nature of the Work

Most communications equipment operators work as *switchboard operators* for a wide variety of businesses, such as hospitals, hotels, and personnel-supply services. Switchboard operators operate private branch exchange (PBX) switchboards to relay incoming, outgoing, and interoffice calls, usually for a single organization. They also may handle other clerical duties, such as supplying information, taking messages, and announcing visitors. Technological improvements have automated many of the tasks handled by switchboard operators. New systems automatically connect outside calls to the correct destination, and voice mail systems take messages without the assistance of an operator.

Some communications equipment operators work as *telephone operators*, assisting customers in making telephone calls. Although most calls are connected automatically, callers sometimes require the assistance of an operator. *Central office operators* help customers complete local and long-distance calls. *Directory assistance operators* provide customers with information such as phone numbers or area codes.

When callers dial "0," they usually reach a central office operator, also known as a *local, long distance*, or *call completion operator*. Most of these operators work for telephone companies, and many of their responsibilities have been automated. For example, callers can make international, collect, and credit card calls without the assistance of a central office operator. Other tasks previously handled by these operators, such as billing calls to third parties or monitoring the cost of a call, also have been automated.

Callers still need a central office operator for a limited number of tasks. These include placing person-to-person calls or interrupting busy lines if an emergency warrants the disruption. When natural disasters occur, such as storms or earthquakes, central office operators provide callers with emergency phone contacts. They also assist callers having difficulty with automated phone systems. An operator monitoring an automated system for placing collect calls, for example, may intervene if a caller needs assistance with the system.

Directory assistance operators provide callers with information such as telephone numbers or area codes. Most directory assistance operators work for telephone companies; increasingly, they also work for companies that provide business services. Automated systems now handle many of the responsibilities once performed by directory assistance operators. The systems prompt callers for a listing, and may even connect the call after providing the phone number. However, directory assistance operators monitor many of the calls received by automated systems. The operators listen to recordings of the customer's request, and then key information into electronic directories to access the correct phone numbers. Directory assistance operators also provide personal assistance to customers having difficulty using the automated system.

Other communications equipment operators include workers who operate telegraphic typewriter, telegraph key, facsimile machine, and related equipment to transmit and receive signals and messages. They prepare messages according to prescribed formats, and verify and correct errors in messages. As part of their job, they also may adjust equipment for proper operation.

Working Conditions

Most communications equipment operators work in pleasant, well-lighted surroundings. Because telephone operators spend much time seated at keyboards and video monitors, employers often provide workstations designed to decrease glare and other physical discomforts. Such improvements reduce the incidence of eyestrain, back discomfort, and injury due to repetitive motion.

Switchboard operators generally work the same hours as other clerical employees at their company. In most organizations, full-time operators work regular business hours over a 5-day workweek. Work schedules are more irregular in hotels, hospitals, and other organizations that require round-the-clock operator services. In these companies, switchboard operators may work in the evenings and on holidays and weekends.

Central office and directory assistance operators must be accessible to customers 24 hours a day and, therefore, work a variety of shifts. Some operators work split shifts; that is, they are on duty during peak calling periods in the late morning and early evening and off duty during the intervening hours. Telephone companies normally assign shifts by seniority, allowing the most experienced operators first choice of schedules. As a result, entry-level operators may have less desirable schedules, including late evening, split-shift, and weekend work. Telephone company operators may work overtime during emergencies.

Approximately 1 in 5 communications equipment operators works part-time. Because of the irregular nature of telephone operator schedules, many employers seek part-time workers for those shifts that are difficult to fill.

An operator's work may be quite repetitive and the pace hectic during peak calling periods. To maintain operator efficiency, supervisors at telephone companies often monitor operator performance, including the amount of time spent on each call. The rapid pace of the job and frequent monitoring may cause stress. To reduce job-related stress, some workplaces attempt to create a more stimulating and less rigid work environment.

Employment

Communications equipment operators held about 339,000 jobs in 2000. About 3 out of 4 worked as switchboard operators. Employment was distributed as follows:

Switchboard operators	259,000
Telephone operators	54,000
All other communications equipment operators	26,000

Most switchboard operators worked for services establishments, such as personnel-supply services, hospitals, and hotels and motels.

Training, Other Qualifications, and Advancement

Communications equipment operators receive their training on the job. At large telephone companies, entry-level central office and directory assistance operators may receive both classroom and on-the-job instruction that can last several weeks. At small telephone companies, operators usually receive shorter, less formal training. These operators may be paired with experienced personnel who provide hands-on instruction. Switchboard operators also may receive short-term, informal training, sometimes provided by the manufacturer of their switchboard equipment.

New employees train in equipment operation and procedures designed to maximize efficiency. They are familiarized with company policies, including the expected level of customer service. Instructors monitor both the time and quality of trainees' responses to customer requests. Supervisors may continue to closely monitor new employees after their initial training session is complete.

Employers generally require a high school diploma for operator positions. Applicants should have strong reading, spelling, and numerical skills; clear speech; and good hearing. Computer literacy and typing skills also are important, and familiarity with a foreign language is helpful because of the increasing diversity of the population. Most companies emphasize customer service skills. They seek operators who will remain courteous to customers while working at a fast pace.

After 1 or 2 years on the job, communications equipment operators may advance to other positions within a company. Many enter clerical occupations in which their operator experience is valuable, such as customer service representatives, dispatchers, and receptionists. Operators with a more technical background may advance into positions installing and repairing equipment. Promotion to supervisory positions also is possible.

Job Outlook

Employment of communications equipment operators is projected to decline through 2010, largely due to new laborsaving communications technologies and to consolidation of telephone operator jobs into fewer locations, often staffed by personnel-supply services firms. Virtually all job openings will result from the need to replace communications equipment operators who transfer to other occupations or leave the labor force.

Developments in communications technologies, specifically the ease and accessibility of voice recognition systems, will continue to have a significant impact on the demand for communications equipment operators. The decline in employment will be sharpest among directory assistance operators; smaller decreases will occur for switchboard operators. Voice recognition technology allows automated phone systems to recognize human speech. Callers speak directly to the system, which interprets the speech and then connects the call. Because voice recognition systems do not require callers to input data on a telephone keypad, they are easier to use than touch-tone systems. The systems also can understand increasingly sophisticated vocabulary and grammatical structures; however, many companies will continue to employ operators so that those callers having problems can access a "live" employee, if desired.

Electronic communications through the Internet or e-mail, for example, provides alternatives to telephone communications and requires no operators. Internet directory assistance services are reducing the need for directory assistance operators. Local phone companies currently have the most reliable phone directory data; however, Internet services provide information such as addresses and maps, in addition to phone numbers. As telephones and computers converge, the convenience of Internet directory assistance is expected to attract many customers, reducing the need for telephone operators to provide this service.

Consolidations among telephone companies also will reduce the need for operators. As communications technologies improve and long-distance prices fall, telephone companies will contract out and consolidate telephone operator jobs. Operators will be employed at fewer locations and will serve larger customer populations.

Earnings

Median hourly earnings of switchboard operators, including answering service, were $9.71 in 2000. The middle 50 percent earned between $8.02 and $11.71. The lowest 10 percent earned less than $6.87, and the highest 10 percent earned more than $13.76. Median hourly earnings in the industries employing the largest numbers of switchboard operators in 2000 are shown in the following:

Offices and clinics of medical doctors	$9.74
Hospitals	9.54
Hotels and motels	9.16
Personnel supply services	9.02
Miscellaneous business services	8.66

Median hourly earnings of telephone operators in 2000 were $13.46. The middle 50 percent earned between $9.40 and $16.76. The lowest 10 percent earned less than $7.23, and the highest 10 percent earned more than $19.57.

Some telephone operators working at telephone companies are members of the Communications Workers of America (CWA) or the International Brotherhood of Electrical Workers (IBEW). For these operators, union contracts govern wage rates, wage increases, and the time required to advance from one pay step to the next. (It normally takes 4 years to rise from the lowest paying, nonsupervisory operator position to the highest.) Contracts also call for extra pay for work beyond the normal 6½ to 7½ hours a day or 5 days a week, for Sunday and holiday work, and for a pay differential for nightwork and split shifts. Many contracts provide for a 1-week vacation after 6 months of service; 2 weeks after 1 year; 3 weeks after 7 years; 4 weeks after 15 years; and 5 weeks after 25 years. Holidays range from 9 to 11 days a year.

Related Occupations

Other workers who provide information to the general public include dispatchers; hotel, motel, and resort desk clerks; information and record clerks; and reservation and transportation ticket agents and travel clerks.

Sources of Additional Information

For more details about employment opportunities, contact a telephone company, temporary-help agency, or write to:

- Communications Workers of America, 501 3rd St. N.W., Washington, DC 20001. Internet: http://www.cwa-union.org
- International Brotherhood of Electrical Workers, Telecommunications Department, 1125 15th St. N.W., Room 807, Washington, DC 20005

Computer and Information Systems Managers

O*NET 11-3021.00

Significant Points

- Projected job growth stems primarily from rapid growth among computer-related occupations.
- Employers prefer managers with advanced technical knowledge acquired through computer-related work experience and formal education.
- Job opportunities should be best for applicants with a master's degree in business administration with technology as a core component.

Nature of the Work

The need for organizations to incorporate existing and future technologies in order to remain competitive has become a more pressing issue over the last several years. As electronic commerce becomes more common, how and when companies use technology are critical issues. Computer and information systems managers play a vital role in the technological direction of their organizations. They do everything from constructing the business plan to overseeing network and Internet operations.

Computer and information systems managers plan, coordinate, and direct research and design the computer-related activities of firms. They determine technical goals in consultation with top management, and make detailed plans for the accomplishment of these goals. For example, working with their staff, they may develop the overall concepts of a new product or identify computer-related problems standing in the way of project completion.

Computer and information systems managers direct the work of systems analysts, computer programmers, support specialists, and other computer-related workers. These managers plan and coordinate activities such as the installation and upgrading of hardware and software, programming and systems design, the development of computer networks, and the implementation of Internet and intranet sites. They are increasingly involved with the upkeep and maintenance of networks. They analyze the computer and information needs of their organization and determine personnel and equipment requirements. They assign and review the work of their subordinates, and stay abreast of the latest technology in order to purchase necessary equipment.

The duties of computer and information systems managers vary with their specific titles. Chief technology officers, for example, evaluate the newest and most innovative technologies and determine how these can help their organization. The chief technology officer, who often reports to the organization's chief information officer, manages and plans technical standards and tends to the daily information technology issues of their firm. Because of the rapid pace of technological change, chief technology officers must constantly be on the lookout for developments that could benefit their organization. They are responsible for demonstrating to a company how information technology can be used as a competitive weapon that not only cuts costs, but also increases revenue.

Management of information systems (MIS) directors manage information systems and computing resources for entire organizations. They also work under the chief information officer and deal directly with lower-level information technology employees. These managers oversee a variety of user services such as an organization's help desk, which employees can call with questions or problems. MIS directors may also make hardware and software upgrade recommendations based on their experience with an organization's technology.

Computer and information system managers need strong communication skills. They coordinate the activities of their unit with those of other units or organizations. They confer with top executives; financial, production, marketing, and other managers; and contractors and equipment and materials suppliers.

Working Conditions

Computer and information systems managers spend most of their time in an office. Most work at least 40 hours a week and may have to work evenings and weekends to meet deadlines or solve unexpected problems. Some computer and information systems managers may experience considerable pressure in meeting technical goals within short timeframes or tight budgets. As networks continue to expand and more work is done remotely, computer and information system managers have to communicate with and oversee offsite employees using modems, laptops, e-mail, and the Internet.

Like other workers who sit continuously in front of a keyboard, computer and information system managers are susceptible to eyestrain, back discomfort, and hand and wrist problems such as carpal tunnel syndrome.

Employment

Computer and information systems managers held about 313,000 jobs in 2000. About 2 in 5 works in services industries, primarily for firms providing computer and data processing services. Other large employers include insurance and financial services firms, government agencies, and manufacturers.

Training, Other Qualifications, and Advancement

Strong technical knowledge is essential for computer and information systems managers, who must understand and guide the work of their subordinates, yet also explain the work in nontechnical terms to senior management and potential customers. Therefore, these management positions usually require work experience and formal education similar to that of other computer occupations.

Many computer and information systems managers have experience as systems analysts; others may have experience as computer support specialists, programmers, or other information technology professionals. A bachelor's degree is usually required for management positions, although employers often prefer a graduate degree, especially a master's degree in business administration (MBA) with technology as a core component. This degree differs from a traditional MBA in that there is a heavy emphasis on information technology in addition to the standard business curriculum. This becomes important because more computer and information systems managers make not only important technology decisions but also important business decisions for their organizations. A few computer and information systems managers may have only an associate degree, provided they have sufficient experience and were able to learn additional skills on the job.

Computer and information systems managers need a broad range of skills. In addition to technical skills, employers also seek managers with strong business skills. Employers want managers who have experience with the specific software or technology to be used on the job, as well as a background in either consulting or business management. The expansion of electronic commerce has elevated the importance of business insight, because many managers are called upon to make important business decisions. Managers need a keen understanding of people, processes, and customer's needs.

Computer and information systems managers must possess strong interpersonal, communication, and leadership skills because they are required to interact not only with their employees, but also with people inside and outside their organization. They must also possess great team skills to work on group projects and other collaborative efforts. Computer and information systems managers increasingly interact with persons outside their organization, reflecting their emerging role as vital parts of their firm's executive team.

Computer and information systems managers may advance to progressively higher leadership positions in their field. Some may become managers in nontechnical areas such as marketing, human resources, or sales. In high technology firms, managers in nontechnical areas often must possess the same specialized knowledge as do managers in technical areas.

Job Outlook

Employment of computer and information systems managers is expected to increase much faster than the average for all occupations through the year 2010. Technological advancements will increase the employment of computer-related workers; as a result, the demand for managers to direct these workers also will increase. In addition, job openings will result from the need to replace managers who retire or move into other occupations. Opportunities for obtaining a management position will be best for workers possessing an MBA with technology as a core component, advanced technical knowledge, and strong communication and administrative skills.

Rapid growth in employment can be attributed to the explosion in information technology and the fast-paced expansion of the computer and data processing services industry. In order to remain competitive, firms will continue to install sophisticated computer networks and set up more complex Internet and intranet sites. Keeping a computer network running smoothly is essential to almost every organization. Firms will be more willing to hire managers who can accomplish that.

The security of computer networks will continue to increase in importance as more business is conducted over the Internet. Organizations need to understand how their systems are vulnerable and how to protect their infrastructure and Internet sites from hackers, viruses, and other acts of cyber-terrorism. As a result, there will be a high demand for managers proficient in computer security issues.

Due to the explosive growth of electronic commerce and the ability of the Internet to create new relationships with customers, the role of computer and information systems managers will continue to evolve in the future. They will continue to become more vital to their companies and the environments in which they work. The expansion of e-commerce will spur the need for computer and information systems managers with both business savvy and technical proficiency.

Opportunities for those who wish to become computer and information systems managers should be closely related to the growth of the occupations they supervise and the industries in which they are found.

Earnings

Earnings for computer and information systems managers vary by specialty and level of responsibility. Median annual earnings of these managers in 2000 were $78,830. The middle 50 percent earned between $59,640 and $100,820. The lowest 10 percent earned less than $44,090, and the highest 10 percent earned more than $127,460.

According to Robert Half International Consulting, average starting salaries in 2001 for information technology managers ranged from $92,250 to $152,500, depending on the area of specialization. According to a 2001 survey by the National Association of Colleges and Employers, starting salary offers for those with an MBA and a technical undergraduate degree averaged $61,196; for those with a master's degree in management information systems/business data processing, $57,225.

In addition, computer and information systems managers, especially those at higher levels, often receive more benefits—such as expense accounts, stock option plans, and bonuses—than do nonmanagerial workers in their organizations.

Related Occupations

The work of computer and information systems managers is closely related to that of computer programmers, computer software engineers; systems analysts, computer scientists, and database administrators; and computer support specialists and systems administrators. Computer and information systems managers also have some high-level responsibilities similar to those of top executives.

Sources of Additional Information

For information about certification as a computing professional, contact:

- Institute for Certification of Computing Professionals (ICCP), 2350 East Devon Ave., Suite 115, Des Plaines, IL 60018. Internet: http://www.iccp.org

Further information about computer careers is available from:

- Association for Computing Machinery (ACM), 1515 Broadway, New York, NY 10036. Internet: http://www.acm.org
- IEEE Computer Society, Headquarters Office, 1730 Massachusetts Ave. N.W., Washington, DC 20036-1992. Internet: http://www.computer.org
- National Workforce Center for Emerging Technologies, 3000 Landerholm Circle S.E., Bellevue, WA 98007. Internet: http://www.nwcet.org
- IEEE Computer Society, 1730 Massachusetts Ave. N.W., Washington, DC 20036-1992. Internet: http://www.computer.org

Computer, Automated Teller, and Office Machine Repairers

O*NET 49-2011.01, 49-2011.02, 49-2011.03

Significant Points

- Workers receive training in electronics from associate degree programs, the military, vocational schools, equipment manufacturers, or employers.
- Job growth reflects the increasing dependence of business and residential customers on computers and other sophisticated office machines.
- Job prospects will be best for applicants with knowledge of electronics, as well as repair experience; opportunities for computer repairers should be excellent, as employers report difficulty finding qualified applicants.

Nature of the Work

Computer repairers, also known as *data processing equipment repairers,* service mainframe, server, and personal computers; printers; and disc drives. These repairers primarily perform hands-on repair, maintenance, and installation of computers and related equipment. Workers who provide technical assistance, in person or by telephone, to computer system users are known as computer support specialists.

Automated teller machines (ATMs) allow customers to carry out bank transactions without the assistance of a teller. ATMs now provide a growing variety of other services, including stamp, phone card, and ticket sales. *Automated teller machine servicers* repair and service these machines.

Office machine and cash register servicers work on photocopiers, cash registers, mail processing equipment, and fax machines. Newer models of office machinery increasingly include computerized components that allow them to function more effectively than earlier models.

To install large equipment, such as mainframe computers and ATMs, repairers connect the equipment to power sources and communication lines. These lines allow the transmission of information over computer networks. For example, when an ATM dispenses cash, it also transmits the withdrawal information to the customer's bank. Workers also may install operating software and peripheral equipment, checking that all components are configured to correctly function together. The installation of personal computers and other small office machines is less complex and may be handled by the purchaser.

When equipment breaks down, many repairers travel to customers' workplaces or other locations to make the necessary repairs. These workers, known as *field technicians*, often have assigned areas in which they perform preventive maintenance on a regular basis. *Bench technicians* work in repair shops located in stores, factories, or service centers. In small companies, repairers may work in both repair shops and at customer locations.

Computer repairers usually replace defective components instead of repairing them. Replacement is common because components are inexpensive and businesses are reluctant to shut down their computers for time-consuming repairs. Components commonly replaced by computer repairers include video cards, which transmit signals from the computer to the monitor; hard drives, which store data; and network cards, which allow communication over the network. Defective components may be given to bench technicians, who use software programs to diagnose the problem and who may repair the components, if possible.

When ATMs malfunction, computer networks recognize the problem and alert repairers. Common problems include worn magnetic heads on card readers, which prevent the equipment from recognizing customer bankcards; and "pick failures," which prevent the equipment from dispensing the correct amount of cash. Field technicians travel to the locations of ATMs and usually repair equipment by removing and replacing defective components. Broken components are brought to a repair shop where bench technicians perform the necessary repairs. Field technicians perform routine maintenance on a regular basis, replacing worn parts and running diagnostic tests to ensure that the equipment functions properly.

Office machine repairers usually work on machinery at the customer's workplace; customers also may bring small equipment to a repair shop for maintenance. Common malfunctions include paper misfeeds due to worn or dirty parts, and poor copy quality due to problems with lamps, lenses, or mirrors. These malfunctions usually can be resolved simply by cleaning components. Breakdowns also may result from failure of commonly used parts. For example, heavy usage of a photocopier may wear down the printhead, which applies ink to the final copy. In such cases, the repairer usually replaces the part, instead of repairing it.

Workers use a variety of tools for diagnostic tests and repair. To diagnose malfunctions, they use multimeters to measure voltage, current, resistance, and other electrical properties; signal generators to provide test signals; and oscilloscopes to monitor equipment signals. When diagnosing computerized equipment, repairers also use software programs. To repair or adjust equipment, workers use handtools, such as pliers, screwdrivers, soldering irons, and wrenches.

Working Conditions

Repairers usually work in clean, well-lighted surroundings. Because computers and office machines are sensitive to extreme temperatures and to humidity, repair shops usually are air-conditioned and well-ventilated. Field repairers must travel frequently to various locations to install, maintain, or repair customer equipment. ATM repairers may have to perform their jobs in small, confined spaces that house the equipment.

Because computers and ATMs are critical for many organizations to function efficiently, data processing equipment repairers and ATM field technicians often work around the clock. Their schedules may include evening, weekend, and holiday shifts; shifts may be assigned on the basis of seniority. Office machine and cash register servicers usually work regular business hours because the equipment they repair is not as critical.

Although their job is not strenuous, repairers must lift equipment and work in a variety of postures. Repairers of computer monitors need to discharge voltage from the equipment to avoid electrocution. Workers may have to wear protective goggles.

Employment

Computer, automated teller, and office machine repairers held about 172,000 jobs in 2000. Wholesale trade establishments employed slightly less than one-half of the workers in this occupation; most of these establishments were wholesalers of professional and commercial equipment. Many workers were employed in computer and data processing services, as well as in appliance, radio, TV, and music stores. More than 1 in 7 computer, automated teller, and office machine repairers was self-employed.

Training, Other Qualifications, and Advancement

Knowledge of electronics is necessary for employment as a computer, automated teller, or office machine repairer. Employers prefer workers who are certified as repairers or who have training in electronics from associate degree programs, the military, vocational schools, or equipment manufacturers. Employers generally provide some training to new repairers on specific equipment; however, workers are expected to arrive on the job with a basic understanding of equipment repair. Employers may send experienced workers to training sessions to keep up with changes in technology and service procedures.

Most office machine and ATM repairer positions require an associate degree in electronics. A basic understanding of mechanical equipment also is important, as many of the parts that fail in office machines and ATMs are mechanical, such as paper loaders. Entry-level employees at large companies normally receive on-the-job training lasting several months. This may include a week of classroom instruction followed by a period of 2 weeks to several months assisting an experienced repairer.

Field technicians work closely with customers and must have good communications skills and a neat appearance. Employers normally require that field technicians have a driver's license.

Several organizations administer certification programs for electronic or computer equipment repairers. Numerous certifications, including A+, Net+, and Server+, are available through the Computing Technology Industry Association (CompTIA). To receive the certifications, candidates must pass several tests that assess computer repair skills. The International Society of Certified Electronics Technicians (ISCET) and the Electronics Technicians Association (ETA) also administer certification programs. Repairers may specialize in computer repair or a variety of other skills. To receive certification, repairers must pass qualifying exams corresponding to their level of training and experience. Both programs offer associate certifications to entry-level repairers.

Newly hired computer repairers may work on personal computers or peripheral equipment. With experience, they can advance to positions maintaining more sophisticated systems, such as networking equipment and servers. Field repairers of ATMs may advance to bench-technician positions responsible for more complex repairs. Experienced workers may become specialists who help other repairers diagnose difficult problems or work with engineers in designing equipment and developing maintenance procedures. Experienced workers also may move into management positions responsible for supervising other repairers.

Because of their familiarity with equipment, experienced repairers also may move into customer service or sales positions. Some experienced workers open their own repair shops or become wholesalers or retailers of electronic equipment.

Job Outlook

Employment of computer, automated teller, and office machine repairers is expected to grow about as fast as the average for all occupations through 2010. Job growth will be driven by the increasing dependence of business and residential customers on computers and other sophisticated office machines. The need to maintain this equipment in working order will create new jobs for repairers. In addition, openings will result from the need to replace repairers who retire or move into new occupations.

Job prospects will be best for applicants with knowledge of electronics as well as repair experience; opportunities for computer repairers should be excellent, as employers report difficulty finding qualified applicants and as reliance on computers continues to increase. Although computer equipment continues to become less expensive and more reliable, malfunctions still occur and can cause severe problems for users, most of whom lack the knowledge to make repairs. Computers are critical to most businesses today and will become even more so to companies that do business on the Internet and to households that make purchases online.

People also are becoming increasingly reliant on ATMs. Besides bank and retail transactions, ATMs provide an increasing number of other services, such as employee information processing and distribution of government payments. ATM design improvements have increased reliability and simplified repair tasks, reducing the number and extent of repairs. Opportunities for ATM repairers should be available, primarily arising from the need to replace workers who leave the specialty, rather than from employment growth.

Conventional office machines, such as calculators, are inexpensive, and often are replaced instead of repaired. However, digital copiers and other newer office machines are more costly and complex. This equipment often is computerized, designed to work on a network, and able to perform multiple functions. The growing need for repairers to service such sophisticated equipment should result in job opportunities for office machine repairers.

Earnings

Median hourly earnings of computer, automated teller, and office machine repairers were $15.08 in 2000. The middle 50 percent earned between $11.80 and $19.20. The lowest 10 percent earned less than $9.50, and the highest 10 percent earned more than $23.42. Median hourly earnings in the industries employing the largest numbers of computer, automated teller, and office machine repairers in 2000 are shown in the following table:

Professional and commercial equipment	$15.28
Computer and data processing services	15.05
Radio, television, and computer stores	13.16

Related Occupations

Workers in other occupations who repair and maintain electronic equipment include broadcast and sound engineering technicians and radio operators; electronic home entertainment equipment installers and repairers; electrical and electronics installers and repairers; industrial machinery installation, repair, and maintenance workers; and radio and telecommunications equipment installers and repairers.

Sources of Additional Information

For information on certification programs, contact:

- Computing Technology Industry Association, 450 East 22nd St., Suite 230, Lombard, IL 60148-6158. Internet: http://www.comptia.org
- International Society of Certified Electronics Technicians, 3608 Pershing Ave., Fort Worth, TX 76107. Internet: http://www.iscet.org
- Electronics Technicians Association, 502 North Jackson, Greencastle, IN 46135

Computer-Control Programmers and Operators

O*NET 51-4011.01, 51-4012.00

Significant Points

- Workers learn in apprenticeship programs, informally on the job, and in secondary, vocational, or post-secondary schools; many entrants have previously worked as machinists or machine setters, operators, and tenders.
- Job opportunities will be excellent, as employers are expected to continue to have difficulty finding qualified workers.

Nature of the Work

Computer-control programmers and operators use computer numerically controlled (CNC) machines to cut and shape precision products, such as automobile parts, machine parts, and compressors. CNC machines include metal-machining tools such as lathes, multi-axis spindles, and milling machines, but the functions formerly performed by human operators are performed by a computer-control module. CNC machines cut away material from a solid block of metal, plastic, or glass—known as a workpiece—to form a finished part. Computer-control programmers and operators normally produce large quantities of one part, although they may produce small batches or one-of-a-kind items. They use their knowledge of the working properties of metals and their skill with CNC programming to design and carry out the operations needed to make machined products that meet precise specifications.

Before CNC programmers—also referred to as numerical tool and process control programmers—machine a part, they must carefully plan and prepare the operation. First, these workers review three-dimensional computerized engineering diagrams (blueprints) of the part. Next, they calculate where to cut or bore into the workpiece, how fast to feed the metal into the machine, and how much metal to remove. They then select tools and materials for the job and plan the sequence of cutting and finishing operations.

Next, computer-control programmers turn the planned machining operations into a set of instructions. These instructions are translated into a computer program containing a set of commands for the machine to follow. The program is then saved onto a computer, which functions as a server. Computer-control programmers and operators check new programs to ensure that the machinery will function properly and that the output will meet specifications. Because a problem with the program could damage costly machinery and cutting tools, computer simulations may be used to check the program instead of a trial run. If errors are found, the program must be changed and retested until the problem is resolved. In addition, growing connectivity between computer-aided design (CAD) software and CNC machine tools is raising productivity by automatically translating designs into instructions for the computer controller on the machine tool. These new computer-automated manufacturing (CAM) technologies enable programs to be easily modified for use on other jobs with similar specifications, thereby reducing time and effort.

After the programming work is completed, computer-controlled machine tool operators, metal and plastic (CNC operators), perform the necessary machining operations. The CNC operators transfer the commands from the server to the CNC control module using a computer network link. Many advanced control modules are conversational, meaning they ask the operator a series of questions about the nature of the task. Computer-control operators position the metal stock on the CNC machine tool—spindle, lathe, milling machine, or other—set the controls, and let the computer make the cuts. Heavier objects may be loaded with the assistance of other workers, a crane, or a forklift. During the machining process, computer-control operators constantly monitor the readouts from the CNC control module, checking to see if any problems exist. Machine tools have unique characteristics, which can be problematic. During a machining operation, the operator modifies the cutting program to account for any problems encountered. Unique, modified CNC programs are saved for every different machine that performs a task.

CNC operators detect some problems by listening for specific sounds—for example, a dull cutting tool or excessive vibration. Dull cutting tools are removed and replaced. Machine tools rotate at high speeds, which can create problems with harmonic vibrations in the workpiece. Vibrations cause the machine tools to make minor cutting errors, hurting the quality of the product. Computer-control operators listen for vibrations and then adjust the cutting speed to compensate. In older, slower machine tools, the cutting speed would be reduced to eliminate the vibrations, but the amount of time needed to finish the product would increase as a result. In newer, high-speed CNC machines, increasing the cutting speed normally eliminates the vibrations and reduces production time. CNC operators also ensure that the workpiece is being properly lubricated and cooled, because the machining of metal products generates a significant amount of heat.

Working Conditions

Most machine shops are clean, well lit, and ventilated. Many computer-controlled machines are totally enclosed, minimizing the exposure of workers to noise, dust, and the lubricants used to cool workpieces during machining. Nevertheless, working around high-speed machine tools presents certain dangers, and workers must follow safety precautions. Computer-controlled machine tool operators, metal and plastic, wear protective equipment such as safety glasses to shield against bits of flying metal and earplugs to dampen machinery noise. They also must exercise caution when handling hazardous coolants and lubricants. The job requires stamina because operators stand most of the day and, at times, may need to lift moderately heavy workpieces.

Numerical tool and process control programmers work in offices that typically are near, but separate from, the shop floor. These work areas usually are clean, well lit, and free of machine noise. Numerical tool and process control programmers occasionally need to enter the shop floor to monitor CNC machining operations. On the shop floor, CNC programmers encounter the same hazards and exercise the same safety precautions as CNC operators.

Most computer-control programmers and operators work a 40-hour week. CNC operators increasingly work evening and weekend shifts as companies justify investments in more expensive machinery by extending hours of operation. Overtime is common during peak production periods.

Employment

Computer-control programmers and operators held about 186,000 jobs in 2000, mostly working in small machining shops or in manufacturing firms that produce durable goods, such as metalworking and industrial machinery, aircraft, or motor vehicles. Although computer-control programmers and operators work in all parts of the country, jobs are most plentiful in the northeast, midwest, and west, where manufacturing is concentrated.

Training, Other Qualifications, and Advancement

Computer-control programmers and operators train in various ways—in apprenticeship programs, informally on the job, and in secondary, vocational, or postsecondary schools. Due to a shortage of qualified applicants, many employers teach introductory courses, which provide a basic understanding of metalworking machines, safety, and blueprint reading. A basic knowledge of computers and electronics also is helpful. Experience with machine tools is extremely important. In fact, many entrants to these occupations have previously worked as machinists or machine setters, operators, and tenders. Persons interested in becoming computer-control programmers or operators should be mechanically inclined and able to work independently and do highly accurate work.

High school or vocational school courses in mathematics, blueprint reading, computer programming, metalworking, and drafting are recommended. Apprenticeship programs consist of shop training and related classroom instruction. In shop training, apprentices learn filing, handtapping, and dowel fitting, as well as the operation of various machine tools. Classroom instruction includes math, physics, programming, blueprint reading, CAD software, safety, and shop practices. Skilled computer-control programmers and operators need an understanding of the machining process, including the complex physics that occur at the cutting point. Thus, most training programs teach CNC operators and programmers to perform operations on manual machines prior to operating CNC machines. A growing number of computer-control programmers and operators receive most of their formal training from community or technical colleges. Less skilled CNC operators may have only 12 weeks of classroom training prior to working on the shop floor.

To boost the skill level of all metalworkers and to create a more uniform standard of competency, a number of training facilities and colleges have recently begun implementing curriculums incorporating national skills standards developed by the National Institute of Metalworking Skills (NIMS). After completing such a curriculum and passing a performance requirement and written exam, a NIMS credential is granted to trainees, providing formal recognition of competency in a metalworking field. Completion of a formal certification program provides expanded career opportunities.

Qualifications for computer-control programmers vary widely depending upon the complexity of the job. Employers often prefer skilled machinists or those with technical school training. For some specialized types of programming, such as that needed to produce complex parts for the aerospace or shipbuilding industries, employers may prefer individuals with a degree in engineering.

For those entering CNC programming directly, a basic knowledge of computers and electronics is necessary, and experience with machine tools is extremely helpful. Classroom training includes an introduction to computer numerical control, the basics of programming, and more complex topics, such as computer-aided manufacturing. Trainees start writing simple programs under the direction of an experienced programmer. Although machinery manufacturers are trying to standardize programming languages, there are numerous languages in use. Because of this, computer-control programmers and operators should be able to learn new programming languages.

As new automation is introduced, computer-control programmers and operators normally receive additional training to update their skills. This training usually is provided by a representative of the equipment manufacturer or a local technical school. Many employers offer tuition reimbursement for job-related courses.

Computer-control programmers and operators can advance in several ways. Experienced CNC operators may become CNC programmers, and some are promoted to supervisory or administrative positions in their firms. A few open their own shops.

Job Outlook

Computer-control programmers and operators should have excellent job opportunities. Due to the limited number of people entering training programs, employers are expected to continue to have difficulty finding workers with the necessary skills and knowledge. Employment of computer-control programmers and operators is projected to grow about as fast as the average for all occupations through 2010. Job growth will be driven by the increasing use of CNC machine tools, but advances in CNC machine tool technology will further simplify minor adjustments, enabling machinists and tool and die makers to perform tasks that previously required computer-control operators. In addition, the demand for computer-control programmers will be negatively affected by the increasing use of software that automatically translates part and product designs into CNC machine tool instructions.

Employment levels of computer-control programmers and operators are influenced by economic cycles—as the demand for machined goods falls, programmers and operators involved in production may be laid off or forced to work fewer hours.

Earnings

Median hourly earnings of computer-controlled machine tool operators, metal and plastic, were about $13.17 in 2000. The middle 50 percent earned between $10.48 and $16.55. The lowest 10 percent earned less than $8.80, whereas the top 10 percent earned more than $20.25. Median hourly earnings in the manufacturing industries employing the largest number of computer-controlled machine tool operators, metal and plastic, in 2000 were as follows:

Metalworking machinery	$15.20
General industrial machinery	15.06
Industrial machinery, not elsewhere classified	13.05
Motor vehicles and equipment	12.05
Miscellaneous plastics products, not elsewhere classified	11.35

Median hourly earnings of numerical tool and process control programmers were $17.70 in 2000. The middle 50 percent earned between $13.81 and $21.74. The lowest 10 percent earned less than $10.39, while the top 10 percent earned more than $26.66.

Related Occupations

Occupations most closely related to computer-control programmers and operators are other metal worker occupations. These include machinists; tool and die makers; machine setters, operators, and tenders—metal and plastic; and welding, soldering, and brazing workers.

Numerical tool and process control programmers apply their knowledge of machining operations, metals, blueprints, and machine programming to write programs that run machine tools. Computer programmers also write detailed programs to meet precise specifications.

Sources of Additional Information

For general information about computer-control programmers and operators, contact:

- Precision Machine Products Association, 6700 West Snowville Rd., Brecksville, OH 44141-3292. Internet: http://www.pmpa.org

For a list of training centers and apprenticeship programs, contact:

- National Tooling and Metalworking Association, 9300 Livingston Rd., Fort Washington, MD 20744. Internet: http://www.ntma.org

For general occupational information, including a list of training programs, contact:

- PMA Educational Foundation, 6363 Oak Tree Blvd., Independence, OH 44131-2500. Internet: http://www.pmaef.org

Computer Hardware Engineers

O*NET 17-2061.00

Significant Points

- Overall job opportunities in engineering are expected to be good.
- A bachelor's degree is required for most entry-level jobs.
- Starting salaries are significantly higher than those of college graduates in other fields.
- Continuing education is critical to keep abreast of the latest technology.

Nature of the Work

Computer hardware engineers research, design, develop, and test computer hardware and supervise its manufacture and installation. Hardware refers to computer chips, circuit boards, computer systems, and related equipment such as keyboards, modems, and printers. (Computer software engineers—often simply called computer engineers—design and develop the software systems that control computers.) The work of computer hardware engineers is very similar to that of electronics engineers, but unlike electronics engineers, computer hardware engineers work with computers and computer-related equipment exclusively. In addition to design and development, computer hardware engineers may supervise the manufacturing and installation of computers and computer-related equipment. The rapid advances in computer technology are largely a result of the research, development, and design efforts of computer hardware engineers. To keep up with technology change, these engineers must continually update their knowledge.

Working Conditions

Most engineers work in office buildings, laboratories, or industrial plants. Others may spend time outdoors at construction sites, mines, and oil and gas exploration and production sites, where they monitor or direct operations or solve onsite problems. Some engineers travel extensively to plants or work sites.

Many engineers work a standard 40-hour week. At times, deadlines or design standards may bring extra pressure to a job. When this happens, engineers may work longer hours and experience considerable stress.

Employment

The number of computer hardware engineers is relatively small as compared with the number of other computer-related workers who work with software or computer applications. Computer hardware engineers held about 60,000 jobs in 2000. About 25 percent were employed in computer and data processing services. About 1 out of 10 worked in computer and office equipment manufacturing, but many also are employed in communications industries and engineering consulting firms.

Training, Other Qualifications, and Advancement

A bachelor's degree in engineering is required for almost all entry-level engineering jobs. College graduates with a degree in a physical science or mathematics occasionally may qualify for some engineering jobs, especially in specialties in high demand. Most engineering degrees are granted in electrical, electronics, mechanical, or civil engineering. However, engineers trained in one branch may work in related branches. This flexibility allows employers to meet staffing needs in new technologies and specialties in which engineers are in short supply. It also allows engineers to shift to fields with better employment prospects or to those that more closely match their interests.

Most engineering programs involve a concentration of study in an engineering specialty, along with courses in both mathematics and science. Most programs include a design course, sometimes accompanied by a computer or laboratory class or both.

In addition to the standard engineering degree, many colleges offer 2- or 4-year degree programs in engineering technology. These programs, which usually include various hands-on laboratory classes that focus on current issues, prepare students for practical design and production work, rather than for jobs which require more theoretical and scientific knowledge. Graduates of 4-year technology programs may get jobs similar to those obtained by graduates with a bachelor's degree in engineering. Engineering technology graduates, however, are not qualified to register as professional engineers under the same terms as graduates with degrees in engineering. Some employers regard technology program graduates as having skills between those of a technician and an engineer.

Graduate training is essential for engineering faculty positions and many research and development programs, but is not required for the majority of entry-level engineering jobs. Many engineers obtain graduate degrees in engineering or business administration to learn new technology and broaden their education. Many high-level executives in government and industry began their careers as engineers.

About 330 colleges and universities offer bachelor's degree programs in engineering that are accredited by the Accreditation Board for Engineering and Technology (ABET), and about 250 colleges offer accredited bachelor's degree programs in engineering technology. ABET accreditation is based on an examination of an engineering program's student achievement, program improvement, faculty, curricular content, facilities, and institutional commitment. Although most institutions offer programs in the major branches of engineering, only a few offer programs in the smaller specialties. Also, programs of the same title may vary in content. For example, some programs emphasize industrial practices, preparing students for a job in industry, whereas others are more theoretical and are designed to prepare students for graduate work. Therefore, students should investigate curricula and check accreditations carefully before selecting a college.

Admissions requirements for undergraduate engineering schools include a solid background in mathematics (algebra, geometry, trigonometry, and calculus) and sciences (biology, chemistry, and physics), and courses in English, social studies, humanities, and computers. Bachelor's degree programs in engineering typically are designed to last 4 years, but many students find that it takes between 4 and 5 years to complete their studies. In a typical 4-year college curriculum, the first 2 years are spent studying mathematics, basic sciences, introductory engineering, humanities, and social sciences. In the last 2 years, most courses are in engineering, usually with a concentration in one branch. Some programs offer a general engineering curriculum; students then specialize in graduate school or on the job.

Some engineering schools and 2-year colleges have agreements whereby the 2-year college provides the initial engineering education, and the engineering school automatically admits students for their last 2 years. In addition, a few engineering schools have arrangements whereby a student spends 3 years in a liberal arts college studying pre-engineering subjects and 2 years in an engineering school studying core subjects, and then receives a bachelor's degree from each school. Some colleges and universities offer 5-year master's degree programs. Some 5- or even 6-year cooperative plans combine classroom study and practical work, permitting students to gain valuable experience and finance part of their education. All 50 states and the District of Columbia usually require licensure for engineers who offer their services directly to the public. Engineers who are licensed are called Professional Engineers (PE). This licensure generally requires a degree from an ABET-accredited engineering program, 4 years of relevant work experience, and successful completion of a state examination. Recent graduates can start the licensing process by taking the examination in two stages. The initial Fundamentals of Engineering (FE) examination can be taken upon graduation. Engineers who pass this examination commonly are called Engineers in Training (EIT) or Engineer Interns (EI). The EIT certification usually is valid for 10 years. After acquiring suitable work experience, EITs can take the second examination, the Principles and Practice of Engineering Exam. Several states have imposed mandatory continuing education requirements for relicensure. Most states recognize licensure from other states.

Engineers should be creative, inquisitive, analytical, and detail oriented. They should be able to work as part of a team and to communicate well, both orally and in writing. Communication abilities are becoming more important because much of their work is becoming more diversified, meaning that engineers interact with specialists in a wide range of fields outside engineering.

Beginning engineering graduates usually work under the supervision of experienced engineers and, in large companies, also may receive formal classroom or seminar-type training. As new engineers gain knowledge and experience, they are assigned more difficult projects with greater independence to develop designs, solve problems, and make decisions. Engineers may advance to become technical specialists or to supervise a staff or team of engineers and technicians. Some may eventually become engineering managers or enter other managerial or sales jobs.

Job Outlook

Computer hardware engineers are expected to have favorable job opportunities. Employment of computer hardware engineers is projected to increase faster than the average for all occupations through 2010, reflecting rapid employment growth in the computer and office equipment industry, which employs the greatest number of computer engineers. Consulting opportunities for computer hardware engineers should grow as businesses need help managing, upgrading, and customizing increasingly complex systems. Growth in embedded systems, a technology that uses computers to control other devices such as appliances or cell phones, also will increase the demand for computer hardware engineers. In addition to job openings arising from employment growth, other vacancies will result from the need to replace workers who move into managerial positions, transfer to other occupations, or leave the labor force.

Earnings

Median annual earnings of computer hardware engineers were $67,300 in 2000. The middle 50 percent earned between $52,960 and $86,280. The lowest 10 percent earned less than $42,620, and the highest 10 percent earned more than $107,360. Median annual earnings in the industries employing the largest numbers of computer hardware engineers in 2000 were:

Computer and office equipment	$75,730
Computer and data processing services	69,490
Electronic components and accessories	67,800
Telephone communication	59,160

Starting salaries for computer engineers with a bachelor's degree can be significantly higher than salaries of bachelor's degree graduates in many other fields. According to the National Association of Colleges and Employers, starting salary offers in 2001 for bachelor's degree candidates in computer engineering averaged $53,924 a year; master's degree candidates averaged $58,026; and PhD candidates averaged $70,140.

Related Occupations

Computer hardware engineers apply the principles of physical science and mathematics in their work. Other workers who use scientific and mathematical principles include architects, except landscape and naval; engineering and natural sciences managers; computer and information systems managers; mathematicians; drafters; engineering technicians; sales engineers; science technicians; and physical and life scientists, including agricultural and food scientists, biological and medical scientists, conservation scientists and foresters, atmospheric scientists, chemists and materials scientists, environmental scientists and geoscientists, and physicists and astronomers.

Sources of Additional Information

For further information about computer hardware engineers, contact:

- IEEE Computer Society, 1730 Massachusetts Ave. N.W., Washington, DC 20036-1992. Internet: http://www.computer.org

High school students interested in obtaining a full package of guidance materials and information (product number SP-01) on a variety of engineering disciplines should contact the Junior Engineering Technical Society by sending $3.50 to:

- JETS-Guidance, 1420 King St., Suite 405, Alexandria, VA 22314-2794. Internet: http://www.jets.org

High school students interested in obtaining information on ABET-accredited engineering programs should contact:

- The Accreditation Board for Engineering and Technology, Inc., 111 Market Place, Suite 1050, Baltimore, MD 21202-4012. Internet: http://www.abet.org

Non-licensed engineers and college students interested in obtaining information on Professional Engineer licensure should contact:

- The National Society of Professional Engineers, 1420 King St., Alexandria, VA 22314-2794. Internet: http://www.nspe.org
- National Council of Examiners for Engineers and Surveying, P.O. Box 1686, Clemson, SC 29633-1686. Internet: http://www.ncees.org

Information on general engineering education and career resources is available from:

- American Society for Engineering Education, 1818 N St. NW, Suite 600, Washington, DC 20036-2479. Internet: http://www.asee.org

Information on obtaining an engineering position with the federal government is available from the Office of Personnel Management (OPM) through a telephone-based system. Consult your telephone directory under U.S. Government for a local number or call (912) 757-3000; Federal Relay Service: (800) 877-8339. The first number is not toll free, and charges may result. Information also is available from the OPM Internet site: http://www.usajobs.opm.gov

Non-high school students and those wanting more detailed information should contact societies representing the individual branches of engineering. Each can provide information about careers in the particular branch. The individual statements elsewhere in this book also provide other information in detail on aerospace; agricultural; biomedical; chemical; civil; electrical and electronics; environmental; industrial, including health and safety; materials; mechanical; mining and geological, including mining safety; nuclear; and petroleum engineering.

Computer Operators

O*NET 43-9011.00

Significant Points

- Employment is expected to decline sharply due to advances in technology.
- Opportunities will be best for operators who have formal computer-related education, are familiar with a variety of operating systems, and keep up-to-date with the latest technology.

Nature of the Work

Computer operators oversee the operation of computer hardware systems, ensuring that these machines are used as efficiently as possible. They may work with mainframes, minicomputers, or networks of personal computers. Computer operators must anticipate problems and take preventive action, as well as solve problems that occur during operations.

The duties of computer operators vary with the size of the installation, the type of equipment used, and the policies of the employer. Generally, operators control the console of either a mainframe digital computer or a group of minicomputers. Working from operating instructions prepared by programmers, users, or operations managers, computer operators set controls on the computer and on peripheral devices required to run a particular job.

Computer operators load equipment with tapes, disks, and paper, as needed. While the computer is running—which may be 24 hours a day for large computers—computer operators monitor the control console and respond to operating and computer messages. Messages indicate the individual specifications of each job being run. If an error message occurs, operators must locate and solve the problem or terminate the program. Operators also maintain logbooks or operating records, listing each job that is run and events, such as machine malfunctions, that occur during their shift. In addition, computer operators may help programmers and systems analysts test and debug new programs.

As the trend toward networking computers accelerates, a growing number of computer operators are working on personal computers (PCs) and minicomputers. In many offices, factories, and other work settings, PCs and minicomputers are connected in networks, often referred to as local area networks (LANs) or multiuser systems. Whereas users in the area operate some of these computers, many require the services of full-time operators. The tasks performed on PCs and minicomputers are very similar to those performed on large computers.

As organizations continue to look for opportunities to increase productivity, automation is expanding into additional areas of computer operations. Sophisticated software, coupled with robotics, enables a computer to perform many routine tasks formerly done by computer operators. Scheduling, loading and downloading programs, mounting tapes, rerouting messages, and running periodic reports can be done without the intervention of an operator. Consequently, these improvements will change what computer operators do in the future. As technology advances, the responsibilities of many computer operators are shifting to areas such as network operations, user support, and database maintenance.

Working Conditions

Computer operators generally work in well-lighted, well-ventilated, comfortable rooms. Because many organizations use their computers 24 hours a day, 7 days a week, computer operators may be required to work evening or night shifts and weekends. Shift assignments usually are made based on seniority. However, increasingly automated operations will lessen the need for shift work, because many companies let the computer take over operations during less desirable working hours. In addition, advances in telecommuting technologies—such as faxes, modems, and e-mail—and data center automation—such as automated tape libraries—enable some operators to monitor batch processes, check systems performance, and record problems for the next shift.

Because computer operators generally spend a lot of time in front of a computer monitor, as well as performing repetitive tasks such as loading and unloading printers, they may be susceptible to eyestrain, back discomfort, and hand and wrist problems.

Employment

Computer operators held about 194,000 jobs in 2000. Most jobs are found in organizations such as wholesale trade establishments, manufacturing companies, business services firms, financial institutions, and government agencies that have data-processing needs requiring large computer installations. A large number of computer operators are employed by service firms in the computer and data-processing services industry, as more companies contract out the operation of their data-processing centers.

Training, Other Qualifications, and Advancement

Computer operators usually receive on-the-job training in order to become acquainted with their employer's equipment and routines. The length of training varies with the job and the experience of the worker. However, previous work experience is the key to obtaining an operator job in many large establishments. Employers generally look for specific, hands-on experience with the type of equipment and related operating systems they use. Additionally, formal computer-related training, perhaps through a community college or technical school, is recommended. Related training also can be obtained through the armed forces and from some computer manufacturers. As computer technology changes and data processing centers become more automated, employers will increasingly require candidates to have formal training and experience for operator jobs. And although not required, a bachelor's degree in a computer-related field can be helpful when one is seeking employment as a computer operator.

Because computer technology changes so rapidly, operators must be adaptable and willing to learn. Analytical and technical expertise also are needed, particularly by operators who work in automated data centers, to deal with unique or high-level problems that a computer is not programmed to handle. Operators must be able to communicate well, and to work effectively with programmers, users, and other operators. Computer operators also must be able to work independently because they may have little or no direct supervision.

A few computer operators may advance to supervisory jobs, although most management positions within data-processing or computer-operations centers require advanced formal education, such as a bachelor's or higher degree. Through on-the-job experience and additional formal education, some computer operators may advance to jobs in areas such as network operations or support. As they gain experience in programming, some operators may advance to jobs as programmers or analysts. A move into these types of jobs is becoming much more difficult, as employers increasingly require candidates for more-skilled computer jobs to possess at least a bachelor's degree.

Job Outlook

Employment of computer operators is expected to decline sharply through the year 2010. Experienced operators are expected to compete for the small number of openings that will arise each year to replace workers who transfer to other occupations or leave the labor force. Opportunities will be best for operators who have formal computer-related education, are familiar with a variety of operating systems, and keep up-to-date with the latest technology.

Advances in technology have reduced both the size and cost of computer equipment, while increasing the capacity for data storage and processing automation. Sophisticated computer hardware and software are now used in practically every industry, in such areas as factory and office automation, telecommunications, medicine, education, and administration. The expanding use of software that automates computer operations gives companies the option of making systems user-friendly, greatly reducing the need for operators. Such improvements require operators to monitor a greater number of operations at the same time and be capable of solving a broader range of problems that may arise. The result is that fewer operators will be needed to perform more highly skilled work.

Computer operators who are displaced by automation may be reassigned to support staffs that maintain personal computer networks or assist other members of the organization. Operators who keep up with changing technology, by updating their skills and enhancing their training, should have the best prospects of moving into other areas such as network administration and technical support. Others may be retrained to perform different job duties, such as supervising an operations center, maintaining automation packages, or analyzing computer operations to recommend ways to increase productivity. In the future, operators who wish to work in the computer field will need to know more about programming, automation software, graphics interface, client/server environments, and open systems, in order to take advantage of changing opportunities.

Earnings

Median annual earnings of computer operators were $27,670 in 2000. The middle 50 percent earned between about $21,280 and $35,320 a year. The top 10 percent earned more than $43,950, and the bottom 10 percent earned less than $17,350. Median annual earnings in the industries employing the largest numbers of computer operators in 2000 are shown in the following:

Computer and data processing services	$28,530
Hospitals	26,550
Commercial banks	22,840
Personnel supply services	22,130
Miscellaneous business services	21,980

The average salary for computer operators employed by the federal government was $37,574 in early 2001.

According to Robert Half International, the average starting salaries for console operators ranged from $28,250 to $40,500 in 2001. Salaries generally are higher in large organizations than in small ones.

Related Occupations

Other occupations involving work with computers include computer software engineers; computer programmers; computer support specialists and systems administrators; and systems analysts, computer scientists, and database administrators. Other occupations in which workers operate electronic office equipment include data entry and information-processing workers, as well as secretaries and administrative assistants.

Sources of Additional Information

For information about work opportunities in computer operations, contact establishments with large computer centers, such as banks, manufacturing and insurance firms, colleges and universities, and data processing service organizations. The local office of the state employment service can supply information about employment and training opportunities.

Computer Programmers

O*NET 15-1021.00

Significant Points

- Employment growth will be considerably slower than that of other computer specialists, due to the spread of pre-packaged software solutions. Three out of 5 computer programmers held at least a bachelor's degree in 2000.
- Prospects should be best for college graduates with knowledge of a variety of programming languages and tools; those with less formal education or its equivalent in work experience should face strong competition for programming jobs.

Nature of the Work

Computer programmers write, test, and maintain the detailed instructions, called programs, that computers must follow to perform their functions. They also conceive, design, and test logical structures for solving problems by computer. Many technical innovations in programming—advanced computing technologies and sophisticated new languages and programming tools—have redefined the role of a programmer and elevated much of the programming work done today. Job titles and descriptions may vary, depending on the organization. In this occupational statement, *computer programmer* refers to individuals whose main job function is programming; this group has a wide range of responsibilities and educational backgrounds.

Computer programs tell the computer what to do, such as which information to identify and access, how to process it, and what equipment to use. Programs vary widely depending upon the type of information to be accessed or generated. For example, the instructions involved in updating financial records are very different from those required to duplicate conditions on board an aircraft for pilots training in a flight simulator. Although simple programs can be written in a few hours, programs that use complex mathematical formulas, whose solutions can only be approximated, or that draw data from many existing systems, may require more than a year of work. In most cases, several programmers work together as a team under a senior programmer's supervision.

Programmers write programs according to the specifications determined primarily by computer software engineers and system analysts. After the design process is complete, it is the job of the programmer to convert that design into a logical series of instructions that the computer can follow. They then code these instructions in a conventional programming language, such as COBOL; an artificial intelligence language, such as Prolog; or one of the most advanced object-oriented languages such as Java, C++, or Smalltalk. Different programming languages are used depending on the purpose of the program. COBOL, for example, is commonly used for business applications, whereas Fortran (short for "formula translation") is used in science and engineering. C++ is widely used for both scientific and business applications. Programmers generally know more than one programming language; and since many languages are similar, they often can learn new languages relatively easily. In practice, programmers often are referred to by the language they know, such as Java programmers, or the type of function they perform or environment in which they work, such as database programmers, mainframe programmers, or Internet programmers.

Many programmers update, repair, modify, and expand existing programs. When making changes to a section of code, called a routine, programmers need to make other users aware of the task the routine is to perform. They do this by inserting comments in the coded instructions, so others can understand the program. Many programmers use computer-assisted software engineering (CASE) tools to automate much of the coding process. These tools enable a programmer to concentrate on writing the unique parts of the program, because the tools automate various pieces of the program being built. CASE tools generate whole sections of code automatically, rather than line by line. This also yields more reliable and consistent programs and increases programmers' productivity by eliminating some routine steps.

Programmers test a program by running it, to ensure the instructions are correct and it produces the desired information. If errors do occur, the programmer must make the appropriate change and recheck the program until it produces the correct results. This process is called debugging. Programmers may continue to fix these problems throughout the life of a program. Programmers working in a mainframe environment may prepare instructions for a computer operator who will run the program. They also may contribute to a manual for users.

Programmers often are grouped into two broad types—applications programmers and systems programmers. *Applications programmers* write programs to handle a specific job, such as a program to track inventory, within an organization. They may also revise existing packaged software. *Systems programmers*, on the other hand, write programs to maintain and control computer systems software, such as operating systems, networked systems, and database systems. These workers make changes in the sets of instructions that determine how the network, workstations, and central processing unit of the system handle the various jobs they have been given and how they communicate with peripheral equipment, such as terminals, printers, and disk drives. Because of their knowledge of the entire computer system, systems programmers often help applications programmers determine the source of problems that may occur with their programs.

Programmers in software development companies may work directly with experts from various fields to create software—either programs designed for specific clients or packaged software for general use—ranging from games and educational software to programs for desktop publishing, financial planning, and spreadsheets. Much of this type of programming is in the preparation of packaged software, which comprises one of the most rapidly growing segments of the computer services industry.

In some organizations, particularly small ones, workers commonly known as *programmer-analysts* are responsible for both the systems analysis and the actual programming work. Advanced programming languages and new object-oriented programming capabilities are increasing the efficiency and productivity of both programmers and users. The transition from a mainframe environment to one that is primarily personal computer (PC) based has blurred the once rigid distinction between the programmer and the user. Increasingly, adept end-users are taking over many of the tasks previously performed by programmers. For example, the growing use of packaged software, like spreadsheet and database management software packages, allows users to write simple programs to access data and perform calculations.

Working Conditions

Programmers generally work in offices in comfortable surroundings. Many programmers may work long hours or weekends, to meet deadlines or fix critical problems that occur during off hours. Given the technology available, telecommuting is becoming common for a wide range of computer professionals—including computer programmers. As computer networks expand, more programmers are able to connect to a customer's computer system remotely to make corrections or fix problems, using modems, e-mail, and the Internet.

Like other workers who spend long periods of time in front of a computer terminal typing at a keyboard, programmers are susceptible to eyestrain, back discomfort, and hand and wrist problems, such as carpal tunnel syndrome.

Employment

Computer programmers held about 585,000 jobs in 2000. Programmers are employed in almost every industry, but the largest concentration is in the computer and data processing services industry, which includes firms that write and sell software. Large numbers of programmers can also be found working for firms that provide engineering and management services, telecommunications companies, manufacturers of computer and office equipment, financial institutions, insurance carriers, educational institutions, and government agencies.

A large number of computer programmers are employed on a temporary or contract basis or work as independent consultants, as companies demand expertise with new programming languages or specialized areas of application. Rather than hiring programmers as permanent employees and then laying them off after a job is completed, employers can contract with temporary help agencies, consulting firms, or directly with programmers themselves. A marketing firm, for example, may only require the services of several programmers to write and debug the software necessary to get a new customer resource management system running. This practice also enables companies to bring in people with a specific set of skills—usually in one of the latest technologies—as it applies to their business needs. Bringing in an independent contractor or consultant with a certain level of experience in a new or advanced programming language, for example, enables an establishment to complete a particular job without having to retrain existing workers. Such jobs may last anywhere from several weeks to a year or longer. There were 22,000 self-employed computer programmers in 2000.

Training, Other Qualifications, and Advancement

While there are many training paths available for programmers, mainly because employers' needs are so varied, the level of education and experience employers seek has been rising, due to the growing number of qualified applicants and the specialization involved with most programming tasks. Bachelor's degrees are commonly required, although some programmers may qualify for certain jobs with 2-year degrees or certificates. Employers are primarily interested in programming knowledge, and computer programmers are able to get certified in a language such as C++ or Java. College graduates who are interested in changing careers or developing an area of expertise also may return to a 2-year community college or technical school for additional training. In the absence of a degree, substantial specialized experience or expertise may be needed. Even with a degree, employers appear to be placing more emphasis on previous experience, for all types of programmers.

About 3 out of 5 computer programmers had a bachelor's degree or higher in 2000 (Table 1). Of these, some hold a degree in computer science, mathematics, or information systems, whereas others have taken special courses in computer programming to supplement their study in fields such as accounting, inventory control, or other areas of business. As the level of education and training required by employers continues to rise, this proportion should increase in the future.

TABLE 1

Highest level of school completed or degree received, computer programmers, 2000

Level	Percent
High school graduate or equivalent or less	11.8
Some college, no degree	17.2
Associate degree	11.0
Bachelor's degree	47.4
Graduate degree	12.8

Required skills vary from job to job, but the demand for various skills generally is driven by changes in technology. Employers using computers for scientific or engineering applications usually prefer college graduates who have degrees in computer or information science, mathematics, engineering, or the physical sciences. Graduate degrees in related fields are required for some jobs. Employers who use computers for business applications prefer to hire people who have had college courses in management information systems (MIS) and business and who possess strong programming skills. Although knowledge of traditional languages still is important, increasing emphasis is placed on newer, object-oriented programming languages and tools, such as C++ and Java. Additionally, employers are seeking persons familiar with fourth and fifth generation languages that involve graphic user interface (GUI) and systems programming. Employers also prefer applicants who have general business skills and experience related to the operations of the firm. Students can improve their employment prospects by participating in a college work-study program or by undertaking an internship.

Most systems programmers hold a 4-year degree in computer science. Extensive knowledge of a variety of operating systems is essential. This includes being able to configure an operating system to work with different types of hardware and adapting the operating system to best meet the needs of a particular organization. Systems programmers also must be able to work with database systems, such as DB2, Oracle, or Sybase, for example.

When hiring programmers, employers look for people with the necessary programming skills who can think logically and pay close attention to detail. The job calls for patience, persistence, and the ability to work on exacting analytical work, especially under pressure. Ingenuity and imagination also are particularly important, when programmers design solutions and test their work for potential failures. The ability to work with abstract concepts and to do technical analysis is especially important for systems programmers, because they work with the software that controls the computer's operation. Because programmers are expected to work in teams and interact directly with users, employers want programmers who are able to communicate with nontechnical personnel.

Entry-level or junior programmers may work alone on simple assignments after some initial instruction or on a team with more experienced programmers. Either way, beginning programmers generally must work under close supervision. Because technology changes so rapidly, programmers must continuously update their training by taking courses sponsored by their employer or software vendors.

For skilled workers who keep up to date with the latest technology, the prospects for advancement are good. In large organizations, programmers may be promoted to lead programmer and be given supervisory responsibilities. Some applications programmers may move into systems programming after they gain experience and take courses in systems software. With general business experience, programmers may become programmer analysts or systems analysts or be promoted to a managerial position. Other programmers, with specialized knowledge and experience with a language or operating system, may work in research and development areas, such as multimedia or Internet technology. As employers increasingly contract out programming jobs, more opportunities should arise for experienced programmers with expertise in a specific area to work as consultants.

Technical or professional certification is a way to demonstrate a level of competency or quality. In addition to language-specific certificates that a programmer can obtain, product vendors or software firms also offer certification and may require professionals who work with their products to be certified. Voluntary certification also is available through other organizations. Professional certification may provide a job seeker a competitive advantage.

Job Outlook

Employment of programmers is expected to grow about as fast as the average for all occupations through 2010. Jobs for both systems and applications programmers should be most plentiful in data processing service firms, software houses, and computer consulting businesses. These types of establishments are part of computer and data processing services, which is projected to be the fastest growing industry in the economy over the 2000–2010 period. As organizations attempt to control costs and keep up with changing technology, they will need programmers to assist in conversions to new computer languages and systems. In addition, numerous job openings will result from the need to replace programmers who leave the labor force or transfer to other occupations such as manager or systems analyst.

Employment of programmers, however, is expected to grow much slower than that of other computer specialists. With the rapid gains in technology, sophisticated computer software now has the capability to write basic code, eliminating the need for more programmers to do this routine work. The consolidation and centralization of systems and applications, developments in packaged software, advanced programming languages and tools, and the growing ability of users to design, write, and implement more of their own programs means more of the programming functions can be transferred to other types of workers. As the level of technological innovation and sophistication increases, programmers should continue to face increasing competition from programming businesses overseas where much routine work can be contracted out at a lower cost.

Nevertheless, employers will continue to need programmers who have strong technical skills and who understand an employer's business and its programming needs. This will mean that programmers will need to keep up with changing programming languages and techniques. Given the importance of networking and the expansion of client/server environments and web-based environments, organizations will look for programmers who can support data communications and help implement electronic commerce and intranet strategies. Demand for programmers with strong object-oriented programming capabilities and technical specialization in areas such as client/server programming, multimedia technology, and graphic user interface (GUI), should arise from the expansion of intranets, extranets, and Internet applications. Programmers also will be needed to create and maintain expert systems and embed these technologies in more and more products.

As programming tasks become increasingly sophisticated and an additional level of skill and experience is demanded by employers, graduates of 2-year programs and people with less than a 2-year degree or its equivalent in work experience should face strong competition for programming jobs. Competition for entry-level positions, however, also can affect applicants with a bachelor's degree. Prospects should be best for college graduates with knowledge of, and experience working with, a variety of programming languages and tools—including C++ and other object-oriented languages like Java, as well as newer, domain-specific languages that apply to computer networking, data base management, and Internet application development. Obtaining vendor or language specific certification also can provide a competitive edge. Because demand fluctuates with employers' needs, job seekers should keep up to date with the latest skills and technologies. Individuals who want to become programmers can enhance their prospects by combining the appropriate formal training with practical work experience.

Earnings

Median annual earnings of computer programmers were $57,590 in 2000. The middle 50 percent earned between $44,850 and $74,500 a year. The lowest 10 percent earned less than $35,020; the highest 10 percent earned more than $93,210. Median annual earnings in the industries employing the largest numbers of computer programmers in 2000 were:

Personnel supply services	$65,780
Professional and commercial equipment	63,780
Computer and data processing services	61,010
Commercial banks	60,180
Management and public relations	57,120

According to the National Association of Colleges and Employers, starting salary offers for graduates with a bachelor's degree in computer programming averaged $48,602 a year in 2001.

According to Robert Half International, average annual starting salaries in 2001 ranged from $58,500 to $90,000 for applications development programmers/developers, and from $54,000 to $77,750 for software development programmers/analysts. Average starting salaries for Internet programmers/analysts ranged from $56,500 to $84,000.

Related Occupations

Other professional workers who deal with data and detail include computer software engineers; systems analysts, computer scientists, and database administrators; statisticians; mathematicians; engineers; financial analysts and personal financial advisors; accountants and auditors; actuaries; and operations research analysts.

Sources of Additional Information

State employment service offices can provide information about job openings for computer programmers. Municipal chambers of commerce are other sources of additional information on an area's largest employers.

For information about certification as a computing professional, contact:

- Institute for Certification of Computing Professionals (ICCP), 2350 East Devon Ave., Suite 115, Des Plaines, IL 60018. Internet: http://www.iccp.org

Further information about computer careers is available from:

- Association for Computing Machinery (ACM), 1515 Broadway, New York, NY 10036. Internet: http://www.acm.org
- IEEE Computer Society, Headquarters Office, 1730 Massachusetts Ave. N.W., Washington, DC 20036-1992. Internet: http://www.computer.org
- National Workforce Center for Emerging Technologies, 3000 Landerholm Circle S.E., Bellevue, WA 98007. Internet: http://www.nwcet.org

Computer Software Engineers

O*NET 15-1031.00, 15-1032.00

Significant Points

- Computer software engineers are projected to be the fastest growing occupation over the 2000–2010 period.
- Very favorable opportunities are expected for college graduates with at least a bachelor's degree in computer engineering or computer science and practical experience working with computers.
- Computer software engineers must continually strive to acquire new skills as computer technology changes rapidly.

Nature of the Work

The explosive impact of computers and information technology on our everyday lives has generated a need to design and develop new computer software systems and to incorporate new technologies in a rapidly growing range of applications. The tasks performed by workers known as computer software engineers evolve rapidly, reflecting new areas of specialization or changes in technology, as well as the preferences and practices of employers. Computer software engineers apply the principles and techniques of computer science, engineering, and mathematical analysis to the design, development, testing, and evaluation of the software and systems that enable computers to perform their many applications.

Software engineers working in applications or systems development analyze users' needs and design, create, and modify general computer applications software or systems. Software engineers can be involved in the design and development of many types of software including software for operating systems, network distribution, and compilers, which convert programs for faster processing. In programming, or coding, software engineers instruct a computer, line by line, how to perform a function. They also solve technical problems that arise. Software engineers must possess strong programming skills, but are more concerned with developing algorithms and analyzing and solving programming problems than with actually writing code.

Computer applications software engineers analyze users' needs and design, create, and modify general computer applications software or specialized utility programs. Different programming languages are used, depending on the purpose of the program. The programming languages most often used are C, C++, and Java, with Fortran and COBOL used less commonly. Some software engineers develop both packaged systems and systems software or create customized applications.

Computer systems software engineers coordinate the construction and mainten ance of a company's computer systems, and plan their future growth. Working with a company, they coordinate each department's computer needs—ordering, inventory, billing, and payroll recordkeeping, for example—and make suggestions about its technical direction. They also might set up the company's intranets, networks that link computers within the organization and ease communication.

Systems software engineers work for companies that configure, implement, and install complete computer systems. They may be members of the marketing or sales staff, where they serve as the primary technical resource for salesworkers and customers. They also may be involved in product sales and in providing their customers with continuing technical support.

Computer software engineers often work as part of a team that designs new hardware, software, and systems. A core team may comprise engineering, marketing, manufacturing, and design people who work together until the product is released.

Working Conditions

Computer software engineers normally work in well-lighted and comfortable offices or computer laboratories in which computer equipment is located. Most software engineers work at least 40 hours a week; however, due to the project-oriented nature of the work, they also may have to work evenings or weekends to meet deadlines or solve unexpected technical problems. And like other workers who sit for hours at a computer typing on a keyboard, software engineers are susceptible to eyestrain, back discomfort, and hand and wrist problems such as carpal tunnel syndrome.

Many computer software engineers interact with customers and coworkers as they strive to improve software for users. Those employed by software vendors and consulting firms, for example, spend much of their time away from their offices, frequently traveling overnight, to meet with customers. They call on customers in businesses ranging from manufacturing plants to financial institutions.

As networks expand, software engineers may be able to use modems, laptops, e-mail, and the Internet to provide more technical support and other services from their main office, connecting to a customer's computer remotely to identify and correct developing problems.

Employment

Computer software engineers held about 697,000 jobs in 2000. About 380,000 were computer software engineers, applications, and about 317,000 were computer software engineers, systems software. Although they are employed in most industries, the largest concentration of computer software engineers, almost 46 percent, is in the computer and data processing services industry. This industry includes firms that develop and produce prepackaged software and firms that provide contractual computer services such as computer programming, systems integration, and information retrieval, including online databases and Internet services. Many computer software engineers also work in other industries, such as government agencies, manufacturers of computers and related electronic equipment, and colleges and universities.

Employers of computer software engineers range from startup companies to established industry leaders. The proliferation of Internet, e-mail, and other communications systems expands electronics to engineering firms traditionally associated with unrelated disciplines. Engineering firms specializing in building bridges and power plants, for example, hire computer software engineers to design and develop new geographic data systems and automated drafting capabilities. Communications firms need computer software engineers to tap into growth in the personal communications market. Major communications companies have many job openings for both computer software applications and systems engineers.

A increasing number of computer software engineers are employed on a temporary or contract basis—many of whom are self-employed, working independently as consultants. Some consultants work for firms that specialize in developing and maintaining client companies' Web sites and intranets. Consulting opportunities for software engineers should grow as businesses need help managing, upgrading, and customizing increasingly complex computer systems. About 49,000 computer software engineers were self-employed in 2000.

Training, Other Qualifications, and Advancement

Most employers prefer to hire persons who have at least a bachelor's degree and broad knowledge and experience with computer systems and technologies. Usual degree concentrations for applications software engineers are computer science or software engineering; for systems software engineers, usual concentrations are computer science or computer information systems. Graduate degrees are preferred for some of the more complex jobs.

Academic programs in software engineering emphasize software and may be offered as a degree option or in conjunction with computer science degrees. Students seeking software engineering jobs enhance their employment opportunities by participating in internship or co-op programs offered through their schools. These experiences provide students with broad knowledge and experience, making them more attractive candidates to employers. Inexperienced college graduates may be hired by large computer and consulting firms that train new hires in intensive, company-based programs. In many firms, mentoring has become part of the evaluation process for new employees.

For systems software engineering jobs that require workers who have a college degree, a bachelor's in computer science or computer information systems is typical. For systems engineering jobs that place less emphasis on workers having a computer-related degree, computer training programs are offered by systems software vendors, including Microsoft, Novell, and Oracle. These training programs usually last from 1 to 4 weeks but are not required in order to sit for a certification exam; several study guides also are available to help prepare for the exams. However, many training authorities feel that program certification alone is not sufficient for most software engineering jobs.

Professional certification is offered by the Institute for Certification of Computing Professionals. This voluntary certification is available to those who have a college degree and at least 2 years of experience. Candidates must pass an examination covering general knowledge and two specialty areas or one specialty area and two computer programming languages. In addition, the Institute of Electrical and Electronics Engineers Computer Society recently announced plans to certify software engineers who pass an examination.

Persons interested in jobs as computer software engineers must have strong problem-solving and analytical skills. They also must be able to communicate effectively with team members, other staff, and the customers they meet. And because they often deal with a number of tasks simultaneously, they must be able to concentrate and pay close attention to detail.

As is the case with most occupations, advancement opportunities for computer software engineers increase with experience. Entry-level computer software engineers are likely to test and verify ongoing designs. As they become more experienced, computer software engineers may be involved in designing and developing software. They eventually may advance to become a project manager, manager of information systems, or chief information officer. Some computer software engineers with several years of experience or expertise find lucrative opportunities working as systems designers or independent consultants or starting their own computer consulting firms.

As technological advances in the computer field continue, employers demand new skills. Computer software engineers must continually strive to acquire new skills if they wish to remain in this extremely dynamic field. To help them keep up with the changing technology, continuing education and professional development seminars are offered by employers and software vendors, colleges and universities, private training institutions, and professional computing societies.

Job Outlook

Computer software engineers are projected to be the fastest growing occupation from 2000 to 2010. Very rapid employment growth in the computer and data processing services industry, which employs the greatest numbers of computer software engineers, should result in very favorable opportunities for those college graduates with at least a bachelor's degree in computer engineering or computer science and practical experience working with computers. Employers will continue to seek computer professionals with strong programming, systems analysis, interpersonal, and business skills.

Employment of computer software engineers is expected to increase much faster than the average for all occupations as businesses and other organizations continue to adopt and integrate new technologies and seek to maximize the efficiency of their computer systems. Competition among businesses will continue to create an incentive for increasingly sophisticated technological innovations, and organizations will need more computer software engineers to implement these new technological changes. In addition to employment growth, many job openings will result annually from the need to replace workers who move into managerial positions, transfer to other occupations, or who leave the labor force.

Demand for computer software engineers will increase as computer networking continues to grow. For example, the expanding integration of Internet technologies and the explosive growth in electronic commerce—doing business on the Internet—have resulted in rising demand for computer software engineers who can develop Internet, intranet, and other Web applications. Likewise, expanding electronic data processing systems in business, telecommunications, government, and other settings continue to become more sophisticated and complex. Growing numbers of systems software engineers will be needed to implement, safeguard, and update systems and resolve problems. Consulting opportunities for computer software engineers also should continue to grow as businesses increasingly need help to manage, upgrade, and customize their increasingly complex computer systems.

Earnings

Median annual earnings of computer software engineers, applications, who worked full time in 2000 were about $67,670. The middle 50 percent earned between $53,390 and $85,490. The lowest 10 percent earned less than $42,710, and the highest 10 percent earned more than $106,680. Median annual earnings in the industries employing the largest numbers of computer applications software engineers in 2000 were:

Computer and office equipment	$74,300
Computer and data processing services	69,520
Engineering and architectural services	68,790
Professional and commercial equipment	64,920
Management and public relations	62,660

Median annual earnings of computer software engineers, systems software, who worked full time in 2000 were about $69,530. The middle 50 percent earned between $54,460 and $86,520. The lowest 10 percent earned less than $43,600, and the highest 10 percent earned more than $105,240. Median annual earnings in the industries employing the largest numbers of computer systems software engineers in 2000 were:

Computer and office equipment	$74,600
Computer and data processing services	70,150
Telephone communication	68,930
Engineering and architectural services	68,030
Commercial banks	65,620

According to the National Association of Colleges and Employers, starting salary offers for graduates with a bachelor's degree in computer engineering averaged $53,924 in 2001, and those with a master's degree averaged $58,026.

According to Robert Half International, starting salaries for software engineers in software development ranged from $62,800 to $92,000 in 2001.

In addition to typical benefits, computer software engineers may be provided with profit sharing, stock options, and a company car with a mileage allowance.

Related Occupations

Other workers who extensively use mathematics and logic include systems analysts, computer scientists, and database administrators; computer programmers; financial analysts and personal financial advisors; computer hardware engineers; statisticians; mathematicians; management analysts; actuaries; and operations research analysts.

Sources of Additional Information

Additional information on a career in computer software engineering is available from:

- Association for Computing Machinery (ACM), 1515 Broadway, New York, NY 10036. Internet: http://www.acm.org
- IEEE Computer Society, Headquarters Office, 1730 Massachusetts Ave. N.W., Washington, DC 20036-1992. Internet: http://www.computer.org
- National Workforce Center for Emerging Technologies, 3000 Landerholm Circle S.E., Bellevue, WA 98007. Internet: http://www.nwcet.org

Further information about the Certified Computing Professional designation is available from:

- Institute for Certification of Computing Professionals (ICCP), 2350 East Devon Ave., Suite 115, Des Plaines, IL 60018. Internet: http://www.iccp.org

Computer Support Specialists and Systems Administrators

O*NET 15-1041.00, 15-1071.00

Significant Points

- Computer support specialists and systems administrators are projected to be among the fastest growing occupations over the 2000–2010 period.
- Job prospects should best for college graduates who are up to date with the latest skills and technologies; certifications and practical experience are essential for persons without degrees.

Nature of the Work

In the last decade, computers have become an integral part of everyday life, used for a variety of reasons at home, in the workplace, and at schools. And almost every computer user encounters a problem occasionally, whether it is the disaster of a crashing hard drive or the annoyance of a forgotten password. The explosion of computer use has created a high demand for specialists to provide advice to users, as well as day-to-day administration, maintenance, and support of computer systems and networks.

Computer support specialists provide technical assistance, support, and advice to customers and other users. This group includes *technical support specialists* and *help-desk technicians*. These troubleshooters interpret problems and provide technical support for hardware, software, and systems. They answer phone calls, analyze problems using automated diagnostic programs, and resolve recurrent difficulties. Support specialists may work either within a company that uses computer systems or directly for a computer hardware or software vendor. Increasingly, these specialists work for help-desk or support services firms, where they provide computer support on a contract basis to clients.

Technical support specialists are troubleshooters, providing valuable assistance to their organization's computer users. Because many nontechnical employees are not computer experts, they often run into computer problems they cannot resolve on their own. Technical support specialists install, modify, clean, and repair computer hardware and software. They also may work on monitors, keyboards, printers, and mice.

Technical support specialists answer phone calls from their organizations' computer users and may run automatic diagnostics programs to resolve problems. They also may write training manuals and train computer users how to properly use the new computer hardware and software. In addition, technical support specialists oversee the daily performance of their company's computer systems and evaluate software programs for usefulness.

Help-desk technicians assist computer users with the inevitable hardware and software questions not addressed in a product's instruction manual. Help-desk technicians field telephone calls and e-mail messages from customers seeking guidance on technical problems. In responding to these requests for guidance, help-desk technicians must listen carefully to the customer, ask questions to diagnose the nature of the problem, and then patiently walk the customer through the problem-solving steps.

Help-desk technicians deal directly with customer issues, and companies value them as a source of feedback on their products. These technicians are consulted for information about what gives customers the most trouble as well as their concerns. Most computer support specialists start out at the help desk.

Network or *computer systems administrators* design, install, and support an organization's LAN, WAN, network segment, Internet, or Intranet system. They provide day-to-day onsite administrative support for software users in a variety of work environments, including professional offices, small businesses, government, and large corporations. They maintain network hardware and software, analyze problems, and monitor the network to ensure availability to system users. These workers gather data to identify customer needs and then use that information to identify, interpret, and evaluate system and network requirements. Administrators also may plan, coordinate, and implement network security measures.

Systems administrators are the information technology employees responsible for the efficient use of networks by organizations. They ensure that the design of an organization's computer site allows all the components, including computers, the network, and software, to fit together and work properly. Furthermore, they monitor and adjust performance of existing networks and continually survey the current computer site to determine future network needs. Administrators also troubleshoot problems as reported by users and automated network monitoring systems and make recommendations for enhancements in the construction of future servers and networks.

In some organizations, *computer security specialists* may plan, coordinate, and implement the organization's information security. These and other growing specialty occupations reflect the increasing emphasis on client-server applications, the expansion of Internet and Intranet applications, and the demand for more end-user support.

Working Conditions

Computer support specialists and systems administrators normally work in well lit, comfortable offices or computer laboratories. They usually work about 40 hours a week, but that may include evening or weekend work if the employer requires computer support over extended hours. Overtime may be necessary when unexpected technical problems arise. Like other workers who type on a keyboard for long periods, computer support specialists and systems administrators are susceptible to eyestrain, back discomfort, and hand and wrist problems such as carpal tunnel syndrome.

Due to the heavy emphasis on helping all types of computer users, computer support specialists and systems administrators constantly interact with customers and fellow employees as they answer questions and give valuable advice. Those who work as consultants are away from their offices much of the time, sometimes spending months working in a client's office.

As computer networks expand, more computer support specialists and systems administrators may be able to connect to a customer's computer remotely using modems, laptops, e-mail, and the Internet to provide technical support to computer users. This capability would reduce or eliminate travel to the customer's workplace. Systems administrators also can administer and configure networks and servers remotely, although it not as common as with computer support specialists.

Employment

Computer support specialists and systems administrators held about 734,000 jobs in 2000. Of these, about 506,000 were computer support specialists and about 229,000 were network and computer systems administrators. Although they worked in a wide range of industries, about one-third of all computer support specialists and systems administrators were employed in business services industries, principally computer and data processing services. Other industries that employed substantial numbers of these workers include banks, government agencies, insurance companies, educational institutions, and wholesale and retail vendors of computers, office equipment, appliances, and home electronic equipment. Many computer support specialists also worked for manufacturers of computers and other office equipment and for firms making electronic components and other accessories.

Employers of computer support specialists and systems administrators range from start-up companies to established industry leaders. With the continued development of the Internet, telecommunications, and e-mail, industries not typically associated with computers—such as construction—increasingly need computer-related workers. Small and large firms across all industries are expanding or developing computer systems, creating an immediate need for computer support specialists and systems administrators.

Training, Other Qualifications, and Advancement

Due to the wide range of skills required, there are a multitude of ways workers can become a computer support specialist or a systems administrator. While there is no universally accepted way to prepare for a job as a computer support specialist, many employers prefer to hire persons with some formal college education. A bachelor's degree in computer science or information systems is a prerequisite for some jobs; however, other jobs may require only a computer-related associate degree. For systems administrators, many employers seek applicants with bachelor's degrees, although not necessarily in a computer-related field.

Many companies are becoming more flexible about requiring a college degree for support positions because of the explosive demand for specialists. However, certification and practical experience demonstrating these skills will be essential for applicants without a degree. Completion of a certification training program, offered by a variety of vendors and product makers, may help some people to qualify for entry-level positions. Relevant computer experience may substitute for formal education.

Beginning computer support specialist start out at an organization dealing directly with customers or in-house users. Then, they may advance into more responsible positions in which they use what they learn from customers to improve the design and efficiency of future products. Job promotions usually depend more on performance than on formal education. Eventually, some computer support specialists become programmers, designing products rather than assisting users. Computer support specialists at hardware and software companies often enjoy great upward mobility; advancement sometimes comes within months of initial employment.

Entry-level network and computer systems administrators are involved in routine maintenance and monitoring of computer systems, typically working behind the scenes in an organization. After gaining experience and expertise, they often are able to advance into more senior-level positions in which they take on more responsibilities. For example, senior network and computer systems administrators may present recommendations to management on matters related to a company's network. They also may translate the needs of an organization into a set of technical requirements, based on the available technology. As with support specialists, administrators may become software engineers, actually involved in the designing of the system or network, not just the day-to-day administration.

Persons interested in becoming a computer support specialist or systems administrator must have strong problem-solving, analytical, and communication skills because troubleshooting and helping others are a vital part of the job. The constant interaction with other computer personnel, customers, and employees require computer support specialists and systems administrators to communicate effectively on paper, via e-mail, or in person. Strong writing skills are useful when preparing manuals for employees and customers.

As technology continues to improve, computer support specialists and systems administrators must keep their skills current and acquire new ones. Many employers, hardware and software vendors, colleges and universities, and private training institutions offer continuing education programs. Professional development seminars offered by computing services firms also can enhance one's skills.

Job Outlook

Computer support specialists and systems administrators are projected to be among the fastest growing occupations over the 2000–2010 period. Employment is expected to increase much faster than the average for all occupations as organizations continue to adopt and integrate increasingly sophisticated technology. Job growth will continue to be driven by rapid gains in computer and data processing services, which is projected to be the fastest growing industry in the U.S. economy.

The falling prices of computer hardware and software should help businesses expand their computing applications and integrate new technology into their operations. To maintain a competitive edge and operate more efficiently, firms will continue to demand computer specialists who are knowledgeable about the latest technologies and able to apply them to meet the needs of the organization.

Demand for computer support specialists is expected to increase because of the rapid pace of improved technology. As computers and software become more complex, support specialists will be needed to provide technical assistance to customers and other users. Consulting opportunities for computer support specialists also should continue to grow as businesses increasingly need help managing, upgrading, and customizing more complex computer systems.

Demand for systems administrators will grow as a result of the upsurge in electronic commerce and as computer applications continue to expand. Companies are looking for workers knowledgeable in the function and administration of networks. Such employees have become increasingly hard to find as systems administration has moved from being a separate function within corporations to one which forms a crucial element of business in an increasingly high-technology economy.

The growth of electronic commerce means more establishments use the Internet to conduct their business online. This translates into a need for information technology specialists who can help organizations use technology to communicate with employees, clients, and consumers. Explosive growth in these areas also is expected to fuel demand for specialists knowledgeable about network, data, and communications security.

Job prospects should be best for college graduates who are up to date with the latest skills and technologies, particularly if they have supplemented their formal education with some relevant work experience. Employers will continue to seek computer specialists who possess a strong background in fundamental computer skills combined with good interpersonal and communication skills. Due to the rapid growth in demand for computer support specialists and systems administrators, those who have strong computer skills but do not have a bachelor's degree should continue to qualify for some entry-level positions. However, certifications and practical experience are essential for persons without degrees.

Earnings

Median annual earnings of computer support specialists were $36,460 in 2000. The middle 50 percent earned between $27,680 and $48,440. The lowest 10 percent earned less than $21,260, and the highest 10 percent earned more than $63,480. Median annual earnings in the industries employing the largest numbers of computer support specialists in 2000 were:

Professional and commercial equipment	$42,970
Computer and data processing services	37,860
Personnel supply services	34,080
Colleges and universities	32,830
Miscellaneous business services	21,070

Median annual earnings of network and computer systems administrators were $51,280 in 2000. The middle 50 percent earned between $40,450 and $65,140. The lowest 10 percent earned less than $32,450, and the highest 10 percent earned more than $81,150. Median annual earnings in the industries employing the largest number of network and computer systems administrators in 2000 were:

Computer and data processing services	$54,400
Telephone communication	52,620
Management and public relations	51,340
Elementary and secondary schools	45,450
Colleges and universities	44,010

According to Robert Half International, starting salaries in 2001 ranged from $30,500 to $56,000 for help-desk support staff, and from $48,000 to $61,000 for more senior computer support specialists. For systems administrators, starting salaries in 2001 ranged from $50,250 to $70,750.

Related Occupations

Other computer-related occupations include computer programmers; computer software engineers; systems analysts, computer scientists, and database administrators; and operations research analysts.

Sources of Additional Information

For additional information about a career as a computer support specialist, contact:

- Association of Computer Support Specialists, 218 Huntington Rd., Bridgeport, CT 06608. Internet: http://www.acss.org
- Association of Support Professionals, 66 Mt. Auburn St., Watertown, MA 02472

For additional information about a career as a systems administrator, contact:

- System Administrators Guild, 2560 9th St., Suite 215, Berkeley, CA 94710. Internet: http://www.sage.org

Further information about computer careers is available from:

- National Workforce Center for Emerging Technologies, 3000 Landerholm Circle S.E., Bellevue, WA 98007. Internet: http://www.nwcet.org

Conservation Scientists and Foresters

O*NET 19-1031.01, 19-1031.02, 19-1031.03, 19-1032.00

Significant Points

- Nearly 3 out of 4 work for federal, state, or local governments.
- A bachelor's degree in forestry, range management, or a related field is the minimum educational requirement.

- Projected average employment growth will stem from continuing emphasis on environmental protection and responsible land management.

Nature of the Work

Forests and range lands supply wood products, livestock forage, minerals, and water; serve as sites for recreational activities; and provide habitats for wildlife. Conservation scientists and foresters manage, develop, use, and help to protect these and other natural resources.

Foresters manage forested lands for a variety of purposes. Those working in private industry may manage company forestland or procure timber from private landowners. Company forests are usually managed to produce a sustainable supply or wood for company mills. Procurement foresters contact local forest owners and gain permission to take inventory of the type, amount, and location of all standing timber on the property, a process known as timber cruising. Foresters then appraise the timber's worth, negotiate its purchase, and draw up a contract for procurement. Next, they subcontract with loggers or pulpwood cutters for tree removal, aid in road layout, and maintain close contact with the subcontractor's workers and the landowner to ensure that the work meets the landowner's requirements, as well as federal, state, and local environmental specifications. Forestry consultants often act as agents for the forest owner, performing these duties and negotiating timber sales with industrial procurement foresters.

Throughout the forest management and procurement processes, foresters consider the economics as well as the environmental impact on natural resources. To do this, they determine how best to conserve wildlife habitats, creek beds, water quality, and soil stability and how best to comply with environmental regulations. Foresters must balance the desire to conserve forested ecosystems for future generations with the need to use forest resources for recreational or economic purposes.

Through a process called regeneration, foresters also supervise the planting and growing of new trees. They choose and prepare the site, using controlled burning, bulldozers, or herbicides to clear weeds, brush, and logging debris. They advise on the type, number, and placement of trees to be planted. Foresters then monitor the seedlings to ensure healthy growth and to determine the best time for harvesting. If they detect signs of disease or harmful insects, they consult with forest pest management specialists to decide on the best course of treatment. Foresters who work for federal and state governments manage public forests and parks and work with private landowners to protect and manage forest land outside of the public domain. They may also design campgrounds and recreation areas.

Foresters use a number of tools to perform their jobs. Clinometers measure the height, diameter tapes measure the diameter, and increment borers and bark gauges measure the growth of trees so that timber volumes can be computed and growth rates estimated. Photogrammetry and remote sensing (aerial photographs and other imagery taken from airplanes and satellites) often are used for mapping large forest areas and for detecting widespread trends of forest and land use. Computers are used extensively, both in the office and in the field, for the storage, retrieval, and analysis of information required to manage the forest land and its resources.

Range managers, also called range *conservationists*, range *ecologists*, or range *scientists*, study, manage, improve, and protect range lands to maximize their use without damaging the environment. Range lands cover about 1 billion acres of the United States, mostly in western states and Alaska. They contain many natural resources, including grass and shrubs for animal grazing, wildlife habitats, water from vast watersheds, recreation facilities, and valuable mineral and energy resources. Range managers help ranchers attain optimum livestock production by determining the number and kind of animals to graze, the grazing system to use, and the best season for grazing. At the same time, however, they maintain soil stability and vegetation for other uses such as wildlife habitats and outdoor recreation. They also plan and implement revegetation of disturbed sites.

Soil conservationists provide technical assistance to farmers, ranchers, forest managers, state and local agencies, and others concerned with the conservation of soil, water, and related natural resources. They develop programs designed to make the most productive use of land without damaging it. Soil conservationists visit areas with erosion problems, find the source of the problem, and help landowners and managers develop management practices to combat it.

Foresters and conservation scientists often specialize in one area, such as forest resource management, urban forestry, wood technology, or forest economics.

Working Conditions

Working conditions vary considerably. Although some of the work is solitary, foresters and conservation scientists also deal regularly with landowners, loggers, forestry technicians and aides, farmers, ranchers, government officials, special interest groups, and the public in general. Some foresters and conservation scientists work regular hours in offices or labs. Others may split their time between field work and office work, while independent consultants and especially new, less experienced workers spend the majority of their time outdoors overseeing or participating in hands-on work.

The work can be physically demanding. Some foresters and conservation scientists work outdoors in all types of weather, sometimes in isolated areas. Other foresters may need to walk long distances through densely wooded land to carry out their work. Foresters also may work long hours fighting fires. Conservation scientists often are called in to prevent erosion after a forest fire, and they provide emergency help after floods, mudslides, and tropical storms.

Employment

Conservation scientists and foresters held about 29,000 jobs in 2000. The federal government employed nearly 4 out of 10 workers, many in the U.S. Department of Agriculture (USDA). Foresters were concentrated in the USDA's Forest Service; soil conservationists in the USDA's Natural Resource Conservation Service. Most range managers worked in the Department of the Interior's Bureau of Land Management, the USDA's Natural Resource Conservation Service or Forest Service. About 1 out of 4 conservation scientists and foresters worked for state governments, and nearly 1 out of 10 worked for local governments. The remainder worked in private industry, mainly in research and testing services, the forestry industry, and logging and lumber companies and sawmills. Some were self-employed as consultants for private landowners, federal and state governments, and forestry-related businesses.

Although conservation scientists and foresters work in every state, employment of foresters is concentrated in the western and southeastern states, where many national and private forests and parks, and most of the lumber and pulpwood-producing forests, are located. Range managers work almost entirely in the western states, where most of the rangeland is located. Soil conservationists, on the other hand, are employed in almost every county in the country.

Training, Other Qualifications, and Advancement

A bachelor's degree in forestry is the minimum educational requirement for professional careers in forestry. In the federal government, a combination of experience and appropriate education occasionally may substitute for a 4-year forestry degree, but job competition makes this difficult.

Sixteen states have mandatory licensing or voluntary registration requirements that a forester must meet in order to acquire the title "professional forester" and practice forestry in the state. Licensing or registration requirements vary by state, but usually entail completing a 4-year degree in forestry and a minimum period of training, and passing an exam.

Foresters who wish to perform specialized research or teach should have an advanced degree, preferably a PhD.

Most land-grant colleges and universities offer bachelor's or higher degrees in forestry; about 100 of these programs are accredited by the Society of American Foresters. Curriculums stress science, mathematics, communications skills, and computer science, as well as technical forestry subjects. Courses in forest economics and business administration supplement the student's scientific and technical knowledge. Forestry curriculums increasingly include courses on best management practices, wetland analysis, water and soil quality, and wildlife conservation, in response to the growing focus on protecting forested lands during timber harvesting operations. Prospective foresters should have a strong grasp of policy issues and of increasingly numerous and complex environmental regulations, which affect many forestry-related activities. Many colleges require students to complete a field session either in a camp operated by the college or in a cooperative work-study program with a federal or state agency or private industry. All schools encourage students to take summer jobs that provide experience in forestry or conservation work.

A bachelor's degree in range management or range science is the usual minimum educational requirement for range managers; graduate degrees usually are required for teaching and research positions. In 2000, about 40 colleges and universities offered degrees in range management or range science or in a closely related discipline with a range management or range science option. A number of other schools offered some courses in range management or range science. Specialized range management courses combine plant, animal, and soil sciences with principles of ecology and resource management. Desirable electives include economics, forestry, hydrology, agronomy, wildlife, animal husbandry, computer science, and recreation.

Very few colleges and universities offer degrees in soil conservation. Most soil conservationists have degrees in environmental studies, agronomy, general agriculture, hydrology, or crop or soil science; a few have degrees in related fields such as wildlife biology, forestry, and range management. Programs of study usually include 30 semester hours in natural resources or agriculture, including at least 3 hours in soil science.

In addition to meeting the demands of forestry and conservation research and analysis, foresters and conservation scientists generally must enjoy working outdoors, be physically hardy, and be willing to move to where the jobs are. They also must work well with people and have good communications skills.

Recent forestry and range management graduates usually work under the supervision of experienced foresters or range managers. After gaining experience, they may advance to more responsible positions. In the federal government, most entry-level foresters work in forest resource management. An experienced federal forester may supervise a ranger district, and may advance to forest supervisor, to regional forester, or to a top administrative position in the national headquarters. In private industry, foresters start by learning the practical and administrative aspects of the business and acquiring comprehensive technical training. They are then introduced to contract writing, timber harvesting, and decision-making. Some foresters work their way up to top managerial positions within their companies. Foresters in management usually leave the fieldwork behind, spending more of their time in an office, working with teams to develop management plans and supervising others. After gaining several years of experience, some foresters may become consulting foresters, working alone or with one or several partners. They contract with state or local governments, private landowners, private industry, or other forestry consulting groups.

Soil conservationists usually begin working within one county or conservation district and, with experience, may advance to the area, state, regional, or national level. Also, soil conservationists can transfer to related occupations such as farm or ranch management advisor or land appraiser.

Job Outlook

Employment of conservation scientists and foresters is expected to grow more slowly than the average for all occupations through 2010. Growth should be strongest in state and local governments and in research and testing services, where demand will be spurred by a continuing emphasis on environmental protection and responsible land management. Job opportunities are expected to be best for soil conservationists and other conservation scientists as government regulations, such as those regarding the management of stormwater and coastlines, have created demand for persons knowledgeable about runoff and erosion on farms and in cities and suburbs. Soil and water quality experts also will be needed as states attempt to improve water quality by preventing pollution by agricultural producers and industrial plants.

Fewer opportunities for conservation scientists and foresters are expected in the federal government, partly due to budgetary constraints. Also, federal land management agencies, such as the Forest Service, have de-emphasized their timber programs and increasingly focused on wildlife, recreation, and sustaining ecosystems, thereby increasing demand for other life and social scientists relative to foresters. However, a large number of foresters are expected to retire or leave the government for other reasons, resulting in many job openings between 2000 and 2010. In addition, a small number of new jobs will result from the need for range and soil conservationists to provide technical assistance to owners of grazing land through the Natural Resource Conservation Service.

Reductions in timber harvesting on public lands, most of which are located in the northwest and California, also will dampen job growth for private industry foresters in these regions. Opportunities will be better for foresters in the southeast, where much forested land is privately owned. Rising demand for timber on private lands will increase the need for forest management plans that maximize production while sustaining the ecosystem for future growth. Salaried foresters working for private industry—such as paper companies, sawmills, and pulp wood mills—and consulting foresters will be needed to provide technical assistance and management plans to landowners.

Research and testing firms have increased their hiring of conservation scientists and foresters in recent years. The firms need professionals to prepare environmental impact statements and erosion and sediment control plans, monitor water quality near logging sites, and advise on tree harvesting practices required by federal, state, or local regulations. Hiring in these firms should continue during the 2000–2010 period, although at a slower rate than over the last 10 years.

Earnings

Median annual earnings of conservation scientists in 2000 were $47,140. The middle 50 percent earned between $37,610 and $56,040. The lowest 10 percent earned less than $30,240, and the highest 10 percent earned more than $68,300.

Median annual earnings of foresters in 2000 were $43,640. The middle 50 percent earned between $34,760 and $53,740. The lowest 10 percent earned less than $27,330, and the highest 10 percent earned more than $65,960.

In 2001, most bachelor's degree graduates entering the federal government as foresters, range managers, or soil conservationists started at $23,776 or $30,035, depending on academic achievement. Those with a master's degree could start at $30,035 or $42,783. Holders of doctorates could start at $52,162 or, in research positions, at $61,451. Beginning salaries were slightly higher in selected areas where the prevailing local pay level was higher. In 2001, the average federal salary for foresters in nonsupervisory, supervisory, and managerial positions was $55,006; for soil conservationists, $53,591; for rangeland managers, $50,715, and for forest products technologists, $71,572.

According to the National Association of Colleges and Employers, graduates with a bachelor's degree in conservation and renewable natural resources received an average starting salary offer of $28,571 in 2001.

In private industry, starting salaries for students with a bachelor's degree were comparable with starting salaries in the federal government, but starting salaries in state and local governments were usually lower.

Conservation scientists and foresters who work for federal, state, and local governments and large private firms generally receive more generous benefits than do those working for smaller firms.

Related Occupations

Conservation scientists and foresters manage, develop, and protect natural resources. Other workers with similar responsibilities include agricultural engineers; environmental engineers; agricultural and food scientists; biological scientists; environmental scientists and geoscientists; and farmers, ranchers, and agricultural managers.

Sources of Additional Information

For information about the forestry profession and lists of schools offering education in forestry, send a self-addressed, stamped business envelope to:

- Society of American Foresters, 5400 Grosvenor Lane, Bethesda, MD 20814. Internet: http://www.safnet.org

For information about career opportunities in forestry in the federal government, contact:

- Chief, U.S. Forest Service, U.S. Department of Agriculture, P.O. Box 96090, Washington, DC 20090-6090. Internet: http://www.fs.fed.us

Information about a career as a range manager, as well as a list of schools offering training, is available from:

- Society for Range Management, 445 Union Blvd., Suite 230, Lakewood, CO 80228-1259. Internet: http://srm.org

Court Reporters

O*NET 23-2091.00

Significant Points

- Court reporters usually need a 2- or 4-year postsecondary school degree.
- Demand for realtime and broadcast captioning and translating will result in employment growth of court reporters.
- Job opportunities should be best for those with certification from the National Court Reporters Association.

Nature of the Work

Court reporters typically take verbatim reports of speeches, conversations, legal proceedings, meetings, and other events when written accounts of spoken words are necessary for correspondence, records, or legal proof. Court reporters not only play a critical role in judicial proceedings, but every meeting where the spoken word must be preserved as a written transcript. They are responsible for ensuring a complete, accurate, and secure legal record. In addition to preparing and protecting the legal record, many court reporters assist judges and trial attorneys in a variety of ways, such as organizing and searching for information in the official record or making suggestions to judges and attorneys regarding courtroom administration and procedure. Increasingly, court reporters are providing closed-captioning and realtime translating services to the deaf and hard-of-hearing community.

Court reporters document all statements made in official proceedings using a stenotype machine, which allows them to press multiple keys at a time to record combinations of letters representing sounds, words, or phrases. These symbols are then recorded on computer disks or CD-ROM, which are then translated and displayed as text in a process called computer-aided transcription (CAT). In all cases, accuracy is crucial because there is only one person creating an official transcript. In a judicial setting, for example, appeals often depend on the court reporter's transcript.

Stenotype machines used for realtime captioning are linked directly to the computer. As the reporter keys in the symbols, they instantly appear as text on the screen. This process, called Communications Access Realtime Translation (CART), is used in courts, classrooms, meetings, and for closed captioning for the hearing-impaired on television.

Court reporters are responsible for a number of duties both before and after transcribing events. First, they must create and maintain the computer dictionary that they use to translate stenographic strokes into written text. They may customize the dictionary with word parts, words, or terminology specific to the proceeding, program, or event—such as a religious service—they plan to transcribe. After documenting proceedings, court reporters must edit their CART translation for correct grammar, accurate identification of proper names and places, and to ensure the record or testimony is distinguishable. They usually prepare written transcripts, make copies, and provide transcript information to court, counsel, parties, and the public upon request. They also develop procedures for easy storage and retrieval of all stenographic notes and files in paper or digital format.

Although many court reporters record official proceedings in the courtroom, the majority of them work outside the courtroom. Freelance reporters, for example, take depositions for attorneys in offices and document proceedings of meetings, conventions, and other private activities. Others capture the proceedings in government agencies of all levels, from the U.S. Congress to state and local governing bodies. Court reporters who specialize in captioning live television programming for people with hearing loss are commonly known as *stenocaptioners*. They work for television networks or cable stations captioning news, emergency broadcasts, sporting events, and other programming. With CART and broadcast captioning, the level of understanding gained by a person with hearing loss depends entirely on the skill of the stenocaptioner. In an emergency situation, such as a tornado or hurricane, peoples' safety may depend entirely on the information provided in the form of captioning.

Medical transcriptionists, have similar duties, but with a different focus. They translate and edit recorded dictation by physicians and other healthcare providers regarding patient assessment and treatment.

Working Conditions

The majority of court reporters work in comfortable settings, such as in offices of attorneys, courtrooms, legislatures, and conventions. An increasing number of court reporters work from home-based offices as independent contractors.

Work in this occupation presents few hazards, although sitting in the same position for long periods can be tiring, and workers can suffer wrist, back, neck, or eye problems due to strain and risk repetitive motion injuries such as carpal tunnel syndrome. Also, the pressure to be accurate and fast also can be stressful.

Many official court reporters work a standard 40-hour week. Self-employed court reporters usually work flexible hours—including part-time, evenings, weekends, or on an on-call basis.

Employment

Court reporters held about 18,000 jobs in 2000. Of those who worked for a wage or salary, about 11,000 worked for state and local governments, a reflection of the large number of court reporters working in courts, legislatures, and various agencies. Most of the rest worked as independent contractors or employees of court reporting agencies. About 13 percent were self-employed.

Training, Other Qualifications, and Advancement

Court reporters usually complete a 2- or 4-year training program, offered by about 160 postsecondary vocational and technical schools and colleges. Currently, the National Court Reporters Association (NCRA) has approved about 86 programs, all of which offer courses in computer-aided transcription and real-time reporting. NCRA-approved programs require students to capture a minimum of 225 words per minute. Court reporters in the federal government must capture at least 225 words a minute.

Some states require court reporters to be notary publics or to be a Certified Court Reporter (CCR); reporters must pass a state certification test administered by a board of examiners to earn this designation. The National Court Reporters Association confers the entry-level designation, Registered Professional Reporter (RPR), upon those who pass a four-part examination and participate in mandatory continuing education programs. Although voluntary, the RPR designation is recognized as a mark of distinction in this field. A reporter may obtain additional certifications that demonstrate higher levels of competency. The NCRA also offers a designation called Certified Realtime Reporter (CRR). This designation promotes and recognizes competence in the specialized skill of converting the spoken word into the written word instantaneously. Reporters, working as stenocaptioners or CART providers, use realtime skills to produce captions for the deaf and hard-of-hearing viewers.

Court reporters must have excellent listening skills, as well as good English grammar and punctuation skills. They must also be aware of business practices and current events, especially the correct spelling of names of people, places, and events that may be mentioned in a broadcast or in court proceedings. For those who work in courtrooms, an expert knowledge of legal terminology and criminal and appellate procedure is essential. Because stenographic capturing of proceedings requires a computerized stenography machine, court reporters must be knowledgeable about computer hardware and software applications.

With experience and education, court reporters can advance to administrative and management positions, consulting, or teaching.

Job Outlook

Employment of court reporters is projected to grow about as fast as the average for all occupations through 2010. Demand for court reporter services will be spurred by the continuing need for accurate transcription of proceedings in courts and in pretrial depositions, and by the growing need to create captions of live or prerecorded television and provide other realtime translating services for the deaf and hard-of-hearing community.

Federal legislation mandates that a by 2006, all new television programming must be captioned for the deaf and hard-of-hearing. Additionally, the American with Disabilities Act gives deaf and hard-of-hearing students in colleges and universities the right to request access to realtime translation in their classes. Both of these factors are expected to increase demand for trained stenographic court reporters to provide realtime captioning and Communications Access Realtime Translation (CART) services. Although these services are transcript-free and differ from traditional court reporting, which uses computer-aided transcription to turn spoken words into permanent text, they require the same skills that court reporters learn in their training.

Despite increasing numbers of civil and criminal cases, budget constraints are expected to limit the ability of federal, state, and local courts to expand, also limiting the demand for traditional court reporting services in courtrooms and other legal venues. Also, in efforts to keep costs down, many courtrooms have installed tape recorders to maintain records of proceedings. Despite the use of audiotape and videotape technology, court reporters can quickly turn spoken words into readable, searchable, permanent text so they will continue to be needed to produce written legal transcripts and proceedings for publication.

The Internet is expected to affect how reporting services are provided as online video technology improves and more meetings, college classes, and even depositions take place on the Internet. Court reporters will be in demand online to provide instantaneous text of those meetings in a searchable, easy-to-access medium.

Job opportunities should be best for those with certification from the National court Reporters Association.

Earnings

Court reporters had median annual earnings of $39,660 in 2000. The middle 50 percent earned between $28,630 and $51,740. The lowest paid 10 percent earned less than $18,750, and the highest paid 10 percent earned over $69,060. Median annual earnings in 2000 were $37,640 for court reporters working in local government.

Compensation methods for court reporters vary, depending on the type of reporting jobs, the experience of the individual reporter, the level of certification achieved and the region. Official court reporters earn a salary and a per-page fee for transcripts. Many salaried court reporters supplement their income by doing additional freelance work. Freelance court reporters are paid per job and receive a per-page fee for transcripts. Communication access realtime translation providers are paid hourly. Stenocaptioners are paid a salary and benefits if they work as employees of a captioning company; stenocaptioners working as independent contractors are paid hourly.

According to a National Court Reporters Association survey of its members, average annual earnings for court reporters were $61,830 in 1999.

Related Occupations

A number of other workers type, record information, and process paperwork. Among these are secretaries and administrative assistants, medical transcriptionists, receptionists and information clerks, and human resources assistants, except payroll and timekeeping. Other workers who provide legal support include paralegals and legal assistants.

Sources of Additional Information

State employment service offices can provide information about job openings for court reporters. For information about careers, training, and certification in court reporting, contact:

- National Court Reporters Association, 8224 Old Courthouse Rd., Vienna, VA 22182. Internet: http://www.ncraonline.org and http://www.bestfuture.com
- United States Court Reporters Association, 1904 Marvel Lane, Liberty, MO 64068. Internet: http://www.uscra.org

Data Entry and Information Processing Workers

O*NET 43-9021.00, 43-9022.00

Significant Points

- Workers can acquire their skills through high schools, community colleges, business schools, or self-teaching aids such as books, tapes, or Internet tutorial applications.
- Overall employment is projected to decline due to the proliferation of personal computers and other technologies; however, the need to replace workers who leave this large occupation each year should produce many job openings.
- Those with expertise in appropriate computer software applications should have the best job prospects.

Nature of the Work

Organizations need to process a rapidly growing amount of information. Data entry and information processing workers help ensure this work is handled smoothly and efficiently. By typing texts, entering data into a computer, operating a variety of office machines, and performing other clerical duties, these workers help organizations keep up with the rapid changes of the "Information Age."

Word processors and *typists* usually set up and prepare reports, letters, mailing labels, and other text material. *Typists* make neat, typed copies of materials written by other clerical, professional, or managerial workers. They may begin as entry-level workers by typing headings on form letters, addressing envelopes, or preparing standard forms on typewriters or computers. As they gain experience, they often are assigned tasks requiring a higher degree of accuracy and independent judgment. Senior typists may work with highly technical material, plan and type complicated statistical tables, combine and rearrange materials from different sources, or prepare master copies.

Most keyboarding is now done on word processing equipment—usually a personal computer or part of a larger computer system—which normally includes a keyboard, video display terminal, and printer, and may have "add-on" capabilities such as optical character recognition readers. *Word processors* use this equipment to record, edit, store, and revise letters, memos, reports, statistical tables, forms, and other printed materials. Although it is becoming less common, some word processing workers are employed in centralized word processing teams that handle the transcription and typing for several departments.

In addition to the duties mentioned above, word processors and typists often perform other office tasks, such as answering telephones, filing, and operating copiers or other office machines. Job titles of these workers often vary to reflect these duties. Clerk typists, for example, combine typing with filing, sorting mail, answering telephones, and other general office work. Notereaders transcribe stenotyped notes of court proceedings into standard formats.

Data entry keyers usually input lists of items, numbers, or other data into computers or complete forms that appear on a computer screen. They may also manipulate existing data, edit current information, or proofread new entries to a database for accuracy. Some examples of data sources include customers' personal information, medical records, and membership lists. Usually this information is used internally by a company and may be reformatted before use by other departments or by customers.

Keyers use various types of equipment to enter data. Many keyers use a machine that converts the information they type to magnetic impulses on tapes or disks for entry into a computer system. Others prepare materials for printing or publication by using data entry composing machines. Some keyers operate online terminals or personal computers. Data entry keyers increasingly also work with non-keyboard forms of data entry such as scanners and electronically transmitted files. When using these new character recognition systems, data entry keyers often enter only those data that cannot be recognized by machines. In some offices, keyers also operate computer peripheral equipment such as printers and tape readers, act as tape librarians, and perform other clerical duties.

Working Conditions

Data entry and information processing workers usually work a standard 40-hour week in clean offices. They sit for long periods and sometimes must contend with high noise levels caused by various office machines. These workers are susceptible to repetitive strain injuries, such as carpal tunnel syndrome and neck, back, and eyestrain. To help prevent these from occurring, many offices have scheduled exercise breaks, ergonomically designed keyboards, and workstations that allow workers to stand or sit as they wish.

Employment

Data entry and information processing workers held about 806,000 jobs in 2000 and were employed in every sector of the economy; 509,000 were data entry keyers, and 297,000 were word processors and typists. Some workers telecommute by working from their homes on personal computers linked by telephone lines to those in the main office. This enables them to type material at home while still being able to produce printed copy in their offices.

About 1 out of 3 data entry and information processing workers held jobs in firms providing business services, including temporary help, word processing, and computer and data processing. Nearly 1 out of 5 worked in federal, state, and local government agencies.

Training, Other Qualifications, and Advancement

Employers generally hire high school graduates who meet their requirements for keyboarding speed. Increasingly, employers also expect applicants to have word processing or data entry training or experience. Spelling, punctuation, and grammar skills are important, as is familiarity with standard office equipment and procedures.

Students acquire skills in keyboarding and in the use of word processing, spreadsheet, and database management computer software packages through high schools, community colleges, business schools, temporary help agencies, or self-teaching aids such as books, tapes, or Internet tutorials applications.

For many people, a job as a data entry and information-processing worker is their first job after graduating from high school or after a period of full-time family responsibilities. This work frequently serves as a steppingstone to higher paying jobs with increased responsibilities. Large companies and government agencies usually have training programs to help administrative employees upgrade their skills and advance to other positions. It is common for data entry and information processing workers to transfer to other administrative jobs, such as secretary, administrative assistant, statistical clerk, or to be promoted to a supervisory job in a word processing or data entry center.

Job Outlook

Overall employment of data entry and information processing workers is projected to decline through 2010. Nevertheless, the need to replace those who transfer to other occupations or leave this large occupation for other reasons will produce numerous job openings each year. Job prospects will be most favorable for those with the best technical skills—in particular, expertise in appropriate computer software applications. Data entry and information processing workers must be willing to continuously upgrade their skills to remain marketable.

Although data entry and information processing workers are all affected by productivity gains stemming from organizational restructuring and the implementation of new technologies, projected growth differs among these workers. Employment of word processors and typists is expected to decline due to the proliferation of personal computers which allows other workers to perform duties formerly assigned to word processors and typists. Most professionals and managers, for example, now use desktop personal computers to do their own word processing. Because technologies affecting data entry keyers tend to be costlier to implement, however, these workers will be less affected by technology and should experience slower-than-average growth.

Employment growth of data entry keyers will still be dampened by productivity gains, as various data capturing technologies, such as bar code scanners, voice recognition technologies, and sophisticated character recognition readers, become more prevalent. These technologies can be applied to a variety of business transactions, such as inventory tracking, invoicing, and order placement. Moreover, as telecommunications technology improves, many organizations will increasingly take advantage of computer networks that allow data to be transmitted electronically, thereby avoiding the reentry of data. These technologies will allow more data to be entered automatically into computers, reducing the demand for data entry keyers.

In addition to technology, employment of data entry and information processing workers will be adversely affected as businesses increasingly contract out their work. Many organizations have reduced or even eliminated permanent in-house staff, for example, in favor of temporary-help and staffing services firms. Some large data entry and information processing firms increasingly employ workers in nations with low wages to enter data. As international trade barriers continue to fall and telecommunications technology improves, this transfer will mean reduced demand for data entry keyers in the United States.

Earnings

Median annual earnings of word processors and typists in 2000 were $24,710. The middle 50 percent earned between $20,070 and $29,500. The lowest 10 percent earned less than $16,410, while the highest 10 percent earned more than $35,410. The salaries of these workers vary by industry and by region. In 2000, median annual earnings in the industries employing the largest numbers of word processors and typists were:

Local government	$25,710
State government	24,850
Federal government	23,890
Elementary and secondary schools	23,300
Personnel supply services	22,720

Median annual earnings of data entry keyers in 2000 were $21,300. The middle 50 percent earned between $17,850 and $25,820. The lowest 10 percent earned less than $15,140, and the highest 10 percent earned more than $30,910. In 2000, median annual earnings in the industries employing the largest numbers of data entry keyers were:

Federal government	$27,260
Accounting, auditing, and bookkeeping	22,310
Computer and data processing services	20,480
Commercial banks	20,410
Personnel supply services	20,070

In the federal government, clerk-typists and data entry keyers without work experience started at $16,015 a year in 2001. Beginning salaries were slightly higher in selected areas where the prevailing local pay level was higher. The average annual salary for all clerk-typists in the federal government was $24,934 in 2001.

Related Occupations

Data entry and information processing workers must transcribe information quickly. Other workers who deliver information in a timely manner are dispatchers and communications equipment operators. Data entry and information processing workers also must be comfortable working with office automation, and in this regard they are similar to court reporters, medical records and health information technicians, secretaries and administrative assistants, and computer operators.

Sources of Additional Information

For information about job opportunities for data entry and information processing workers, contact the nearest office of the state employment service.

Dental Assistants

O*NET 31-9091.00

Significant Points

- Rapid employment growth and substantial replacement needs should result in good job opportunities.
- Dentists are expected to hire more assistants to perform routine tasks so that they may devote their own time to more profitable procedures.
- Infection control is a crucial responsibility of dental assistants. Proper infection control protects patients and members of the dental health team.

Nature of the Work

Dental assistants perform a variety of patient care, office, and laboratory duties. They work chairside as dentists examine and treat patients. They make patients as comfortable as possible in the dental chair, prepare them for treatment, and obtain dental records. Assistants hand instruments and materials to dentists, and keep patients' mouths dry and clear by using suction or other devices. Assistants also sterilize and disinfect instruments and equipment, prepare tray setups for dental procedures, and instruct patients on postoperative and general oral healthcare.

Some dental assistants prepare materials for making impressions and restorations, expose radiographs, and process dental X-ray film as directed by a dentist. They also may remove sutures, apply anesthetics to gums or cavity-preventive agents to teeth, remove excess cement used in the filling process, and place rubber dams on the teeth to isolate them for individual treatment.

Those with laboratory duties make casts of the teeth and mouth from impressions taken by dentists, clean and polish removable appliances, and make temporary crowns. Dental assistants with office duties schedule and confirm appointments, receive patients, keep treatment records, send bills, receive payments, and order dental supplies and materials.

Dental assistants should not be confused with dental hygienists, who are licensed to perform different clinical tasks.

Working Conditions

Dental assistants work in a well-lighted, clean environment. Their work area usually is near the dental chair so that they can arrange instruments, materials, and medication and hand them to the dentist when needed. Dental assistants wear gloves, masks, eyewear, and protective clothing to protect themselves and their patients from infectious diseases. Following safety procedures also minimizes the risks associated with the use of radiographic equipment.

Almost half of dental assistants have a 35- to 40-hour workweek, which may include work on Saturdays or evenings.

Employment

Dental assistants held about 247,000 jobs in 2000. Almost 2 out of 5 worked part-time, sometimes in more than one dental office.

Virtually all dental assistants work in a private dental office. A small number work in dental schools, private and government hospitals, state and local public health departments, or clinics.

Training, Other Qualifications, and Advancement

Most assistants learn their skills on the job, although some are trained in dental assisting programs offered by community and junior colleges, trade schools, technical institutes, or the armed forces. Assistants must be a dentist's "third hand"; therefore, dentists look for people who are reliable, can work well with others, and have good manual dexterity. High school students interested in a career as a dental assistant should take courses in biology, chemistry, health, and office practices.

The American Dental Association's Commission on Dental Accreditation approved 248 dental assisting training programs in 2000. Programs include classroom, laboratory, and preclinical instruction in dental assisting skills and related theory. In addition, students gain practical experience in dental schools, clinics, or dental offices. Most programs take 1 year or less to complete and lead to a certificate or diploma. Two-year programs offered in community and junior colleges lead to an associate degree. All programs require a high school diploma or its equivalent, and some require a typing or science course for admission. Some private vocational schools offer 4- to 6-month courses in dental assisting, but the Commission on Dental Accreditation does not accredit these.

Some states regulate the duties dental assistants may complete through licensure or registration. Licensure or registration may require passing a written or practical examination. States offering licensure or registration have a variety of schools offering courses—approximately 10 to 12 months in length—that meet their state's requirements. Some states require continuing education to maintain licensure or registration. A few states allow dental assistants to perform any function delegated to them by the dentist.

Individual states have adopted different standards for dental assistants who perform certain medical duties, such as radiological procedures. Completion of the Radiation Health and Safety examination offered by the Dental Assisting National Board, Inc. (DANB) meets those standards in 31 states. Some states require the completion of a state-approved course in radiology as well.

Certification is available through DANB and is recognized or required in 20 states. Other organizations offer registration, most often at the state level. Certification is an acknowledgment of an assistant's qualifications and professional competence, and may be an asset when seeking employment. Candidates may qualify to take the DANB certification examination by graduating from an accredited training program or by having 2 years of full-time, or 4 years of part-time, experience as a dental assistant. In addition, applicants must have current certification in cardiopulmonary resuscitation. Recertification is offered annually for applicants who have earned continuing education credits.

Without further education, advancement opportunities are limited. Some dental assistants become office managers, dental assisting instructors, or dental product sales representatives. Others go back to school to become dental hygienists. For many, this entry-level occupation provides basic training and experience and serves as a steppingstone to more highly skilled and higher paying jobs.

Job Outlook

Job prospects for dental assistants should be good. Employment is expected to grow much faster than the average for all occupations through the year 2010. In addition, numerous job openings will occur due to the need to replace assistants who transfer to other occupations, retire, or leave the labor force for other reasons. Many opportunities are for entry-level positions offering on-the-job training.

Population growth and greater retention of natural teeth by middle-aged and older people will fuel demand for dental services. Older dentists, who are less likely to employ assistants, will leave and be replaced by recent graduates, who are more likely to use one, or even two. In addition, as dentists' workloads increase, they are expected to hire more assistants to perform routine tasks, so that they may devote their own time to more profitable procedures.

Earnings

Median hourly earnings of dental assistants were $12.49 in 2000. The middle 50 percent earned between $9.99 and $15.51 an hour. The lowest 10 percent earned less than $8.26, and the highest 10 percent earned more than $18.57 an hour.

Benefits vary substantially by practice setting and may be contingent upon full-time employment. According to the American Dental Association's 1999 Workforce Needs Assessment Survey, almost all full-time dental assistants employed by private practitioners received paid vacation. The survey also found that 9 out of 10 full- and part-time dental assistants received dental coverage.

Related Occupations

Workers in other occupations supporting health practitioners include medical assistants, occupational therapist assistants and aides, pharmacy aides, pharmacy technicians, physical therapist assistants and aides, and veterinary technologists, technicians, and assistants.

Sources of Additional Information

Information about career opportunities and accredited dental assistant programs is available from:

- Commission on Dental Accreditation, American Dental Association, 211 E. Chicago Ave., Suite 1814, Chicago, IL 60611. Internet: http://www.ada.org

For information on becoming a Certified Dental Assistant and a list of state boards of dentistry, contact:

- Dental Assisting National Board, Inc., 676 North Saint Clair, Suite 1880, Chicago, IL 60611. Internet: http://www.danb.org

For general information about continuing education for dental assistants, contact:

- American Dental Assistants Association, 203 North LaSalle St., Suite 1320, Chicago, IL 60601. Internet: http://www.dentalassistant.org

Dental Hygienists

O*NET 29-2021.00

Significant Points

- Dental hygienists are projected to be one of the 30 fastest growing occupations.
- Population growth and greater retention of natural teeth will stimulate demand for dental hygienists.
- Opportunities for part-time work and flexible schedules are common.

Nature of the Work

Dental hygienists remove soft and hard deposits from teeth, teach patients how to practice good oral hygiene, and provide other preventive dental care. Hygienists examine patients' teeth and gums, recording the presence of diseases or abnormalities. They remove calculus, stains, and plaque from teeth; take and develop dental X rays; and apply cavity-preventive agents such as fluorides and pit and fissure sealants. In some states, hygienists administer anesthetics; place and carve filling materials, temporary fillings, and periodontal dressings; remove sutures; perform root-planing as a periodontal therapy; and smooth and polish metal restorations. Although hygienists may not diagnose diseases, they can prepare clinical and laboratory diagnostic tests for the dentist to interpret. Hygienists sometimes work chairside with the dentist during treatment.

Dental hygienists also help patients develop and maintain good oral health. For example, they may explain the relationship between diet and oral health, or even the link between oral health and such serious conditions as heart disease and stroke. They also inform patients how to select toothbrushes and show them how to brush and floss their teeth.

Dental hygienists use hand and rotary instruments and ultrasonics to clean and polish teeth, X-ray machines to take dental pictures, syringes with needles to administer local anesthetics, and models of teeth to explain oral hygiene.

Working Conditions

Flexible scheduling is a distinctive feature of this job. Full-time, part-time, evening, and weekend schedules are widely available.

Dentists frequently hire hygienists to work only 2 or 3 days a week, so hygienists may hold jobs in more than one dental office.

Dental hygienists work in clean, well-lighted offices. Important health safeguards include strict adherence to proper radiological procedures, and use of appropriate protective devices when administering anesthetic gas. Dental hygienists also wear safety glasses, surgical masks, and gloves to protect themselves from infectious diseases.

Employment

Dental hygienists held about 147,000 jobs in 2000. Because multiple jobholding is common in this field, the number of jobs exceeds the number of hygienists. More than half of all dental hygienists worked part-time—less than 35 hours a week.

Almost all dental hygienists work in private dental offices. Some work in public health agencies, hospitals, and clinics.

Training, Other Qualifications, and Advancement

Dental hygienists must be licensed by the state in which they practice. To qualify for licensure, a candidate must graduate from an accredited dental hygiene school and pass both a written and clinical examination. The American Dental Association Joint Commission on National Dental Examinations administers the written examination accepted by all states and the District of Columbia. State or regional testing agencies administer the clinical examination. In addition, most states require an examination on legal aspects of dental hygiene practice. Alabama allows candidates to take its examinations if they have been trained through a state-regulated on-the-job program in a dentist's office.

In 2000, the Commission on Dental Accreditation accredited about 256 programs in dental hygiene. Although some programs lead to a bachelor's degree, most grant an associate degree. A dozen universities offer master's degree programs in dental hygiene or a related area.

An associate degree is sufficient for practice in a private dental office. A bachelor's or master's degree usually is required for research, teaching, or clinical practice in public or school health programs.

About half of the dental hygiene programs prefer applicants who have completed at least 1 year of college. However, requirements vary from one school to another. Schools offer laboratory, clinical, and classroom instruction in subjects such as anatomy, physiology, chemistry, microbiology, pharmacology, nutrition, radiography, histology (the study of tissue structure), periodontology (the study of gum diseases), pathology, dental materials, clinical dental hygiene, and social and behavioral sciences.

Dental hygienists should work well with others and must have good manual dexterity because they use dental instruments within a patient's mouth, with little room for error. High school students interested in becoming a dental hygienist should take courses in biology, chemistry, and mathematics.

Job Outlook

Employment of dental hygienists is expected to grow much faster than the average for all occupations through 2010, in response to increasing demand for dental care and the greater substitution of the services of hygienists for those previously performed by dentists. Job prospects are expected to remain very good unless the number of dental hygienist program graduates grows much faster than during the last decade, and results in a much larger pool of qualified applicants.

Population growth and greater retention of natural teeth will stimulate demand for dental hygienists. Older dentists, who are less likely to employ dental hygienists, will leave and be replaced by recent graduates, who are more likely to do so. In addition, as dentists' workloads increase, they are expected to hire more hygienists to perform preventive dental care such as cleaning, so that they may devote their own time to more profitable procedures.

Earnings

Median hourly earnings of dental hygienists were $24.68 in 2000. The middle 50 percent earned between $20.46 and $29.72 an hour. The lowest 10 percent earned less than $15.53, and the highest 10 percent earned more than $35.39 an hour.

Earnings vary by geographic location, employment setting, and years of experience. Dental hygienists who work in private dental offices may be paid on an hourly, daily, salary, or commission basis.

Benefits vary substantially by practice setting, and may be contingent upon full-time employment. According to the American Dental Association's 1999 Workforce Needs Assessment Survey, almost all full-time dental hygienists employed by private practitioners received paid vacation. The survey also found that 9 out of 10 full- and part-time dental hygienists received dental coverage. Dental hygienists who work for school systems, public health agencies, the federal government, or state agencies usually have substantial benefits.

Related Occupations

Workers in other occupations supporting health practitioners in an office setting include dental assistants, medical assistants, occupational therapist assistants and aides, physical therapist assistants and aides, physician assistants, and registered nurses.

Sources of Additional Information

For information on a career in dental hygiene and the educational requirements to enter this occupation, contact:

- Division of Professional Development, American Dental Hygienists' Association, 444 N. Michigan Ave., Suite 3400, Chicago, IL 60611. Internet: http://www.adha.org

For information about accredited programs and educational requirements, contact:

- Commission on Dental Accreditation, American Dental Association, 211 E. Chicago Ave., Suite 1814, Chicago, IL 60611. Internet: http://www.ada.org

The State Board of Dental Examiners in each state can supply information on licensing requirements.

Dental Laboratory Technicians

O*NET 51-9081.00

Significant Points

- Employment should increase slowly, as the public's improving dental health requires fewer dentures but more bridges and crowns.
- Dental laboratory technicians need artistic aptitude for detailed and precise work, a high degree of manual dexterity, and good vision.

Nature of the Work

Dental laboratory technicians fill prescriptions from dentists for crowns, bridges, dentures, and other dental prosthetics. First, dentists send a specification of the item to be fabricated, along with an impression (mold) of the patient's mouth or teeth. Then, dental laboratory technicians, also called dental technicians, create a model of the patient's mouth by pouring plaster into the impression and allowing it to set. Next, they place the model on an apparatus that mimics the bite and movement of the patient's jaw. The model serves as the basis of the prosthetic device. Technicians examine the model, noting the size and shape of the adjacent teeth, as well as gaps within the gumline. Based upon these observations and the dentist's specifications, technicians build and shape a wax tooth or teeth model, using small hand instruments called wax spatulas and wax carvers. They use this wax model to cast the metal framework for the prosthetic device.

After the wax tooth has been formed, dental technicians pour the cast and form the metal and, using small hand-held tools, prepare the surface to allow the metal and porcelain to bond. They then apply porcelain in layers, to arrive at the precise shape and color of a tooth. Technicians place the tooth in a porcelain furnace to bake the porcelain onto the metal framework, and then adjust the shape and color, with subsequent grinding and addition of porcelain to achieve a sealed finish. The final product is nearly an exact replica of the lost tooth or teeth.

In some laboratories, technicians perform all stages of the work, whereas in other labs, each technician does only a few. Dental laboratory technicians can specialize in one of five areas: orthodontic appliances, crowns and bridges, complete dentures, partial dentures, or ceramics. Job titles can reflect specialization in these areas. For example, technicians who make porcelain and acrylic restorations are called *dental ceramists*.

Working Conditions

Dental laboratory technicians generally work in clean, well-lighted, and well-ventilated areas. Technicians usually have their own workbenches, which can be equipped with Bunsen burners, grinding and polishing equipment, and hand instruments, such as wax spatulas and wax carvers.

The work is extremely delicate and time consuming. Salaried technicians usually work 40 hours a week, but self-employed technicians frequently work longer hours.

Employment

Dental laboratory technicians held about 43,000 jobs in 2000. Most jobs were in commercial dental laboratories, which usually are small, privately owned businesses with fewer than five employees. However, some laboratories are large; a few employ more than 50 technicians.

Some dental laboratory technicians work in dentists' offices. Others work for hospitals providing dental services, including U.S. Department of Veterans Affairs' hospitals. Some technicians work in dental laboratories in their homes, in addition to their regular job.

Training, Other Qualifications, and Advancement

Most dental laboratory technicians learn their craft on the job. They begin with simple tasks, such as pouring plaster into an impression, and progress to more complex procedures, such as making porcelain crowns and bridges. Becoming a fully trained technician requires an average of 3 to 4 years, depending upon the individual's aptitude and ambition, but it may take a few years more to become an accomplished technician.

Training in dental laboratory technology also is available through community and junior colleges, vocational-technical institutes, and the armed forces. Formal training programs vary greatly both in length and in the level of skill they impart.

In 2000, 30 programs in dental laboratory technology were accredited by the Commission on Dental Accreditation in conjunction with the American Dental Association (ADA). These programs provide classroom instruction in dental materials science, oral anatomy, fabrication procedures, ethics, and related subjects. In addition, each student is given supervised practical experience in a school or an associated dental laboratory. Accredited programs normally take 2 years to complete and lead to an associate degree.

Graduates of 2-year training programs need additional hands-on experience to become fully qualified. Each dental laboratory owner operates in a different way, and classroom instruction does not necessarily expose students to techniques and procedures favored by individual laboratory owners. Students who have taken enough courses to learn the basics of the craft usually are considered good candidates for training, regardless of whether they have completed a formal program. Many employers will train someone without any classroom experience.

The National Board for Certification, an independent board established by the National Association of Dental Laboratories, offers certification in dental laboratory technology. Certification, which is voluntary, can be obtained in five specialty areas: crowns and bridges, ceramics, partial dentures, complete dentures, and orthodontic appliances.

In large dental laboratories, technicians may become supervisors or managers. Experienced technicians may teach or may take jobs with dental suppliers in such areas as product development, marketing, and sales. Still, for most technicians, opening one's own laboratory is the way toward advancement and higher earnings.

A high degree of manual dexterity, good vision, and the ability to recognize very fine color shadings and variations in shape are necessary. An artistic aptitude for detailed and precise work also is important. High school students interested in becoming dental laboratory technicians should take courses in art, metal and wood shop, drafting, and sciences. Courses in management and business may help those wishing to operate their own laboratories.

Job Outlook

Job opportunities for dental laboratory technicians should be favorable, despite very slow growth in the occupation. Employers have difficulty filling trainee positions, probably because entry-level salaries are relatively low and because the public is not familiar with the occupation.

Although job opportunities are favorable, slower-than-average growth in the employment of dental laboratory technicians is expected through the year 2010, due to changes in dental care. The overall dental health of the population has improved because of fluoridation of drinking water, which has reduced the incidence of dental cavities, and greater emphasis on preventive dental care since the early 1960s. As a result, full dentures will be less common, as most people will need only a bridge or crown. However, during the last few years, demand has arisen from an aging public that is growing increasingly interested in cosmetic prostheses. For example, many dental laboratories are filling orders for composite fillings that are the same shade of white as natural teeth to replace older, less attractive fillings.

Earnings

Median hourly earnings of dental laboratory technicians were $12.94 in 2000. The middle 50 percent earned between $9.83 and $16.82 an hour. The lowest 10 percent earned less than $7.78, and the highest 10 percent earned more than $21.47 an hour. Median hourly earnings of dental laboratory technicians in 2000 were $12.88 in offices and clinics of dentists and $12.87 in medical and dental laboratories.

Technicians in large laboratories tend to specialize in a few procedures, and, therefore, tend to be paid a lower wage than those employed in small laboratories that perform a variety of tasks.

Related Occupations

Dental laboratory technicians fabricate artificial teeth, crowns and bridges, and orthodontic appliances, following specifications and instructions provided by dentists. Other workers who make and repair medical devices include dispensing opticians, ophthalmic laboratory technicians, orthotists and prosthetists, and precision instrument and equipment repairers.

Sources of Additional Information

For a list of accredited programs in dental laboratory technology, contact:

- Commission on Dental Accreditation, American Dental Association, 211 E. Chicago Ave., Chicago, IL 60611. Internet: http://www.ada.org

For information on requirements for certification, contact:

- National Board for Certification in Dental Technology, 1530 Metropolitan Blvd., Tallahassee, FL 32308. Internet: http://www.nadl.org/html/certification.html

For information on career opportunities in commercial laboratories, contact:

- National Association of Dental Laboratories, 1530 Metropolitan Blvd., Tallahassee, FL 32308. Internet: http://www.nadl.org

General information on grants and scholarships is available from dental technology schools.

Designers

O*NET 27-1021.00, 27-1022.00, 27-1023.00, 27-1024.00, 27-1025.00, 27-1026.00, 27-1027.01, 27-1027.02

Significant Points

- Three out of 10 designers are self-employed—almost 5 times the proportion for all professional and related occupations.
- Creativity is crucial in all design occupations; most designers need a bachelor's degree, and candidates with a master's degree hold an advantage.
- Keen competition is expected for most jobs, despite projected faster-than-average employment growth, because many talented individuals are attracted to careers as designers.

Nature of the Work

Designers are people with a desire to create. They combine practical knowledge with artistic ability to turn abstract ideas into formal designs for the merchandise we buy, the clothes we wear, the publications we read, and the living and office space we inhabit. Designers usually specialize in a particular area of design, such as automobiles, industrial or medical equipment, or home appliances; clothing and textiles; floral arrangements; publications, logos, signage, or movie or TV credits; interiors of homes or office buildings; merchandise displays; or movie, television, and theater sets.

The first step in developing a new design or altering an existing one is to determine the needs of the client, the ultimate function for which the design is intended, and its appeal to customers. When creating a design, designers often begin by researching the desired design characteristics, such as size, shape, weight, color, materials used, cost, ease of use, fit, and safety.

Designers then prepare sketches, by hand or with the aid of a computer, to illustrate the vision for the design. After consulting with the client, an art or design director, or a product development team, designers create detailed designs using drawings, a structural model, computer simulations, or a full-scale prototype. Many designers increasingly are using computer-aided design (CAD) tools to create and better visualize the final product. Computer models allow greater ease and flexibility in exploring a greater number of design alternatives, thus reducing design costs and cutting the time it takes to deliver a product to market. Industrial designers use computer-aided industrial design (CAID) tools to create designs and machine-readable instructions that communicate with automated production tools.

Designers sometimes supervise assistants who carry out their creations. Designers who run their own businesses also may devote a considerable amount of time to developing new business contacts, reviewing equipment and space needs, and performing administrative tasks, such as reviewing catalogues and ordering samples. Design encompasses a number of different fields. Many designers specialize in a particular area of design, whereas others work in more than one area.

Commercial and industrial designers, including designers of commercial products and equipment, develop countless manufactured products, including airplanes; cars; children's toys; computer equipment; furniture; home appliances; and medical, office, and recreational equipment. They combine artistic talent with research on product use, customer needs, marketing, materials, and production methods to create the most functional and appealing design that will be competitive with others in the marketplace. Industrial designers typically concentrate in an area of sub-specialization such as kitchen appliances, auto interiors, or plastic-molding machinery.

Fashion designers design clothing and accessories. Some high-fashion designers are self-employed and design for individual clients. Other high-fashion designers cater to specialty stores or high-fashion department stores. These designers create original garments, as well as those that follow established fashion trends. Most fashion designers, however, work for apparel manufacturers, creating designs of men's, women's, and children's fashions for the mass market.

Floral designers cut and arrange live, dried, or artificial flowers and foliage into designs, according to the customer's order. They trim flowers and arrange bouquets, sprays, wreaths, dish gardens, and terrariums. They usually work from a written order indicating the occasion, customer preference for color and type of flower, price, the time at which the floral arrangement or plant is to be ready, and the place to which it is to be delivered. The variety of duties performed by floral designers depends on the size of the shop and the number of designers employed. In a small operation, floral designers may own their shops and do almost everything, from growing and purchasing flowers to keeping financial records.

Graphic designers use a variety of print, electronic, and film media to create designs that meet clients' commercial needs. Using computer software, they develop the overall layout and design of magazines, newspapers, journals, corporate reports, and other publications. They also may produce promotional displays and marketing brochures for products and services, design distinctive company logos for products and businesses, and develop signs and signage systems—called environmental graphics—for business and government. An increasing number of graphic designers develop material to appear on Internet home pages. Graphic designers also produce the credits that appear before and after television programs and movies.

Interior designers plan the space and furnish the interiors of private homes, public buildings, and business or institutional facilities, such as offices, restaurants, retail establishments, hospitals, hotels, and theaters. They also plan the interiors when existing structures are renovated or expanded. Most interior designers specialize. For example, some may concentrate in residential design, and others may further specialize by focusing on particular rooms, such as kitchens or baths. With a client's tastes, needs, and budget in mind, interior designers prepare drawings and specifications for non-load bearing interior construction, furnishings, lighting, and finishes. Increasingly, designers use computers to plan layouts, which can easily be changed to include ideas received from the client. Interior designers also design lighting and architectural details—such as crown molding, built-in bookshelves, or cabinets—coordinate colors, and select furniture, floor coverings, and window treatments. Interior designers must design space to conform to federal, state, and local laws, including building codes. Designs for public areas also must meet accessibility standards for the disabled and elderly.

Merchandise displayers and window dressers, or *visual merchandisers,* plan and erect commercial displays, such as those in windows and interiors of retail stores or at trade exhibitions. Those who work on building exteriors erect major store decorations, including building and window displays, and spot lighting. Those who design store interiors outfit store departments, arrange table displays, and dress mannequins. In large retail chains, store layouts typically are designed corporately, through a central design department. To retain the chain's visual identity and ensure that a particular image or theme is promoted in each store, designs are distributed to individual stores by e-mail, downloaded to computers equipped with the appropriate design software, and adapted to meet individual store size and dimension requirements.

Set and exhibit designers create sets for movie, television, and theater productions and design special exhibition displays. Set designers study scripts, confer with directors and other designers, and conduct research to determine the appropriate historical period, fashion, and architectural styles. They then produce sketches or scale models to guide in the construction of the actual sets or exhibit spaces. Exhibit designers work with curators, art and museum directors, and trade show sponsors to determine the most effective use of available space.

Working Conditions

Working conditions and places of employment vary. Designers employed by manufacturing establishments, large corporations, or design firms generally work regular hours in well-lighted and comfortable settings. Self-employed designers tend to work longer hours.

Designers who work on a contract, or job, basis frequently adjust their workday to suit their clients' schedules, meeting with them during evening or weekend hours when necessary. Designers may transact business in their own offices or studios or in clients' homes or offices, or they may travel to other locations, such as showrooms, design centers, clients' exhibit sites, and manufacturing facilities. Designers who are paid by the assignment are under pressure to please clients and to find new ones to maintain a constant income. All designers face frustration at times when their designs are rejected or when they cannot be as creative as they wish. With the increased use of computers in the workplace and the advent of Internet Web sites, more designers conduct business, research design alternatives, and purchase supplies electronically than ever before.

Occasionally, industrial designers may work additional hours to meet deadlines. Similarly, graphic designers usually work regular hours, but may work evenings or weekends to meet production schedules. In contrast, set and exhibit designers work long and irregular hours; often, they are under pressure to make rapid changes. Merchandise displayers and window trimmers who spend most of their time designing space typically work in office-type settings; however, those who also construct and install displays spend much of their time doing physical labor, such as those tasks performed by a carpenter or someone constructing and moving stage scenery. Fashion designers may work long hours to meet production deadlines or prepare for fashion shows. In addition, fashion designers may be required to travel to production sites across the United States and overseas. Interior designers generally work under deadlines and may work extra hours to finish a job. Also, they regularly carry heavy, bulky sample books to meetings with clients. Floral designers usually work regular hours in a pleasant work environment, but holiday, wedding, and funeral orders often require overtime.

Employment

Designers held about 492,000 jobs in 2000. About one-third were self-employed. Employment was distributed as follows:

Graphic designers	190,000
Floral designers	102,000
Merchandise displayers and window trimmers	76,000
Commercial and industrial designers	50,000
Interior designers	46,000
Fashion designers	16,000
Set and exhibit designers	12,000

Designers work in a number of different industries, depending on their design specialty. Most industrial designers, for example, work for engineering or architectural consulting firms or for large corporations. Most salaried interior designers work for furniture and home furnishings stores, interior designing services, and architectural firms. Others are self-employed and do freelance work—full time or part-time—in addition to a salaried job in another occupation.

Set and exhibit designers work for theater companies; film and television production companies; and museums, art galleries, and convention and conference centers. Fashion designers generally work for textile, apparel, and pattern manufacturers; wholesale distributors of clothing, furnishings, and accessories; or for fashion salons, high-fashion department stores, and specialty shops. Most floral designers work for retail flower shops or in floral departments located inside grocery and department stores.

Training, Other Qualifications, and Advancement

Creativity is crucial in all design occupations. People in this field must have a strong sense of the esthetic—an eye for color and detail, a sense of balance and proportion, and an appreciation for beauty. Despite the advancement of computer-aided design, sketching ability remains an important advantage in most types of design, especially fashion design. A good portfolio—a collection of examples of a person's best work—often is the deciding factor in getting a job.

A bachelor's degree is required for most entry-level design positions, except for floral design and visual merchandising. Esthetic ability is important for floral design and visual merchandising, but formal preparation typically is not necessary. Many candidates in industrial design pursue a master's degree to better compete for open positions.

Interior design is the only design field subject to government regulation. According to the American Society for Interior Designers, 19 states and the District of Columbia require interior designers to be licensed or registered. Passing the National Council for Interior Design qualification examination is required for licensure. To take the exam, one must complete at least 2 years of postsecondary education in design, at least 2 years of practical work experience in the field, plus additional related education or experience to total at least 6 years of combined education and experience in design. Because licensing is not mandatory in all states, membership in a professional association is an indication of an interior designer's qualifications and professional standing and can aid in obtaining clients.

In fashion design, employers seek individuals with a 2- or 4-year degree who are knowledgeable in the areas of textiles, fabrics, and ornamentation, as well as trends in the fashion world. Set and exhibit designers typically have college degrees in design. A Master of Fine Arts (MFA) degree from an accredited university program further establishes one's design credentials. Membership in the United Scenic Artists, Local 829, is a nationally recognized standard of achievement for scenic designers.

Most floral designers learn their skills on the job. When employers hire trainees, they generally look for high school graduates who have a flair for arranging and a desire to learn. Completion of formal training, however, is an asset for floral designers, particularly for advancement to the chief floral designer level. Vocational and technical schools offer programs in floral design, usually lasting less than a year, while 2- and 4-year programs in floriculture, horticulture, floral design, or ornamental horticulture are offered by community and junior colleges, and colleges and universities.

Formal training for some design professions also is available in 2- and 3-year professional schools that award certificates or associate degrees in design. Graduates of 2-year programs normally qualify as assistants to designers. The bachelor of fine arts degree is granted at 4-year colleges and universities. The curriculum in these schools includes art and art history, principles of design, designing and sketching, and specialized studies for each of the individual design disciplines, such as garment construction, textiles, mechanical and architectural drawing, computerized design, sculpture, architecture, and basic engineering. A liberal arts education, with courses in merchandising, business administration, marketing, and psychology, along with training in art, is recommended for designers who want to freelance. Additionally, persons with training or experience in architecture qualify for some design occupations, particularly interior design.

Because computer-aided design is increasingly common, many employers expect new designers to be familiar with its use as a design tool. For example, industrial designers extensively use computers in the aerospace, automotive, and electronics industries. Interior designers use computers to create numerous versions of interior space designs—images can be inserted, edited, and replaced easily and without added cost—making it possible for a client to see and choose among several designs.

The National Association of Schools of Art and Design currently accredits about 200 postsecondary institutions with programs in art and design; most of these schools award a degree in art. Some award degrees in industrial, interior, textile, graphic, or fashion design. Many schools do not allow formal entry into a bachelor's degree program until a student has successfully finished a year of basic art and design courses. Applicants may be required to submit sketches and other examples of their artistic ability.

The Foundation for Interior Design Education Research also accredits interior design programs and schools. Currently, there are more than 120 accredited professional programs in the United States and Canada, primarily located in schools of art, architecture, and home economics.

Individuals in the design field must be creative, imaginative, persistent, and able to communicate their ideas in writing, visually, and verbally. Because tastes in style and fashion can change quickly, designers need to be well-read, open to new ideas and influences, and quick to react to changing trends. Problem-solving skills and the ability to work independently and under pressure are important traits. People in this field need self-discipline to start projects on their own, to budget their time, and to meet deadlines and production schedules. Good business sense and sales ability also are important, especially for those who freelance or run their own business.

Beginning designers usually receive on-the-job training, and normally need 1 to 3 years of training before they can advance to higher-level positions. Experienced designers in large firms may advance to chief designer, design department head, or other supervisory positions. Some designers become teachers in design schools and colleges and universities. Many faculty members continue to consult privately or operate small design studios to complement their classroom activities. Some experienced designers open their own firms.

Job Outlook

Despite projected faster-than-average employment growth, designers in most fields—with the exception of floral design—are expected to face keen competition for available positions. Many talented individuals are attracted to careers as designers. Individuals with little or no formal education in design, as well as those who lack creativity and perseverance, will find it very difficult to establish and maintain a career in design. Floral design should be the least competitive of all design fields because of the relatively low pay and limited opportunities for advancement, as well as the relatively high job turnover of floral designers in retail flower shops.

Overall, the employment of designers is expected to grow faster than the average for all occupations through the year 2010. In addition to those that result from employment growth, many job openings will arise from the need to replace designers who leave the field. Increased demand for industrial designers will stem from the continued emphasis on product quality and safety; the demand for new products that are easy and comfortable to use; the development of high-technology products in medicine, transportation, and other fields; and growing global competition among businesses. Demand for graphic designers should increase because of the rapidly increasing demand for Web-based graphics and the expansion of the video entertainment market, including television, movies, videotape, and made-for-Internet outlets. Rising demand for professional design of private homes, offices, restaurants and other retail establishments, and institutions that care for the rapidly growing elderly population should spur employment growth of interior designers. Demand for fashion designers should remain strong, because many consumers continue to demand new fashions and apparel styles.

Earnings

Median annual earnings for commercial and industrial designers were $48,780 in 2000. The middle 50 percent earned between $36,460 and $64,120. The lowest 10 percent earned less than $27,290, and the highest 10 percent earned more than $77,790.

Median annual earnings for fashion designers were $48,530 in 2000. The middle 50 percent earned between $34,800 and $73,780. The lowest 10 percent earned less than $24,710, and the highest 10 percent earned more than $103,970. Median annual earnings were $52,860 in apparel, piece goods, and notions—the industry employing the largest numbers of fashion designers.

Median annual earnings for floral designers were $18,360 in 2000. The middle 50 percent earned between $14,900 and $22,110. The lowest 10 percent earned less than $12,570, and the highest 10 percent earned more than $27,860. Median annual earnings were $20,160 in grocery stores and $17,760 in miscellaneous retail stores, including florists.

Median annual earnings for graphic designers were $34,570 in 2000. The middle 50 percent earned between $26,560 and $45,130. The lowest 10 percent earned less than $20,480, and the highest 10 percent earned more than $58,400. Median annual earnings in the industries employing the largest numbers of graphic designers were as follows:

Management and public relations	$37,570
Advertising	37,080
Mailing, reproduction, and stenographic services	36,130
Commercial printing	29,730
Newspapers	28,170

Median annual earnings for interior designers were $36,540 in 2000. The middle 50 percent earned between $26,800 and $51,140. The lowest 10 percent earned less than $19,840, and the highest 10 percent earned more than $66,470. Median annual earnings were $40,710 in engineering and architectural services and $34,890 in furniture and home furnishings stores.

Median annual earnings of merchandise displayers and window dressers were $20,930 in 2000. The middle 50 percent earned between $16,770 and $26,840. The lowest 10 percent earned less than $13,790, and the highest 10 percent earned more than $31,130. Median annual earnings were $22,210 in groceries and related products and $18,820 in department stores.

Median annual earnings for set and exhibit designers were $31,440 in 2000. The middle 50 percent earned between $21,460 and $42,800. The lowest 10 percent earned less than $13,820, and the highest 10 percent earned more than $57,400.

According to the Industrial Designers Society of America, the median base salary, excluding deferred compensation, bonuses, royalties, and commissions, for an industrial designer with 1 to 2 years of experience was about $36,500 in 2000. Staff designers with 5 years of experience earned $45,000, whereas senior designers with 8 years of experience earned $64,000. Industrial designers in managerial, executive, or ownership positions earned substantially more—up to $600,000 annually; however, the $80,000 to $180,000 range was more representative.

The American Institute of Graphic Arts (AIGA) reported 1999 median earnings for graphic designers with increasing levels of responsibility. Staff-level graphic designers earned $36,000, while senior designers, who may supervise junior staff or have some decision-making authority that reflects their knowledge of graphic design, earned $50,000. Solo designers, who freelance or work independently of a company, reported median earnings of $50,000. Design directors, the creative heads of design firms or in-house corporate design departments, earned $80,000. Graphic designers with business responsibilities for the operation of a firm as owners, partners, or principals earned $90,000.

Related Occupations

Workers in other occupations who design or arrange objects, materials, or interiors to enhance their appearance and function include artists and related workers; architects, except landscape and naval; engineers, landscape architects, and photographers. Some computer-related occupations require design skills, including computer software engineers and desktop publishers.

Sources of Additional Information

For general information about art and design and a list of accredited college-level programs, contact:

- National Association of Schools of Art and Design, 11250 Roger Bacon Dr., Suite 21, Reston, VA 20190. Internet: http://www.arts-accredit.org/nasad/default.htm

For information on industrial design careers and a list of academic programs in industrial design, write to:

- Industrial Designers Society of America, 1142 Walker Rd., Great Falls, VA 22066. Internet: http://www.idsa.org

For information about graphic design careers, contact:

- American Institute of Graphic Arts, 164 Fifth Ave., New York, NY 10010. Internet: http://www.aiga.org

For information on degree, continuing education, and licensure programs in interior design and interior design research, contact:

- American Society for Interior Designers, 608 Massachusetts Ave. NE, Washington, DC 20002-6006. Internet: http://www.asid.org

For information on degree, continuing education, and licensure programs, and general information on the interior design profession, contact:

- International Interior Design Association, 997 Merchandise Mart, Chicago, IL 60654. Internet: http://www.iida.org

For a list of schools with accredited programs in interior design, contact:

- Foundation for Interior Design Education Research, 146 Monroe Center N.W., Suite 1318, Grand Rapids, MI 49503. Internet: http://www.fider.org

For information about careers in floral design, contact:

- Society of American Florists, 1601 Duke St., Alexandria, VA 22314

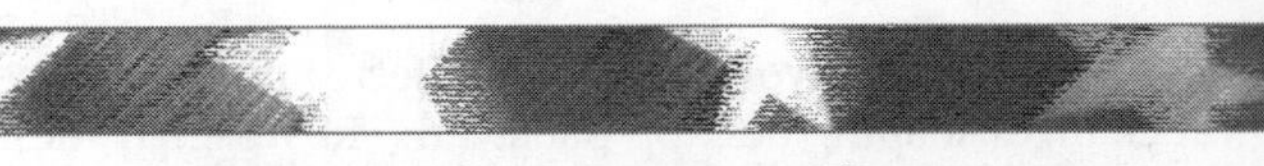

Desktop Publishers

O*NET 43-9031.00

Significant Points

- Desktop publishers rank among the 10 fastest growing occupations.
- Most jobs are in firms that handle commercial or business printing, and in newspaper plants.
- Although formal training is not always required, those with certification or degrees will have the best job opportunities.

Nature of the Work

Using computer software, desktop publishers format and combine text, numerical data, photographs, charts, and other visual graphic elements to produce publication-ready material. Depending on the nature of a particular project, desktop publishers may write and edit text, create graphics to accompany text, convert photographs and drawings into digital images and then manipulate those images. They also design page layouts, create proposals, develop presentations and advertising campaigns, typeset and do color separation, and translate electronic information onto film or other traditional forms. Materials produced by desktop publishers include books, business cards, calendars, magazines, newsletters and newspapers, packaging, slides, and tickets. As companies have brought the production of marketing, promotional, and other kinds of materials in-house, they increasingly have employed people who can produce such materials.

Desktop publishers use a keyboard to enter and select formatting specifics such as size and style of type, column width, and spacing, and store them in the computer. The computer then displays and arranges columns of type on a video display terminal or computer monitor. An entire newspaper, catalog, or book page, complete with artwork and graphics, can be created on the screen exactly as it will appear in print. Operators transmit the pages for production either into film and then into printing plates, or directly into plates.

Desktop publishing is a rapidly changing field that encompasses a number of different kinds of jobs. Personal computers enable desktop publishers to perform publishing tasks that would otherwise require complicated equipment and human effort. Advances in computer software and printing technology continue to change and enhance desktop publishing work. Instead of receiving simple typed text from customers, desktop publishers get the material on a computer disk. Other innovations in desktop publishing work include digital color page makeup systems, electronic page layout systems, and off-press color proofing systems. And because most materials today often are published on the Internet, desktop publishers may need to know electronic publishing technologies, such as Hypertext Markup Language (HTML), and may be responsible for converting text and graphics to an Internet-ready format.

Typesetting and page layout have been affected by the technological changes shaping desktop publishing. Increasingly, desktop publishers use computers to do much of the typesetting and page layout work formerly done by prepress workers, posing new challenges for the printing industry. The old "hot type" method of text composition—which used molten lead to create individual letters, paragraphs, and full pages of text—is nearly extinct. Today, composition work is primarily done with computers. Improvements in desktop publishing software also allow customers to do much more of their own typesetting.

Desktop publishers use scanners to capture photographs, images or art as digital data that can be incorporated directly into electronic page layouts or further manipulated using computer software. The desktop publisher then can correct for mistakes or compensate for deficiencies in the original color print or transparency. Digital files are used to produce printing plates. Like photographers and multimedia artists and animators, desktop publishers also can create special effects or other visual images using film, video, computers, or other electronic media.

Depending on the establishment employing these workers, desktop publishers also may be referred to as publications specialists, electronic publishers, DTP operators, desktop publishing editors, electronic prepress technicians, electronic publishing specialists, image designers, typographers, compositors, layout artists, and Web publications designers.

Working Conditions

Desktop publishers usually work in clean, air-conditioned office areas with little noise. Desktop publishers usually work an 8-hour day, 5 days a week. Some workers—particularly those self-employed—work night shifts, weekends, and holidays.

Desktop publishers often are subject to stress and the pressures of short deadlines and tight work schedules. Like other workers who spend long hours working in front of a computer monitor, they may be susceptible to eyestrain, back discomfort, and hand and wrist problems.

Employment

Desktop publishers held about 38,000 jobs in 2000. Nearly all worked in the printing and publishing industries. About 1,000 desktop publishers were self-employed.

Most desktop publishing jobs were found in firms that handle commercial or corporate printing, and in newspaper plants. Commercial printing firms print a wide range of products (newspaper inserts, catalogs, pamphlets, and advertisements); business form establishments print material such as sales receipts. A large number of desktop publishers also were found in printing trade services firms. Establishments in printing trade services typically perform custom compositing, platemaking, and related prepress services. Others work printing or publishing materials "in-house" or "in-plant" for business services firms, government agencies, hospitals, or universities, typically in a reproduction or publications department that operates within the organization.

The printing and publishing industry is one of the most geographically dispersed in the United States, and desktop publishing jobs are found throughout the country. However, job prospects may be best in large metropolitan cities.

Training, Other Qualifications, and Advancement

Most workers qualify for jobs as desktop publishers by taking classes or completing certificate programs at vocational schools, universities and colleges, or via the Internet. Programs range in length, but the average nondegree certification training program takes approximately 1 year. However, some desktop publishers train on the job to develop the necessary skills. The length of training on the job varies by company. An internship or part-time desktop publishing assignment is another way to gain experience as a desktop publisher.

Students interested in pursuing a career in desktop publishing also may obtain an associate degree in applied science or a bachelor's degree in graphic arts, graphic communications or graphic design. Graphic arts programs are a good way to learn about desktop publishing software used to format pages, assign type characteristics, and import text and graphics into electronic page layouts to produce printed materials such as advertisements, brochures, newsletters, and forms. Applying this knowledge of graphic arts techniques and computerized typesetting usually are intended for students who may eventually move into management positions, while 2-year associate degree programs are designed to train skilled workers. Students also develop finely tuned skills in typography, print mediums, packaging, branding and identity, Web design and motion graphics. These programs teach print and graphic design fundamentals and provide an extensive background in imaging, prepress, print reproduction, and emerging media. Courses in other aspects of printing also are available at vocational-technical institutes, industry-sponsored update and retraining programs, and private trade and technical schools.

Although formal training is not always required, those with certification or degrees will have the best job opportunities. Most employers prefer to hire people who have at least a high school diploma, possess good communication skills, basic computer skills, and a strong work ethic. Desktop publishers should be able to deal courteously with people because in small shops they may have to take customer orders. They also may add, subtract, multiply, divide, and compute ratios to estimate job costs. Persons interested in working for firms using advanced printing technology need to know the basics of electronics and computers.

Desktop publishers need good manual dexterity, and they must be able to pay attention to detail and work independently. Good eyesight, including visual acuity, depth perception, field of view, color vision, and the ability to focus quickly, also are assets. Artistic ability often is a plus. Employers also seek persons who are even-tempered and adaptable—important qualities for workers who often must meet deadlines and learn how to operate new equipment.

Workers with limited training and experience may start as helpers. They begin with instruction from an experienced desktop publisher and advance based on their demonstrated mastery of skills at each level. All workers should expect to be retrained from time to time to handle new, improved software and equipment. As workers gain experience, they advance to positions with greater responsibility. Some move into supervisory or management positions. Other desktop publishers may start their own company or work as an independent consultant, while those with more artistic talent and further education may find opportunities in graphic design or commercial art.

Job Outlook

Employment of desktop publishers is expected to grow much faster than the average for all occupations through 2010, as more page layout and design work is performed in-house using computers and sophisticated publishing software. Desktop publishing is replacing much of the prepress work done by compositors and typesetters, enabling organizations to reduce costs while increasing production speeds. Many new jobs for desktop publishers are expected to emerge in commercial printing and publishing establishments. However, more companies also are turning to in-house desktop publishers, as computers with elaborate text and graphics capabilities have become common, and desktop publishing software has become cheaper and easier to use. In addition to employment growth, many job openings for desktop publishers also will result from the need to replace workers who move into managerial positions, transfer to other occupations, or who leave the labor force.

Printing and publishing costs represent a significant portion of a corporation's expenses, no matter the industry, and corporations are finding it more profitable to print their own newsletters and other reports than to send them out to trade shops. Desktop publishing reduces the time needed to complete a printing job, and allows commercial printers to make inroads into new markets that require fast turnaround.

Most employers prefer to hire experienced desktop publishers. As more people gain desktop publishing experience, however, competition for jobs may increase. Among persons without experience, opportunities should be best for those with computer backgrounds who are certified or who have completed postsecondary programs in desktop publishing or graphic design. Many employers prefer graduates of these programs because the comprehensive training they receive helps them learn the page layout process and adapt more rapidly to new software and techniques.

Earnings

Earnings for desktop publishers vary according to level of experience, training, location, and size of firm. Median annual earnings of desktop publishers were $30,600 in 2000. The middle 50 percent earned between $22,890 and $40,210. The lowest 10 percent earned less than $17,800, and the highest 10 percent earned more than $50,920 a year. Median annual earnings in the industries employing the largest numbers of these workers in 2000 are shown in the following:

Commercial printing	$30,940
Newspapers	24,520

Related Occupations

Desktop publishers use artistic and editorial skills in their work. These skills also are essential for artists and related workers; designers; news analysts, reporters, and correspondents; public relations specialists; writers and editors; and prepress technicians and workers.

Sources of Additional Information

Details about apprenticeship and other training programs may be obtained from local employers such as newspapers and printing shops, or from local offices of the state employment service.

For information on careers and training in printing, desktop publishing, and graphic arts, write to:

- Graphic Communications Council, 1899 Preston White Dr., Reston, VA 20191. Internet: http://www.npes.org
- Graphic Arts Technical Foundation, 200 Deer Run Rd., Sewickley, PA 15143. Internet: http://www.gatf.org

For information on benefits and compensation in desktop publishing, write to:

- Printing Industries of America, Inc., 100 Daingerfield Rd., Alexandria, VA 22314. Internet: http://www.gain.org

Diagnostic Medical Sonographers

O*NET 29-2032.00

Significant Points

- Sonographers should experience favorable job opportunities as ultrasound becomes an increasingly attractive alternative to radiologic procedures.
- More than half of all sonographers are employed by hospitals, and most of the remainder work in physicians' offices and clinics, including diagnostic imaging centers.
- Beginning in 2005, an associate or higher degree from an accredited program will be required for registration.

Nature of the Work

Diagnostic imaging embraces several procedures that aid in diagnosing ailments, the most familiar being the X ray. Another increasingly common diagnostic imaging method, called magnetic resonance imaging (MRI), uses giant magnets and radio waves rather than radiation to create an image. Not all imaging technologies use ionizing radiation or radio waves, however. Sonography, or ultrasonography, is the use of sound waves to generate an image used for assessment and diagnosis of various medical conditions. Many people associate sonography with obstetrics and the viewing of the fetus in the womb. But this technology has many other applications in the diagnosis and treatment of medical conditions.

Diagnostic medical sonographers, also known as *ultrasonographers*, use special equipment to direct nonionizing, high frequency sound waves into areas of the patient's body. Sonographers operate the equipment, which collects reflected echoes and forms an image that may be videotaped, transmitted, or photographed for interpretation and diagnosis by a physician.

Sonographers begin by explaining the procedure to the patient and recording any additional medical history that may be relevant to the condition being viewed. They then select appropriate equipment settings and direct the patient to move into positions that will provide the best view. To perform the exam, sonographers use a transducer, which transmits sound waves in a cone- or rectangle-shaped beam. Although techniques vary based on the area being examined, sonographers usually spread a special gel on the skin to aid the transmission of sound waves.

Viewing the screen during the scan, sonographers look for subtle visual cues that contrast healthy areas from unhealthy ones. They decide whether the images are satisfactory for diagnostic purposes and select which ones to show to the physician.

Diagnostic medical sonographers may specialize in obstetric and gynecologic sonography (the female reproductive system), abdominal sonography (the liver, kidneys, gallbladder, spleen, and pancreas), neurosonography (the brain), or ophthalmologic sonography (the eyes). In addition, sonographers also may specialize in vascular technology or echocardiography.

Obstetric and gynecologic sonographers specialize in the study of the female reproductive system. This includes one of the more well known uses of sonography: examining the fetus of a pregnant woman to track its growth and health.

Abdominal sonographers inspect a patient's abdominal cavity to help diagnose and treat conditions involving primarily the gallbladder, bile ducts, kidneys, liver, pancreas, and spleen. Abdominal sonographers also are able to scan parts of the heart, although diagnosis of the heart using ultrasound usually is done by echocardiographers.

Neurosonographers use ultrasound technology to focus on the nervous system, including the brain. In neonatal care, neurosonographers study and diagnose neurological and nervous system disorders in premature infants. They also may scan blood vessels to check for abnormalities indicating a stroke in infants diagnosed with sickle cell anemia. Like other sonographers, neurosonographers operate transducers to perform the ultrasound, but use different frequencies and beam shapes than obstetric and abdominal sonographers.

Ophthalmologic sonographers use ultrasound to study the eyes. Ultrasound aids in the insertion of prosthetic lenses by allowing accurate measurement of the eyes. Ophthalmologic ultrasound also helps diagnose and track tumors, blood supply conditions, separated retinas, and other ailments of the eye and the surrounding tissue. Ophthalmologic sonographers use high frequency transducers made exclusively to study the eyes, which are much smaller than those used in other specialties.

In addition to working directly with patients, diagnostic medical sonographers keep patient records and adjust and maintain equipment. They also may prepare work schedules, evaluate equipment purchases, or manage a sonography or diagnostic imaging department.

Working Conditions

Most full-time sonographers work about 40 hours a week; they may have evening weekend hours and times when they are on call and must be ready to report to work on short notice.

Sonographers typically work in healthcare facilities that are clean and well lit. Some travel to patients in large vans equipped with sophisticated diagnostic equipment. Sonographers are on their feet for long periods and may have to lift or turn disabled patients. They work at diagnostic imaging machines but may also do some procedures at patients' bedsides.

Employment

Diagnostic medical sonographers held about 33,000 jobs in 2000. More than half of all sonographer jobs are in hospitals. Most of the rest are in physicians' offices and clinics, primarily in offices specializing in obstetrics and in diagnostic imaging centers. According to the 2000 Sonography Benchmark Survey conducted by the Society of Diagnostic Medical Sonographers (SDMS), about three out of four sonographers worked in urban areas.

Training, Other Qualifications, and Advancement

There are several avenues for entry into the field of diagnostic medical sonography. Sonographers may train in hospitals, vocational-technical institutions, colleges and universities, and the armed forces. Some training programs prefer applicants with a background in science or experience in other health professions, but also will consider high school graduates with courses in math and science, as well as applicants with liberal arts backgrounds.

Colleges and universities offer formal training in both 2- and 4-year programs, culminating in an associate or bachelor's degree. Two-year programs are most prevalent. Course work includes classes in anatomy, physiology, instrumentation, basic physics, patient care, and medical ethics. The Joint Review Committee on Education for Diagnostic Medical Sonography accredits most formal training programs (76 programs in 1999).

Some health workers, such as obstetric nurses and radiologic technologists, seek to increase their marketability by cross-training in fields such as sonography. Many take 1-year programs resulting in a certificate. Additionally, sonographers specializing in one discipline often seek competency in others; for example, obstetric sonographers might seek training in and exposure to abdominal sonography to broaden their opportunities.

While no state requires licensure in diagnostic medical sonography, the American Registry of Diagnostic Medical Sonographers (ARDMS) certifies the competency of sonographers through registration. Because registration provides an independent, objective measure of an individual's professional standing, many employers prefer to hire registered sonographers. Registration with ARDMS requires passing a general physics and instrumentation examination, in addition to passing an exam in a specialty such as obstetrics/gynecology, abdominal, or neurosonography.

While formal education is not necessary to take the exams, an associate or bachelor's degree from an accredited program is preferred. Beginning in 2005, ARDMS will consider for registration only those holding an associate or higher degree. To keep their registration current, sonographers must complete 30 hours of continuing education every 3 years to stay abreast of advances in the occupation and in technology.

Sonographers need good communication and interpersonal skills because they must be able to explain technical procedures and results to their patients, some of whom may be nervous about the exam or the problems it may reveal. They also should have some background in math and science, especially when they must perform mathematical and scientific calculations in analyses for diagnosis.

Job Outlook

Employment of diagnostic medical sonographers is expected to grow faster than the average for all occupations through 2010 as the population grows and ages, increasing the demand for diagnostic imaging and therapeutic technology. Some job openings also will arise from the need to replace sonographers who leave the occupation.

Ultrasound is becoming an increasingly attractive alternative to radiologic procedures as patients seek safer treatment methods. Because ultrasound—unlike most diagnostic imaging methods—does not involve radiation, harmful side effects and complications from repeated use are rarer for both the patient and the sonographer. Sonographic technology is expected to evolve rapidly and to spawn many new ultrasound procedures, such as 3D-ultrasonography for use in obstetric and ophthalmologic diagnosis. However, high costs may limit the rate at which some promising new technologies are adopted.

Hospitals will remain the principal employer of diagnostic medical sonographers. However, employment is expected to grow more rapidly in offices and clinics of physicians, including diagnostic imaging centers. Health facilities such as these are expected to grow very rapidly through 2010 due to the strong shift toward outpatient care, encouraged by third-party payers and made possible by technological advances that permit more procedures to be performed outside the hospital.

Earnings

Median annual earnings of diagnostic medical sonographers were $44,820 in 2000. The middle 50 percent earned between $38,390 and $52,750 a year. The lowest 10 percent earned less than $32,470, and the highest 10 percent earned more than $59,310. Median annual earnings of diagnostic medical sonographers in 2000 were $43,950 in hospitals and $46,190 in offices and clinics of medical doctors.

Related Occupations

Diagnostic medical sonographers operate sophisticated equipment to help physicians and other health practitioners diagnose and treat patients. Workers in related occupations include cardiovascular technologists and technicians, clinical laboratory technologists and technicians, nuclear medicine technologists, radiologic technologists and technicians, and respiratory therapists.

Sources of Additional Information

For more information on a career as a diagnostic medical sonographer, contact:

- Society of Diagnostic Medical Sonographers, 12770 Coit Rd., Suite 708, Dallas, TX 75251. Internet: http://www.sdms.org
- The American Registry of Diagnostic Medical Sonographers, 600 Jefferson Plaza, Suite 360, Rockville, MD 20852-1150. Internet: http://www.ardms.org

For a current list of accredited education programs in diagnostic medical sonography, write to:

- The Joint Review Committee on Education in Diagnostic Medical Sonography, 1248 Harwood Rd., Bedford, TX 76021-4244. Internet: http://www.caahep.org

Diesel Service Technicians and Mechanics

O*NET 49-3031.00

Significant Points

- A career as a diesel service technician or mechanic offers relatively high wages and the challenge of skilled repair work.
- Opportunities are expected to be good for persons who complete formal training programs.
- National certification is the recognized standard of achievement for diesel service technicians and mechanics.

Nature of the Work

The diesel engine is the workhorse powering the nation's trucks and buses, because it delivers more power and is more durable than its gasoline-burning counterpart. Diesel-powered engines also are becoming more prevalent in light vehicles, including pickups and other work trucks.

Diesel service technicians and mechanics, also known as *bus and truck mechanics and diesel engine specialists*, repair and maintain the diesel engines that power transportation equipment such as heavy trucks, buses, and locomotives. Some diesel technicians and mechanics also work on heavy vehicles and mobile equipment such as bulldozers, cranes, road graders, farm tractors, and combines. A small number of technicians repair diesel-powered passenger automobiles, light trucks, or boats.

Technicians who work for organizations that maintain their own vehicles spend most of their time doing preventive maintenance, to ensure that equipment will operate safely. These workers also eliminate unnecessary wear on and damage to parts that could result in costly breakdowns. During a routine maintenance check on a vehicle, technicians follow a checklist that includes inspection of brake systems, steering mechanisms, wheel bearings, and other important parts. Following inspection, technicians repair or adjust parts that do not work properly or remove and replace parts that cannot be fixed.

Increasingly, technicians must be flexible, in order to adapt to customer needs and new technologies. It is common for technicians to handle all kinds of repairs, from working on a vehicle's electrical system one day, to doing major engine repairs the next. Diesel maintenance is becoming increasingly complex, as more electronic components are used to control engine operation. For example, microprocessors regulate and manage fuel timing, increasing engine efficiency. In modern shops, diesel service technicians use hand-held computers to diagnose problems and adjust engine functions. Technicians must continually learn about new techniques and advanced materials.

Diesel service technicians use a variety of tools in their work, including power tools, such as pneumatic wrenches, to remove bolts quickly; machine tools, such as lathes and grinding machines, to rebuild brakes; welding and flame-cutting equipment to remove and repair exhaust systems; and jacks and hoists to lift and move large parts. Common handtools—screwdrivers, pliers, and wrenches—are used to work on small parts and get at hard-to-reach places. Diesel service technicians and mechanics also use a variety of computerized testing equipment to pinpoint and analyze malfunctions in electrical systems and engines.

In large shops, technicians generally receive their assignments from shop supervisors or service managers. Most supervisors and managers are experienced technicians who also assist in diagnosing problems and maintaining quality standards. Technicians may work as a team or be assisted by an apprentice or helper when doing heavy work, such as removing engines and transmissions.

Working Conditions

Diesel technicians usually work indoors, although they occasionally make repairs to vehicles on the road. Diesel technicians may lift heavy parts and tools, handle greasy and dirty parts, and stand or lie in awkward positions to repair vehicles and equipment. Minor cuts, burns, and bruises are common, although serious accidents can usually be avoided if the shop is kept clean and orderly and safety procedures are followed. Technicians normally work in well-lighted, heated, and ventilated areas; however, some shops are drafty and noisy. Many employers provide lockers and shower facilities.

Employment

Diesel service technicians and mechanics held about 285,000 jobs in 2000. About 25 percent serviced buses, trucks, and other diesel-powered equipment for customers of vehicle and equipment dealers, automotive rental and leasing agencies, or independent automotive repair shops. About 20 percent worked for local and long-distance trucking companies, and another 19 percent maintained the buses, trucks, and other equipment of buslines, public transit companies, school systems, or federal, state, and local governments. The remaining technicians maintained vehicles and other equipment for manufacturing, construction, or other companies. A relatively small number were self-employed. Nearly every section of the country employs diesel service technicians and mechanics, although most work in towns and cities where trucking companies, buslines, and other fleet owners have large operations.

Training, Other Qualifications, and Advancement

Although many persons qualify for diesel service technician and mechanic jobs through years of on-the-job training, authorities recommend completion of a formal diesel engine training program. Employers prefer to hire graduates of formal training programs because these workers often have a head start in training and are able to quickly advance to the journey level.

Many community colleges and trade and vocational schools offer programs in diesel repair. These programs, lasting 6 months to 2 years, lead to a certificate of completion or an associate degree. Programs vary in the degree of hands-on training they provide on equipment. Some offer about 30 hours per week on equipment, whereas others offer more lab or classroom instruction. Training provides a foundation in the latest diesel technology and instruction in the service and repair of the vehicles and equipment that technicians will encounter on the job. Training programs also improve the skills needed to interpret technical manuals and to communicate with coworkers and customers. In addition to the hands-on aspects of the training, many institutions teach communication skills, customer service, basic understanding of physics, and logical thought. Increasingly, employers work closely with representatives of training programs, providing instructors with the latest equipment, techniques, and tools and offering jobs to graduates.

Whereas most employers prefer to hire persons who have completed formal training programs, some technicians and mechanics continue to learn their skills on the job. Unskilled beginners usually are assigned tasks such as cleaning parts, fueling and lubricating vehicles, and driving vehicles into and out of the shop. Beginners usually are promoted to trainee positions, as they gain experience and as vacancies become available. In some shops, beginners with experience in automobile service start as trainee technicians.

Most trainees perform routine service tasks and make minor repairs after a few months' experience. These workers advance to increasingly difficult jobs as they prove their ability and competence. After technicians master the repair and service of diesel engines, they learn to work on related components, such as brakes, transmissions, and electrical systems. Generally, technicians with at least 3 to 4 years of on-the-job experience will qualify as journey-level diesel technicians. Completion of a formal training program speeds advancement to the journey level.

For unskilled entry-level jobs, employers usually look for applicants who have mechanical aptitude and strong problem-solving skills, and who are at least 18 years of age and in good physical condition. Nearly all employers require completion of high school. Courses in automotive repair, electronics, English, mathematics, and physics provide a strong educational background for a career as a diesel service technician or mechanic. Technicians need a state commercial driver's license to test-drive trucks or buses on public roads. Practical experience in automobile repair at a gasoline service station, in the armed forces, or as a hobby is also valuable.

Employers often send experienced technicians and mechanics to special training classes conducted by manufacturers and vendors, in which workers learn the latest technology and repair techniques. Technicians constantly receive updated technical manuals and service procedures outlining changes in techniques and standards for repair. It is essential for technicians to read, interpret, and comprehend service manuals, in order to keep abreast of engineering changes.

Voluntary certification by the National Institute for Automotive Service Excellence (ASE) is recognized as the standard of achievement for diesel service technicians and mechanics. Technicians may be certified as Master Heavy-Duty Truck Technicians or in specific areas of heavy-duty truck repair, such as gasoline engines, drive trains, brakes, suspension and steering, electrical and electronic systems, or preventive maintenance and inspection.

For certification in each area, a technician must pass one or more of the ASE-administered exams and present proof of 2 years of relevant hands-on work experience. Two years of relevant formal training from a high school, vocational or trade school, or community or junior college program may be substituted for up to 1 year of the work experience requirement. To remain certified, technicians must retest every 5 years. This ensures that service technicians and mechanics keep up with changing technology.

Diesel service technicians and mechanics may opt for ASE certification as schoolbus technicians. The certification identifies and recognizes technicians with the knowledge and skills required to diagnose, service, and repair different subsystems of schoolbuses. The ASE School Bus Technician Test Series includes seven certification exams: Body Systems and Special Equipment (S1), Diesel Engines (S2), Drive Train (S3), Brakes (S4), Suspension and Steering (S5), Electrical/Electronic Systems (S6), and Air Conditioning Systems and Controls (S7). Whereas several of these tests parallel existing ASE truck tests, each one is designed to test knowledge of systems specific to schoolbuses. In order to become ASE-certified in schoolbus repair, technicians must pass one or more of the exams and present proof of 2 years of relevant hands-on work experience. Technicians who pass tests S1 through S6 become ASE-Certified Master School Bus Technicians.

The most important work possessions of technicians and mechanics are their handtools. Technicians and mechanics usually provide their own tools, and many experienced workers have thousands of dollars invested in them. Employers typically furnish expensive power tools, computerized engine analyzers, and other diagnostic equipment; but individual workers ordinarily accumulate handtools with experience.

Experienced technicians and mechanics with leadership ability may advance to shop supervisor or service manager. Technicians and mechanics with sales ability sometimes become sales representatives. Some open their own repair shops.

Job Outlook

Employment of diesel service technicians and mechanics is expected to increase about as fast as the average for all occupations through the year 2010. Besides openings resulting from employment growth, opportunities will be created by the need to replace workers who retire or transfer to other occupations.

Employment of diesel service technicians and mechanics is expected to grow as freight transportation by truck increases. Additional trucks will be needed to keep pace with the increasing volume of freight shipped nationwide. Trucks also serve as intermediaries for other forms of transportation, such as rail and air. Due to the greater durability and economy of the diesel engine relative to the gasoline engine, buses and trucks of all sizes are expected to be increasingly powered by diesels. In addition, diesel service technicians will be needed to maintain and repair the growing number of schoolbuses in operation.

Careers as diesel service technicians attract many because of relatively high wages and the challenge of skilled repair work. Opportunities should be good for persons who complete formal training in diesel mechanics at community and junior colleges and vocational and technical schools. Applicants without formal training may face stiffer competition for entry-level jobs.

Most persons entering this occupation can expect steady work, because changes in economic conditions have little effect on the diesel repair business. During a financial downturn, however, some employers may be reluctant to hire new workers.

Earnings

Median hourly earnings of bus and truck mechanics and diesel engine specialists, including incentive pay, were $15.55 in 2000. The middle 50 percent earned between $12.33 and $19.30 an hour. The lowest 10 percent earned less than $9.88, and the highest 10 percent earned more than $22.63 an hour. Median hourly earnings in the industries employing the largest numbers of bus and truck mechanics and diesel engine specialists in 2000 were as follows:

Local government	$17.93
Motor vehicles, parts, and supplies	15.48
Automotive repair shops	14.74
Trucking and courier services, except air	14.65
Elementary and secondary schools	14.63

Because many experienced technicians employed by truck fleet dealers and independent repair shops receive a commission related to the labor cost charged to the customer, weekly earnings depend on the amount of work completed. Beginners usually earn from 50 to 75 percent of the rate of skilled workers and receive increases, as they become more skilled, until they reach the rates of skilled service technicians.

The majority of service technicians work a standard 40-hour week, although some work longer hours, particularly if they are self-employed. A growing number of shops have expanded their hours to better perform repairs and routine service when needed, or as a convenience to customers. Those employed by truck and bus firms providing service around the clock may work evenings, nights, and weekends. These technicians usually receive a higher rate of pay for working non-traditional hours.

Many diesel service technicians and mechanics are members of labor unions, including the International Association of Machinists and Aerospace Workers; the Amalgamated Transit Union; the International Union, United Automobile, Aerospace and Agricultural Implement Workers of America; the Transport Workers Union of America; the Sheet Metal Workers' International Association; and the International Brotherhood of Teamsters.

Related Occupations

Diesel service technicians and mechanics repair trucks, buses, and other diesel-powered equipment. Related technician and mechanic occupations include aircraft and avionics equipment mechanics and service technicians, automotive service technicians and mechanics, heavy vehicle and mobile equipment service technicians and mechanics, and small engine mechanics.

Sources of Additional Information

More details about work opportunities for diesel service technicians and mechanics may be obtained from local employers such as trucking companies, truck dealers, or bus lines; locals of the unions previously mentioned; and local offices of your state employment service. Local state employment service offices also may have information about training programs. State boards of postsecondary career schools also have information on licensed schools with training programs for diesel service technicians and mechanics.

For general information about a career as a diesel service technician or mechanic, write:

- Detroit Diesel, Personnel Director, MS B39, 13400 West Outer Dr., Detroit, MI 48239

Information on how to become a certified medium/heavy-duty diesel technician or bus technician is available from:

- ASE, 101 Blue Seal Dr. S.E., Suite 101, Leesburg, VA 20175. Internet: http://www.asecert.org

For a directory of accredited private trade and technical schools with training programs for diesel service technicians and mechanics, contact:

- Accrediting Commission of Career Schools and Colleges of Technology, 2101 Wilson Blvd., Suite 302, Arlington, VA 22201. Internet: http://www.accsct.org
- National Automotive Technicians Education Foundation, 101 Blue Seal Dr. S.E., Suite 101, Leesburg, VA 20175. Internet: http://www.natef.org

For a directory of public training programs for diesel service technicians and mechanics, contact:

- SkillsUSA-VICA, P.O. Box 3000, 14001 James Monroe Hwy., Leesburg, VA 22075. Internet: http://www.skillsusa.org

Dietitians and Nutritionists

O*NET 29-1031.00

Significant Points

- Employment of dietitians is expected to grow about as fast as the average for all occupations through the year 2010 as a result of increasing emphasis on disease prevention through improved health habits.
- Dietitians and nutritionists need at least a bachelor's degree in dietetics, foods and nutrition, food service systems management, or a related area.

Nature of the Work

Dietitians and nutritionists plan food and nutrition programs, and supervise the preparation and serving of meals. They help prevent and treat illnesses by promoting healthy eating habits and suggesting diet modifications, such as less salt for those with high blood pressure or reduced fat and sugar intake for those who are overweight.

Dietitians run food service systems for institutions such as hospitals and schools, promote sound eating habits through education, and conduct research. Major areas of practice include clinical, community, management, and consultant dietetics.

Clinical dietitians provide nutritional services for patients in institutions such as hospitals and nursing homes. They assess patients' nutritional needs, develop and implement nutrition programs, and evaluate and report the results. They also confer with doctors and other healthcare professionals in order to coordinate medical and nutritional needs. Some clinical dietitians specialize in the management of overweight patients, care of the critically ill, or of renal (kidney) and diabetic patients. In addition, clinical dietitians in nursing homes, small hospitals, or correctional facilities also may manage the food service department.

Community dietitians counsel individuals and groups on nutritional practices designed to prevent disease and promote good health. Working in places such as public health clinics, home health agencies, and health maintenance organizations, they evaluate individual needs, develop nutritional care plans, and instruct individuals and their families. Dietitians working in home health agencies provide instruction on grocery shopping and food preparation to the elderly, individuals with special needs, and children.

Increased interest in nutrition has led to opportunities in food manufacturing, advertising, and marketing, in which dietitians analyze foods, prepare literature for distribution, or report on issues such as the nutritional content of recipes, dietary fiber, or vitamin supplements.

Management dietitians oversee large-scale meal planning and preparation in healthcare facilities, company cafeterias, prisons, and schools. They hire, train, and direct other dietitians and food service workers; budget for and purchase food, equipment, and supplies; enforce sanitary and safety regulations; and prepare records and reports.

Consultant dietitians work under contract with healthcare facilities or in their own private practice. They perform nutrition screenings for their clients, and offer advice on diet-related concerns such as weight loss or cholesterol reduction. Some work for wellness programs, sports teams, supermarkets, and other nutrition-related businesses. They may consult with Food Service Managers, providing expertise in sanitation, safety procedures, menu development, budgeting, and planning.

Working Conditions

Most dietitians work a regular 40-hour week, although some work weekends. Many dietitians work part-time.

Dietitians and nutritionists usually work in clean, well-lighted, and well-ventilated areas. However, some dietitians work in warm, congested kitchens. Many dietitians and nutritionists are on their feet for much of the workday.

Employment

Dietitians and nutritionists held about 49,000 jobs in 2000. More than half were in hospitals, nursing homes, or offices and clinics of physicians.

State and local governments provided about 1 job in 10—mostly in health departments and other public health related areas. Other jobs were in restaurants, social service agencies, residential care facilities, diet workshops, physical fitness facilities, school systems, colleges and universities, and the federal government—mostly in the U.S. Department of Veterans Affairs. Some dietitians and nutritionists were employed by firms that provide food services on contract to such facilities as colleges and universities, airlines, correctional facilities, and company cafeterias.

Some dietitians were self-employed, working as consultants to facilities such as hospitals and nursing homes, or providing dietary counseling to individual clients.

Training, Other Qualifications, and Advancement

High school students interested in becoming a dietitian or nutritionist should take courses in biology, chemistry, mathematics, health, and communications. Dietitians and nutritionists need at least a bachelor's degree in dietetics, foods and nutrition, food service systems management, or a related area. College students in these majors take courses in foods, nutrition, institution management, chemistry, biochemistry, biology, microbiology, and physiology. Other suggested courses include business, mathematics, statistics, computer science, psychology, sociology, and economics.

Twenty-seven of the 41 states with laws governing dietetics require licensure, 13 require certification, and 1 requires registration. The Commission on Dietetic Registration of the American Dietetic Association (ADA) awards the Registered Dietitian credential to those who pass a certification exam after completing their academic coursework and supervised experience. Because practice requirements vary by state, interested candidates should determine the requirements of the state in which they want to work before sitting for any exam.

As of 2001, there were 234 bachelor's and master's degree programs approved by the ADA's Commission on Accreditation for Dietetics Education (CADE). Supervised practice experience can be acquired in two ways. The first requires completion of an ADA-accredited coordinated program. As of 2001, there were 51 accredited programs, which combined academic and supervised practice experience and generally lasted 4 to 5 years. The second option requires completion of 900 hours of supervised practice experience in any of the 258 CADE-accredited/approved internships. Internships and may be full-time programs lasting 6 to 12 months, or part-time programs lasting 2 years. Students interested in research, advanced clinical positions, or public health may need an advanced degree.

Experienced dietitians may advance to assistant, associate, or director of a dietetic department, or become self-employed. Some dietitians specialize in areas such as renal or pediatric dietetics. Others may leave the occupation to become sales representatives for equipment, pharmaceutical, or food manufacturers.

Job Outlook

Employment of dietitians is expected to grow about as fast as the average for all occupations through 2010 as a result of increasing emphasis on disease prevention through improved dietary habits. A growing and aging population will increase the demand for meals and nutritional counseling in nursing homes, schools, prisons, community health programs, and home healthcare agencies. Public interest in nutrition and the emphasis on health education and prudent lifestyles will also spur demand, especially in management. In addition to employment growth, job openings also will result from the need to replace experienced workers who leave the occupation.

The number of dietitian positions in hospitals is expected to grow slowly as hospitals continue to contract out food service operations. On the other hand, employment is expected to grow fast in contract providers of food services, social services agencies, and offices and clinics of physicians.

Employment growth for dietitians and nutritionists may be somewhat constrained by some employers substituting other workers such as health educators, food service managers, and dietetic technicians. Growth also is constrained by limitations on insurance reimbursement for dietetic services.

Earnings

Median annual earnings of dietitians and nutritionists were $38,450 in 2000. The middle 50 percent earned between $31,070 and $45,950 a year. The lowest 10 percent earned less than $23,680, and the highest 10 percent earned more than $54,940 a year. Median annual earnings in hospitals, the industry employing the largest numbers of dietitians and nutritionists, were $39,450.

According to the American Dietetic Association, median annual income for registered dietitians in 1999 varied by practice area as follows: $48,810 in consultation and business, $48,370 in food and nutrition management, $47,040 in education and research, $37,990 in community nutrition, and $37,565 in clinical nutrition. Salaries also vary by years in practice, educational level, geographic region, and size of community.

Related Occupations

Workers in other occupations who may apply the principles of food and nutrition include food service managers, health educators, and registered nurses.

Sources of Additional Information

For a list of academic programs, scholarships, and other information about dietetics, contact:

- The American Dietetic Association, 216 West Jackson Blvd., Suite 800, Chicago, IL 60606-6995. Internet: http://www.eatright.org

Drafters

O*NET 17-3011.01, 17-3011.02, 17-3012.01, 17-3012.02, 17-3013.00

Significant Points

- The type and quality of postsecondary drafting programs vary considerably; prospective students should be careful in selecting a program.
- Opportunities should be best for individuals who have at least 2 years of postsecondary training in drafting and considerable skill and experience using computer-aided drafting (CAD) systems.
- Demand for particular drafting specializations varies geographically, depending on the needs of local industry.

Nature of the Work

Drafters prepare technical drawings and plans used by production and construction workers to build everything from manufactured products, such as toys, toasters, industrial machinery, or spacecraft, to structures, such as houses, office buildings, or oil and gas pipelines. Their drawings provide visual guidelines, showing the technical details of the products and structures and specifying dimensions, materials to be used, and procedures and processes to be followed. Drafters fill in technical details, using drawings, rough sketches, specifications, codes, and calculations previously made by engineers, surveyors, architects, or scientists. For example, they use their knowledge of standardized building techniques to draw in the details of a structure. Some drafters use their knowledge of engineering and manufacturing theory and standards to draw the parts of a machine in order to determine design elements, such as the number and kind of fasteners needed to assemble it. They use technical handbooks, tables, calculators, and computers to do this.

Traditionally, drafters sat at drawing boards and used pencils, pens, compasses, protractors, triangles, and other drafting devices to prepare a drawing manually. Most drafters now use computer-aided drafting (CAD) systems to prepare drawings. Consequently, some drafters are referred to as CAD operators.

CAD systems employ computer workstations to create a drawing on a video screen. The drawings are stored electronically so that revisions or duplications can be made easily. These systems also permit drafters to easily and quickly prepare variations of a design. Although drafters use CAD extensively, it is only a tool. Persons who produce technical drawings using CAD still function as drafters, and need the knowledge of traditional drafters—relating to drafting skills and standards—in addition to CAD skills. Despite the near-universal use of CAD systems, manual drafting still is used in certain applications.

Drafting work has many specialties, and titles may denote a particular discipline of design or drafting. *Aeronautical drafters* prepare engineering drawings detailing plans and specifications used for the manufacture of aircraft, missiles, and related parts.

Architectural drafters draw architectural and structural features of buildings and other structures. They may specialize by the type of structure, such as residential or commercial, or by the kind of material used, such as reinforced concrete, masonry, steel, or timber.

Civil drafters prepare drawings and topographical and relief maps used in major construction or civil engineering projects, such as highways, bridges, pipelines, flood control projects, and water and sewage systems.

Electrical drafters prepare wiring and layout diagrams used by workers who erect, install, and repair electrical equipment and wiring in communication centers, powerplants, electrical distribution systems, and buildings.

Electronic drafters draw wiring diagrams, circuitboard assembly diagrams, schematics, and layout drawings used in the manufacture, installation, and repair of electronic devices and components.

Mechanical drafters prepare detail and assembly drawings of a wide variety of machinery and mechanical devices, indicating dimensions, fastening methods, and other requirements.

Process piping or *pipeline drafters* prepare drawings used for layout, construction, and operation of oil and gas fields, refineries, chemical plants, and process piping systems.

Working Conditions

Drafters usually work in comfortable offices furnished to accommodate their tasks. They may sit at adjustable drawing boards or drafting tables when doing manual drawings, although most drafters work at computer terminals much of the time. Because they spend long periods in front of computer terminals doing detailed work, drafters may be susceptible to eyestrain, back discomfort, and hand and wrist problems.

Employment

Drafters held about 213,000 jobs in 2000. More than 40 percent of drafters worked in engineering and architectural services firms that design construction projects or do other engineering work on a contract basis for organizations in other industries. Another 29 percent worked in durable goods manufacturing industries, such as machinery, electrical equipment, and fabricated metals. The remainder were mostly employed in the construction; government; transportation, communications, and utilities; and personnel-supply services industries. About 10,000 were self-employed in 2000.

Training, Other Qualifications, and Advancement

Employers prefer applicants who have completed postsecondary school training in drafting, which is offered by technical institutes, community colleges, and some 4-year colleges and universities. Employers are most interested in applicants who have well-developed drafting and mechanical drawing skills; a knowledge of drafting standards, mathematics, science, and engineering technology; and a solid background in computer-aided drafting and design techniques. In addition, communication and problem-solving skills are important.

Individuals planning careers in drafting should take courses in math, science, computer technology, design or computer graphics, and any high school drafting courses available. Mechanical ability and visual aptitude also are important. Prospective drafters should be able to draw three-dimensional objects as well as draw freehand. They also should do detailed work accurately and neatly. Artistic ability is helpful in some specialized fields, as is knowledge of manufacturing and construction methods. In addition, prospective drafters should have good interpersonal skills because they work closely with engineers, surveyors, architects, other professionals, and sometimes customers.

Training and coursework differ somewhat within the drafting specialties. The initial training for each specialty is similar. All incorporate math and communication skills, for example, but coursework relating to the specialty varies. In an electronics drafting program, for example, students learn the ways that electronic components and circuits are depicted in drawings.

Entry-level or junior drafters usually do routine work under close supervision. After gaining experience, intermediate-level drafters progress to more difficult work with less supervision. They may be required to exercise more judgment and perform calculations when preparing and modifying drawings. Drafters may eventually advance to senior drafter, designer, or supervisor. Many employers pay for continuing education and, with appropriate college degrees, drafters may go on to become engineering technicians, engineers, or architects.

Many types of publicly and privately operated schools provide some form of drafting training. The kind and quality of programs vary considerably. Therefore, prospective students should be careful in selecting a program. They should contact prospective employers regarding their preferences and ask schools to provide information about the kinds of jobs obtained by graduates, type and condition of instructional facilities and equipment, and faculty qualifications.

Technical institutes offer intensive technical training but less general education than junior and community colleges. Certificates or diplomas based on completion of a certain number of course hours may be rewarded. Many technical institutes offer 2-year associate degree programs, which are similar to, or part of, the programs offered by community colleges or state university systems. Other technical institutes are run by private, often for-profit, organizations, sometimes called proprietary schools. Their programs vary considerably in both length and type of courses offered.

Community colleges offer curriculums similar to those in technical institutes but include more courses on theory and liberal arts. Often, there is little or no difference between technical institute and community college programs. However, courses taken at community colleges are more likely to be accepted for credit at 4-year colleges than are those at technical institutes. After completing a 2-year associate degree program, graduates may obtain jobs as drafters or continue their education in a related field at 4-year colleges. Four-year colleges usually do not offer drafting training, but college courses in engineering, architecture, and mathematics are useful for obtaining a job as a drafter.

Area vocational-technical schools are postsecondary public institutions that serve local students and emphasize training needed by local employers. Many offer introductory drafting instruction. Most require a high school diploma, or its equivalent, for admission.

Technical training obtained in the armed forces also can be applied in civilian drafting jobs. Some additional training may be necessary, depending on the technical area or military specialty.

The American Design Drafting Association (ADDA) has established a certification program for drafters. Although drafters usually are not required to be certified by employers, certification demonstrates that the understanding of nationally recognized practices and knowledge standards have been met. Individuals who wish to become certified must pass the Drafter Certification Test, which is administered periodically at ADDA-authorized test sites. Applicants are tested on their knowledge and understanding of basic drafting concepts such as geometric construction, working drawings, and architectural terms and standards.

Job Outlook

Employment of drafters is expected to grow about as fast as the average for all occupations through 2010. Industrial growth and increasingly complex design problems associated with new products and manufacturing processes will increase the demand for drafting services. Further, drafters are beginning to break out of the traditional drafting role and increasingly do work traditionally performed by engineers and architects, thus increasing the need for drafters. However, the greater use of CAD equipment by drafters, as well as by architects and engineers, should limit demand for lesser-skilled drafters. In addition to those created by employment growth, many job openings are expected to arise as drafters move to other occupations or leave the labor force.

Opportunities should be best for individuals who have at least 2 years of postsecondary training in a drafting program that provides strong technical skills, and who have considerable skill and experience using CAD systems. CAD has increased the complexity of drafting applications while enhancing the productivity of drafters. It also has enhanced the nature of drafting by creating more possibilities for design and drafting. As technology continues to advance, employers will look for drafters with a strong background in fundamental drafting principles, a higher level of technical sophistication, and an ability to apply this knowledge to a broader range of responsibilities.

Demand for particular drafting specialties varies throughout the country because employment usually is contingent upon the needs of local industry. Employment of drafters remains highly concentrated in industries that are sensitive to cyclical changes in the economy, such as engineering and architectural services and durable-goods manufacturing. During recessions, drafters may be laid off. However, a growing number of drafters should continue to be employed on a temporary or contract basis, as more companies turn to the personnel-supply services industry to meet their changing needs.

Earnings

Earnings for drafters vary by specialty and level of responsibility. Median hourly earnings of architectural and civil drafters were $16.93 in 2000. The middle 50 percent earned between $13.79 and $20.86. The lowest 10 percent earned less than $11.18, and the highest 10 percent earned more than $26.13. Median hourly earnings of architectural and civil drafters in engineering and architectural services in 2000 were $16.75.

Median hourly earnings of electrical and electronics drafters were $18.37 in 2000. The middle 50 percent earned between $14.19 and $23.76. The lowest 10 percent earned less than $11.30, and the highest 10 percent earned more than $29.46. In engineering and architectural services, the average hourly earnings for electrical and electronics drafters were $17.30.

Median hourly earnings of mechanical drafters were $18.19 in 2000. The middle 50 percent earned between $14.43 and $23.20. The lowest 10 percent earned less than $11.70, and the highest 10 percent earned more than $28.69. The average hourly earnings for mechanical drafters in engineering and architectural services were $16.98.

Related Occupations

Other workers who prepare or analyze detailed drawings and make precise calculations and measurements include architects, except landscape and naval; landscape architects; designers; engineers; engineering technicians; science technicians; and surveyors, cartographers, photogrammetrists, and surveying technicians.

Sources of Additional Information

Information on schools offering programs in drafting and related fields is available from:

- Accrediting Commission of Career Schools and Colleges of Technology, 2101 Wilson Blvd., Suite 302, Arlington, VA 22201. Internet: http://www.accsct.org

Information about certification is available from:

- American Design Drafting Association, P.O. Box 11937, Columbia, SC 29211. Internet: http://www.adda.org

Economists and Market and Survey Researchers

O*NET 19-3011.00, 19-3021.00, 19-3022.00

Significant Points

- Demand for qualified market and survey researchers should be strong.
- Candidates who hold an advanced degree will have the best employment prospects and advancement opportunities.

Nature of the Work

Economists. Economists study how society distributes scarce resources such as land, labor, raw materials, and machinery to produce goods and services. They conduct research, collect and analyze data, monitor economic trends, and develop forecasts. They research issues such as energy costs, inflation, interest rates, imports, or employment levels.

Most economists are concerned with practical applications of economic policy. They use their understanding of economic relationships to advise businesses and other organizations, including insurance companies, banks, securities firms, industry and trade associations, labor unions, and government agencies. Economists use mathematical models to help predict answers to questions such as the nature and length of business cycles, the effects of a specific rate of inflation on the economy, or the effects of tax legislation on unemployment levels.

Economists devise methods and procedures for obtaining the data they need. For example, sampling techniques may be used to conduct a survey, and various mathematical modeling techniques may be used to develop forecasts. Preparing reports, including tables and charts, on research results is an important part of an economist's job. Presenting economic and statistical concepts in a clear and meaningful way is particularly important for economists whose research is directed toward making policies for an organization.

Economists who work for government agencies may assess economic conditions in the United States or abroad, in order to estimate the economic effects of specific changes in legislation or public policy. They may study areas such as how the dollar's fluctuation against foreign currencies affects import and export levels. The majority of government economists work in the area of agriculture, labor, or quantitative analysis; however, economists work in almost every area of government. For example, economists in the U.S. Department of Commerce study production, distribution, and consumption of commodities produced overseas, while economists employed with the U.S. Bureau of Labor Statistics analyze data on the domestic economy such as prices, wages, employment, productivity, and safety and health. An economist working in state or local government might analyze data on the growth of school-aged populations, prison growth, and employment and unemployment rates, in order to project future spending needs.

Market Research Analysts. Market, or marketing, research analysts are concerned with the potential sales of a product or service. They analyze statistical data on past sales to predict future sales. They gather data on competitors and analyze prices, sales, and methods of marketing and distribution. Like economists, market research analysts devise methods and procedures for obtaining the data they need. They often design telephone, personal, or mail interview surveys to assess consumer preferences. Trained interviewers, under the market research analyst's direction, usually conduct the surveys.

After compiling the data, market research analysts evaluate it and make recommendations to their client or employer based upon their findings. They provide a company's management with information needed to make decisions on the promotion, distribution, design, and pricing of products or services. The information may also be used to determine the advisability of adding new lines of merchandise, opening new branches, or otherwise diversifying the company's operations. Analysts may conduct opinion research to determine public attitudes on various issues, which may help political or business leaders and others assess public support for their electoral prospects or advertising policies.

Survey Researchers. Survey researchers design and conduct surveys. They use surveys to collect information that is used for research, making fiscal or policy decisions, and measuring policy effectiveness, for example. As with market research analysts, survey researchers may use a variety of mediums to conduct surveys, such as the Internet, personal or telephone interviews, or mail questionnaires. They also may supervise interviewers who conduct surveys in person or over the telephone.

Survey researchers design surveys in many different formats, depending upon the scope of research and method of collection. Interview surveys, for example, are common because they can increase survey participation rates. Survey researchers may consult with economists, statisticians, market research analysts, or other data users in order to design surveys. They also may present survey results to clients.

Working Conditions

Economists and market and survey researchers have structured work schedules. They often work alone, writing reports, preparing statistical charts, and using computers, but they also may be an integral part of a research team. Most work under pressure of deadlines and tight schedules, which may require overtime. Their routine may be interrupted by special requests for data, as well as by the need to attend meetings or conferences. Frequent travel may be necessary.

Employment

Economists and market and survey researchers held about 134,000 jobs in 2000. Private industry provided about 9 out of

10 jobs for salaried workers, particularly economic and marketing research firms, management consulting firms, banks, securities and commodities brokers, and computer and data processing companies. A wide range of government agencies provided the remaining jobs, primarily for economists. The U.S. Departments of Labor, Agriculture, and Commerce are the largest federal employers of economists. A number of economists and market and survey researchers combine a full-time job in government, academia, or business with part-time or consulting work in another setting.

Employment of economists and market and survey researchers is concentrated in large cities. Some work abroad for companies with major international operations, for U.S. Government agencies, and for international organizations like the World Bank and the United Nations.

Besides the jobs described above, many economists and market and survey researchers held faculty positions in colleges and universities. Economics and marketing faculties have flexible work schedules, and may divide their time among teaching, research, consulting, and administration.

Training, Other Qualifications, and Advancement

Graduate education is required for many private sector economist and market and survey research jobs, and for advancement to more responsible positions. Economics includes many specialties at the graduate level, such as advanced economic theory, econometrics, international economics, and labor economics. Students should select graduate schools strong in specialties in which they are interested. Undergraduate economics majors can choose from a variety of courses, ranging from microeconomics, macroeconomics, and econometrics, to more philosophical courses, such as the history of economic thought.

In the federal government, candidates for entry-level economist positions must have a bachelor's degree with a minimum of 21 semester hours of economics and 3 hours of statistics, accounting, or calculus.

Market and survey researchers may earn advanced degrees in economics, business administration, marketing, statistics, or some closely related discipline. Some schools help graduate students find internships or part-time employment in government agencies, economic consulting firms, financial institutions, or marketing research firms prior to graduation.

In addition to courses in business, marketing, and consumer behavior, marketing majors should take other liberal arts and social science courses, including economics, psychology, English, and sociology. Because of the importance of quantitative skills to economists and market and survey researchers, courses in mathematics, statistics, econometrics, sampling theory and survey design, and computer science are extremely helpful.

Whether working in government, industry, research organizations, marketing, or consulting firms, economists and market and survey researchers with bachelor degrees usually qualify for most entry-level positions as a research assistant, administrative or management trainee, marketing interviewer, or any of a number of professional sales jobs. A master's degree usually is required to qualify for more responsible research and administrative positions. Many businesses, research and consulting firms, and government agencies seek individuals who have strong computer and quantitative skills and can perform complex research. A PhD is necessary for top economist or marketing positions in many organizations. Many corporation and government executives have a strong background in economics or marketing.

A master's degree is usually the minimum requirement for a job as an instructor in junior and community colleges. In most colleges and universities, however, a PhD is necessary for appointment as an instructor. A PhD and extensive publications in academic journals are required for a professorship, tenure, and promotion.

Aspiring economists and market and survey researchers should gain experience gathering and analyzing data, conducting interviews or surveys, and writing reports on their findings while in college. This experience can prove invaluable later in obtaining a full-time position in the field, since much of their work, in the beginning, may center on these duties. With experience, economists and market and survey researchers eventually are assigned their own research projects.

Those considering careers as economists or market and survey researchers should be able to pay attention to details because much time is spent on precise data analysis. Patience and persistence are necessary qualities since economists and market and survey researchers must spend long hours on independent study and problem solving. At the same time, they must work well with others, especially market and survey researchers, who often oversee interviews for a wide variety of individuals. Economists and market and survey researchers must be able to present their findings, both orally and in writing, in a clear, concise manner.

Job Outlook

Employment of economists and market and survey researchers is expected to grow about as fast as the average for all occupations through 2010. Many job openings are likely to result from the need to replace experienced workers who transfer to other occupations, retire, or leave the labor force for other reasons. Employment growth of economists is expected to be as fast as average over the projection period, while growth for market research analysts and survey researchers is expected to be faster than average.

Opportunities for economists should be best in private industry, especially in research, testing, and consulting firms, as more companies contract out for economic research services. The growing complexity of the global economy, competition, and increased reliance on quantitative methods for analyzing the current value of future funds, business trends, sales, and purchasing should spur demand for economists. The growing need for economic analyses in virtually every industry should result in additional jobs for economists. Employment of economists in the federal government should decline more slowly than other occupations in the federal workforce. Slow employment growth is expected among economists in state and local government.

Candidates who meet state certification requirements may become high school economics teachers. The demand for secondary school economics teachers is expected to grow, as economics becomes an increasingly important and popular course.

Demand for qualified market research analysts should be healthy because of an increasingly competitive economy. Marketing research provides organizations valuable feedback from purchasers, allowing companies to evaluate consumer satisfaction and more effectively plan for the future. As companies seek to expand their market and consumers become better informed, the need for marketing professionals will increase.

Opportunities for market research analysts with graduate degrees should be good in a wide range of employment settings, particularly in marketing research firms, as companies find it more profitable to contract out for marketing research services rather than support their own marketing department. Other organizations, including financial services organizations, healthcare institutions, advertising firms, manufacturing firms producing consumer goods, and insurance companies may offer job opportunities for market research analysts.

Opportunities for survey researchers should be strong as the demand for market and opinion research increase. Employment opportunities will be especially favorable in commercial market and opinion research as an increasingly competitive economy requires businesses to more effectively and efficiently allocate advertising funds.

An advanced degree coupled with a strong background in economic theory, mathematics, statistics, and econometrics provides the basis for acquiring any specialty within the economics and market and survey research field. Those skilled in quantitative techniques and their application to economic modeling and forecasting, coupled with good communications skills, should have the best job opportunities.

Bachelor's degree holders may face competition for the limited number of positions for which they qualify. They will qualify for a number of other positions, however, where they can take advantage of their economic knowledge in conducting research, developing surveys, or analyzing data. Many graduates with bachelor's degrees will find good jobs in industry and business as management or sales trainees, or administrative assistants. Bachelor's degree holders with good quantitative skills and a strong background in mathematics, statistics, survey design, and computer science also may be hired by private firms as research assistants or interviewers.

PhD degree holders in economics and marketing should have good opportunities in most areas such as industry and consulting firms. However, PhD holders are likely to face keen competition for tenured teaching positions in colleges and universities.

Earnings

Median annual earnings of economists were $64,830 in 2000. The middle 50 percent earned between $47,370 and $87,890. The lowest 10 percent earned less than $35,690, and the highest 10 percent earned more than $114,580.

The federal government recognizes education and experience in certifying applicants for entry-level positions. The entrance salary for economists having a bachelor's degree was about $21,900 a year in 2001; however, those with superior academic records could begin at $27,200. Those having a master's degree could qualify for positions at an annual salary of $33,300. Those with a PhD could begin at $40,200, while some individuals with experience and an advanced degree could start at $48,200. Starting salaries were slightly higher in selected areas where the prevailing local pay was higher. The average annual salary for economists employed by the federal government was $74,090 a year in 2001.

Median annual earnings of market research analysts in 2000 were $51,190. The middle 50 percent earned between $37,030 and $71,660. The lowest 10 percent earned less than $27,570, and the highest 10 percent earned more than $96,360. Median annual earnings in the industries employing the largest numbers of market research analysts in 2000 were as follows:

Computer and data processing services	$61,320
Management and public relations	44,580
Research and testing services	43,660

Median annual earnings of survey researchers in 2000 were $26,200. The middle 50 percent earned between $17,330 and $47,820. The lowest 10 percent earned less than $15,050, and the highest 10 percent earned more than $71,790. Median annual earnings of survey researchers in 2000 were $52,470 in computer and data processing services and $18,780 in research and testing services.

Related Occupations

Economists are concerned with understanding and interpreting financial matters, among other subjects. Other jobs in this area include actuaries; budget analysts; financial analysts and personal financial advisors; financial managers; insurance underwriters; loan counselors and officers; and purchasing managers, buyers, and purchasing agents.

Market research analysts do research to find out how well the market receives products or services. This may include planning, implementation, and analysis of surveys to determine people's needs and preferences. Other jobs using these skills include psychologists, sociologists, and urban and regional planners.

Sources of Additional Information

For information on careers in business economics, contact:

- National Association for Business Economics, 1233 20th St. NW, Suite 505, Washington, DC 20036

For information about careers and salaries in market and survey research, contact:

- Marketing Research Association, 1344 Silas Deane Hwy., Suite 306, Rocky Hill, CT 06067-0230. Internet: http://www.mra-net.org
- Council of American Survey Research Organizations, 3 Upper Devon, Port Jefferson, NY 11777. Internet: http://www.casro.org

Information on obtaining a position as an economist with the federal government is available from the Office of Personnel Management (OPM) through a telephone-based system. Consult your telephone directory under U.S. Government for a local number or call (912) 757-3000; Federal Relay Service: (800) 877-8339. The first number is not toll free, and charges may result. Information also is available from the OPM Internet site: http://www.usajobs.opm.gov

Electrical and Electronics Engineers

O*NET 17-2071.00, 17-2072.00

Significant Points

- Overall job opportunities in engineering are expected to be good.
- A bachelor's degree is required for most entry-level jobs.
- Starting salaries are significantly higher than those of college graduates in other fields.
- Continuing education is critical to keep abreast of the latest technology.

Nature of the Work

From geographical information systems that can continuously provide the location of a vehicle to giant electric power generators, electrical and electronics engineers are responsible for a wide range of technologies. Electrical and electronics engineers design, develop, test, and supervise the manufacture of electrical and electronic equipment. Some of this equipment includes power generating, controlling, and transmission devices used by electric utilities; and electric motors, machinery controls, lighting, and wiring in buildings, automobiles, aircraft, radar and navigation systems, and broadcast and communications systems. Many electrical and electronics engineers also work in areas closely related to computers. However, engineers whose work is related exclusively to computer hardware are considered computer hardware engineers.

Electrical and electronics engineers specialize in different areas such as power generation, transmission, and distribution; communications; and electrical equipment manufacturing, or a subdivision of these areas—industrial robot control systems or aviation electronics, for example. Electrical and electronics engineers design new products, write performance requirements, and develop maintenance schedules. They also test equipment, solve operating problems, and estimate the time and cost of engineering projects.

Working Conditions

Most engineers work in office buildings, laboratories, or industrial plants. Others may spend time outdoors at construction sites, mines, and oil and gas exploration and production sites, where they monitor or direct operations or solve onsite problems. Some engineers travel extensively to plants or work sites.

Many engineers work a standard 40-hour week. At times, deadlines or design standards may bring extra pressure to a job. When this happens, engineers may work longer hours and experience considerable stress.

Employment

Electrical and electronics engineers held about 288,000 jobs in 2000, making their occupation the largest branch of engineering. Most jobs were in engineering and business consulting firms, government agencies, and manufacturers of electrical and electronic and computer and office equipment, industrial machinery, and professional and scientific instruments. Transportation, communications, and utilities firms as well as personnel supply services and computer and data processing services firms accounted for most of the remaining jobs.

California, Texas, New York, and New Jersey—states with many large electronics firms—employ nearly one-third of all electrical and electronics engineers.

Training, Other Qualifications, and Advancement

A bachelor's degree in engineering is required for almost all entry-level engineering jobs. College graduates with a degree in a physical science or mathematics occasionally may qualify for some engineering jobs, especially in specialties in high demand. Most engineering degrees are granted in electrical, electronics, mechanical, or civil engineering. However, engineers trained in one branch may work in related branches. This flexibility allows employers to meet staffing needs in new technologies and specialties in which engineers are in short supply. It also allows engineers to shift to fields with better employment prospects or to those that more closely match their interests.

Most engineering programs involve a concentration of study in an engineering specialty, along with courses in both mathematics and science. Most programs include a design course, sometimes accompanied by a computer or laboratory class or both.

In addition to the standard engineering degree, many colleges offer 2- or 4-year degree programs in engineering technology. These programs, which usually include various hands-on laboratory classes that focus on current issues, prepare students for practical design and production work, rather than for jobs which require more theoretical and scientific knowledge. Graduates of 4-year technology programs may get jobs similar to those obtained by graduates with a bachelor's degree in engineering. Engineering technology graduates, however, are not qualified to register as professional engineers under the same terms as graduates with degrees in engineering. Some employers regard technology program graduates as having skills between those of a technician and an engineer.

Graduate training is essential for engineering faculty positions and many research and development programs, but is not required for the majority of entry-level engineering jobs. Many engineers obtain graduate degrees in engineering or business administration to learn new technology and broaden their education. Many high-level executives in government and industry began their careers as engineers.

About 330 colleges and universities offer bachelor's degree programs in engineering that are accredited by the Accreditation Board for Engineering and Technology (ABET), and about 250 colleges offer accredited bachelor's degree programs in engineering technology. ABET accreditation is based on an examination of an engineering program's student achievement, program improvement, faculty, curricular content, facilities, and institutional commitment. Although most institutions offer programs in the major branches of engineering, only a few offer programs in the smaller specialties. Also, programs of the same title may vary in content. For example, some programs emphasize industrial practices, preparing students for a job in industry, whereas others are more theoretical and are designed to prepare students for graduate work. Therefore, students should investigate curricula and check accreditations carefully before selecting a college.

Admissions requirements for undergraduate engineering schools include a solid background in mathematics (algebra, geometry, trigonometry, and calculus) and sciences (biology, chemistry, and physics), and courses in English, social studies, humanities, and computers. Bachelor's degree programs in engineering typically are designed to last 4 years, but many students find that it takes between 4 and 5 years to complete their studies. In a typical 4-year college curriculum, the first 2 years are spent studying mathematics, basic sciences, introductory engineering, humanities, and social sciences. In the last 2 years, most courses are in engineering, usually with a concentration in one branch. Some programs offer a general engineering curriculum; students then specialize in graduate school or on the job.

Some engineering schools and 2-year colleges have agreements whereby the 2-year college provides the initial engineering education, and the engineering school automatically admits students for their last 2 years. In addition, a few engineering schools have arrangements whereby a student spends 3 years in a liberal arts college studying pre-engineering subjects and 2 years in an engineering school studying core subjects, and then receives a bachelor's degree from each school. Some colleges and universities offer 5-year master's degree programs. Some 5- or even 6-year cooperative plans combine classroom study and practical work, permitting students to gain valuable experience and finance part of their education. All 50 states and the District of Columbia usually require licensure for engineers who offer their services directly to the public. Engineers who are licensed are called Professional Engineers (PE). This licensure generally requires a degree from an ABET-accredited engineering program, 4 years of relevant work experience, and successful completion of a state examination. Recent graduates can start the licensing process by taking the examination in two stages. The initial Fundamentals of Engineering (FE) examination can be taken upon graduation. Engineers who pass this examination commonly are called Engineers in Training (EIT) or Engineer Interns (EI). The EIT certification usually is valid for 10 years. After acquiring suitable work experience, EITs can take the second examination, the Principles and Practice of Engineering Exam. Several states have imposed mandatory continuing education requirements for relicensure. Most states recognize licensure from other states. Many electrical engineers are licensed as PEs.

Engineers should be creative, inquisitive, analytical, and detail oriented. They should be able to work as part of a team and to communicate well, both orally and in writing. Communication abilities are becoming more important because much of their work is becoming more diversified, meaning that engineers interact with specialists in a wide range of fields outside engineering.

Beginning engineering graduates usually work under the supervision of experienced engineers and, in large companies, also may receive formal classroom or seminar-type training. As new engineers gain knowledge and experience, they are assigned more difficult projects with greater independence to develop designs, solve problems, and make decisions. Engineers may advance to become technical specialists or to supervise a staff or team of engineers and technicians. Some may eventually become engineering managers or enter other managerial or sales jobs.

Job Outlook

Electrical and electronics engineering graduates should have favorable job opportunities. The number of job openings resulting from employment growth and the need to replace electrical engineers who transfer to other occupations or leave the labor force is expected to be in rough balance with the supply of graduates. Employment of electrical and electronics engineers is expected to grow about as fast as the average for all occupations through 2010.

Projected job growth stems largely from increased demand for electrical and electronic goods, including advanced communications equipment, defense-related electronic equipment, and consumer electronics products. The need for electronics manufacturers to invest heavily in research and development to remain competitive and gain a scientific edge will provide openings for graduates who have learned the latest technologies. Opportunities for electronics engineers in defense-related firms should improve as aircraft and weapons systems are upgraded with improved navigation, control, guidance, and targeting systems. However, job growth is expected to be fastest in services industries—particularly consulting firms that provide electronic engineering expertise.

Continuing education is important for electrical and electronics engineers. Engineers who fail to keep up with the rapid changes in technology risk becoming more susceptible to layoffs or, at a minimum, more likely to be passed over for advancement.

Earnings

Median annual earnings of electrical engineers were $64,910 in 2000. The middle 50 percent earned between $51,700 and

$80,600. The lowest 10 percent earned less than $41,740, and the highest 10 percent earned more than $94,490. Median annual earnings in the industries employing the largest numbers of electrical engineers in 2000 were:

Computer and office equipment	$69,700
Measuring and controlling devices	67,570
Search and navigation equipment	67,330
Electronic components and accessories	65,830
Engineering and architectural services	65,040

Median annual earnings of electronics engineers, except computer, were $64,830 in 2000. The middle 50 percent earned between $52,430 and $79,960. The lowest 10 percent earned less than $43,070, and the highest 10 percent earned more than $94,330. Median annual earnings in the industries employing the largest numbers of electronics engineers in 2000 were:

Federal government	$70,890
Search and navigation equipment	68,930
Electronic components and accessories	63,890
Electrical goods	62,860
Telephone communication	57,710

According to a 2001 salary survey by the National Association of Colleges and Employers, bachelor's degree candidates in electrical and electronics engineering received starting offers averaging $51,910 a year; master's degree candidates averaged $63,812; and PhD candidates averaged $79,241.

Related Occupations

Electrical and electronics engineers apply the principles of physical science and mathematics in their work. Other workers who use scientific and mathematical principles include architects, except landscape and naval; engineering and natural sciences managers; computer and information systems managers; mathematicians; drafters; engineering technicians; sales engineers; science technicians; and physical and life scientists, including agricultural and food scientists, biological and medical scientists, conservation scientists and foresters, atmospheric scientists, chemists and materials scientists, environmental scientists and geoscientists, and physicists and astronomers.

Sources of Additional Information

Information on electrical and electronics engineers is available from:

- Institute of Electrical and Electronics Engineers, 445 Hoes Lane, Piscatway, NJ 08855-1331. Internet: http://www.ieee.org

High school students interested in obtaining a full package of guidance materials and information (product number SP-01) on a variety of engineering disciplines should contact the Junior Engineering Technical Society by sending $3.50 to:

- JETS-Guidance, 1420 King St., Suite 405, Alexandria, VA 22314-2794. Internet: http://www.jets.org

High school students interested in obtaining information on ABET-accredited engineering programs should contact:

- The Accreditation Board for Engineering and Technology, Inc., 111 Market Place, Suite 1050, Baltimore, MD 21202-4012. Internet: http://www.abet.org

Non-licensed engineers and college students interested in obtaining information on Professional Engineer licensure should contact:

- The National Society of Professional Engineers, 1420 King St., Alexandria, VA 22314-2794. Internet: http://www.nspe.org
- National Council of Examiners for Engineers and Surveying, P.O. Box 1686, Clemson, SC 29633-1686. Internet: http://www.ncees.org

Information on general engineering education and career resources is available from:

- American Society for Engineering Education, 1818 N St. N.W., Suite 600, Washington, DC 20036-2479. Internet: http://www.asee.org

Information on obtaining an engineering position with the federal government is available from the Office of Personnel Management (OPM) through a telephone-based system. Consult your telephone directory under U.S. Government for a local number or call (912) 757-3000; Federal Relay Service: (800) 877-8339. The first number is not toll free, and charges may result. Information also is available from the OPM Internet site: http://www.usajobs.opm.gov

Non-high school students and those wanting more detailed information should contact societies representing the individual branches of engineering. Each can provide information about careers in the particular branch. The individual statements elsewhere in this book also provide other information in detail on aerospace; agricultural; biomedical; chemical; civil; computer hardware; environmental; industrial, including health and safety; materials; mechanical; mining and geological, including mining safety; nuclear; and petroleum engineering.

Electrical and Electronics Installers and Repairers

O*NET 49-2092.01, 49-2092.02, 49-2092.03, 49-2092.04, 49-2092.05, 49-2092.06, 49-2093.00, 49-2094.00, 49-2095.00, 49-2096.00

Significant Points

- Knowledge of electrical equipment and electronics is necessary for employment; many applicants complete 1 to 2 years at vocational schools and community colleges, although some less skilled repairers may have only a high school diploma.
- Projected employment growth will be slower than average, but varies by occupational specialty.
- Job opportunities will be best for applicants with a thorough knowledge of electrical and electronic equipment, as well as repair experience.

Nature of the Work

Businesses and other organizations depend on complex electronic equipment for a variety of functions. Industrial controls automatically monitor and direct production processes on the factory floor. Transmitters and antennae provide communications links for many organizations. Electric power companies use electronic equipment to operate and control generating plants, substations, and monitoring equipment. The federal government uses radar and missile control systems to provide for the national defense and to direct commercial air traffic. These complex pieces of electronic equipment are installed, maintained, and repaired by electrical and electronics installers and repairers.

Electrical equipment and electronics equipment are two distinct types of industrial equipment, although much equipment contains both electrical and electronic components. In general, electrical portions of equipment provide the power for the equipment while electronic components control the device, although many types of equipment still are controlled with electrical devices. Electronic sensors monitor the equipment and the manufacturing process, providing feedback to the programmable logic control (PLC) that controls the equipment. The PLC processes the information provided by the sensors and makes adjustments to optimize output. To adjust the output the PLC sends signals to the electrical, hydraulic, and pneumatic devices that power the machine—changing feed rates, pressures, and other variables in the manufacturing process. Many installers and repairers, known as *field technicians*, travel to factories or other locations to repair equipment. These workers often have assigned areas where they perform preventive maintenance on a regular basis. When equipment breaks down, field technicians go to a customer's site to repair the equipment. *Bench technicians* work in repair shops located in factories and service centers. They work on components that cannot be repaired on the factory floor.

Some industrial electronic equipment is self-monitoring and alerts repairers to malfunctions. When equipment breaks down, repairers first check for common causes of trouble, such as loose connections or obviously defective components. If routine checks do not locate the trouble, repairers may refer to schematics and manufacturers' specifications that show connections and provide instructions on how to locate problems. Automated electronic control systems are increasing in complexity, making diagnosing problems more challenging. Repairers use software programs and testing equipment to diagnose malfunctions. They use multimeters, which measure voltage, current, and resistance; advanced multimeters also measure capacitance, inductance, and current gain of transistors. They also use signal generators that provide test signals, and oscilloscopes that graphically display signals. Repairers use handtools such as pliers, screwdrivers, soldering irons, and wrenches to replace faulty parts and to adjust equipment.

Because component repair is complex and factories cannot allow production equipment to stand idle, repairers on the factory floor usually remove and replace defective units, such as circuit boards, instead of fixing them. Defective units are discarded or returned to the manufacturer or to a specialized shop for repair. Bench technicians at these locations have the training, tools, and parts to thoroughly diagnose and repair circuit boards or other complex components. These workers also locate and repair circuit defects, such as poorly soldered joints, blown fuses, or malfunctioning transistors.

Electrical and electronics installers often fit older manufacturing equipment with new automated control devices. Older manufacturing machines are frequently in good working order, but are limited by inefficient control systems that lack replacement parts. Installers replace old electronic control units with new PLCs. Setting up and installing a new PLC involves connecting it to different sensors and electrically powered devices (electric motors, switches, pumps) and writing a computer program to operate the PLC. Electronics installers coordinate their efforts with other workers installing and maintaining equipment.

Electronic equipment installers and repairers, motor vehicles have a significantly different job. They install, diagnose, and repair communications, sound, security, and navigation equipment in motor vehicles. Most installation work involves either new alarm or sound systems. New sound systems vary significantly in cost and complexity of installation. Replacing a head unit (radio) with a new computer disc (CD) player is quite simple, requiring removing a few screws and connecting a few wires. Installing a new sound system with a subwoofer, amplifier, and fuses is far more complicated. The installer builds a box, of fiberglass or wood, designed to hold the subwoofer and to fit in the unique dimensions of the automobile. Installing sound-deadening material, which often is necessary with more powerful speakers, requires an installer to remove many parts of a car (seats, carpeting, interiors of doors), add sound-absorbing material in empty spaces, and reinstall the interior parts. They also run new speaker and electrical cables. Additional electrical power may require additional fuses; a new electrical line to be run from the battery, through a newly drilled hole in the fire wall into the interior of the vehicle; or an additional or more powerful alternator and/or battery.

Repairing automotive electronic equipment is similar to other electronic installation and repair work. Multimeters are used to diagnose the source of the problem. Many parts often are removed and replaced, rather than repaired. Many repairs are quite simple, only requiring a fuse to be replaced. Motor vehicle installers and repairs work with an increasingly complex range of electronic equipment, including DVD players, VCRs, satellite navigation equipment, passive security tracking systems, and active security systems.

Working Conditions

Many electrical and electronics installers and repairers work on factory floors where they are subject to noise, dirt, vibration, and heat. Bench technicians work primarily in repair shops where the surroundings are relatively quiet, comfortable, and well-lighted. Field technicians spend much time on the road, traveling to different customer locations.

Because electronic equipment is critical to industries and other organizations, repairers work around the clock. Their schedules may include evening, weekend, and holiday shifts; shifts may be assigned on the basis of seniority.

Installers and repairers may have to do heavy lifting and work in a variety of positions. They must follow safety guidelines and often wear protective goggles and hardhats. When working on ladders or on elevated equipment, repairers must wear harnesses to prevent falls. Before repairing a piece of machinery, these workers must follow procedures to insure that others cannot start the equipment during the repair process. They also must take precautions against electric shock by locking off power to the unit under repair.

Electronic equipment installers and repairers, motor vehicles normally work indoors in well-ventilated and well-lighted repair shops. Minor cuts and bruises are common, but serious accidents usually are avoided when safety practices are observed.

Employment

Electrical and electronics installers and repairers held about 171,000 jobs in 2000. The following table breaks down employment by occupational specialty:

Electrical and electronics repairers, commercial and industrial equipment	90,000
Electric motor, power tools, and related repairers	37,000
Electrical and electronics repairers, powerhouse, substation, and relay	18,000
Electrical and electronics installers and repairers, transportation equipment	14,000
Electronic equipment installers and repairers, motor vehicles	13,000

Many repairers worked for wholesale trade companies, general electrical work companies, the federal government, electrical repair shops, and manufacturers of electronic components and accessories and communications equipment.

Training, Other Qualifications, and Advancement

Knowledge of electrical equipment and electronics is necessary for employment. Many applicants gain this training through programs lasting 1 to 2 years at vocational schools and community colleges, although some less skilled repairers may have only a high school diploma. Entry-level repairers may work closely with more experienced technicians who provide technical guidance.

Installers and repairers should have good eyesight and color perception in order to work with the intricate components used in electronic equipment. Field technicians work closely with customers and should have good communications skills and a neat appearance. Employers also may require that field technicians have a driver's license.

The International Society of Certified Electronics Technicians (ISCET) and the Electronics Technicians Association (ETA) administer certification programs for electronics installation and repair technicians. Repairers may specialize—in industrial electronics, for example. To receive certification, repairers must pass qualifying exams corresponding to their level of training and experience. Both programs offer associate certifications to entry-level repairers.

Experienced repairers with advanced training may become specialists or troubleshooters who help other repairers diagnose difficult problems. Workers with leadership ability may become supervisors of other repairers. Some experienced workers open their own repair shops.

Job Outlook

Job opportunities should be best for applicants with a thorough knowledge of electrical equipment and electronics, as well as repair experience. Overall employment of electrical and electronics installers and repairers is expected to grow more slowly than the average for all occupations over the 2000–2010 period, but varies by occupational specialty. In addition to employment growth, many job openings should result from the need to replace workers who transfer to other occupations or leave the labor force.

Average employment growth is projected for electrical and electronics installers and repairers of transportation equipment. Commercial and industrial electronic equipment will become more sophisticated and used more frequently, as businesses strive to lower costs by increasing and improving automation. Companies will install electronic controls, robots, sensors, and other equipment to automate processes such as assembly and testing. As prices decline, applications will be found across a number of industries, including services, utilities, and construction, as well as manufacturing. Improved equipment reliability should not constrain employment growth, however; companies increasingly will rely on repairers, because any malfunction that idles commercial and industrial equipment is costly.

Employment of electronics installers and repairers of motor vehicles also is expected to grow about as fast as average. Motor vehicle manufacturers will install more and better sound, security, entertainment, and navigation systems in new vehicles, limiting employment growth for after-market electronic equipment installers. However, repairing the new electronic systems should help drive employment growth.

On the other hand, employment of electric motor, power tool, and related repairers is expected to grow more slowly than average. Improvements in electrical and electronic equipment design should limit job growth by simplifying repair tasks. More parts are being designed to be easily disposable, further reducing employment growth.

Employment of electrical and electronics installers and repairers, powerhouse, substation, and relay is expected to decline slightly. Consolidation and privatization in utilities industries should improve productivity, reducing employment. Newer equipment will be more reliable and easier to repair, further limiting employment.

Earnings

Median hourly earnings of electrical and electronics repairers, commercial and industrial equipment were $17.75 in 2000. The middle 50 percent earned between $13.92 and $21.32. The lowest 10 percent earned less than $10.90, and the highest 10 percent earned more than $25.78.

Median hourly earnings of electric motor, power tool, and related repairers were $15.80 in 2000. The middle 50 percent earned between $11.91 and $20.04. The lowest 10 percent earned less than $9.13, and the highest 10 percent earned more than $25.17.

Median hourly earnings of electrical and electronics repairers, powerhouse, substation, and relay were $23.34 in 2000. The middle 50 percent earned between $19.07 and $26.21. The lowest 10 percent earned less than $14.79, and the highest 10 percent earned more than $29.00.

Median hourly earnings of electrical and electronics repairers, transportation equipment were $16.93 in 2000. The middle 50 percent earned between $12.25 and $21.54. The lowest 10 percent earned less than $9.60, and the highest 10 percent earned more than $25.76.

Median hourly earnings of electronics installers and repairers, motor vehicles were $12.06 in 2000. The middle 50 percent earned between $9.60 and $15.25. The lowest 10 percent earned less than $7.98, and the highest 10 percent earned more than $18.69.

Related Occupations

Workers in other occupations who install and repair electronic equipment include broadcast and sound engineering technicians and radio operators; computer, automated teller, and office machine repairers; electronic home entertainment equipment installers and repairers; and radio and telecommunications equipment installers and repairers. Industrial machinery installation, repair, and maintenance workers also install, maintain, and repair industrial machinery.

Sources of Additional Information

For information on careers and certification, contact:

- International Society of Certified Electronics Technicians, 3608 Pershing Ave., Fort Worth, TX 76107-4527. Internet: http://www.iscet.org
- Electronics Technicians Association, 502 North Jackson, Greencastle, IN 46135

Electronic Home Entertainment Equipment Installers and Repairers

O*NET 49-2097.00

Significant Points

- Employment is expected to decline because it often is cheaper to replace than to repair equipment.
- Job opportunities will be best for applicants with knowledge of electronics and related hands-on experience.

Nature of the Work

Electronic home entertainment equipment installers and repairers, also called *service technicians*, repair a variety of equipment, including televisions and radios, stereo components, video and audio disc players, video cameras, and videocassette recorders. They also repair home security systems, intercom equipment, and home theater equipment, which consist of large-screen televisions and sophisticated, surround-sound systems.

Customers usually bring small, portable equipment to repair shops for servicing. Repairers at these locations, known as *bench technicians*, are equipped with a full array of electronic tools and parts. When larger, less mobile equipment breaks down, customers may pay repairers to come to their homes. These repairers, known as *field technicians*, travel with a limited set of tools and parts, and attempt to complete the repair at the customer's location. If the repair is complex, technicians may bring defective components back to the repair shop for a thorough diagnosis and repair.

When equipment breaks down, repairers check for common causes of trouble, such as dirty or defective components. Many repairs consist of simply cleaning and lubricating equipment. For example, cleaning the tape heads on a videocassette recorder will prevent tapes from sticking to the equipment. If routine checks do not locate the trouble, repairers may refer to schematics and manufacturers' specifications that provide instructions on how to locate problems. Repairers use a variety of test equipment to diagnose and identify malfunctions. They use multimeters to detect short circuits, failed capacitors, and blown fuses by measuring the voltage, current, and resistance. They use color bar and dot generators to provide onscreen test patterns, signal generators to test signals, and oscilloscopes and digital storage scopes to measure complex waveforms produced by electronic equipment. Repairs may involve removing and replacing a failed capacitor, transistor, or fuse. Repairers use handtools such as pliers, screwdrivers, soldering irons, and wrenches to replace faulty parts. They also make adjustments to equipment, such as focusing and converging the picture of a television set or balancing the audio on a surround-sound system.

Improvements in technology have miniaturized and digitized many audio and video recording devices. Miniaturization has made repairwork significantly more difficult, as both the components and acceptable tolerances are smaller. For example, an analog video camera operates at 1800 revolutions per minute (rpm), while a digital video camera may operate at 9000 rpm. Components now are mounted on the surface of circuit boards, instead of plugged into slots, requiring more precise soldering when a new part is installed. Improved technologies also have lowered the price of electronic home entertainment equipment. As a result, customers often replace broken equipment instead of repairing it.

Working Conditions

Most repairers work in well-lighted electrical repair shops. Field technicians, however, spend much time traveling in service vehicles and working in customers' residences.

Repairers may have to work in a variety of positions and carry heavy equipment. Although the work of repairers is comparatively safe, they must take precautions against minor burns and electric shock. Because television monitors carry high voltage even when turned off, repairers need to discharge the voltage before servicing such equipment.

Employment

Electronic home entertainment equipment installers and repairers held about 37,000 jobs in 2000. Most repairers work in stores that sell and service electronic home entertainment products, or in electrical repair shops and service centers. About 1 in 6 electronic home entertainment equipment installers and repairers is self-employed.

Training, Other Qualifications, and Advancement

Employers prefer applicants who have basic knowledge and skills in electronics. Applicants should be familiar with schematics and have some hands-on experience repairing electronic equipment. Many applicants gain these skills at vocational training programs and community colleges. Training programs should include both a hands-on and theoretical education in digital consumer electronics. Entry-level repairers may work closely with more experienced technicians, who provide technical guidance.

Field technicians work closely with customers and must have good communications skills and a neat appearance. Employers also may require that field technicians have a driver's license.

The International Society of Certified Electronics Technicians (ISCET) and the Electronics Technicians Association (ETA) administer certification programs for electronics technicians. Repairers may specialize in a variety of skill areas, including consumer electronics. To receive certification, repairers must pass qualifying exams corresponding to their level of training and experience. Both programs offer associate certifications to entry-level repairers.

Experienced repairers with advanced training may become specialists or troubleshooters, who help other repairers diagnose difficult problems. Workers with leadership ability may become supervisors of other repairers. Some experienced workers open their own repair shops.

Job Outlook

Employment of electronic home entertainment equipment installers and repairers is expected to decline through 2010, due to decreased demand for repair work. Some job openings will occur, however, as repairers retire or gain higher paying jobs in other occupations requiring electronics experience. Opportunities will be best for applicants with hands-on experience and knowledge of electronics.

The need for repairers is declining because home entertainment equipment is less expensive than in the past. As technological developments have lowered equipment prices and improved reliability, the demand for repair services has decreased. When malfunctions do occur, it often is cheaper for consumers to replace equipment rather than to pay for repairs.

Employment of repairers will continue to decline despite the introduction of sophisticated digital equipment, such as DVDs, digital televisions, and digital camcorders. So long as the price of such equipment remains high, purchasers will be willing to hire repairers when malfunctions occur. However, the need for repairers to maintain this costly equipment will not be great enough to offset the overall decline in demand for their services.

Earnings

Median hourly earnings of electronic home entertainment equipment installers and repairers were $12.72 in 2000. The middle 50 percent earned between $9.90 and $16.63. The lowest 10 percent earned less than $7.84, and the highest 10 percent earned more than $20.72. Median hourly earnings in the industries employing the largest numbers of electronic home entertainment equipment repairers in 2000 are shown in the following table:

Electrical repair shops	$12.30
Radio, television, and computer stores	11.67

Related Occupations

Other workers who repair and maintain electronic equipment include broadcast and sound engineering technicians and radio operators; computer, automated teller, and office machine repairers; electrical and electronics installers and repairers; and radio and telecommunications equipment installers and repairers.

Sources of Additional Information

For information on careers and certification, contact:

- The International Society of Certified Electronics Technicians, 3608 Pershing Ave., Fort Worth, TX 76107. Internet: http://www.iscet.org
- Electronics Technicians Association, 502 North Jackson, Greencastle, IN 46135

Emergency Medical Technicians and Paramedics

O*NET 29-2041.00

Significant Points

- Job stress is common due to irregular hours and treating patients in life-or-death situations.
- Formal training and certification are required but state requirements vary.
- Employment is projected to grow faster than average as paid emergency medical technician positions replace unpaid volunteers.

Nature of the Work

People's lives often depend on the quick reaction and competent care of emergency medical technicians (EMTs) and paramedics, EMTs with additional advanced training to perform more difficult pre-hospital medical procedures. Incidents as varied as automobile accidents, heart attacks, drownings, childbirth, and gunshot wounds all require immediate medical attention. EMTs and paramedics provide this vital attention as they care for and transport the sick or injured to a medical facility.

Depending on the nature of the emergency, EMTs and paramedics typically are dispatched to the scene by a 911 operator and often work with police and fire department personnel. Once they arrive, they determine the nature and extent of the patient's condition while trying to ascertain whether the patient has pre-existing medical problems. Following strict rules and guidelines, they give appropriate emergency care and, when necessary, transport the patient. Some paramedics are trained to treat patients with minor injuries on the scene of an accident or at their home without transporting them to a medical facility. Emergency treatments for more complicated problems are carried out under the direction of medical doctors by radio preceding or during transport.

EMTs and paramedics may use special equipment such as backboards to immobilize patients before placing them on stretchers and securing them in the ambulance for transport to a medical facility. Usually, one EMT or paramedic drives while the other monitors the patient's vital signs and gives additional care as needed. Some EMTs work as part of the flight crew of helicopters that transport critically ill or injured patients to hospital trauma centers.

At the medical facility, EMTs and paramedics help transfer patients to the emergency department, report their observations and actions to staff, and may provide additional emergency treatment. After each run, EMTs and paramedics replace used supplies and check equipment. If a transported patient had a contagious disease, EMTs and paramedics decontaminate the interior of the ambulance and report cases to the proper authorities.

Beyond these general duties, the specific responsibilities of EMTs and paramedics depend on their level of qualification and training. To determine this, the National Registry of Emergency Medical Technicians (NREMT) registers emergency medical service (EMS) providers at four levels: First Responder, EMT-Basic, EMT-Intermediate, and EMT-Paramedic. Some states, however, do their own certification and use numeric ratings from 1 to 4 to distinguish levels of proficiency.

The lowest level workers—First Responders—are trained to provide basic emergency medical care because they tend to be the first persons to arrive at the scene of an incident. Many firefighters, police officers, and other emergency workers have this level of training. The EMT-Basic, also known as EMT-1, represents the first component of the emergency medical technician system. An EMT-1 is trained to care for patients on accident scenes and on transport by ambulance to the hospital under medical direction. The EMT-1 has the emergency skills to assess a patient's condition and manage respiratory, cardiac, and trauma emergencies.

The EMT-Intermediate (EMT-2 and EMT-3) has more advanced training that allows administration of intravenous fluids, use of manual defibrillators to give lifesaving shocks to a stopped heart, and use of advanced airway techniques and equipment to assist patients experiencing respiratory emergencies. EMT-Paramedics (EMT-4) provide the most extensive pre-hospital care. In addition to the procedures already described, paramedics may administer drugs orally and intravenously, interpret electrocardiograms (EKGs), perform endotracheal intubations, and use monitors and other complex equipment.

Working Conditions

EMTs and paramedics work both indoors and outdoors, in all types of weather. They are required to do considerable kneeling, bending, and heavy lifting. These workers risk noise-induced hearing loss from sirens and back injuries from lifting patients. In addition, EMTs and paramedics may be exposed to diseases such as Hepatitis-B and AIDS, as well as violence from drug overdose victims or mentally unstable patients. The work is not only physically strenuous, but also stressful, involving life-or-death situations and suffering patients. Nonetheless, many people find the work exciting and challenging and enjoy the opportunity to help others.

EMTs and paramedics employed by fire departments work about 50 hours a week. Those employed by hospitals frequently work between 45 and 60 hours a week, and those in private ambulance services, between 45 and 50 hours. Some of these workers, especially those in police and fire departments, are on call for extended periods. Because emergency services function 24 hours a day, EMTs and paramedics have irregular working hours that add to job stress.

Employment

EMTs and paramedics held about 172,000 jobs in 2000. Most career EMTs and paramedics work in metropolitan areas. There are many more volunteer EMTs and paramedics, especially in smaller cities, towns, and rural areas. They volunteer for fire departments, emergency medical services (EMS), or hospitals and may respond to only a few calls for service per month, or may answer the majority of calls, especially in smaller communities. EMTs and paramedics work closely with firefighters, who often are certified as EMTs as well and act as first responders.

Full- and part-time paid EMTs and paramedics were employed in a number of industries. About 4 out of 10 worked in local and suburban transportation, as employees of private ambulance services. About 3 out of 10 worked in local government for fire departments, public ambulance services and EMS. Another 2 out 10 were found in hospitals, where they worked full time within the medical facility or responded to calls in ambulances or helicopters to transport critically ill or injured patients. The remainder worked in various industries providing emergency services.

Training, Other Qualifications, and Advancement

Formal training and certification is needed to become an EMT or paramedic. All 50 states possess a certification procedure. In 38 states and the District of Columbia, registration with the National Registry of Emergency Medical Technicians (NREMT) is required at some or all levels of certification. Other states administer their own certification examination or provide the option of taking the NRMET examination. To maintain certification, EMTs and paramedics must reregister, usually every 2 years. In order to re-register, an individual must be working as an EMT or paramedic and meet a continuing education requirement.

Training is offered at progressive levels: EMT-Basic, also known as EMT-1; EMT-Intermediate, or EMT-2 and EMT-3; and EMT-paramedic, or EMT-4. The EMT-Basic represents the first level of skills required to work in the emergency medical system. Coursework typically emphasizes emergency skills such as managing respiratory, trauma, and cardiac emergencies and patient assessment. Formal courses are often combined with time in an emergency room or ambulance. The program also provides instruction and practice in dealing with bleeding, fractures, airway obstruction, cardiac arrest, and emergency childbirth. Students learn to use and maintain common emergency equipment, such as backboards, suction devices, splints, oxygen delivery systems, and stretchers. Graduates of approved EMT basic training programs who pass a written and practical examination administered by the state certifying agency or the NREMT earn the title of Registered EMT-Basic. The course also is a prerequisite for EMT-Intermediate and EMT-Paramedic training.

EMT-Intermediate training requirements vary from state to state. Applicants can opt to receive training in EMT-Shock Trauma, where the caregiver learns to start intravenous fluids and give certain medications, or in EMT-Cardiac, which includes learning heart rhythms and administering advanced medications. Training commonly includes 35 to 55 hours of additional instruction beyond EMT-Basic coursework and covers patient assessment, as well as the use of advanced airway devices and intravenous fluids. Prerequisites for taking the EMT-Intermediate examination include registration as an EMT-Basic, required classroom work, and a specified amount of clinical experience.

The most advanced level of training for this occupation is EMT-Paramedic. At this level, the caregiver receives additional training in body function and more advanced skills. The Paramedic Technology program usually lasts up to 2 years and results in an associate degree in applied science. Such education prepares the graduate to take the NREMT examination and become certified as an EMT-Paramedic. Extensive related coursework and clinical and field experience is required. Due to the longer training requirement, almost all EMT-Paramedics are in paid positions. Refresher courses and continuing education are available for EMTs and paramedics at all levels.

EMTs and paramedics should be emotionally stable, have good dexterity, agility, and physical coordination, and be able to lift and carry heavy loads. They also need good eyesight (corrective lenses may be used) with accurate color vision.

Advancement beyond the EMT-Paramedic level usually means leaving fieldwork. An EMT-Paramedic can become a supervisor, operations manager, administrative director, or executive director of emergency services. Some EMTs and paramedics become instructors, dispatchers, or physician assistants, while others move into sales or marketing of emergency medical equipment. A number of people become EMTs and paramedics to assess their interest in healthcare and then decide to return to school and become registered nurses, physicians, or other health workers.

Job Outlook

Employment of emergency medical technicians and paramedics is expected to grow faster than the average for all occupations through 2010. Population growth and urbanization will increase the demand for full-time paid EMTs and paramedics rather than for volunteers. In addition, a large segment of the population—the aging baby boomers—will further spur demand for EMT services, as they become more likely to have medical emergencies. There will still be demand for part-time, volunteer EMTs and paramedics in rural areas and smaller metropolitan areas. In addition to job growth, openings will occur because of replacement needs; some workers leave because of stressful working conditions, limited advancement potential, and the modest pay and benefits in the private sector.

Most opportunities for EMTs and paramedics are expected to arise in hospitals and private ambulance services. Competition will be greater for jobs in local government, including fire, police, and independent third service rescue squad departments, where salaries and benefits tend to be slightly better. Opportunities will be best for those who have advanced certifications, such as EMT-Intermediate and EMT-Paramedic, as clients and patients demand higher levels of care before arriving at the hospital.

Earnings

Earnings of EMTs and paramedics depend on the employment setting and geographic location as well as the individual's training and experience. Median annual earnings of EMTs and paramedics were $22,460 in 2000. The middle 50 percent earned between $17,930 and $29,270. The lowest 10 percent earned less than $14,660, and the highest 10 percent earned more than $37,760. Median annual earnings in the industries employing the largest numbers of EMTs and paramedics in 2000 were:

Local government	$24,800
Hospitals	23,590
Local and suburban transportation	20,950

Those in emergency medical services who are part of fire or police departments receive the same benefits as firefighters or police officers. For example, many are covered by pension plans that provide retirement at half pay after 20 or 25 years of service or if disabled in the line of duty.

Related Occupations

Other workers in occupations that require quick and level-headed reactions to life-or-death situations are air traffic controllers, firefighting occupations, physician assistants, police and detectives, and registered nurses.

Sources of Additional Information

General information about emergency medical technicians and paramedics is available from:

- National Association of Emergency Medical Technicians, 408 Monroe St., Clinton, MS 39056. Internet: http://www.naemt.org
- National Registry of Emergency Medical Technicians, P.O. Box 29233, Columbus, OH 43229. Internet: http://www.nremt.org
- National Highway Transportation Safety Administration, EMS Division, 400 7th St. S.W., NTS-14, Washington, DC. Internet: http://www.nhtsa.dot.gov/people/injury/ems

Engineering and Natural Sciences Managers

O*NET 11-9041.00, 11-9121.00

Significant Points

- Most engineering and natural sciences managers have previous experience as engineers, scientists, or mathematicians.
- Employers prefer managers with advanced technical knowledge and strong communication and administrative skills.

Nature of the Work

Engineering and natural sciences managers plan, coordinate, and direct research, design, and production activities. They may supervise engineers, scientists, and technicians, along with support personnel. These managers use advanced technical knowledge of engineering and science to oversee a variety of activities. They determine scientific and technical goals within broad outlines provided by top executives. These goals may include improving manufacturing processes, advancing scientific research, or redesigning aircraft. Managers make detailed plans to accomplish these goals—for example, they may develop the overall concepts of a new product or identify technical problems preventing the completion of a project.

To perform effectively, they also must possess knowledge of administrative procedures, such as budgeting, hiring, and supervision. These managers propose budgets for projects and programs and determine staff, training, and equipment purchases. They hire and assign scientists, engineers, and support personnel to carry out specific parts of each project. They also supervise the work of these employees, review their output, and establish administrative procedures and policies—including environmental standards, for example.

In addition, these managers use communication skills extensively. They spend a great deal of time coordinating the activities of their unit with those of other units or organizations. They confer with higher levels of management; with financial, production, marketing, and other managers; and with contractors and equipment and materials suppliers.

Engineering managers supervise people who design and develop machinery, products, systems, and processes; or direct and coordinate production, operations, quality assurance, testing, or maintenance in industrial plants. Many are plant engineers, who direct and coordinate the design, installation, operation, and maintenance of equipment and machinery in industrial plants. Others manage research and development teams that produce new products and processes or improve existing ones.

Natural sciences managers oversee the work of life and physical scientists, including agricultural scientists, chemists, biologists, geologists, medical scientists, and physicists. These managers direct research and development projects and coordinate activities such as testing, quality control, and production. They may work on basic research projects or on commercial activities. Science managers sometimes conduct their own research in addition to managing the work of others.

Working Conditions

Engineering and natural sciences managers spend most of their time in an office. Some managers, however, also may work in laboratories, where they may be exposed to the same conditions as research scientists, or in industrial plants, where they may be exposed to the same conditions as production workers. Most managers work at least 40 hours a week and may work much longer on occasion to meet project deadlines. Some may experience considerable pressure to meet technical or scientific goals on a short deadline or within a tight budget.

Employment

Engineering and natural sciences managers held about 324,000 jobs in 2000. Nearly 3 out of 10 worked in services industries, primarily for firms providing computer and data processing, engineering and architectural, or research and testing services. Manufacturing industries employed one-third. Manufacturing industries with the largest employment include those producing industrial machinery and equipment, electronic and other electrical equipment, transportation equipment, instruments, and chemicals. Other large employers include government agencies and transportation, communications, and utilities companies.

Training, Other Qualifications, and Advancement

Strong technical knowledge is essential for engineering and natural sciences managers, who must understand and guide the work of their subordinates and explain the work in nontechnical terms to senior management and potential customers. Therefore, these management positions usually require work experience and formal education similar to that of engineers, scientists, or mathematicians.

Most engineering managers begin their careers as engineers, after completing a bachelor's degree in the field. To advance to higher level positions, engineers generally must assume management responsibility. To fill management positions, employers seek engineers who possess administrative and communications skills in addition to technical knowledge in their specialty. Many engineers gain these skills by obtaining a master's degree in engineering management or a master's degree in business administration (MBA). Employers often pay for such training. In large firms, some courses required in these degree programs may be offered onsite. Engineers who prefer to manage in technical areas should get a master's degree in engineering management, whereas those interested in nontechnical management should get an MBA.

Many science managers begin their careers as scientists, such as chemists, biologists, geologists, or mathematicians. Most scientists or mathematicians engaged in basic research have a PhD. Some in applied research and other activities may have a bachelor's or master's degree. Science managers must be specialists in the work they supervise. In addition, employers prefer managers with good communication and administrative skills. Graduate programs allow scientists to augment their undergraduate training with instruction in other fields, such as management or computer technology. Given the rapid pace of scientific developments, science managers must continuously upgrade their knowledge.

Engineering and natural sciences managers may advance to progressively higher leadership positions within their discipline. Some may become managers in nontechnical areas such as marketing, human resources, or sales. In high technology firms, managers in nontechnical areas often must possess the same specialized knowledge as managers in technical areas. For example, employers in an engineering firm may prefer to hire experienced engineers as sales workers because only someone with specialized engineering knowledge can market the firm's complex services.

Job Outlook

Employment of engineering and natural sciences managers is expected to increase more slowly than the average for all occupations through the year 2010—in line with projected employment growth in engineering and most sciences. However, many additional jobs will result from the need to replace managers who retire or move into other occupations. Opportunities for obtaining a management position will be best for workers with advanced technical knowledge and strong communication and administrative skills.

The job outlook for engineering and natural sciences managers should be closely related to the growth of the occupations they supervise and the industries in which they are found. For example, opportunities for managers should be better in rapidly growing areas of engineering, such as electrical, computer, and biomedical engineering than in more slowly growing areas of engineering or physical science. In addition, many employers are finding it more efficient to contract engineering and science management services to outside companies and consultants, creating good opportunities for managers in management services and management consulting firms.

Earnings

Earnings for engineering and natural sciences managers vary by specialty and level of responsibility. Median annual earnings of engineering managers were $84,070 in 2000. The middle 50 percent earned between $66,420 and $105,630. The lowest 10 percent earned less than $52,350, and the highest 10 percent earned more than $130,350. Median annual earnings in the industries employing the largest numbers of engineering managers in 2000 were as follows:

Electronic components and accessories	$98,940
Computer and data processing services	98,890
Aircraft and parts	88,620
Federal government	83,840
Engineering and architectural services	83,390

Median annual earnings of natural sciences managers were $75,880 in 2000. The middle 50 percent earned between $56,320 and $100,760. The lowest 10 percent earned less than $43,110, and the highest 10 percent earned more than $128,090. Median annual earnings in the industries employing the largest numbers of natural sciences managers in 2000 were as follows:

Research and testing services	$87,070
Federal government	74,780

A survey of manufacturing firms, conducted by Abbot, Langer & Associates, found that engineering department managers and superintendents earned a median annual income of $85,154 in 1999, and research and development managers earned $84,382.

In addition, engineering and natural sciences managers, especially those at higher levels, often receive more benefits—such as expense accounts, stock option plans, and bonuses—than nonmanagerial workers do in their organizations.

Related Occupations

The work of engineering and natural sciences managers is closely related to that of engineers; mathematicians; and physical and life scientists, including agricultural and food scientists, biological and medical scientists, conservation scientists and foresters, atmospheric scientists, chemists and materials scientists, environmental scientists and geoscientists, and physicists and astronomers. It also is related to the work of other managers, especially top executives.

Sources of Additional Information

High school students interested in obtaining a full package of guidance materials and information (product number SP-01) on a variety of engineering disciplines should contact the Junior Engineering Technical Society by sending $3.50 to:

- JETS-Guidance, 1420 King St., Suite 405, Alexandria, VA 22314-2794. Internet: http://www.jets.org

High school students interested in obtaining information on ABET-accredited engineering programs should contact:

- The Accreditation Board for Engineering and Technology, Inc., 111 Market Place, Suite 1050, Baltimore, MD 21202-4012. Internet: http://www.abet.org

Information on general engineering education and career resources is available from:

- American Society for Engineering Education, 1818 N St. N.W., Suite 600, Washington, DC 20036-2479. Internet: http://www.asee.org

Information on obtaining an engineering position with the federal government is available from the Office of Personnel Management (OPM) through a telephone-based system. Consult your telephone directory under U.S. Government for a local number or call (912) 757-3000; Federal Relay Service: (800) 877-8339. The first number is not toll free, and charges may result. Information also is available from the OPM Internet site: http://www.usajobs.opm.gov

Non-high school students and those wanting more detailed information should contact societies representing the individual branches of engineering. Each can provide information about careers in the particular branch. The individual statements elsewhere in this book also provide other information in detail on aerospace; agricultural; biomedical; chemical; civil; computer hardware; electrical and electronics; environmental; industrial, including health and safety; materials; mechanical; mining and geological, including mining safety; nuclear; and petroleum engineering.

Information on careers in agricultural science is available from:

- American Society of Agronomy, Crop Science Society of America, Soil Science Society of America, 677 S. Segoe Rd., Madison, WI 53711-1086
- Food and Agricultural Careers for Tomorrow, Purdue University, 1140 Agricultural Administration Bldg., West Lafayette, IN 47907-1140

For information on careers in food technology, write to:

- Institute of Food Technologists, Suite 300, 221 N. LaSalle St., Chicago IL 60601-1291

For information on careers in the biological sciences, contact:

- American Institute of Biological Sciences, Suite 200, 1444 I St. N.W., Washington, DC 20005. Internet: http://www.aibs.org

For information about the forestry profession and lists of schools offering education in forestry, send a self-addressed, stamped business envelope to:

- Society of American Foresters, 5400 Grosvenor Lane, Bethesda, MD 20814. Internet: http://www.safnet.org

Information about careers in meteorology is available from:

- American Meteorological Society, 45 Beacon St., Boston, MA 02108. Internet: http://www.ametsoc.org/AMS

For general information on materials science, contact:

- Materials Research Society (MRS), 506 Keystone Dr., Warrendale, PA 15086-7573. Internet: http://www.mrs.org

Information on training and career opportunities for geologists is available from:

- American Geological Institute, 4220 King St., Alexandria, VA 22302-1502. Internet: http://www.agiweb.org
- Geological Society of America, P.O. Box 9140, Boulder, CO 80301-9140. Telephone: (717) 447-2020. Internet: http://www.geosociety.org
- American Association of Petroleum Geologists, P.O. Box 979, Tulsa, OK 74101. Internet: http://www.aapg.org

General information on career opportunities in physics is available from:

- American Institute of Physics, Career Services Division and Education and Employment Division, One Physics Ellipse, College Park, MD 20740-3843. Internet: http://www.aip.org
- The American Physical Society, One Physics Ellipse, College Park, MD 20740-3844. Internet: http://www.aps.org

Engineering Technicians

O*NET 17-3021.00, 17-3022.00, 17-3023.01, 17-3023.02, 17-3023.03, 17-3024.00, 17-3025.00, 17-3026.00, 17-3027.00

Significant Points

- Electrical and electronic engineering technicians make up about 45 percent of all engineering technicians.
- Because the type and quality of training programs vary considerably, prospective students should carefully investigate training programs before enrolling.
- Opportunities will be best for individuals with an associate degree or extensive job training in engineering technology.

Nature of the Work

Engineering technicians use the principles and theories of science, engineering, and mathematics to solve technical problems in research and development, manufacturing, sales, construction, inspection, and maintenance. Their work is more limited in scope and more practically oriented than that of scientists and engineers. Many engineering technicians assist engineers and scientists, especially in research and development. Others work in quality control—inspecting products and processes, conducting tests, or collecting data. In manufacturing, they may assist in product design, development, or production. Many workers who repair or maintain various types of electrical, electronic, or mechanical equipment often are called technicians.

Engineering technicians who work in research and development build or set up equipment, prepare and conduct experiments, collect data, calculate or record the results, and help engineers or scientists in other ways, such as making prototype versions of newly designed equipment. They also assist in design work, often using computer-aided design equipment.

Most engineering technicians specialize in certain areas, learning skills and working in the same disciplines as engineers. Occupational titles, therefore, tend to follow the same structure as those of engineers.

Aerospace engineering and operations technicians install, construct, maintain, and test systems used to test, launch, or track aircraft and space vehicles. They may calibrate test equipment and determine the cause of equipment malfunctions. Using computer and communications systems, aerospace engineering and operations technicians often record and interpret test data.

Chemical engineering technicians usually are employed in industries producing pharmaceuticals, chemicals, and petroleum products, among others. They work in laboratories as well as processing plants. They help develop new chemical products and processes, test processing equipment and instrumentation, gather data, and monitor quality.

Civil engineering technicians help civil engineers plan and build highways, buildings, bridges, dams, wastewater treatment systems, and other structures, and perform related surveys and studies. Some estimate construction costs and specify materials to be used, and some may even prepare drawings or perform land-surveying duties. Others may set up and monitor instruments used to study traffic conditions.

Electrical and electronics engineering technicians help design, develop, test, and manufacture electrical and electronic equipment such as communication equipment, radar, industrial and medical measuring or control devices, navigational equipment, and computers. They may work in product evaluation and testing, using measuring and diagnostic devices to adjust, test, and repair equipment.

Electrical and electronic engineering technology is also applied to a wide variety of systems such as communications and process controls. *Electromechanical engineering technicians* combine fundamental principles of mechanical engineering technology with knowledge of electrical and electronic circuits to design, develop, test, and manufacture electrical and computer-controlled mechanical systems.

Environmental engineering technicians work closely with environmental engineers and scientists in developing methods and devices used in the prevention, control, or correction of environmental hazards. They inspect and maintain equipment affecting air pollution and recycling. Some inspect water and wastewater treatment systems to ensure that pollution control requirements are met.

Industrial engineering technicians study the efficient use of personnel, materials, and machines in factories, stores, repair shops, and offices. They prepare layouts of machinery and equipment, plan the flow of work, make statistical studies, and analyze production costs.

Mechanical engineering technicians help engineers design, develop, test, and manufacture industrial machinery, consumer products, and other equipment. They may assist in product tests—by setting up instrumentation for auto crash tests, for example. They may make sketches and rough layouts, record data, make computations, analyze results, and write reports. When planning production, mechanical engineering technicians prepare layouts and drawings of the assembly process and of parts to be manufactured. They estimate labor costs, equipment life, and plant space. Some test and inspect machines and equipment in manufacturing departments or work with engineers to eliminate production problems.

Working Conditions

Most engineering technicians work at least 40 hours a week in laboratories, offices, or manufacturing or industrial plants, or on construction sites. Some may be exposed to hazards from equipment, chemicals, or toxic materials.

Employment

Engineering technicians held about 519,000 jobs in 2000. About 233,000 of these were electrical and electronics engineering technicians. About 35 percent of all engineering technicians worked in durable goods manufacturing, mainly in the electrical and electronic equipment, industrial machinery and equipment, instruments and related products, and transportation equipment industries. Another 26 percent worked in service industries, mostly in engineering or business services companies that do engineering work on contract for government, manufacturing firms, or other organizations.

In 2000, the federal government employed about 23,000 engineering technicians. The major employer was the Department of Defense, followed by the Departments of Transportation, Agriculture, and Interior, the Tennessee Valley Authority, and the National Aeronautics and Space Administration. State governments employed about 22,000, and local governments, about 21,000.

Training, Other Qualifications, and Advancement

Although it may be possible to qualify for a few engineering technician jobs without formal training, most employers prefer to hire someone with at least a 2-year associate degree in engineering technology. Training is available at technical institutes, community colleges, extension divisions of colleges and universities, public and private vocational-technical schools, and the armed forces. Persons with college courses in science, engineering, and mathematics may qualify for some positions but may need additional specialized training and experience. Although employers usually do not require engineering technicians to be certified, such certification may provide job seekers a competitive advantage.

Prospective engineering technicians should take as many high school science and math courses as possible to prepare for postsecondary programs in engineering technology. Most 2-year associate degree programs accredited by the Technology Accreditation Commission of the Accreditation Board for Engineering and Technology (TAC/ABET) require, at a minimum, college algebra and trigonometry, and one or two basic science courses. Depending on the specialty, more math or science may be required.

The type of technical courses required also depends on the specialty. For example, prospective mechanical engineering technicians may take courses in fluid mechanics, thermodynamics, and mechanical design; electrical engineering technicians may take classes in electric circuits, microprocessors, and digital electronics; and those preparing to work in environmental engineering technology need courses in environmental regulations and safe handling of hazardous materials.

Because many engineering technicians may assist in design work, creativity is desirable. Good communication skills and the ability to work well with others also is important because these workers often are part of a team of engineers and other technicians.

Engineering technicians usually begin by performing routine duties under the close supervision of an experienced technician, technologist, engineer, or scientist. As they gain experience, they are given more difficult assignments with only general supervision. Some engineering technicians eventually become supervisors.

Many publicly and privately operated schools provide technical training; the type and quality of programs vary considerably. Therefore, prospective students should be careful in selecting a program. They should contact prospective employers regarding their preferences and ask schools to provide information about the kinds of jobs obtained by graduates, instructional facilities and equipment, and faculty qualifications. Graduates of ABET-accredited programs usually are recognized to have achieved an acceptable level of competence in the mathematics, science, and technical courses required for this occupation.

Technical institutes offer intensive technical training through application and practice, but less theory and general education than community colleges. Many offer 2-year associate degree programs, and are similar to or part of a community college or state university system. Other technical institutes are run by private, often for-profit, organizations, sometimes called proprietary schools. Their programs vary considerably in length and types of courses offered, although some are 2-year associate degree programs.

Community colleges offer curriculums that are similar to those in technical institutes, but that may include more theory and liberal arts. Often there may be little or no difference between technical institute and community college programs, as both offer associate degrees. After completing the 2-year program, some graduates get jobs as engineering technicians, while others continue their education at 4-year colleges. However, there is a difference between an associate degree in pre-engineering and one in engineering technology. Students who enroll in a 2-year pre-engineering program may find it very difficult to find work as an engineering technician should they decide not to enter a 4-year engineering program, because pre-engineering programs usually focus less on hands-on applications and more on academic preparatory work. Conversely, graduates of 2-year engineering technology programs may not receive credit for many of the courses they have taken if they choose to transfer to a 4-year engineering program. Colleges with these 4-year programs usually do not offer engineering technician training, but college courses in science, engineering, and mathematics are useful for obtaining a job as an engineering technician. Many 4-year colleges offer bachelor's degrees in engineering technology, but graduates of these programs often are hired to work as technologists or applied engineers, not technicians.

Area vocational-technical schools, another source of technical training, include postsecondary public institutions that serve local students and emphasize training needed by local employers. Most require a high school diploma or its equivalent for admission.

Other training in technical areas may be obtained in the armed forces. Many military technical training programs are highly regarded by employers. However, skills acquired in military programs are often narrowly focused, so they may not be useful in civilian industry, which often requires broader training. Therefore, some additional training may be needed, depending on the acquired skills and the kind of job.

The National Institute for Certification in Engineering Technologies (NICET) has established a voluntary certification program for engineering technicians. Certification is available at various levels, each level combining a written examination in 1 of more than 30 specialties with a certain amount of job-related experience, a supervisory evaluation, and a recommendation.

Job Outlook

Opportunities will be best for individuals with an associate degree or extensive job training in engineering technology. As technology becomes more sophisticated, employers continue to look for technicians who are skilled in new technology and require a minimum of additional job training. An increase in the number of jobs affecting public health and safety should create job opportunities for certified engineering technicians.

Overall employment of engineering technicians is expected to increase about as fast as the average for all occupations through 2010. As production of technical products continues to grow, competitive pressures will force companies to improve and update manufacturing facilities and product designs more rapidly than in the past. However, the growing availability and use of advanced technologies, such as computer-aided design and drafting and computer simulation, will continue to increase productivity and limit job growth. In addition to growth, many job openings will stem from the need to replace technicians who retire or leave the labor force.

Like engineers, employment of engineering technicians is influenced by local and national economic conditions. As a result, the employment outlook varies with industry and specialization. Some types of engineering technicians, such as civil engineering and aerospace engineering and operations technicians, experience greater cyclical fluctuations in employment than do others. Increasing demand for more sophisticated electrical and electronic products, as well as the expansion of these products and systems into all areas of industry and manufacturing processes, will contribute to average growth in the largest specialty—electrical and electronics engineering technicians. At the same time, new specializations will contribute to growth among all other engineering technicians; fire protection engineering, water quality control, and environmental technology are some of many new specialties for which demand is increasing.

Earnings

Median annual earnings of electrical and electronics engineering technicians were $40,020 in 2000. The middle 50 percent earned between $31,570 and $49,680. The lowest 10 percent earned less than $25,210, and the highest 10 percent earned more than $58,320. Median annual earnings in the industries employing the largest numbers of electrical and electronics engineering technicians in 2000 are shown below.

Federal government	$50,000
Telephone communication	45,640
Engineering and architectural services	40,690
Electrical goods	38,120
Electronic components and accessories	35,500

Median annual earnings of civil engineering technicians were $35,990 in 2000. The middle 50 percent earned between $27,810 and $44,740. The lowest 10 percent earned less than $21,830, and the highest 10 percent earned more than $54,770. Median annual earnings in the industries employing the largest numbers of civil engineering technicians in 2000 are shown below.

Local government	$39,080
Engineering and architectural services	36,670
State government	32,160

In 2000, the average annual salary for aerospace engineering and operations technicians in the aircraft and parts industry was $53,340, and the average annual salary for environmental engineering technicians in engineering and architectural services was $29,960. The average annual salary for industrial engineering technicians in computer and data processing services and electric components and accessories was $73,320 and $36,300, respectively. In engineering and architectural services, the average annual salary for mechanical engineering technicians was $40,580.

Related Occupations

Engineering technicians apply scientific and engineering principles usually acquired in postsecondary programs below the baccalaureate level. Similar occupations include science technicians; drafters; surveyors, cartographers, photogrammetrists, and surveying technicians; and broadcast and sound engineering technicians and radio operators.

Sources of Additional Information

For $3.50, a full package of guidance materials and information (product number SP-01) on a variety of engineering technician and technology careers is available from:

- Junior Engineering Technical Society (JETS), 1420 King St., Suite 405, Alexandria, VA 22314-2794. Internet: http://www.jets.org

Information on ABET-accredited engineering technology programs is available from:

- Accreditation Board for Engineering and Technology, Inc., 111 Market Place, Suite 1050, Baltimore, MD 21202. Internet: http://www.abet.org

Information on certification of engineering technicians is available from:

- National Institute for Certification in Engineering Technologies (NICET), 1420 King St., Alexandria, VA 22314-2794. Internet: http://www.nicet.org

Engineers

O*NET 17-2011.00, 17-2021.00, 17-2031.00, 17-2041.00, 17-2051.00, 17-2061.00, 17-2071.00, 17-2072.00, 17-2081.00, 17-2111.01, 17-2111.02, 17-2111.03, 17-2112.00, 17-2131.00, 17-2141.00, 17-2151.00, 17-2161.00, 17-2171.00

Significant Points

- Overall job opportunities in engineering are expected to be good, but to vary by specialty.
- A bachelor's degree is required for most entry-level jobs.
- Starting salaries are significantly higher than those of college graduates in other fields are.
- Continuing education is critical to keep abreast of the latest technology.

Nature of the Work

Engineers apply the theories and principles of science and mathematics to research and develop economical solutions to technical problems. Their work is the link between perceived social needs and commercial applications. Engineers design products, machinery to build those products, factories in which those products are made, and the systems that ensure the quality of the products and efficiency of the workforce and manufacturing process. Engineers design, plan, and supervise the construction of buildings, highways, and transit systems. They develop and implement improved ways to extract, process, and use raw materials, such as petroleum and natural gas. They develop new materials that both improve the performance of products and take advantage of advances in technology. They harness the power of the sun, the earth, atoms, and electricity for use in supplying the nation's power needs, and create millions of products using power. They analyze the impact of the products they develop or the systems they design on the environment and people using them. Engineering knowledge is applied to improving many things, including the quality of healthcare, the safety of food products, and the efficient operation of financial systems.

Engineers consider many factors when developing a new product. For example, in developing an industrial robot, engineers determine precisely what function the robot needs to perform; design and test the robot's components; fit the components together in an integrated plan; and evaluate the design's overall effectiveness, cost, reliability, and safety. This process applies to many different products, such as chemicals, computers, gas turbines, helicopters, and toys.

In addition to design and development, many engineers work in testing, production, or maintenance. These engineers supervise production in factories, determine the causes of breakdowns, and test manufactured products to maintain quality. They also estimate the time and cost to complete projects. Some move into engineering management or into sales. In sales, an engineering background enables them to discuss technical aspects and assist in product planning, installation, and use.

Most engineers specialize. More than 25 major specialties are recognized by professional societies, and the major branches have numerous subdivisions. Some examples include structural, environmental, and transportation engineering, which are subdivisions of civil engineering; and ceramic, metallurgical, and polymer engineering, which are subdivisions of materials engineering. Engineers also may specialize in one industry, such as motor vehicles, or in one field of technology, such as turbines or semiconductor materials.

Some branches of engineering have established college programs, such as architectural engineering (the design of a building's internal support structure) and marine engineering (the design and installation of ship machinery and propulsion systems).

Engineers in each branch have a base of knowledge and training that can be applied in many fields. Electronics engineers, for example, work in the medical, computer, communications, and missile guidance fields. Because there are many separate problems to solve in a large engineering project, engineers in one field often work closely with specialists in other scientific, engineering, and business occupations.

Engineers use computers to produce and analyze designs; to simulate and test how a machine, structure, or system operates; and to generate specifications for parts. New communications technologies using computers are changing the way engineers work on designs. Engineers can collaborate on designs with other engineers around the country or even abroad, using the Internet or related communications systems. Many engineers also use computers to monitor product quality and control process efficiency. They spend a great deal of time writing reports and consulting with other engineers, as complex projects often require an interdisciplinary team of engineers. Supervisory engineers are responsible for major components or entire projects.

Working Conditions

Most engineers work in office buildings, laboratories, or industrial plants. Others may spend time outdoors at construction sites, mines, and oil and gas exploration and production sites, where they monitor or direct operations or solve onsite problems. Some engineers travel extensively to plants or work sites.

Many engineers work a standard 40-hour week. At times, deadlines or design standards may bring extra pressure to a job. When this happens, engineers may work longer hours and experience considerable stress.

Employment

In 2000, engineers held 1.5 million jobs. The following table shows the distribution of employment by engineering specialty.

Specialty	Employment	Percent
Total, all engineers	1,465,000	100
Electrical and electronics	288,000	20
Civil	232,000	16
Mechanical	221,000	15
Industrial, including health and safety	198,000	14
Computer hardware	60,000	4
Environmental	52,000	4
Aerospace	50,000	3
Chemical	33,000	2
Materials	33,000	2
Nuclear	14,000	1
Petroleum	9,000	1
Biomedical	7,200	<1
Mining and geological, including mining safety	6,500	<1
Marine engineers and naval architects	5,100	<1
Agricultural	2,400	<1
All other engineers	253,000	17

Almost half of all wage and salary engineering jobs were found in manufacturing industries, such as transportation equipment, electrical and electronic equipment, industrial machinery, and instruments and related products. About 401,000 wage and salary jobs were in services industries, primarily in engineering and architectural services, research and testing services, and business services, where firms designed construction projects or did other engineering work on a contractual basis. Engineers also worked in the construction and transportation, communications and utilities industries.

Federal, state, and local governments employed about 179,000 engineers in 2000. Almost half of these were in the federal government, mainly in the Departments of Defense, Transportation, Agriculture, Interior, and Energy, and in the National Aeronautics and Space Administration. Most engineers in state and local government agencies worked in highway and public works departments. In 2000, about 43,000 engineers were self-employed, many as consultants.

Engineers are employed in every state, in small and large cities, and in rural areas. Some branches of engineering are concentrated in particular industries and geographic areas, as discussed later in this chapter.

Training, Other Qualifications, and Advancement

A bachelor's degree in engineering is required for almost all entry-level engineering jobs. College graduates with a degree in a physical science or mathematics occasionally may qualify for some engineering jobs, especially in specialties in high demand. Most engineering degrees are granted in electrical, electronics, mechanical, or civil engineering. However, engineers trained in one branch may work in related branches. For example, many aerospace engineers have training in mechanical engineering. This flexibility allows employers to meet staffing needs in new technologies and specialties in which engineers are in short supply. It also allows engineers to shift to fields with better employment prospects or to those that more closely match their interests.

Most engineering programs involve a concentration of study in an engineering specialty, along with courses in both mathematics and science. Most programs include a design course, sometimes accompanied by a computer or laboratory class or both.

In addition to the standard engineering degree, many colleges offer 2- or 4-year degree programs in engineering technology. These programs, which usually include various hands-on laboratory classes that focus on current issues, prepare students for practical design and production work, rather than for jobs which require more theoretical and scientific knowledge. Graduates of 4-year technology programs may get jobs similar to those obtained by graduates with a bachelor's degree in engineering. Engineering technology graduates, however, are not qualified to register as professional engineers under the same terms as graduates with degrees in engineering. Some employers regard technology program graduates as having skills between those of a technician and an engineer.

Graduate training is essential for engineering faculty positions and many research and development programs, but is not required for the majority of entry-level engineering jobs. Many engineers obtain graduate degrees in engineering or business administration to learn new technology and broaden their education. Many high-level executives in government and industry began their careers as engineers.

About 330 colleges and universities offer bachelor's degree programs in engineering that are accredited by the Accreditation Board for Engineering and Technology (ABET), and about 250 colleges offer accredited bachelor's degree programs in engineering technology. ABET accreditation is based on an examination of an engineering program's student achievement, program improvement, faculty, curricular content, facilities, and institutional commitment. Although most institutions offer programs in the major branches of engineering, only a few offer programs in the smaller specialties. Also, programs of the same title may vary in content. For example, some programs emphasize industrial practices, preparing students for a job in industry, whereas others are more theoretical and are designed to prepare students for graduate work. Therefore, students should investigate curricula and check accreditations carefully before selecting a college.

Admissions requirements for undergraduate engineering schools include a solid background in mathematics (algebra, geometry, trigonometry, and calculus) and sciences (biology, chemistry, and physics), and courses in English, social studies, humanities, and computers. Bachelor's degree programs in engineering typically are designed to last 4 years, but many students find that it takes between 4 and 5 years to complete their studies. In a typical 4-year college curriculum, the first 2 years are spent studying mathematics, basic sciences, introductory engineering, humanities, and social sciences. In the last 2 years, most courses are in engineering, usually with a concentration in one branch. For example, the last 2 years of an aerospace program might include courses in fluid mechanics, heat transfer, applied aerodynamics, analytical mechanics, flight vehicle design, trajectory dynamics, and aerospace propulsion systems. Some programs offer a general engineering curriculum; students then specialize in graduate school or on the job.

Some engineering schools and 2-year colleges have agreements whereby the 2-year college provides the initial engineering education, and the engineering school automatically admits students for their last 2 years. In addition, a few engineering schools have arrangements whereby a student spends 3 years in a liberal arts college studying pre-engineering subjects and 2 years in an engineering school studying core subjects, and then receives a bachelor's degree from each school. Some colleges and universities offer 5-year master's degree programs. Some 5- or even 6-year cooperative plans combine classroom study and practical work, permitting students to gain valuable experience and finance part of their education. All 50 states and the District of Columbia usually require licensure for engineers who offer their services directly to the public. Engineers who are licensed are called Professional Engineers (PE). This licensure generally requires a degree from an ABET-accredited engineering program, 4 years of relevant work experience, and successful completion of a state examination. Recent graduates can start the licensing process by taking the examination in two stages. The initial Fundamentals of Engineering (FE) examination can be taken upon graduation. Engineers who pass this examination commonly are called Engineers in Training (EIT) or Engineer Interns (EI). The EIT certification usually is valid for 10 years. After acquiring suitable work experience, EITs can take the second examination, the Principles and Practice of Engineering Exam. Several states have imposed mandatory continuing education requirements for relicensure. Most states recognize licensure from other states.

Engineers should be creative, inquisitive, analytical, and detail-oriented. They should be able to work as part of a team and to communicate well, both orally and in writing. Communication abilities are becoming more important because much of their work is becoming more diversified, meaning that engineers interact with specialists in a wide range of fields outside engineering.

Beginning engineering graduates usually work under the supervision of experienced engineers and, in large companies, also may receive formal classroom or seminar-type training. As new engineers gain knowledge and experience, they are assigned more difficult projects with greater independence to develop designs, solve problems, and make decisions. Engineers may advance to become technical specialists or to supervise a staff or team of engineers and technicians. Some may eventually become engineering managers or enter other managerial or sales jobs.

Job Outlook

Overall engineering employment is expected to increase more slowly than the average for all occupations. However, overall job opportunities in engineering are expected to be good through 2010 because the number of engineering degrees granted is not expected to increase significantly over the 2000–2010 period. Projected employment growth and, thus, job opportunities vary by specialty, ranging from a decline in employment of mining and geological engineers to faster-than-average growth among environmental engineers. Competitive pressures and advancing technology will force companies to improve and update product designs and to optimize their manufacturing processes. Employers will rely on engineers to further increase productivity, as investment in plant and equipment increases to expand output of goods and services. New computer and communications systems have improved the design process, enabling engineers to produce and analyze various product designs much more rapidly than in the past and to collaborate on designs with other engineers throughout the world. Despite these widespread applications, computer technology is not expected to limit employment opportunities. Finally, additional engineers will be needed to improve or build new roads, bridges, water and pollution control systems, and other public facilities.

Many engineering jobs are related to developing technologies used in national defense. Because defense expenditures—particularly expenditures for aircraft, missiles, and other weapons systems—are not expected to return to previously high levels, job outlook may not be as favorable for engineers working in defense-related fields although defense expenditures are expected to increase.

The number of bachelor's degrees awarded in engineering began declining in 1987 and has continued to stay at about the same level through much of the 1990s. The total number of graduates from engineering programs is not expected to increase significantly over the projection period.

Although only a relatively small proportion of engineers leaves the profession each year, many job openings will arise from replacement needs. A greater proportion of replacement openings is created by engineers who transfer to management, sales, or other professional occupations than by those who leave the labor force.

Most industries are less likely to lay off engineers than other workers. Many engineers work on long-term research and development projects or in other activities that continue even during economic slowdowns. In industries such as electronics and aerospace, however, large cutbacks in defense expenditures and government research and development funds, as well as the trend toward contracting out engineering work to engineering services firms, have resulted in significant layoffs for engineers.

It is important for engineers, like those working in other technical occupations, to continue their education throughout their careers because much of their value to their employer depends on their knowledge of the latest technology. Although the pace of technological change varies by engineering specialty and industry, advances in technology have significantly affected every engineering discipline. Engineers in high-technology areas, such as advanced electronics or information technology, may find that technical knowledge can become obsolete rapidly. Even those who continue their education are vulnerable to layoffs if the particular technology or product in which they have specialized becomes obsolete. By keeping current in their field, engineers are able to deliver the best solutions and greatest value to their employers. Engineers who have not kept current in their field may find themselves passed over for promotions or vulnerable to layoffs, should they occur. On the other hand, it often is these high-technology areas that offer the greatest challenges, the most interesting work, and the highest salaries. Therefore, the choice of engineering specialty and employer involves an assessment not only of the potential rewards but also of the risk of technological obsolescence.

Earnings

Earnings will vary by engineering speciality. Starting salaries are significantly higher than those of college graduates in other fields.

Related Occupations

Engineers apply the principles of physical science and mathematics in their work. Other workers who use scientific and mathematical principles include architects, except landscape and naval; engineering and natural sciences managers; computer and information systems managers; mathematicians; drafters; engineering technicians; sales engineers; science technicians; and physical and life scientists, including agricultural and food scientists, biological and medical scientists, conservation scientists and foresters, atmospheric scientists, chemists and materials scientists, environmental scientists and geoscientists, and physicists and astronomers.

Sources of Additional Information

High school students interested in obtaining a full package of guidance materials and information (product number SP-01) on a variety of engineering disciplines should contact the Junior Engineering Technical Society by sending $3.50 to:

- JETS-Guidance, 1420 King St., Suite 405, Alexandria, VA 22314-2794. Internet: http://www.jets.org

High school students interested in obtaining information on ABET-accredited engineering programs should contact:

- The Accreditation Board for Engineering and Technology, Inc., 111 Market Place, Suite 1050, Baltimore, MD 21202-4012. Internet: http://www.abet.org

Non-licensed engineers and college students interested in obtaining information on Professional Engineer licensure should contact:

- The National Society of Professional Engineers, 1420 King St., Alexandria, VA 22314-2794. Internet: http://www.nspe.org

- National Council of Examiners for Engineers and Surveying, P.O. Box 1686, Clemson, SC 29633-1686. Internet: http://www.ncees.org

Information on general engineering education and career resources is available from:

- American Society for Engineering Education, 1818 N St. N.W., Suite 600, Washington, DC 20036-2479. Internet: http://www.asee.org

Information on obtaining an engineering position with the federal government is available from the Office of Personnel Management (OPM) through a telephone-based system. Consult your telephone directory under U.S. Government for a local number or call (912) 757-3000; Federal Relay Service: (800) 877-8339. The first number is not toll free, and charges may result. Information also is available from the OPM Internet site: http://www.usajobs.opm.gov

Non-high school students and those wanting more detailed information should contact societies representing the individual branches of engineering. Each can provide information about careers in the particular branch. The individual statements elsewhere in this book also provide other information in detail on aerospace; agricultural; biomedical; chemical; civil; computer hardware; electrical and electronics; environmental; industrial, including health and safety; materials; mechanical; mining and geological, including mining safety; nuclear; and petroleum engineering.

Environmental Engineers

O*NET 17-2081.00

Significant Points

- Overall job opportunities in engineering are expected to be good.
- A bachelor's degree is required for most entry-level jobs.
- Starting salaries are significantly higher than those of college graduates in other fields.
- Continuing education is critical to keep abreast of the latest technology.

Nature of the Work

Using the principles of biology and chemistry, environmental engineers develop methods to solve problems related to the environment. They are involved in water and air pollution control, recycling, waste disposal, and public health issues. Environmental engineers conduct hazardous-waste management studies, evaluate the significance of the hazard, offer analysis on treatment and containment, and develop regulations to prevent mishaps. They design municipal sewage and industrial wastewater systems. They analyze scientific data, research controversial projects, and perform quality control checks.

Environmental engineers are concerned with local and worldwide environmental issues. They study and attempt to minimize the effects of acid rain, global warming, automobile emissions, and ozone depletion. They also are involved in the protection of wildlife.

Many environmental engineers work as consultants, helping their clients comply with regulations and clean up hazardous sites, including brownfields, which are abandoned urban or industrial sites that may contain environmental hazards.

Working Conditions

Most engineers work in office buildings, laboratories, or industrial plants. Others may spend time outdoors at construction sites, mines, and oil and gas exploration and production sites, where they monitor or direct operations or solve onsite problems. Some engineers travel extensively to plants or work sites.

Many engineers work a standard 40-hour week. At times, deadlines or design standards may bring extra pressure to a job. When this happens, engineers may work longer hours and experience considerable stress.

Employment

Environmental engineers held about 52,000 jobs in 2000. More than one-third worked in engineering and management services, and about 16,000 were employed in federal, state, and local government agencies. Most of the rest worked in various manufacturing industries.

Training, Other Qualifications, and Advancement

A bachelor's degree in engineering is required for almost all entry-level engineering jobs. College graduates with a degree in a physical science or mathematics occasionally may qualify for some engineering jobs, especially in specialties in high demand. Most engineering degrees are granted in electrical, electronics, mechanical, or civil engineering. However, engineers trained in one branch may work in related branches. This flexibility allows employers to meet staffing needs in new technologies and specialties in which engineers are in short supply. It also allows engineers to shift to fields with better employment prospects or to those that more closely match their interests.

Most engineering programs involve a concentration of study in an engineering specialty, along with courses in both mathematics and science. Most programs include a design course, sometimes accompanied by a computer or laboratory class or both.

In addition to the standard engineering degree, many colleges offer 2- or 4-year degree programs in engineering technology. These programs, which usually include various hands-on laboratory classes that focus on current issues, prepare students for practical design and production work, rather than for jobs which require more theoretical and scientific knowledge. Graduates of 4-year technology programs may get jobs similar to those obtained by graduates with a bachelor's degree in engineering. Engineering technology graduates, however, are not qualified to register as professional engineers under the same terms as graduates with degrees in engineering. Some employers regard technology program graduates as having skills between those of a technician and an engineer.

Graduate training is essential for engineering faculty positions and many research and development programs, but is not required for the majority of entry-level engineering jobs. Many engineers obtain graduate degrees in engineering or business administration to learn new technology and broaden their education. Many high-level executives in government and industry began their careers as engineers.

About 330 colleges and universities offer bachelor's degree programs in engineering that are accredited by the Accreditation Board for Engineering and Technology (ABET), and about 250 colleges offer accredited bachelor's degree programs in engineering technology. ABET accreditation is based on an examination of an engineering program's student achievement, program improvement, faculty, curricular content, facilities, and institutional commitment. Although most institutions offer programs in the major branches of engineering, only a few offer programs in the smaller specialties. Also, programs of the same title may vary in content. For example, some programs emphasize industrial practices, preparing students for a job in industry, whereas others are more theoretical and are designed to prepare students for graduate work. Therefore, students should investigate curricula and check accreditations carefully before selecting a college.

Admissions requirements for undergraduate engineering schools include a solid background in mathematics (algebra, geometry, trigonometry, and calculus) and sciences (biology, chemistry, and physics), and courses in English, social studies, humanities, and computers. Bachelor's degree programs in engineering typically are designed to last 4 years, but many students find that it takes between 4 and 5 years to complete their studies. In a typical 4-year college curriculum, the first 2 years are spent studying mathematics, basic sciences, introductory engineering, humanities, and social sciences. In the last 2 years, most courses are in engineering, usually with a concentration in one branch. Some programs offer a general engineering curriculum; students then specialize in graduate school or on the job.

Some engineering schools and 2-year colleges have agreements whereby the 2-year college provides the initial engineering education, and the engineering school automatically admits students for their last 2 years. In addition, a few engineering schools have arrangements whereby a student spends 3 years in a liberal arts college studying pre-engineering subjects and 2 years in an engineering school studying core subjects, and then receives a bachelor's degree from each school. Some colleges and universities offer 5-year master's degree programs. Some 5- or even 6-year cooperative plans combine classroom study and practical work, permitting students to gain valuable experience and finance part of their education. All 50 states and the District of Columbia usually require licensure for engineers who offer their services directly to the public. Engineers who are licensed are called Professional Engineers (PE). This licensure generally requires a degree from an ABET-accredited engineering program, 4 years of relevant work experience, and successful completion of a state examination. Recent graduates can start the licensing process by taking the examination in two stages. The initial Fundamentals of Engineering (FE) examination can be taken upon graduation. Engineers who pass this examination commonly are called Engineers in Training (EIT) or Engineer Interns (EI). The EIT certification usually is valid for 10 years. After acquiring suitable work experience, EITs can take the second examination, the Principles and Practice of Engineering Exam. Several states have imposed mandatory continuing education requirements for relicensure. Most states recognize licensure from other states.

Engineers should be creative, inquisitive, analytical, and detail oriented. They should be able to work as part of a team and to communicate well, both orally and in writing. Communication abilities are becoming more important because much of their work is becoming more diversified, meaning that engineers interact with specialists in a wide range of fields outside engineering.

Beginning engineering graduates usually work under the supervision of experienced engineers and, in large companies, also may receive formal classroom or seminar-type training. As new engineers gain knowledge and experience, they are assigned more difficult projects with greater independence to develop designs, solve problems, and make decisions. Engineers may advance to become technical specialists or to supervise a staff or team of engineers and technicians. Some may eventually become engineering managers or enter other managerial or sales jobs.

Job Outlook

Employment of environmental engineers is expected to increase faster than the average for all occupations through 2010. More environmental engineers will be needed to meet environmental regulations and to develop methods of cleaning up existing hazards. A shift in emphasis toward preventing problems rather than controlling those that already exist, as well as increasing public health concerns, also will spur demand for environmental engineers. However, political factors determine the job outlook for environmental engineers more than that for other engineers. Looser environmental regulations would reduce job opportunities; stricter regulations would enhance opportunities.

Even though employment of environmental engineers should be less affected by economic conditions than that of most other types of engineers, a significant economic downturn could reduce the emphasis on environmental protection, reducing employment opportunities. Environmental engineers need to keep abreast of a range of environmental issues to ensure steady employment because their area of focus may change frequently—for example, from hazardous-waste site cleanup to the prevention of water pollution.

Earnings

Median annual earnings of environmental engineers were $57,780 in 2000. The middle 50 percent earned between $45,740 and $71,280. The lowest 10 percent earned less than $37,210, and the highest 10 percent earned more than $87,290. Median annual earnings in the industries employing the largest numbers of environmental engineers in 2000 were:

Engineering and architectural services	$53,580
State government	53,210
Management and public relations	52,100

According to a 2001 salary survey by the National Association of Colleges and Employers, bachelor's degree candidates in environmental engineering received starting offers averaging $51,167 a year.

Related Occupations

Environmental engineers apply the principles of physical science and mathematics in their work. Other workers who use scientific and mathematical principles include architects, except landscape and naval; engineering and natural sciences managers; computer and information systems managers; mathematicians; drafters; engineering technicians; sales engineers; science technicians; and physical and life scientists, including agricultural and food scientists, biological and medical scientists, conservation scientists and foresters, atmospheric scientists, chemists and materials scientists, environmental scientists and geoscientists, and physicists and astronomers.

Sources of Additional Information

Further information about environmental engineers can be obtained from:

- American Academy of Environmental Engineers, 130 Holiday Court, Suite 100, Annapolis, MD 21401. Internet: http://www.enviro-engrs.org

High school students interested in obtaining a full package of guidance materials and information (product number SP-01) on a variety of engineering disciplines should contact the Junior Engineering Technical Society by sending $3.50 to:

- JETS-Guidance, 1420 King St., Suite 405, Alexandria, VA 22314-2794. Internet: http://www.jets.org

High school students interested in obtaining information on ABET-accredited engineering programs should contact:

- The Accreditation Board for Engineering and Technology, Inc., 111 Market Place, Suite 1050, Baltimore, MD 21202-4012. Internet: http://www.abet.org

Non-licensed engineers and college students interested in obtaining information on Professional Engineer licensure should contact:

- The National Society of Professional Engineers, 1420 King St., Alexandria, VA 22314-2794. Internet: http://www.nspe.org
- National Council of Examiners for Engineers and Surveying, P.O. Box 1686, Clemson, SC 29633-1686. Internet: http://www.ncees.org

Information on general engineering education and career resources is available from:

- American Society for Engineering Education, 1818 N St. N.W., Suite 600, Washington, DC 20036-2479. Internet: http://www.asee.org

Information on obtaining an engineering position with the federal government is available from the Office of Personnel Management (OPM) through a telephone-based system. Consult your telephone directory under U.S. Government for a local number or call (912) 757-3000; Federal Relay Service: (800) 877-8339. The first number is not toll free, and charges may result. Information also is available from the OPM Internet site: http://www.usajobs.opm.gov

Non-high school students and those wanting more detailed information should contact societies representing the individual branches of engineering. Each can provide information about careers in the particular branch. The individual statements elsewhere in this book also provide other information in detail on aerospace; agricultural; biomedical; chemical; civil; computer hardware; electrical and electronics; industrial, including health and safety; materials; mechanical; mining and geological, including mining safety; nuclear; and petroleum engineering.

Environmental Scientists and Geoscientists

O*NET 19-2041.00, 19-2042.01, 19-2043.00

Significant Points

- Work at remote field sites is common.
- A bachelor's degree in geology or geophysics is adequate for entry-level jobs; better jobs with good advancement potential usually require at least a master's degree.
- A PhD degree is required for most research positions in colleges and universities and in government.

Nature of the Work

Environmental scientists and geoscientists use their knowledge of the physical makeup and history of the earth to locate water, mineral, and energy resources; protect the environment; predict future geologic hazards; and offer advice on construction and land use projects.

Environmental scientists conduct research to identify and abate or eliminate sources of pollutants that affect people, wildlife, and their environments. They analyze and report measurements and observations of air, water, soil, and other sources to make recommendations on how best to clean and preserve the environment. They often use their skills and knowledge to design and monitor waste disposal sites, preserve water supplies, and reclaim contaminated land and water to comply with federal environmental regulations.

Geoscientists study the composition, structure, and other physical aspects of the earth. By using sophisticated instruments and analyses of the earth and water, geoscientists study the earth's geologic past and present in order to make predictions about its future. For example, they may study the earth's movements to try to predict when and where the next earthquake or volcano will occur and the probable impact on surrounding areas to minimize the damage. Many geoscientists are involved in the search for oil and gas, while others work closely with environmental scientists in preserving and cleaning up the environment.

Geoscientists usually study, and are subsequently classified in, one of several closely related fields of geoscience, including geology, geophysics, and oceanography. *Geologists* study the composition, processes, and history of the earth. They try to find out how rocks were formed and what has happened to them since formation. They also study the evolution of life by analyzing plant and animal fossils. *Geophysicists* use the principles of physics, mathematics, and chemistry to study not only the earth's surface, but also its internal composition; ground and surface waters; atmosphere; oceans; and its magnetic, electrical, and gravitational forces. *Oceanographers* use their knowledge of geology and geophysics, in addition to biology and chemistry, to study the world's oceans and coastal waters. They study the motion and circulation of the ocean waters and their physical and chemical properties, and how these properties affect coastal areas, climate, and weather.

Geoscientists can spend a large part of their time in the field identifying and examining rocks, studying information collected by remote sensing instruments in satellites, conducting geological surveys, constructing field maps, and using instruments to measure the earth's gravity and magnetic field. For example, they often perform seismic studies, which involve bouncing energy waves off buried rock layers, to search for oil and gas or understand the structure of subsurface rock layers. Seismic signals generated by earthquakes are used to determine the earthquake's location and intensity.

In laboratories, geologists and geophysicists examine the chemical and physical properties of specimens. They study fossil remains of animal and plant life or experiment with the flow of water and oil through rocks. Some geoscientists use two- or three-dimensional computer modeling to portray water layers and the flow of water or other fluids through rock cracks and porous materials. They use a variety of sophisticated laboratory instruments, including X-ray diffractometers, which determine the crystal structure of minerals, and petrographic microscopes, for the study of rock and sediment samples.

Geoscientists working in mining or the oil and gas industry sometimes process and interpret data produced by remote sensing satellites to help identify potential new mineral, oil, or gas deposits. Seismic technology also is an important exploration tool. Seismic waves are used to develop a three-dimensional picture of underground or underwater rock formations. Seismic reflection technology may also reveal unusual underground features that sometimes indicate accumulations of natural gas or petroleum, facilitating exploration and reducing the risks associated with drilling in previously unexplored areas.

Numerous subdisciplines or specialties fall under the two major disciplines of geology and geophysics that further differentiate the type of work geoscientists do. For example, *petroleum geologists* explore for oil and gas deposits by studying and mapping the subsurface of the ocean or land. They use sophisticated geophysical instrumentation, well log data, and computers to interpret geological information. *Engineering geologists* apply geologic principles to the fields of civil and environmental engineering, offering advice on major construction projects and assisting in environmental remediation and natural hazard reduction projects. *Mineralogists* analyze and classify minerals and precious stones according to composition and structure and study their environment in order to find new mineral resources. *Paleontologists* study fossils found in geological formations to trace the evolution of plant and animal life and the geologic history of the earth. *Stratigraphers* study the formation and layering of rocks to understand the environment in which they were formed. *Volcanologists* investigate volcanoes and volcanic phenomena to try to predict the potential for future eruptions and possible hazards to human health and welfare.

Geophysicists may specialize in areas such as geodesy, seismology, or magnetic geophysics. *Geodesists* study the size and shape of the earth, its gravitational field, tides, polar motion, and rotation. *Seismologists* interpret data from seismographs and other geophysical instruments to detect earthquakes and locate earthquake-related faults. *Geochemists* study the nature and distribution of chemical elements in ground water and earth materials. *Geomagnetists* measure the earth's magnetic field and use measurements taken over the past few centuries to devise theoretical models to explain the earth's origin. *Paleomagnetists* interpret fossil magnetization in rocks and sediments from the continents and oceans, to record the spreading of the sea floor, the wandering of the continents, and the many reversals of polarity that the earth's magnetic field has undergone through time. Other geophysicists study atmospheric sciences and space physics.

Hydrology is closely related to the disciplines of geology and geophysics. *Hydrologists* study the quantity, distribution, circulation, and physical properties of underground and surface waters. They study the form and intensity of precipitation, its rate of infiltration into the soil, its movement through the earth, and its return to the ocean and atmosphere. The work they do is particularly important in environmental preservation, remediation, and flood control.

Oceanography also has several subdisciplines. *Physical oceanographers* study the ocean tides, waves, currents, temperatures, density, and salinity. They study the interaction of various forms of energy, such as light, radar, sound, heat, and wind with the sea, in addition to investigating the relationship between the sea, weather, and climate. Their studies provide the maritime fleet with up-to-date oceanic conditions. *Chemical oceanographers* study the distribution of chemical compounds and chemical interactions that occur in the ocean and sea floor. They may investigate how pollution affects the chemistry of the ocean. *Geological and geophysical oceanographers* study the topographic features and the physical makeup of the ocean floor. Their knowledge can help oil and gas producers find these minerals on the bottom of the ocean. *Biological oceanographers*, often called marine biologists, study the distribution and migration patterns of the many diverse forms of sea life in the ocean.

Working Conditions

Some geoscientists spend the majority of their time in an office, but many others divide their time between fieldwork and office or laboratory work. Geologists often travel to remote field sites by helicopter or four-wheel drive vehicles and cover large areas on foot. An increasing number of exploration geologists and geophysicists work in foreign countries, sometimes in remote areas and under difficult conditions. Oceanographers may spend considerable time at sea on academic research ships. Fieldwork often requires working long hours, but workers are usually rewarded by longer than normal vacations. Environmental scientists and geoscientists in research positions with the federal government or in colleges and universities often are required to design programs and write grant proposals in order to continue their data collection and research. Environmental scientists and geoscientists in consulting jobs face similar pressures to market their skills and write proposals to maintain steady work. Travel often is required to meet with prospective clients or investors.

Employment

Environmental scientists and geoscientists held about 97,000 jobs in 2000. Environmental scientists accounted for 64,000 of the total; geoscientists, 25,000; and hydrologists, 8,000. Many more individuals held environmental science and geoscience faculty positions in colleges and universities, but they are considered college and university faculty.

Among salaried geoscientists, nearly 1 in 3 were employed in engineering and management services, and slightly more than 1 in 5 worked for oil and gas extraction companies or metal mining companies. The federal government employed about 3,100 geoscientists, including geologists, geophysicists, and oceanographers in 2000, mostly within the U.S. Department of the Interior for the U.S. Geological Survey (USGS), and the U.S. Department of Defense. More than 2,600 worked for state agencies, such as state geological surveys and state departments of conservation. About 1 geoscientist in 25 was self-employed; most were consultants to industry or government.

For environmental scientists, about 2 in 5 were employed in state and local governments, about 1 in 8 in management and public relations, 1 in 10 in engineering and architectural services, and 1 in 10 in the federal government. A small number were self-employed.

Nearly 1 in 3 hydrologist worked in the federal government in 2000. Another 1 in 5 worked in management and public relations, 1 in 6 in engineering and architectural services, and 1 in 6 for state governments.

Training, Other Qualifications, and Advancement

A bachelor's degree in geology or geophysics is adequate for some entry-level geoscientist jobs, but more job opportunities and better jobs with good advancement potential usually require at least a master's degree in geology or geophysics. Environmental scientists require at least a bachelor's degree in hydrogeology; environmental, civil, or geological engineering; or geochemistry or geology, but employers usually prefer candidates with master's degrees. A master's degree is required for most entry-level research positions in colleges and universities, federal agencies, and state geological surveys. A PhD is necessary for most high-level research positions.

Hundreds of colleges and universities offer a bachelor's degree in geology; fewer schools offer programs in geophysics, hydrogeology, or other geosciences. Other programs offering related training for beginning geological scientists include geophysical technology, geophysical engineering, geophysical prospecting, engineering geology, petroleum geology, geohydrology, and geochemistry. In addition, several hundred universities award advanced degrees in geology or geophysics.

Traditional geoscience courses emphasizing classical geologic methods and topics (such as mineralogy, petrology, paleontology, stratigraphy, and structural geology) are important for all geoscientists and make up the majority of college training. Persons studying physics, chemistry, biology, mathematics, engineering, or computer science may also qualify for some environmental science and geoscience positions if their coursework includes study in geology. Those students interested in working in the environmental or regulatory fields, either in environmental consulting firms or for federal or state governments, should take courses in hydrology, hazardous waste management, environmental legislation, chemistry, fluid mechanics, and geologic logging. An understanding of environmental regulations and government permit issues is also valuable for those planning to work in mining and oil and gas extraction. Hydrologists and environmental scientists should have some knowledge of the potential liabilities associated with some environmental work. Computer skills are essential for prospective environmental scientists and geoscientists; students who have some experience with computer modeling, data analysis and integration, digital mapping, remote sensing, and geographic information systems (GIS) will be the most prepared entering the job market. A knowledge of the Global Positioning System (GPS)—a locator system that uses satellites—also is very helpful. Some employers seek applicants with field experience, so a summer internship may be beneficial to prospective geoscientists.

Environmental scientists and geoscientists must have excellent interpersonal skills, because they usually work as part of a team with other scientists, engineers, and technicians. Strong oral and written communication skills also are important, because writing technical reports and research proposals, as well as communicating research results to others, are important aspects of the work. Because many jobs require foreign travel, knowledge of a second language is becoming an important attribute to employers. Geoscientists must be inquisitive and able to think logically and have an open mind. Those involved in fieldwork must have physical stamina.

Environmental scientists and geoscientists often begin their careers in field exploration or as research assistants or technicians in laboratories or offices. They are given more difficult assignments as they gain experience. Eventually, they may be promoted to project leader, program manager, or another management and research position.

Job Outlook

Employment of environmental scientists and hydrologists is expected to grow faster than the average for all occupations through

2010, while employment of geoscientists is expected to grow about as fast as the average. The need to replace environmental scientists and geoscientists who retire will result in many job openings over the next decade. Driving the growth of environmental scientists and geoscientists will be the continuing need for companies and organizations to comply with environmental laws and regulations, particularly those regarding groundwater contamination and flood control. However, oil company mergers and stagnant or declining government funding for research may affect the hiring of petroleum geologists and geoscientists involved in research. Instead, increased construction and exploration for oil and natural gas abroad may require geoscientists to work overseas unless additional sites in the United States are opened for exploration.

In the past, employment of geologists and some other geoscientists has been cyclical and largely affected by the price of oil and gas. When prices were low, oil and gas producers curtailed exploration activities and laid off geologists. When prices were up, companies had the funds and incentive to renew exploration efforts and hire geoscientists in large numbers. In recent years, a growing worldwide demand for oil and gas and new exploration and recovery techniques—particularly in deep water and previously inaccessible sites—have returned some stability to the petroleum industry, with a few companies increasing their hiring of geoscientists. Growth in this area, though, will be limited due to increasing efficiencies in finding oil and gas. Geoscientists who speak a foreign language and who are willing to work abroad should enjoy the best opportunities.

The need for companies to comply with environmental laws and regulations is expected to contribute to the demand for environmental scientists and some geoscientists, especially hydrologists and engineering geologists. Issues of water conservation, deteriorating coastal environments, and rising sea levels also will stimulate employment growth of these workers. As the population increases and moves to more environmentally sensitive locations, environmental scientists and hydrologists will be needed to assess building sites for potential geologic hazards and to address issues of pollution control and waste disposal. Hydrologists and environmental scientists also will be needed to conduct research on hazardous waste sites to determine the impact of hazardous pollutants on soil and groundwater so engineers can design remediation systems. The need for environmental scientists and geoscientists who understand both the science and engineering aspects of waste remediation is growing. An expected increase in highway building and other infrastructure projects will be an additional source of jobs for engineering geologists.

Employment of environmental scientists and geoscientists is more sensitive to changes in governmental energy or environmental policy than employment of other scientists. If environmental regulations are rescinded or loosened, job opportunities will shrink. On the other hand, increased exploration for energy sources will result in improved job opportunities for geoscientists.

Jobs with the federal and state governments and with organizations dependent on federal funds for support will experience little growth over the next decade, unless budgets increase significantly. The federal government is expected to increasingly outsource environmental services to private consulting firms. This lack of funding will affect mostly geoscientists performing basic research.

Earnings

Median annual earnings of environmental scientists were $44,180 in 2000. The middle 50 percent earned between $34,570 and $58,490. The lowest 10 percent earned less than $28,520, and the highest 10 percent earned more than $73,790.

Median annual earnings of geoscientists were $56,230 in 2000. The middle 50 percent earned between $43,320 and $77,180. The lowest 10 percent earned less than $33,910, and the highest 10 percent earned more than $106,040.

Median annual earnings of hydrologists were $55,410 in 2000. The middle 50 percent earned between $43,740 and $68,500. The lowest 10 percent earned less than $35,910, and the highest 10 percent earned more than $85,260.

Median annual earnings in the industries employing the largest number of environmental scientists in 2000 were as follows:

Federal government	$59,590
Engineering and architectural services	43,920
Management and public relations	43,900
Local government	42,880
State government	39,330

According to the National Association of Colleges and Employers, beginning salary offers in 2001 for graduates with bachelor's degrees in geology and the geological sciences averaged about $35,568 a year; graduates with a master's degree averaged $41,100; graduates with a doctoral degree averaged $57,500.

In 2001, the federal government's average salary for geologists in managerial, supervisory, and nonsupervisory positions was $70,763; for geophysicists, $79,660; for hydrologists, $64,810; and for oceanographers, $71,881.

The petroleum, mineral, and mining industries offer higher salaries, but less job security, than other industries. These industries are vulnerable to recessions and changes in oil and gas prices, among other factors, and usually release workers when exploration and drilling slow down.

Related Occupations

Many geoscientists work in the petroleum and natural gas industry. This industry also employs many other workers in the scientific and technical aspects of petroleum and natural gas exploration and extraction, including engineering technicians, science technicians, petroleum engineers, and surveyors, cartographers and photogrammetrists. Also, some physicists, chemists, and atmospheric scientists—as well as mathematicians and systems analysts, computer scientists, and database administrators—perform related work in both petroleum and natural gas exploration and extraction and in environment-related activities.

Sources of Additional Information

Information on training and career opportunities for geologists is available from:

- American Geological Institute, 4220 King St., Alexandria, VA 22302-1502. Internet: http://www.agiweb.org
- Society of America, P.O. Box 9140, Boulder, CO 80301-9140. Telephone: (717) 447-2020. Internet: http://www.geosociety.org
- American Association of Petroleum Geologists, P.O. Box 979, Tulsa, OK 74101. Internet: http://www.aapg.org

Information on training and career opportunities for geophysicists is available from:

- American Geophysical Union, 2000 Florida Ave. NW, Washington, DC 20009. Telephone: (202) 777-7512. Internet: http://www.agu.org
- Society of Exploration Geophysicists, 8801 South Yale, Tulsa, OK 74137. Telephone: (918) 497-5500. Internet: http://www.seg.org

A packet of free career information, and a list of education and training programs in oceanography and related fields priced at $6.00, is available from:

- Marine Technology Society, 1828 L St. N.W., Suite 906, Washington, DC 20036. Internet: http://www.mtsociety.org

Information on acquiring a job as a geologist, geophysicist, hydrologist, or oceanographer with the federal government may be obtained through a telephone-based system from the Office of Personnel Management. Consult your telephone directory under U.S. Government for a local number. You can also call (912) 757-3000 or the Federal Relay Service (800) 877-8339. This number is not toll free, and charges may result. Information also is available from the Internet site: http://www.usajobs.opm.gov

Heating, Air-Conditioning, and Refrigeration Mechanics and Installers

O*NET 49-9021.01, 49-9021.02

Significant Points

- Opportunities should be very good for mechanics and installers with technical school or formal apprenticeship training.
- Mechanics and installers need a basic understanding of microelectronics because they increasingly install and service equipment with electronic controls.

Nature of the Work

What would those living in Chicago do without heating, those in Miami do without air-conditioning, or blood banks all over the country do without refrigeration? Heating and air-conditioning systems control the temperature, humidity, and the total air quality in residential, commercial, industrial, and other buildings. Refrigeration systems make it possible to store and transport food, medicine, and other perishable items. *Heating, air-conditioning, and refrigeration mechanics and installers*—also called *technicians*—install, maintain, and repair such systems.

Heating, air-conditioning, and refrigeration systems consist of many mechanical, electrical, and electronic components such as motors, compressors, pumps, fans, ducts, pipes, thermostats, and switches. In central heating systems, for example, a furnace heats air that is distributed throughout the building via a system of metal or fiberglass ducts. Technicians must be able to maintain, diagnose, and correct problems throughout the entire system. To do this, they adjust system controls to recommended settings and test the performance of the entire system using special tools and test equipment.

Although they are trained to do both, technicians often specialize in either installation or maintenance and repair. Some specialize in one type of equipment—for example, oil burners, solar panels, or commercial refrigerators. Technicians may work for large or small contracting companies or directly for a manufacturer or wholesaler. Those working for smaller operations tend to do both installation and servicing, and work with heating, cooling, and refrigeration equipment.

Heating and air-conditioning mechanics install, service, and repair heating and air-conditioning systems in both residences and commercial establishments. *Furnace installers*, also called *heating equipment technicians*, follow blueprints or other specifications to install oil, gas, electric, solid-fuel, and multiple-fuel heating systems. *Air-conditioning mechanics* install and service central air-conditioning systems. After putting the equipment in place, they install fuel and water supply lines, air ducts and vents, pumps, and other components. They may connect electrical wiring and controls and check the unit for proper operation. To ensure the proper functioning of the system, furnace installers often use combustion test equipment such as carbon dioxide and oxygen testers.

After a furnace has been installed, heating equipment technicians often perform routine maintenance and repairwork to keep the system operating efficiently. During the fall and winter, for example, when the system is used most, they service and adjust burners and blowers. If the system is not operating properly, they check the thermostat, burner nozzles, controls, or other parts to diagnose and then correct the problem.

During the summer, when the heating system is not being used, heating equipment technicians do maintenance work, such as replacing filters, ducts, and other parts of the system that may accumulate dust and impurities during the operating season. During the winter, air-conditioning mechanics inspect the systems and do required maintenance, such as overhauling compressors.

Refrigeration mechanics install, service, and repair industrial and commercial refrigerating systems and a variety of refrigeration equipment. They follow blueprints, design specifications, and manufacturers' instructions to install motors, compressors, condensing units, evaporators, piping, and other components. They connect this equipment to the ductwork, refrigerant lines, and electrical power source. After making the connections, they charge the system with refrigerant, check it for proper operation, and program control systems.

When heating, air-conditioning, and refrigeration mechanics service equipment, they must use care to conserve, recover, and recycle chlorofluorocarbon (CFC) and hydrochlorofluoro-carbon (HCFC) refrigerants used in air-conditioning and refrigeration systems. The release of CFCs and HCFCs contributes to the depletion of the stratospheric ozone layer, which protects plant and animal life from ultraviolet radiation. Technicians conserve the refrigerant by making sure that there are no leaks in the system; they recover it by venting the refrigerant into proper cylinders; and they recycle it for reuse with special filter-dryers.

Heating, air-conditioning, and refrigeration mechanics and installers are adept at using a variety of tools, including hammers, wrenches, metal snips, electric drills, pipe cutters and benders, measurement gauges, and acetylene torches, to work with refrigerant lines and air ducts. They use voltmeters, thermometers, pressure gauges, manometers, and other testing devices to check air flow, refrigerant pressure, electrical circuits, burners, and other components.

New technology, in the form of cellular "Web" phones that allow technicians to tap into the Internet, may soon affect the way technicians diagnose problems. Computer hardware and software have been developed that allows heating, venting, and refrigeration units to automatically contact the maintenance establishment when problems arise. The maintenance establishment can then notify the mechanic in the field via cellular phone. The mechanic can then access the Internet to "talk" with the unit needing maintenance. While this technology is cutting-edge and not yet widespread, its potential for cost-savings may spur its acceptance.

Other craft workers sometimes install or repair cooling and heating systems. For example, on a large air-conditioning installation job, especially where workers are covered by union contracts, ductwork might be done by sheet metal workers and duct installers; electrical work by electricians; and installation of piping, condensers, and other components by pipelayers, plumbers, pipefitters, and steamfitters. Home appliance repairers usually service room air conditioners and household refrigerators.

Working Conditions

Heating, air-conditioning, and refrigeration mechanics and installers work in homes, stores of all kinds, hospitals, office buildings, and factories—anywhere there is climate-control equipment. They may be assigned to specific job sites at the beginning of each day, or if they are making service calls, they may be dispatched to jobs by radio, telephone, or pagers. Increasingly, employers are using cell phones to coordinate technicians' schedules.

Technicians may work outside in cold or hot weather or in buildings that are uncomfortable because the air-conditioning or heating equipment is broken. In addition, technicians might have to work in awkward or cramped positions and sometimes are required to work in high places. Hazards include electrical shock, burns, muscle strains, and other injuries from handling heavy equipment. Appropriate safety equipment is necessary when handling refrigerants because contact can cause skin damage, frostbite, or blindness. Inhalation of refrigerants when working in confined spaces is also a possible hazard.

The majority of mechanics and installers work more than a 40-hour week, particularly during peak seasons when they often work overtime or irregular hours. Maintenance workers, including those who provide maintenance services under contract, often work evening or weekend shifts, and are on call. Most employers try to provide a full workweek the year round by scheduling both installation and maintenance work, and many manufacturers and contractors now provide or even require service contracts. In most shops that service both heating and air-conditioning equipment, employment is very stable throughout the year.

Employment

Heating, air-conditioning, and refrigeration mechanics and installers held about 243,000 jobs in 2000; approximately one third of these worked for cooling and heating contractors. The remainder were employed in a variety of industries throughout the country, reflecting a widespread dependence on climate-control systems. Some worked for fuel oil dealers, refrigeration and air-conditioning service and repair shops, schools, and department stores that sell heating and air-conditioning systems. Local governments, the federal government, hospitals, office buildings, and other organizations that operate large air-conditioning, refrigeration, or heating systems employed others. Approximately 1 of every 5 mechanics and installers was self-employed.

Training, Other Qualifications, and Advancement

Because of the increasing sophistication of heating, air-conditioning, and refrigeration systems, employers prefer to hire those with technical school or apprenticeship training. A sizable number of mechanics and installers, however, still learn the trade informally on the job.

Many secondary and postsecondary technical and trade schools, junior and community colleges, and the armed forces offer 6-month to 2-year programs in heating, air-conditioning, and refrigeration. Students study theory, design, and equipment construction, as well as electronics. They also learn the basics of installation, maintenance, and repair.

Apprenticeship programs are frequently run by joint committees representing local chapters of the Air-Conditioning Contractors of America, the Mechanical Contractors Association of America, the National Association of Plumbing-Heating-Cooling Contractors, and locals of the Sheet Metal Workers' International Association or the United Association of Journeymen and Apprentices of the Plumbing and Pipefitting Industry of the United States and Canada. Other apprenticeship programs are sponsored by local chapters of the Associated Builders and Contractors and the National Association of Home Builders. Formal apprenticeship programs normally last 3 to 5 years and combine on-the-job training with classroom instruction. Classes include subjects such as the use and care of tools, safety practices, blueprint reading, and the theory and design of heating, ventilation, air-conditioning, and refrigeration systems. Applicants for these programs must have a high school diploma or equivalent.

Those who acquire their skills on the job usually begin by assisting experienced technicians. They may begin performing simple tasks such as carrying materials, insulating refrigerant lines, or cleaning furnaces. In time, they move on to more difficult tasks, such as cutting and soldering pipes and sheet metal and checking electrical and electronic circuits.

Courses in shop math, mechanical drawing, applied physics and chemistry, electronics, blueprint reading, and computer applications provide a good background for those interested in entering this occupation. Some knowledge of plumbing or electrical work is also helpful. A basic understanding of microelectronics is becoming more important because of the increasing use of this technology in solid-state equipment controls. Because technicians frequently deal directly with the public, they should be courteous and tactful, especially when dealing with an aggravated customer. They also should be in good physical condition because they sometimes have to lift and move heavy equipment.

All technicians who purchase or work with refrigerants must be certified in their proper handling. To become certified to purchase and handle refrigerants, technicians must pass a written examination specific to the type of work in which they specialize. The three possible areas of certification are: Type I—servicing small appliances, Type II—high pressure refrigerants, and Type III—low pressure refrigerants. Exams are administered by organizations approved by the U.S. Environmental Protection Agency, such as trade schools, unions, contractor associations, or building groups.

Several organizations have begun to offer basic self-study, classroom, and Internet courses for individuals with limited experience. In addition to understanding how systems work, technicians must be knowledgeable about refrigerant products, and legislation and regulation that govern their use. The industry recently announced the adoption of one standard for certification of experienced technicians—the Air-Conditioning Excellence program, which is offered through North American Technician Excellence, Inc. (NATE).

Advancement usually takes the form of higher wages. Some technicians, however, may advance to positions as supervisor or service manager. Others may move into areas such as sales and marketing. Still others may become building superintendents, cost estimators, or, with the necessary certification, teachers. Those with sufficient money and managerial skill can open their own contracting business.

Job Outlook

Job prospects for highly skilled heating, air-conditioning, and refrigeration mechanics and installers are expected to be very good, particularly for those with technical school or formal apprenticeship training to install, remodel, and service new and existing systems. In addition to job openings created by employment growth, thousands of openings will result from the need to replace workers who transfer to other occupations or leave the labor force.

Employment of heating, air-conditioning, and refrigeration mechanics and installers is expected to increase faster than the average for all occupations through the year 2010. As the population and economy grow, so does the demand for new residential, commercial, and industrial climate-control systems. Technicians who specialize in installation work may experience periods of unemployment when the level of new construction activity declines, but maintenance and repair work usually remains relatively stable. People and businesses depend on their climate control systems and must keep them in good working order, regardless of economic conditions.

Renewed concern for energy conservation should continue to prompt the development of new energy-saving heating and air-conditioning systems. An emphasis on better energy management should lead to the replacement of older systems and the installation of newer, more efficient systems in existing homes and buildings. Also, demand for maintenance and service work should increase as businesses and home owners strive to keep systems operating at peak efficiency. Regulations prohibiting the discharge of CFC and HCFC refrigerants took effect in 1993, and regulations banning CFC production became effective in 2000. Consequently, these regulations should continue to result in demand for technicians to replace many existing systems, or modify them to use new environmentally safe refrigerants. In addition, the continuing focus on improving indoor air quality should contribute to the growth of jobs for heating, air-conditioning, and refrigeration technicians. Also, growth of business establishments that use refrigerated equipment—such as supermarkets and convenience stores—will contribute to a growing need for technicians.

Earnings

Median hourly earnings of heating, air-conditioning, and refrigeration mechanics and installers were $15.76 in 2000. The middle 50 percent earned between $12.25 and $19.92 an hour. The lowest 10 percent earned less than $9.71, and the top 10 percent earned more than $24.58. Median hourly earnings in the industries employing the largest numbers of heating, air-conditioning, and refrigeration mechanics and installers in 2000 were as follows:

Hardware, plumbing, and heating equipment	$16.83
Elementary and secondary schools	16.45
Fuel dealers	16.40
Colleges and universities	16.12
Electrical repair shops	15.16
Plumbing, heating, and air-conditioning	15.08

Apprentices usually begin at about 50 percent of the wage rate paid to experienced workers. As they gain experience and improve their skills, they receive periodic increases until they reach the wage rate of experienced workers.

Heating, air-conditioning, and refrigeration mechanics and installers enjoy a variety of employer-sponsored benefits. In addition to typical benefits like health insurance and pension plans, some employers pay for work-related training and provide uniforms, company vans, and tools.

More than 1 out of every 5 heating, air-conditioning, and refrigeration mechanics and installers is a member of a union. The unions to which the greatest numbers of mechanics and installers belong are the Sheet Metal Workers' International Association and the United Association of Journeymen and Apprentices of the Plumbing and Pipefitting Industry of the United States and Canada.

Related Occupations

Heating, air-conditioning, and refrigeration mechanics and installers work with sheet metal and piping, and repair machinery, such as electrical motors, compressors, and burners. Other workers who have similar skills are boilermakers; home appliance repairers; electricians; sheet metal workers; and pipelayers, plumbers, pipefitters, and steamfitters.

Sources of Additional Information

For more information about opportunities for training, certification, and employment in this trade, contact local vocational and technical schools; local heating, air-conditioning, and refrigeration contractors; a local of the unions previously mentioned; a local joint union-management apprenticeship committee; a local chapter of the Associated Builders and Contractors; or the nearest office of the state employment service or apprenticeship agency.

For information on career opportunities, training, and technician certification, contact:

- Air-Conditioning Contractors of America (ACCA), Suite 300, 2800 Shirlington Rd., Arlington, VA 22206. Internet: http://www.acca.org
- Refrigeration Service Engineers Society (RSES), 1666 Rand Rd., Des Plaines, IL 60016-3552
- National Association of Plumbing-Heating-Cooling Contractors (PHCC), 180 S. Washington St., P.O. Box 6808, Falls Church, VA 22046. Internet: http://www.naphcc.org
- Northamerican Heating, Refrigeration, and Air-conditioning Wholesalers Association (NHRAW), 1389 Dublin Road, PO Box 16790, Columbus, OH 43216-6790. Internet: http://www.nhraw.org

For information on technician testing and certification, contact:

- North American Technician Excellence (NATE), Suite 300, 8201 Greensboro Dr., McLean, VA 22102. Internet: http://www.natex.org

For information on career opportunities and training, write to:

- Associated Builders and Contractors, Suite 800, 1300 North 17th St., Rosslyn, VA 22209. Internet: http://www.abc.org
- Home Builders Institute, National Association of Home Builders, 1201 15th St. N.W., 6th Floor, Washington, DC 20005. Internet: http://www.hbi.org
- Mechanical Contractors Association of America, 1385 Piccard Dr., Rockville, MD 20850-4329. Internet: http://www.mcca.org
- Air-Conditioning and Refrigeration Institute, 4301 North Fairfax Dr., Suite 425, Arlington, VA 22203. Internet: http://www.coolcareers.org

Heavy Vehicle and Mobile Equipment Service Technicians and Mechanics

O*NET 49-3041.00, 49-3042.00, 49-3043.00

Significant Points

- Opportunities should be good for persons with formal postsecondary training in diesel or heavy equipment mechanics, especially if they also have training in basic electronics and hydraulics.
- This occupation offers relatively high wages and the challenge of skilled repair work.
- Skill in using computerized diagnostic equipment is becoming more important.

Nature of the Work

Heavy vehicles and mobile equipment are indispensable to many industrial activities, from construction to railroads. Various types of equipment move materials, till land, lift beams, and dig earth to pave the way for development and production. *Heavy vehicle and mobile equipment service technicians and mechanics* repair and maintain engines and hydraulic, transmission, and electrical systems powering farm equipment, cranes, bulldozers, and railcars.

Service technicians perform routine maintenance checks on diesel engines and fuel, brake, and transmission systems to ensure peak performance, safety, and longevity of the equipment. Maintenance checks and comments from equipment operators usually alert technicians to problems. With many types of modern heavy and mobile equipment, technicians can plug hand-held diagnostic computers into onboard computers to diagnose any component needing adjustment or repair. After locating the problem, these technicians rely on their training and experience to use the best possible technique to solve the problem. If necessary, they may partially dismantle the component to examine parts for damage or excessive wear. Then, using hand-held tools, they repair, replace, clean, and lubricate parts, as necessary. In some cases, technicians calibrate systems by typing codes into the onboard computer. After reassembling the component and testing it for safety, they put it back into the equipment and return the equipment to the field.

Many types of heavy and mobile equipment use hydraulics to raise and lower movable parts, such as scoops, shovels, log forks, and scraper blades. When hydraulic components malfunction, technicians examine them for hydraulic fluid leaks, ruptured hoses, or worn gaskets on fluid reservoirs. Occasionally, the equipment requires extensive repairs, such as replacing a defective hydraulic pump.

In addition to routine maintenance checks, service technicians perform a variety of other repairs. They diagnose electrical problems and adjust or replace defective components. They also disassemble and repair undercarriages and track assemblies. Occasionally, technicians weld broken equipment frames and structural parts, using electric or gas welders.

It is common for technicians in large shops to specialize in one or two types of repair. For example, a shop may have individual specialists in major engine repair, transmission work, electrical systems, and suspension or brake systems. The technology used in heavy equipment is becoming more sophisticated with the increased use of electronic and computer-controlled components. Training in electronics is essential for these technicians to make engine adjustments and diagnose problems. Training in the use of hand-held computers also is necessary, because computers help technicians diagnose problems and adjust component functions.

Service technicians use a variety of tools in their work. They use power tools, such as pneumatic wrenches to remove bolts quickly, machine tools like lathes and grinding machines to rebuild brakes, welding and flame-cutting equipment to remove and repair exhaust systems, and jacks and hoists to lift and move large parts. They also use common handtools—screwdrivers, pliers, and wrenches—to work on small parts and to get at hard-to-reach places. Service technicians may use a variety of computerized testing equipment to pinpoint and analyze malfunctions in electrical systems and other essential systems. For example, they use tachometers and dynamometers to locate engine malfunctions. Service technicians also use ohmmeters, ammeters, and voltmeters when working on electrical systems.

Mobile heavy equipment mechanics and service technicians keep construction and surface mining equipment such bulldozer, cranes, crawlers, draglines, graders, excavators, and other equipment in working order. They typically work for equipment wholesale distribution and leasing firms, large construction and mining companies, local and federal governments, or other organizations operating and maintaining heavy machinery and equipment fleets. Service technicians employed by the federal government may work on tanks and other armored equipment.

Farm equipment mechanics service, maintain, and repair farm equipment as well as smaller lawn and garden tractors sold to suburban homeowners. What typically was a general repairer's job around the farm has evolved into a specialized technical career. Farmers have increasingly turned to farm equipment dealers to service and repair their equipment because the machinery has grown in complexity. Modern equipment uses more electronics and hydraulics making it difficult to perform repairs without some specialized training.

Farm equipment mechanics work mostly on equipment brought into the shop for repair and adjustment. During planting and harvesting seasons, they may travel to farms to make emergency repairs to minimize delays in farm operations.

Railcar repairers specialize in servicing railroad locomotives and other rolling stock, streetcars and subway cars, or mine cars. Most work for railroads, public and private transit companies, and underground mine operators.

Working Conditions

Service technicians usually work indoors, although many make repairs at the work site. Technicians often lift heavy parts and tools, handle greasy and dirty parts, and stand or lie in awkward positions, to repair vehicles and equipment. Minor cuts, burns, and bruises are common; but serious accidents are normally avoided when the shop is kept clean and orderly and safety practices are observed. Technicians usually work in well-lighted, heated, and ventilated areas. However, some shops are drafty and noisy. Many employers provide uniforms, locker rooms, and shower facilities.

When heavy and mobile equipment breaks down at a construction site, it may be too difficult or expensive to bring it into a repair shop, so the shop often sends a field service technician to the job site to make repairs. Field service technicians work outdoors and spend much of their time away from the shop. Generally, more experienced service technicians specialize in field service. They usually drive trucks specially equipped with replacement parts and tools. On occasion, they must travel many miles to reach disabled machinery. Field technicians normally earn a higher wage than their counterparts, because they are required to make on-the-spot decisions necessary to serve their customers.

The hours of work for farm equipment mechanics vary according to the season of the year. During the busy planting and harvesting seasons, mechanics often work 6 or 7 days a week, 10 to 12 hours daily. In slow winter months, however, mechanics may work fewer than 40 hours a week.

Employment

Heavy vehicle and mobile equipment service technicians and mechanics held about 185,000 jobs in 2000. About 130,000 were mobile heavy equipment mechanics; 41,000 were farm equipment mechanics; and 14,000 were railcar repairers. Heavy and mobile equipment dealers and distributors employed more than 40 percent. About 11 percent were employed by federal, state, and local governments; and nearly 9 percent worked for construction contractors. Other service technicians worked for agricultural production and services, mine operators, public utilities, or heavy equipment rental and leasing companies. Still others repaired equipment for machinery manufacturers, airlines, railroads, steel mills, or oil and gas field companies. Less than 4 percent of service technicians were self-employed.

Nearly every section of the country employs heavy and mobile equipment service technicians and mechanics, although most work in towns and cities where equipment dealers, equipment rental and leasing companies, and construction companies have repair facilities.

Training, Other Qualifications, and Advancement

Although many persons qualify for service technician jobs through years of on-the-job training, most employers prefer that applicants complete a formal diesel or heavy equipment mechanic training program after graduating from high school. They seek persons with mechanical aptitude who are knowledgeable about the fundamentals of diesel engines, transmissions, electrical systems, and hydraulics. Additionally, the constant change in equipment technology makes it necessary for technicians to be flexible and have the capacity to learn new skills quickly.

Many community colleges and vocational schools offer programs in diesel technology. Some tailor programs to heavy equipment mechanics. These programs educate the student in the basics of analysis and diagnostic techniques, electronics, and hydraulics. The increased use of electronics and computers makes training in the fundamentals of electronics essential for new heavy and mobile equipment mechanics. Some 1- to 2-year programs lead to a certificate of completion, whereas others lead to an associate degree in diesel or heavy equipment mechanics. These programs provide a foundation in the components of diesel and heavy equipment technology. These programs also enable trainee technicians to advance more rapidly to the journey, or experienced worker, level.

A combination of formal and on-the-job training prepares trainee technicians with the knowledge to efficiently service and repair equipment handled by a shop. Most beginners perform routine service tasks and make minor repairs, after a few months' experience. They advance to harder jobs, as they prove their ability and competence. After trainees master the repair and service of diesel engines, they learn to work on related components, such as brakes, transmissions, and electrical systems. Generally, a service technician with at least 3 to 4 years of on-the-job experience is accepted as fully qualified.

Many employers send trainee technicians to training sessions conducted by heavy equipment manufacturers. These sessions, which typically last up to 1 week, provide intensive instruction in the repair of a manufacturer's equipment. Some sessions focus on particular components found in the manufacturer's equipment, such as diesel engines, transmissions, axles, and electrical systems. Other sessions focus on particular types of equipment, such as crawler-loaders and crawler-dozers. As they progress, trainees may periodically attend additional training sessions. When appropriate, experienced technicians attend training sessions to gain familiarity with new technology or equipment.

High school courses in automobile repair, physics, chemistry, and mathematics provide a strong foundation for a career as a service technician or mechanic. It is also essential for technicians to be able to read and interpret service manuals to keep abreast of engineering changes. Experience working on diesel engines and heavy equipment acquired in the armed forces also is valuable.

Voluntary certification by the National Institute for Automotive Service Excellence (ASE) is recognized as the standard of achievement for heavy and mobile equipment diesel service technicians. Technicians may be certified as a Master Heavy-Duty Diesel Technician or in 1 or more of 6 different areas of heavy-duty equipment repair: brakes, gasoline engines, diesel engines, drive trains, electrical systems, and suspension and steering. For certification in each area, technicians must pass a written examination and have at least 2 years' experience. High school, vocational or trade school, or community or junior college training in gasoline or diesel engine repair may substitute for up to 1 year's experience. To remain certified, technicians must retest every 5 years. This ensures that service technicians keep up with changing technology. However, there are currently no certification programs for other heavy vehicle and mobile equipment repair specialties.

The most important work possessions of technicians are their handtools. Service technicians typically buy their own handtools, and many experienced technicians have thousands of dollars invested in them. Employers typically furnish expensive power tools, computerized engine analyzers, and other diagnostic equipment; but handtools are normally accumulated with experience.

Experienced technicians may advance to field service jobs, where they have a greater opportunity to tackle problems independently and earn additional pay. Technicians with leadership ability may become shop supervisors or service managers. Some technicians open their own repair shops or invest in a franchise.

Job Outlook

Opportunities for heavy vehicle and mobile equipment service technicians and mechanics should be good for persons who have completed formal training programs in diesel or heavy equipment mechanics. Persons without formal training are expected to encounter growing difficulty entering these jobs.

Employment of heavy vehicle and mobile equipment service technicians and mechanics is expected to grow slower than the average for all occupations through the year 2010. Most job openings will arise from the need to replace experienced repairers who retire. Employers report difficulty finding candidates with formal postsecondary training to fill available service technician positions because many young people with mechanic training prefer to take jobs as automotive service technicians, diesel service technicians, or industrial machinery repairers—jobs that offer relatively higher earnings and a wider variety of locations in which to work.

Increasing numbers of service technicians will be required to support growth in the construction industry, equipment dealers, and rental and leasing companies. Because of the nature of construction activity, demand for service technicians follows the nation's economic cycle. As the economy expands, construction activity increases, resulting in the use of more mobile heavy equipment. More equipment is needed to grade construction sites, excavate basements, and lay water and sewer lines, increasing the need for periodic service and repair. In addition, the construction and repair of highways and bridges also requires more technicians to service equipment. Also, as equipment becomes more complex, repairs increasingly must be made by specially trained technicians. Job openings for farm equipment mechanics and railcar repairers are mostly expected to arise due to replacement needs.

Construction and mining are particularly sensitive to changes in the level of economic activity; therefore, heavy and mobile equipment may be idled during downturns. In addition, winter is traditionally the slow season for construction and farming activity, particularly in cold regions. Few technicians may be needed during periods when equipment is used less; however, employers usually try to retain experienced workers. Employers may be reluctant to hire inexperienced workers during slow periods though.

Earnings

Median hourly earnings of mobile heavy equipment mechanics were $16.32 in 2000. The middle 50 percent earned between $13.32 and $19.86. The lowest 10 percent earned less than $10.93, and the highest 10 percent earned more than $23.29. Median hourly earnings in the industries employing the largest numbers of mobile heavy equipment mechanics in 2000 were as follows:

Federal government	$18.67
Local government	17.09
Machinery, equipment, and supplies	16.05
Miscellaneous equipment rental and leasing	15.95
Heavy construction, except highway	15.54

Median hourly earnings of farm equipment mechanics were $12.38 in 2000. The middle 50 percent earned between $9.99 and $15.29. The lowest 10 percent earned less than $8.15, and the highest 10 percent earned more than $18.23.

Median hourly earnings of railcar repairers were $16.19 in 2000. The middle 50 percent earned between $12.31 and $19.34. The lowest 10 percent earned less than $9.78, and the highest 10 percent earned more than $21.19.

About one-fourth of all service technicians and mechanics are members of unions including the International Association of Machinists and Aerospace Workers, the International Union of Operating Engineers, and the International Brotherhood of Teamsters.

Related Occupations

Workers in related repair occupations include aircraft and avionics equipment mechanics and service technicians; automotive service technicians and mechanics; diesel service technicians and mechanics; heating, air-conditioning, and refrigeration mechanics and installers; and small engine mechanics.

Sources of Additional Information

More details about job openings for heavy vehicle and mobile equipment service technicians and mechanics may be obtained from local heavy and mobile equipment dealers and distributors, construction contractors, and government agencies. Local offices of the state employment service also may have information on job openings and training programs.

For general information about a career as a heavy vehicle and mobile equipment service technician or mechanic, contact:

- The Equipment Maintenance Counsel, P.O. Box 1368, Glenwood Springs, CO 81602. Internet: http://www.equipment.org
- Specialized Carriers and Rigging Association, 2750 Prosperity Ave., Suite 620, Fairfax, VA 22031-4312
- The AED Foundation (Associated Equipment Dealers affiliate), 615 W. 22nd St., Oak Brook, IL 60523. Internet: http://www.aednet.org/aed_foundation

For a directory of public training programs in heavy and mobile equipment mechanics, contact:

- SkillsUSA-VICA, P.O. Box 3000, 1401 James Monroe Hwy., Leesburg, VA 22075. Internet: http://www.skillsusa.org

A list of certified diesel service technician training programs can be obtained from:

- National Automotive Technician Education Foundation (NATEF), 101 Blue Seal Dr. S.E., Suite 101, Leesburg, VA 20175. Internet: http://www.natef.org

Information on certification as a heavy-duty diesel service technician is available from:

- ASE, 101 Blue Seal Dr. S.E., Suite 101, Leesburg, VA 20175. Internet: http://www.asecert.org

Industrial Engineers, Including Health and Safety

O*NET 17-2111.01, 17-2111.02, 17-2111.03, 17-2112.00

Significant Points

- Overall job opportunities in engineering are expected to be good.
- A bachelor's degree is required for most entry-level jobs.
- Starting salaries are significantly higher than those of college graduates in other fields.
- Continuing education is critical to keep abreast of the latest technology.

Nature of the Work

Industrial engineers determine the most effective ways for an organization to use the basic factors of production—people, machines, materials, information, and energy—to make a product or to provide a service. They are the bridge between management goals and operational performance. They are more concerned with increasing productivity through the management of people, methods of business organization, and technology than are engineers in other specialties, who generally work more with products or processes. Although most industrial engineers work in manufacturing industries, they also work in consulting services, healthcare, and communications.

To solve organizational, production, and related problems most efficiently, industrial engineers carefully study the product and its requirements, use mathematical methods such as operations research to meet those requirements, and design manufacturing and information systems. They develop management control systems to aid in financial planning and cost analysis, design production planning and control systems to coordinate activities and ensure product quality, and design or improve systems for the physical distribution of goods and services. Industrial engineers determine which plant location has the best combination of raw materials availability, transportation facilities, and costs. Industrial engineers use computers for simulations and to control various activities and devices, such as assembly lines and robots. They also develop wage and salary administration systems and job evaluation programs. Many industrial engineers move into management positions because the work is closely related.

The work of health and safety engineers is similar to that of industrial engineers in that they are concerned with the entire production process. They promote work site or product safety and health by applying knowledge of industrial processes, as well as mechanical, chemical, and psychological principles. They must be able to anticipate and evaluate hazardous conditions as well as develop hazard control methods. They also must be familiar with the application of health and safety regulations.

Working Conditions

Most engineers work in office buildings, laboratories, or industrial plants. Others may spend time outdoors at construction sites, mines, and oil and gas exploration and production sites, where they monitor or direct operations or solve onsite problems. Some engineers travel extensively to plants or work sites.

Many engineers work a standard 40-hour week. At times, deadlines or design standards may bring extra pressure to a job. When this happens, engineers may work longer hours and experience considerable stress.

Employment

Industrial engineers, including health and safety, held about 198,000 jobs in 2000. More than 65 percent of these jobs were in manufacturing industries. Because their skills can be used in almost any type of organization, industrial engineers are more widely distributed among manufacturing industries than are other engineers.

Their skills can be readily applied outside manufacturing as well. Some work in engineering and management services, utilities, and business services; others work for government agencies or as independent consultants.

Training, Other Qualifications, and Advancement

A bachelor's degree in engineering is required for almost all entry-level engineering jobs. College graduates with a degree in a physical science or mathematics occasionally may qualify for some engineering jobs, especially in specialties in high demand. Most engineering degrees are granted in electrical, electronics, mechanical, or civil engineering. However, engineers trained in one branch may work in related branches. This flexibility allows employers to meet staffing needs in new technologies and specialties in which engineers are in short supply. It also allows engineers to shift to fields with better employment prospects or to those that more closely match their interests.

Most engineering programs involve a concentration of study in an engineering specialty, along with courses in both mathematics and science. Most programs include a design course, sometimes accompanied by a computer or laboratory class or both.

In addition to the standard engineering degree, many colleges offer 2- or 4-year degree programs in engineering technology. These programs, which usually include various hands-on laboratory classes that focus on current issues, prepare students for practical design and production work, rather than for jobs which require more theoretical and scientific knowledge. Graduates of 4-year technology programs may get jobs similar to those obtained by graduates with a bachelor's degree in engineering. Engineering technology graduates, however, are not qualified to register as professional engineers under the same terms as graduates with degrees in engineering. Some employers regard technology program graduates as having skills between those of a technician and an engineer.

Graduate training is essential for engineering faculty positions and many research and development programs, but is not required for the majority of entry-level engineering jobs. Many engineers obtain graduate degrees in engineering or business administration to learn new technology and broaden their education. Many high-level executives in government and industry began their careers as engineers.

About 330 colleges and universities offer bachelor's degree programs in engineering that are accredited by the Accreditation Board for Engineering and Technology (ABET), and about 250 colleges offer accredited bachelor's degree programs in engineering technology. ABET accreditation is based on an examination of an engineering program's student achievement, program improvement, faculty, curricular content, facilities, and institutional commitment. Although most institutions offer programs in the major branches of engineering, only a few offer programs in the smaller specialties. Also, programs of the same title may vary in content. For example, some programs emphasize industrial practices, preparing students for a job in industry, whereas others are more theoretical and are designed to prepare students for graduate work. Therefore, students should investigate curricula and check accreditations carefully before selecting a college.

Admissions requirements for undergraduate engineering schools include a solid background in mathematics (algebra, geometry, trigonometry, and calculus) and sciences (biology, chemistry, and physics), and courses in English, social studies, humanities, and computers. Bachelor's degree programs in engineering typically are designed to last 4 years, but many students find that it takes between 4 and 5 years to complete their studies. In a typical 4-year college curriculum, the first 2 years are spent studying mathematics, basic sciences, introductory engineering, humanities, and social sciences. In the last 2 years, most courses are in engineering, usually with a concentration in one branch. Some programs offer a general engineering curriculum; students then specialize in graduate school or on the job.

Some engineering schools and 2-year colleges have agreements whereby the 2-year college provides the initial engineering education, and the engineering school automatically admits students for their last 2 years. In addition, a few engineering schools have arrangements whereby a student spends 3 years in a liberal arts college studying pre-engineering subjects and 2 years in an engineering school studying core subjects, and then receives a bachelor's degree from each school. Some colleges and universities offer 5-year master's degree programs. Some 5- or even 6-year cooperative plans combine classroom study and practical work, permitting students to gain valuable experience and finance part of their education. All 50 states and the District of Columbia usually require licensure for engineers who offer their services directly to the public. Engineers who are licensed are called Professional Engineers (PE). This licensure generally requires a degree from an ABET-accredited engineering program, 4 years of relevant work experience, and successful completion of a state examination. Recent graduates can start the licensing process by taking the examination in two stages. The initial Fundamentals of Engineering (FE) examination can be taken upon graduation. Engineers who pass this examination commonly are called Engineers in Training (EIT) or Engineer Interns (EI). The EIT certification usually is valid for 10 years. After acquiring suitable work experience, EITs can take the second examination, the Principles and Practice of Engineering Exam. Several states have imposed mandatory continuing education requirements for relicensure. Most states recognize licensure from other states.

Engineers should be creative, inquisitive, analytical, and detail oriented. They should be able to work as part of a team and to communicate well, both orally and in writing. Communication abilities are becoming more important because much of their work is becoming more diversified, meaning that engineers interact with specialists in a wide range of fields outside engineering.

Beginning engineering graduates usually work under the supervision of experienced engineers and, in large companies, also may receive formal classroom or seminar-type training. As new engineers gain knowledge and experience, they are assigned more difficult projects with greater independence to develop designs, solve problems, and make decisions. Engineers may advance to become technical specialists or to supervise a staff or team of engineers and technicians. Some may eventually become engineering managers or enter other managerial or sales jobs.

Job Outlook

Despite industrial growth and more complex business operations, overall employment of industrial engineers, including health and safety, is expected to grow more slowly than the average for all occupations through 2010, reflecting greater use of automation in factories and offices. Employment of industrial engineers is expected to grow more slowly than average while health and safety engineers are expected to grow about as fast as average.

Because the main function of industrial and health and safety engineers is to make a higher quality product as efficiently and as safely as possible, their services should be in demand in the manufacturing sector as firms seek to reduce costs and increase productivity. There also is an increased demand for industrial engineers within the financial services sector, as more emphasis is put on information technology. Also, the growing concern for health and safety within work environments should increase the need for health and safety engineers.

Earnings

Median annual earnings of industrial engineers were $58,580 in 2000. The middle 50 percent earned between $47,530 and $71,050. The lowest 10 percent earned less than $38,140, and the highest 10 percent earned more than $86,370. Median annual earnings in the manufacturing industries employing the largest numbers of industrial engineers in 2000 were:

Motor vehicles and equipment	$63,010
Electronic components and accessories	62,560
Computer and office equipment	62,260
Computer and data processing services	60,510
Aircraft and parts	58,290

Median annual earnings of health and safety engineers were $54,630 in 2000. The middle 50 percent earned between $44,230 and $67,500. The lowest 10 percent earned less than $34,710, and the highest 10 percent earned more than $82,320. In 2000, the median annual earnings of health and safety engineers in railroads were $56,970.

According to a 2001 salary survey by the National Association of Colleges and Employers, bachelor's degree candidates in industrial engineering received starting offers averaging about $48,320 a year; master's degree candidates averaged $56,265 a year; and PhD candidates were initially offered $59,800.

Related Occupations

Industrial engineers apply the principles of physical science and mathematics in their work. Other workers who use scientific and mathematical principles include architects, except landscape and naval; engineering and natural sciences managers; computer and information systems managers; mathematicians; drafters; engineering technicians; sales engineers; science technicians; and physical and life scientists, including agricultural and food scientists, biological and medical scientists, conservation scientists and foresters, atmospheric scientists, chemists and materials scientists, environmental scientists and geoscientists, and physicists and astronomers.

Sources of Additional Information

For further information about industrial engineers, contact:

- Institute of Industrial Engineers, Inc., 25 Technology Park/Atlanta, Norcross, GA 30092. Internet: http://www.iienet.org

General information about safety engineers is available from:

- American Society of Safety Engineers, 1800 E Oakton St., Des Plaines, IL 60018. Internet: http://www.asse.org

Information about certification of safety professionals, including safety engineers, is available from:

- Board of Certified Safety Professionals, 208 Burwash Ave., Savoy, IL 61874. Internet: http://www.bcsp.org

High school students interested in obtaining a full package of guidance materials and information (product number SP-01) on a variety of engineering disciplines should contact the Junior Engineering Technical Society by sending $3.50 to:

- JETS-Guidance, 1420 King St., Suite 405, Alexandria, VA 22314-2794. Internet: http://www.jets.org

High school students interested in obtaining information on ABET-accredited engineering programs should contact:

- The Accreditation Board for Engineering and Technology, Inc., 111 Market Place, Suite 1050, Baltimore, MD 21202-4012. Internet: http://www.abet.org

Non-licensed engineers and college students interested in obtaining information on Professional Engineer licensure should contact:

- The National Society of Professional Engineers, 1420 King St., Alexandria, VA 22314-2794. Internet: http://www.nspe.org
- National Council of Examiners for Engineers and Surveying, P.O. Box 1686, Clemson, SC 29633-1686. Internet: http://www.ncees.org

Information on general engineering education and career resources is available from:

- American Society for Engineering Education, 1818 N St. N.W., Suite 600, Washington, DC 20036-2479. Internet: http://www.asee.org

Information on obtaining an engineering position with the federal government is available from the Office of Personnel Management (OPM) through a telephone-based system. Consult your telephone directory under U.S. Government for a local number or call (912) 757-3000; Federal Relay Service: (800) 877-8339. The first number is not toll free, and charges may result. Information also is available from the OPM Internet site: http://www.usajobs.opm.gov

Non-high school students and those wanting more detailed information should contact societies representing the individual branches of engineering. Each can provide information about careers in the particular branch. The individual statements elsewhere in this book also provide other information in detail on aerospace; agricultural; biomedical; chemical; civil; computer hardware; electrical and electronics; environmental; materials; mechanical; mining and geological, including mining safety; nuclear; and petroleum engineering.

Industrial Machinery Installation, Repair, and Maintenance Workers

O*NET 49-9041.00, 49-9042.00, 49-9043.00, 49-9044.00

Significant Points

- Workers learn their trade through a 4-year apprenticeship program, or through informal on-the-job training supplemented by classroom instruction.
- Despite slower-than-average employment growth resulting from technological advancements in machinery, applicants with broad skills in machine repair should have favorable job prospects.

Nature of the Work

When production workers encounter problems with the machines they operate, they call industrial machinery installation, repair, and maintenance workers. These workers include industrial machinery mechanics, millwrights, and general maintenance and repair and machinery maintenance workers. Their work is important not only because an idle machine will delay production, but also because a machine that is not properly repaired and maintained may damage the final product or injure the operator.

Industrial machinery mechanics repair, install, adjust, or maintain industrial production and processing machinery or refinery and pipeline distribution systems. *Millwrights* install, dismantle, or move machinery and heavy equipment according to layout plans, blueprints, or other drawings. *General maintenance and repair workers* perform work involving the skills of two or more maintenance or craft occupations to keep machines, mechanical equipment, or the structure of an establishment in repair. *Machinery maintenance workers* lubricate machinery, change parts, or perform other routine machinery maintenance.

Much of the work begins when machinery arrives at the job site. New equipment must be unloaded, inspected, and moved into position. To lift and move light machinery, industrial machinery installation, repair, and maintenance workers use rigging and hoisting devices, such as pulleys and cables. In other cases, they require the assistance of hydraulic lift-truck or crane operators to position the machinery. Because industrial machinery installation, repair, and maintenance workers often decide which device to use for moving machinery, they must know the load-bearing properties of ropes, cables, hoists, and cranes.

Industrial machinery installation, repair, and maintenance workers consult with production managers and others to determine the optimal placement of machines in a plant. In some instances, this placement requires building a new foundation. Industrial machinery installation, repair, and maintenance workers either prepare the foundation themselves or supervise its construction, so they must know how to read blueprints and work with building materials, such as concrete, wood, and steel.

When assembling machinery, industrial machinery installation, repair, and maintenance workers fit bearings, align gears and wheels, attach motors, and connect belts, according to the manufacturer's blueprints and drawings. Precision leveling and alignment are important in the assembly process; industrial machinery installation, repair, and maintenance workers must have good mathematical skills, so that they can measure angles, material thickness, and small distances with tools such as squares, calipers, and micrometers. When a high level of precision is required, devices such as lasers and ultrasonic measuring tools may be used. Industrial machinery installation, repair, and maintenance workers also work with hand and power tools, such as cutting torches, welding machines, and soldering guns. Some of these workers use metalworking equipment, such as lathes or grinders, to modify parts to specifications.

Maintenance mechanics must be able to detect and diagnose minor problems and correct them before they become major ones. For example, after hearing a vibration from a machine, the mechanic must decide whether it is due to worn belts, weak motor bearings, or some other problem. Computerized maintenance, vibration analysis techniques, and self-diagnostic systems are making this task easier. Self-diagnostic features on new industrial machinery can determine the cause of a malfunction and, in some cases, alert the mechanic to potential trouble spots before symptoms develop.

After diagnosing the problem, the mechanic disassembles the equipment and repairs or replaces the necessary parts. Once the machine is reassembled, the final step is to test it to ensure that it is running smoothly. When repairing electronically controlled machinery, maintenance mechanics may work closely with electronic repairers or electricians who maintain the machine's electronic parts. However, industrial machinery installation, repair, and maintenance workers increasingly need electronic and computer skills to repair sophisticated equipment on their own.

Although repairing machines is the most important job of industrial machinery installation, repair, and maintenance workers, they also perform preventive maintenance. This includes keeping machines and their parts well oiled, greased, and cleaned. Repairers regularly inspect machinery and check performance. For example, they adjust and calibrate automated manufacturing equipment such as industrial robots, and rebuild components of other industrial machinery. By keeping complete and up-to-date records, mechanics try to anticipate trouble and service equipment before factory production is interrupted.

A wide range of tools may be used when performing repairs or preventive maintenance. Repairers may use a screwdriver and wrench to adjust a motor, or a hoist to lift a printing press off the ground. When replacements for broken or defective parts are not readily available, or when a machine must be quickly returned to production, repairers may sketch a part that can be fabricated by the plant's machine shop. Repairers use catalogs to order replacement parts and often follow blueprints and engineering specifications to maintain and fix equipment.

Installation of new machinery is another responsibility of industrial machinery installation, repair, and maintenance workers. As plants retool and invest in new equipment, they increasingly rely on these workers to properly situate and install the machinery. As employers increasingly seek workers who have a variety of skills, industrial machinery installation, repair, and maintenance workers are taking on new responsibilities.

Working Conditions

Working conditions for repairers who work in manufacturing are similar to those of production workers. These workers are subject to common shop injuries such as cuts and bruises, and use protective equipment such as hardhats, protective glasses, and safety belts. Industrial machinery installation, repair, and maintenance workers also may face additional hazards because they often work on top of a ladder or underneath or above large machinery in cramped conditions. Industrial machinery installation, repair, and maintenance workers may work independently or as part of a team. They must work quickly and precisely, because disabled machinery costs a company time and money.

Because factories and other facilities cannot afford breakdowns of industrial machinery, repairers may be called to the plant at night or on weekends for emergency repairs. Overtime is common among industrial machinery installation, repair, and maintenance workers—more than a third work over 40 hours a week. During power outages, industrial machinery installation, repair, and maintenance workers may be assigned overtime and be required to work in shifts to deal with the emergency.

Employment

Industrial machinery installation, repair, and maintenance workers held about 1.6 million jobs in 2000. Employment was distributed among the following occupations:

Maintenance and repair workers, general	1,251,000
Industrial machinery mechanics	198,000
Maintenance workers, machinery	114,000
Millwrights	72,000

About 1 of every 3 worked in manufacturing industries, primarily food processing, textile mill products, chemicals, fabricated metal products, and primary metals. Others worked for government agencies, public utilities, mining companies, and other establishments in which industrial machinery is used.

Industrial machinery installation, repair, and maintenance workers are found in a wide variety of plants and in every part of the country. However, employment is concentrated in heavily industrialized areas.

Training, Other Qualifications, and Advancement

Most industrial machinery installation, repair, and maintenance workers, including millwrights, learn their trade through a 4-year apprenticeship program combining classroom instruction with on-the-job training. These programs usually are sponsored by a local trade union. Other machinery maintenance workers start as helpers and pick up the skills of the trade informally and by taking courses offered by machinery manufacturers and community colleges.

Trainee repairers learn from experienced repairers how to operate, disassemble, repair, and assemble machinery. Trainees also may work with concrete and receive instruction in related skills, such as carpentry, welding, and sheet metal work. Classroom instruction focuses on subjects such as shop mathematics, blueprint reading, welding, electronics, and computer training.

Most employers prefer to hire those who have completed high school or its equivalency, and who have some vocational training or experience. High school courses in mechanical drawing, mathematics, blueprint reading, physics, computers, and electronics are especially useful.

Mechanical aptitude and manual dexterity are important characteristics for workers in this trade. Good physical conditioning and agility also are necessary because repairers sometimes have to lift heavy objects or climb to reach equipment located high above the floor.

Opportunities for advancement are limited. Industrial machinery installation, repair, and maintenance workers advance either by working with more complicated equipment or by becoming supervisors. The most highly skilled repairers can be promoted to master mechanic or can become machinists or tool and die makers.

Job Outlook

Overall employment of industrial machinery installation, repair, and maintenance workers is projected to grow more slowly than the average for all occupations through 2010. Nevertheless, applicants with broad skills in machine repair should have favorable job prospects. As more firms introduce automated production equipment, industrial machinery installation, repair, and maintenance workers will be needed to ensure that these machines are properly maintained and consistently in operation. However, many new machines are capable of self-diagnosis, increasing their reliability and, thus, reducing the need for repairers. As a result, the majority of job openings will stem from the need to replace repairers who transfer to other occupations or leave the labor force.

As automation of machinery becomes more widespread, there is a greater need for repair work than for the installation of new machinery. Industrial machinery installation, repair, and maintenance workers are becoming more productive through the use of technologies such as hydraulic torque wrenches, ultrasonic measuring tools, and laser shaft alignment, as these technologies allow fewer workers to perform more work. In addition, the demand for industrial machinery installation, repair, and maintenance workers will be adversely affected as lower-paid workers, such as electronics technicians, increasingly assume some installation and maintenance duties.

Unlike many other occupations concentrated in manufacturing industries, industrial machinery installation, repair, and maintenance workers usually are not affected by seasonal changes in production. During slack periods, when some plant workers are laid off, repairers often are retained to do major overhaul jobs. Although these workers may face layoff or a reduced workweek when economic conditions are particularly severe, they usually are less affected than are other workers because machines have to be maintained regardless of production level.

Earnings

Earnings of industrial machinery installation, repair, and maintenance workers vary by industry and geographic region. Median hourly earnings of industrial machinery mechanics were $17.30 in 2000. The middle 50 percent earned between $13.73 and $21.93. The lowest 10 percent earned less than $11.31, and the highest 10 percent earned more than $26.26. Median hourly earnings in the industries employing the largest numbers of industrial machinery mechanics in 2000 are shown in the following table:

Motor vehicles and equipment	$24.28
Electric services	24.12
Plastics materials and synthetics	20.14
Machinery, equipment, and supplies	15.01
Meat products	13.06

Median hourly earnings of general maintenance and repair workers were $13.39 in 2000. The middle 50 percent earned between $10.05 and $17.47. The lowest 10 percent earned less than $7.84, and the highest 10 percent earned more than $21.43. Median hourly earnings in the industries employing the largest numbers of general maintenance and repair workers in 2000 are shown in the following table:

Local government	$13.99
Elementary and secondary schools	13.17
Real estate agents and managers	10.85
Real estate operators and lessors	10.71
Hotels and motels	10.07

Median hourly earnings of millwrights were $19.33 in 2000. The middle 50 percent earned between $15.19 and $23.98. The lowest 10 percent earned less than $12.02, and the highest 10 percent earned more than $27.07. Median hourly earnings in the industries employing the largest numbers of millwrights in 2000 are shown in the following table:

Motor vehicles and equipment	$25.73
Miscellaneous special trade contractors	19.64
Blast furnace and basic steel products	18.85

Median hourly earnings of machinery maintenance workers were $14.89 in 2000. The middle 50 percent earned between $11.54 and $18.79. The lowest 10 percent earned less than $9.20, and the highest 10 percent earned more than $22.74. Median hourly earnings in miscellaneous plastics products, the industry employing the largest numbers of machinery maintenance workers, were $15.28 in 2000.

More than 25 percent of industrial machinery mechanics are union members. More than 67 percent of millwrights belong to labor unions, one of the highest rates of unionization in the economy. Labor unions that represent industrial machinery installation, repair, and maintenance workers include the United Steelworkers of America; the United Automobile, Aerospace and Agricultural Implement Workers of America; the International Association of Machinists and Aerospace Workers; and the International Union of Electronic, Electrical, Salaried, Machine, and Furniture Workers.

Related Occupations

Other occupations that involve repairing machinery include aircraft and avionics equipment mechanics and service technicians; electrical and electronics installers and repairers; coin, vending, and amusement machine servicers and repairers; automotive body and related repairers; automotive service technicians and mechanics; electronic home entertainment equipment installers and repairers; heating, air-conditioning, and refrigeration mechanics and installers; and radio and telecommunications equipment installers and repairers.

Sources of Additional Information

Information about employment and apprenticeship opportunities for industrial machinery installation, repair, and maintenance workers may be obtained from local offices of the state employment service or from:

- United Brotherhood of Carpenters and Joiners of America, 101 Constitution Ave. N.W., Washington, DC 20001
- The National Tooling and Machining Association, 9300 Livingston Rd., Fort Washington, MD 20744. Internet: http://www.ntma.org
- Precision Machined Products Association, 6700 West Snowville Rd., Brecksville, OH 44141. Internet: http://www.pmpa.org
- Associated General Contractors of America, 1957 E St. N.W., Washington, DC 20006. Internet: http://www.agc.org

Job Opportunities in the Armed Forces

O*NET 55-1011.00, 55-1012.00, 55-1013.00, 55-1014.00, 55-1015.00, 55-1016.00, 55-1017.00, 55-1019.99, 55-2011.00, 55-2012.00, 55-2013.00, 55-3011.00, 55-3012.00, 55-3013.00, 55-3014.00, 55-3015.00, 55-3016.00, 55-3017.00, 55-3018.00, 55-3019.99

Significant Points

- Opportunities should be good in all branches of the armed forces for applicants who meet designated standards.
- Most enlisted personnel need at least a high school diploma, while officers need a bachelor's or advanced degree.
- Hours and working conditions can be arduous and vary substantially.
- Some training and duty assignments are hazardous, even in peacetime.

Nature of the Work

Maintaining a strong national defense encompasses such diverse activities as running a hospital, commanding a tank, programming computers, operating a nuclear reactor, or repairing and maintaining a helicopter. The military provides training and work experience in these fields and many others for more than 1.5 million people who serve in the active Army, Navy, Marine Corps, Air Force, and Coast Guard, their Reserve components, and the Air and Army National Guard.

The military distinguishes between enlisted and officer careers. Enlisted personnel, who make up about 85 percent of the armed forces, carry out the fundamental operations of the military in areas such as combat, administration, construction, engineering, healthcare, and human services. Officers, who make up the remaining 15 percent of the armed forces, are the leaders of the military. They supervise and manage activities in every occupational specialty in the military.

The following sections discuss the major occupational groups for enlisted personnel and officers.

Enlisted occupational groups:

Administrative careers include a wide variety of positions. The military must keep accurate information for planning and managing its operations. Paper and electronic records are kept on personnel and on equipment, funds, supplies, and other property of the military. Enlisted administrative personnel record information, type reports, maintain files, and review information to assist military offices. Personnel may work in a specialized area such as finance, accounting, legal, maintenance, supply, or transportation. Some examples of administrative specialists are recruiting specialists, who recruit and place qualified personnel and provide information about military careers to young people, parents, schools, and local communities; training specialists and instructors, who provide the training programs necessary to help people perform their jobs effectively; and personnel specialists, who collect and store information about individuals in the military, including training, job assignment, promotion, and health information.

Combat specialty occupations refer to enlisted specialties, such as infantry, artillery, and special forces, whose members operate weapons or execute special missions during combat situations. Persons in these occupations normally specialize by the type of weapon system or combat operation. These personnel maneuver against enemy forces, and position and fire artillery, guns, and missiles to destroy enemy positions. They also may operate tanks and amphibious assault vehicles in combat or scouting missions. When the military has difficult and dangerous missions to perform, they call upon special operations teams. These elite combat forces stay in a constant state of readiness to strike anywhere in the world on a moment's notice. Special operations forces team members conduct offensive raids, demolitions, intelligence, search and rescue, and other missions from aboard aircraft, helicopters, ships, or submarines.

Construction occupations in the military include personnel who build or repair buildings, airfields, bridges, foundations, dams, bunkers, and the electrical and plumbing components of these structures. Enlisted personnel in construction occupations operate bulldozers, cranes, graders, and other heavy equipment. Construction specialists also may work with engineers and other building specialists as part of military construction teams. Some personnel specialize in areas such as plumbing or electrical wiring. Plumbers and pipefitters install and repair the plumbing and pipe systems needed in buildings and on aircraft and ships. Building electricians install and repair electrical wiring systems in offices, airplane hangars, and other buildings on military bases.

Electronic and electrical equipment repair personnel repair and maintain electronic and electrical equipment used in the military. Repairers normally specialize by type of equipment, such as avionics, computer, optical, communications, or weapons systems. For example, electronic instrument repairers install, test, maintain, and repair a wide variety of electronic systems, including navigational controls and biomedical instruments. Weapons maintenance technicians maintain and repair weapons used by combat forces, most of which have electronic components and systems that assist in locating targets and in aiming and firing weapons.

The military has many *engineering, science, and technical occupations*, whose members require specific knowledge to operate technical equipment, solve complex problems, or provide and interpret information. Enlisted personnel normally specialize in one area, such as space operations, emergency management, environmental health and safety, or intelligence. Space operations specialists use and repair spacecraft ground control command equipment, including electronic systems that track spacecraft location and operation. Emergency management specialists prepare emergency procedures for all types of disasters, such as floods, tornadoes, and earthquakes. Environmental health and safety specialists inspect military facilities and food supplies for the presence of disease, germs, or other conditions hazardous to health and the environment. Intelligence specialists gather and study information using aerial photographs and various types of radar and surveillance systems.

Healthcare personnel assist medical professionals in treating and providing services for men and women in the military. They may work as part of a patient service team in close contact with doctors, dentists, nurses, and physical therapists to provide the necessary support functions within a hospital or clinic. Healthcare specialists normally specialize in a particular area. They may provide emergency medical treatment, operate diagnostic equipment such as X-ray and ultrasound equipment, conduct laboratory tests on tissue and blood samples, maintain pharmacy supplies, or maintain patient records.

Human services specialists help military personnel and their families with social or personal problems, or assist chaplains. For example, caseworkers and counselors work with personnel who may be experiencing social problems, such as drug or alcohol dependence or depression. Religious program specialists assist chaplains with religious services, religious education programs, and administrative duties.

Machine operator and production occupations operate industrial equipment, machinery, and tools to fabricate and repair parts for a variety of items and structures. They may operate engines, turbines, nuclear reactors, and water pumps. Personnel often specialize by type of work performed. Welders and metal workers, for instance, work with various types of metals to repair or form the structural parts of ships, submarines, buildings, or other equipment. Survival equipment specialists inspect, maintain, and repair survival equipment such as parachutes and aircraft life support equipment. Dental and optical laboratory technicians construct and repair dental equipment and eyeglasses for military personnel.

Media and public affairs occupations are involved in the public presentation and interpretation of military information and events. Enlisted media and public affairs personnel take and develop photographs; film, record, and edit audio and video programs; present news and music programs; and produce graphic artwork, drawings, and other visual displays. Other public affairs specialists act as interpreters and translators to convert written or spoken foreign languages into English or other languages.

Service personnel include those who enforce military laws and regulations, provide emergency response to natural and manmade disasters, and maintain food standards. Personnel normally specialize by function. Military police control traffic, prevent crime, and respond to emergencies. Other law enforcement and security specialists investigate crimes committed on military property and guard inmates in military correctional facilities. Firefighters put out, control, and help prevent fires in buildings, on aircraft, and aboard ships. Food service specialists prepare all types of food in dining halls, hospitals, and ships.

Transportation and material handling specialists ensure the safe transport of people and cargo. Most personnel within this occupational group are classified according to mode of transportation, such as aircraft, motor vehicle, or ship. Aircrew members operate equipment on board aircraft during operations. Vehicle drivers operate all types of heavy military vehicles including fuel or water tank trucks, semi-tractor trailers, heavy troop transports, and passenger buses. Quartermasters and boat operators navigate and pilot many types of small watercraft, including tugboats, gunboats, and barges. Cargo specialists load and unload military supplies and material using equipment such as forklifts and cranes.

Vehicle and machinery mechanics conduct preventive and corrective maintenance on aircraft, ships, automotive and heavy equipment, heating and cooling systems, marine engines, and powerhouse station equipment. They typically specialize by the type of equipment that they maintain. For example, aircraft mechanics inspect, service, and repair helicopters and airplanes. Automotive and heavy equipment mechanics maintain and repair vehicles such as jeeps, cars, trucks, tanks, self-propelled missile launchers, and other combat vehicles. They also repair bulldozers, power shovels, and other construction equipment. Heating and cooling mechanics install and repair air-conditioning, refrigeration, and heating equipment. Marine engine mechanics repair and maintain gasoline and diesel engines on ships, boats, and other watercraft. They also repair shipboard mechanical and electrical equipment. Powerhouse mechanics install, maintain, and repair electrical and mechanical equipment in power-generating stations.

Officer occupational groups:

Combat specialty officers plan and direct military operations, oversee combat activities, and serve as combat leaders. This category includes officers in charge of tanks and other armored assault vehicles, artillery systems, special operations forces, and infantry. They normally specialize by type of unit that they lead. Within the unit, they may specialize by the type of weapon system. Artillery and missile system officers, for example, direct personnel as they target, launch, test, and maintain various types of missiles and artillery. Special-operations officers lead their units in offensive raids, demolitions, intelligence gathering, and search and rescue missions.

Engineering, science, and technical officers have a wide range of responsibilities based on their area of expertise. They lead or perform activities in areas such as space operations, environmental health and safety, and engineering. These officers may direct the operations of communications centers or the development of complex computer systems. Environmental health and safety officers study the air, ground, and water to identify and analyze sources of pollution and its effects. They also direct programs to control safety and health hazards in the workplace. Other personnel work as aerospace engineers to design and direct the development of military aircraft, missiles, and spacecraft.

Executive, administrative, and managerial officers oversee and direct military activities in key functional areas such as finance, accounting, health administration, international relations, and supply. Health services administrators, for instance, are responsible for the overall quality of care provided at the hospitals and clinics they operate. They must ensure that each department works together to provide the highest quality of care. Purchasing and contracting managers are another example: they negotiate and monitor contracts for the purchase of the billions of dollars worth of equipment, supplies, and services that the military buys from private industry each year.

Healthcare officers provide health services at military facilities, based on their area of specialization. Officers who assist in examining, diagnosing, and treating patients with illness, injury, or disease include physician assistants and registered nurses. Other healthcare officers provide therapy, rehabilitative treatment, and other services for patients. Physical and occupational therapists plan and administer therapy to help patients adjust to disabilities, regain independence, and return to work. Speech therapists evaluate and treat patients with hearing and speech problems. Dietitians manage food service facilities, and plan meals for hospital patients and for outpatients who need special diets. Pharmacists manage the purchasing, storing, and dispensing of drugs and medicines.

Health diagnosing and treating practitioner officers examine, diagnose, and provide treatment for illnesses, injuries, and disorders. For example, physicians and surgeons in this occupational group provide the majority of medical services to the military and their families. Dentists treat diseases and disorders of the mouth. Optometrists treat vision problems by prescribing eyeglasses or contact lenses. Psychologists provide mental healthcare, and also conduct research on behavior and emotions.

Human services officers perform services in support of the morale and well-being of military personnel and their families. Social workers focus on improving conditions that cause social problems, such as drug and alcohol abuse, racism, and sexism. Chaplains conduct worship services for military personnel and perform other spiritual duties covering beliefs and practices of all religious faiths.

Media and public affairs officers oversee the development, production, and presentation of information or events for the public. These officers may produce and direct motion pictures, videotapes, and television and radio broadcasts that are used for training, news, and entertainment. Some plan, develop, and direct the activities of military bands. Public information officers respond to inquiries about military activities and prepare news releases and reports to keep the public informed.

Officers in *transportation occupations* manage and perform activities related to the safe transport of military personnel and material by air and water. Officers normally specialize by mode of transportation or area of expertise because, in many cases, they must meet licensing and certification requirements. Pilots in the military fly various types of specialized airplanes and helicopters to carry troops and equipment and execute combat missions. Navigators use radar, radio, and other navigation equipment to determine their position and plan their route of travel. Officers on ships and submarines work as a team to manage the various departments aboard their vessels. Ship engineers direct engineering departments aboard ships and submarines, including engine operations, maintenance, repair, heating, and power generation.

Employment

In 2000, more than 1.5 million individuals were on active duty in the armed forces—about 530,500 in the Army, 400,000 in the Navy, 385,000 in the Air Force, 174,000 in the Marine Corps, and 37,000 in the Coast Guard. Table 1 shows the occupational composition of enlisted personnel in 2001, while Table 2 presents similar information for officers.

Military personnel are stationed throughout the United States and in many countries around the world. More than half of all military jobs are located in California, Texas, North Carolina, Virginia, Florida, and Georgia. About 258,000 individuals were stationed outside the United States in 2000, including those assigned to ships at sea. More than 117,000 of these were stationed in Europe, mainly in Germany, and another 101,000 were assigned to east Asia and the Pacific area, mostly in Japan and the Republic of Korea.

TABLE 1

Military Enlisted Personnel by Broad Occupational Category and Branch of Military Service, April 2001

Occupational Group—Enlisted	Army	Air Force	Coast Guard	Marine Corps	Navy	Total, All Services
Administrative occupations	19,862	23,124	2,211	11,560	16,760	73,517
Combat specialty occupations	102,844	1,092	—	33,127	4,242	141,305
Construction occupations	15,815	6,130	881	5,503	5,897	34,226
Electronic and electrical repair occupations	29,628	47,485	1,725	15,828	62,269	156,935
Engineering, science, and technical occupations	43,368	42,018	2,153	25,098	44,979	157,616
Healthcare occupations	26,443	19,140	664	—	24,559	70,806
Human resource development occupations	13,287	12,514	654	5,097	4,557	36,109
Machine operator and precision work occupations	2,881	9,729	4,410	2,275	6,870	26,165
Media and public affairs occupations	7,740	6,683	114	1,974	3,578	20,089
Protective service occupations	21,731	31,123	1,913	5,801	14,780	75,348
Support services occupations	12,651	7,029	1,173	3,062	9,254	33,169
Transportation and material handling occupations	54,555	36,534	10,355	25,911	65,825	193,180
Vehicle machinery mechanic occupations	45,921	37,477	1,626	17,536	48,174	150,734

SOURCE: U.S. Department of Defense, Defense Manpower Data Center East

TABLE 2

Military Officer Personnel by Broad Occupational Category and Branch of Military Service, April 2001

Occupational Group—Officer	Army	Air Force	Coast Guard	Marine Corps	Navy	Total, All Services
Combat specialty occupations	18,714	5,260	38	4,741	4,068	32,821
Engineering, science, and technical occupations	16,095	17,257	1,315	3,027	10,431	48,125
Executive, administrative, and managerial occupations	10,619	8,613	578	2,220	7,163	29,193
Healthcare occupations	10,829	10,383	16	—	8,327	29,555
Human resource development occupations	1,828	2,471	275	659	3,658	8,891
Media and public affairs occupations	601	468	20	152	370	1,611
Protective service occupations	2,063	1,207	981	350	917	5,518
Support services occupations	1,578	1,214	—	44	1,164	4,000
Transportation occupations	12,749	20,846	3,645	6,916	16,774	60,930
Total, by service	75,076	67,719	6,868	18,109	52,872	220,644

SOURCE: U.S. Department of Defense, Defense Manpower Data Center East

Training, Other Qualifications, and Advancement

Enlisted personnel. In order to join the services, enlisted personnel must sign a legal agreement called an enlistment contract, which usually involves a commitment to 8 years of service. Depending on the terms of the contract, 2 to 6 years are spent on active duty and the balance is spent in the reserves. The enlistment contract obligates the service to provide the agreed-upon job, rating, pay, cash bonuses for enlistment in certain occupations, medical and other benefits, occupational training, and continuing education. In return, enlisted personnel must serve satisfactorily for the period specified.

Requirements for each service vary, but certain qualifications for enlistment are common to all branches. In order to enlist, one must be between 17 and 35 years old, be a U.S. citizen or immigrant alien holding permanent resident status, not have a felony record, and possess a birth certificate. Applicants who are aged 17 must have the consent of a parent or legal guardian before entering the service. Coast Guard enlisted personnel must enter active duty before their 28th birthday, while Marine Corps enlisted personnel must not be over the age of 29. Applicants must both pass a written examination—the Armed Services Vocational Aptitude Battery—and meet certain minimum physical standards such as height, weight, vision, and overall health. All branches of the armed forces require high school graduation or its equivalent for certain enlistment options. In 2000, more than 9 out of 10 recruits were high school graduates.

People thinking about enlisting in the military should learn as much as they can about military life before making a decision. This is especially important if you are thinking about making the military a career. Speaking to friends and relatives with military experience is a good idea. Determine what the military can offer you and what it will expect in return. Then, talk to a recruiter, who can determine if you qualify for enlistment, explain the various enlistment options, and tell you which military occupational specialties currently have openings. Bear in mind that the recruiter's job is to recruit promising applicants into his or her branch of military service, so the information that the recruiter gives you is likely to stress the positive aspects of military life in the branch in which he or she serves.

Ask the recruiter for the branch you have chosen to assess your chances of being accepted for training in the occupation of your choice, or, better still, take the aptitude exam to see how well you score. The military uses the aptitude exam as a placement exam, and test scores largely determine an individual's chances of being accepted into a particular training program. Selection for a particular type of training depends on the needs of the service, your general and technical aptitudes, and your personal preference. Because all prospective recruits are required to take the exam, those who do so before committing themselves to enlist have the advantage of knowing in advance whether they stand a good chance of being accepted for training in a particular specialty. The recruiter can schedule you for the Armed Services Vocational Aptitude Battery without any obligation. Many high schools offer the exam as an easy way for students to explore the possibility of a military career, and the test also provides insight into career areas in which the student has demonstrated aptitudes and interests.

If you decide to join the military, the next step is to pass the physical examination and sign an enlistment contract. Negotiating the contract involves choosing, qualifying, and agreeing on a number of enlistment options such as length of active duty time, which may vary according to the enlistment option. Most active duty programs have first-term enlistments of 4 years, although there are some 2-, 3-, and 6-year programs. The contract also will state the date of enlistment and other options, such as bonuses and types of training to be received. If the service is unable to fulfill its part of the contract, such as providing a certain kind of training, the contract may become null and void.

All services offer a "delayed entry program" by which an individual can delay entry into active duty for up to 1 year after enlisting. High school students can enlist during their senior year and enter a service after graduation. Others choose this program because the job training they desire is not currently available but will be within the coming year, or because they need time to arrange personal affairs.

Women are eligible to enter most military specialties—for example, mechanics, missile maintenance technicians, heavy equipment operators, and fighter pilots, as well as medical care, administrative support, and intelligence specialties. Generally, only occupations involving direct exposure to combat are excluded.

People planning to apply the skills gained through military training to a civilian career should first determine how good the prospects are for civilian employment in jobs related to the military specialty that interests them. Second, they should know the prerequisites for the related civilian job. Because many civilian occupations require a license, certification, or minimum level of education, it is important to determine whether military training is sufficient to enter the civilian equivalent or, if not, what additional training will be required. Additional information often can be obtained from school counselors.

Following enlistment, new members of the armed forces undergo recruit training, which is better known as "basic" training. Recruit training provides a 6- to 12-week introduction to military life with courses in military skills and protocol. Days and nights are carefully structured, and include rigorous physical exercise designed to improve strength and endurance and build unit cohesion.

Following basic training, most recruits take additional training at technical schools that prepare them for a particular military occupational specialty. The formal training period generally lasts from 10 to 20 weeks, although training for certain occupations—nuclear power plant operator, for example—may take as long as a year. Recruits not assigned to classroom instruction receive on-the-job training at their first duty assignment.

Many service people get college credit for the technical training they receive on duty, which, combined with off-duty courses, can lead to an associate degree through community college programs such as the Community College of the Air Force. In addition to on-duty training, military personnel may choose from a variety of educational programs. Most military installations have tuition assistance programs for people wishing to take courses during off-duty hours. These may be correspondence courses or degree programs offered by local colleges or universities. Tuition assistance pays up to 75 percent of college costs. Also available are courses designed to help service personnel earn high school equivalency diplomas. Each service branch provides opportunities for full-time study to a limited number of exceptional applicants. Military personnel accepted into these highly competitive programs—in law or medicine, for example—receive full pay, allowances, tuition, and related fees. In return, they must agree to serve an additional amount of time in the service. Other very selective programs enable enlisted personnel to qualify as commissioned officers through additional military training.

Warrant officers. Warrant officers are technical and tactical leaders who specialize in a specific technical area; for example, army aviators make up one group of warrant officers. The Army Warrant Officer Corps constitutes less than 5 percent of the total army. Although the Corps is small in size, its level of responsibility is high. Its members receive extended career opportunities, worldwide leadership assignments, and increased pay and retirement benefits. Selection to attend the Warrant Officer Candidate School is highly competitive and restricted to those with the rank of E5 or higher (see Table 3).

Officers. Officer training in the armed forces is provided through the federal service academies (Military, Naval, Air Force, and Coast Guard); the Reserve Officers Training Corps (ROTC) program offered at many colleges and universities; Officer Candidate School (OCS) or Officer Training School (OTS); the National Guard (state Officer Candidate School programs); the Uniformed Services University of Health Sciences; and other programs. All are very selective and are good options for those wishing to make the military a career. Persons interested in obtaining training through the federal service academies must be single to enter and graduate, while those seeking training through OCS, OTS, or ROTC need not be single. Single parents with one or more minor dependents are not eligible for officer commissioning.

Federal service academies provide a 4-year college program leading to a Bachelor of Science degree. Midshipmen or cadets are provided free room and board, tuition, medical and dental care, and a monthly allowance. Graduates receive regular or reserve commissions and have a 5-year active duty obligation, or more if they are entering flight training.

To become a candidate for appointment as a cadet or midshipman in one of the service academies, applicants are required to obtain a nomination from an authorized source, usually a member of Congress. Candidates do not need to know a member of Congress personally to request a nomination. Nominees must have an academic record of the requisite quality, college aptitude test scores above an established minimum, and recommendations from teachers or school officials; they also must pass a medical examination. Appointments are made from the list of eligible nominees. Appointments to the Coast Guard Academy, however, are based strictly on merit and do not require a nomination.

ROTC programs train students in about 950 Army, 67 Navy and Marine Corps, and 1,000 Air Force units at participating colleges and universities. Trainees take 2 to 5 hours of military instruction a week, in addition to regular college courses. After graduation, they may serve as officers on active duty for a stipulated period. Some may serve their obligation in the Reserves or National Guard. In the last 2 years of a ROTC program, students receive a monthly allowance while attending school, and additional pay for summer training. ROTC scholarships for 2, 3, and 4 years are available on a competitive basis. All scholarships pay for tuition and have allowances for subsistence, textbooks, supplies, and other costs.

College graduates can earn a commission in the armed forces through OCS or OTS programs in the Army, Navy, Air Force, Marine Corps, Coast Guard, and National Guard. These officers generally must serve their obligation on active duty. Those with training in certain health professions may qualify for direct appointment as officers. In the case of persons studying for the health professions, financial assistance and internship opportunities are available from the military in return for specified periods of military service. Prospective medical students can apply to the Uniformed Services University of Health Sciences, which offers free tuition in a program leading to a Doctor of Medicine (MD) degree. In return, graduates must serve for 7 years in either the military or the U.S. Public Health Service. Direct appointments also are available for those qualified to serve in other specialty areas, such as the judge advocate general (legal) or chaplain corps. Flight training is available to commissioned officers in each branch of the armed forces. In addition, the Army has a direct enlistment option to become a warrant officer aviator.

Each service has different criteria for promoting personnel. Generally, the first few promotions for both enlisted and officer personnel come easily; subsequent promotions are much more competitive. Criteria for promotion may include time in service and grade, job performance, a fitness report (supervisor's recommendation), and written examinations. People who are passed over for promotion several times generally must leave the military. Table 3 shows the officer, warrant officer, and enlisted ranks by service.

TABLE 3

Military Rank and Employment for Active Duty Personnel, April 2001

Grade	Rank and Title: Army	Navy and Coast Guard	Air Force	Marine Corps	Total DOD Employment
Commissioned Officers:					
O-10	General	Admiral	General	General	34
O-9	Lieutenant General	Vice Admiral	Lieutenant General	Lieutenant General	118
O-8	Major General	Rear Admiral Upper	Major General	Major General	282
O-7	Brigadier General	Rear Admiral Lower	Brigadier General	Brigadier General	441
O-6	Colonel	Captain	Colonel	Colonel	11,302
O-5	Lieutenant Colonel	Commander	Lieutenant Colonel	Lieutenant Colonel	27,543
O-4	Major	Lieutenant Commander	Major	Major	43,151
O-3	Captain	Lieutenant	Captain	Captain	65,917
O-2	1st Lieutenant	Lieutenant (JG)	1st Lieutenant	1st Lieutenant	24,759
O-1	2nd Lieutenant	Ensign	2nd Lieutenant	2nd Lieutenant	25,303
Warrant Officers:					
W-5	Chief Warrant Officer	Chief Warrant Officer	—	Chief Warrant Officer	476
W-4	Chief Warrant Officer	Chief Warrant Officer	—	Chief Warrant Officer	1,958
W-3	Chief Warrant Officer	Chief Warrant Officer	—	Chief Warrant Officer	3,837
W-2	Chief Warrant Officer	Chief Warrant Officer	—	Chief Warrant Officer	6,350
W-1	Warrant Officer	Warrant Officer	—	Warrant Officer	2,302

(continues)

TABLE 3 (CONTINUED)

Military Rank and Employment for Active Duty Personnel, April 2001

Grade	Rank and Title: Army	Navy and Coast Guard	Air Force	Marine Corps	Total DOD Employment
Enlisted Personnel:					
E-9	Sergeant Major	Master Chief Petty Officer	Chief Master Sergeant	Sergeant Major	10,197
E-8	1st Sergeant/Master Sergeant	Sr. Chief Petty Officer	Senior Master Sergeant	Master Sergeant/1st Sergeant	25,399
E-7	Sergeant First Class	Chief Petty Officer	Master Sergeant	Gunnery Sergeant	97,052
E-6	Staff Sergeant	Petty Officer 1st Class	Technical Sergeant	Staff Sergeant	165,130
E-5	Sergeant	Petty Officer 2nd Class	Staff Sergeant	Sergeant	231,750
E-4	Corporal/Specialist	Petty Officer 3rd Class	Senior Airman	Corporal	247,691
E-3	Private First Class	Seaman	Airman 1st Class	Lance Corporal	207,432
E-2	Private	Seaman Apprentice	Airman	Private 1st Class	96,420
E-1	Private	Seaman Recruit	Airman Basic	Private	60,228

Source: U.S. Department of Defense

Job Outlook

Opportunities should be good for qualified individuals in all branches of the armed forces through 2010. Many military personnel retire with a pension after 20 years of service, while they still are young enough to start a new career. More than 365,000 enlisted personnel and officers must be recruited each year to replace those who complete their commitment or retire. Since the end of the draft in 1973, the military has met its personnel requirements with volunteers. When the economy is good, it is more difficult for all the services to meet their recruitment quotas, while it is much easier to do so during a recession.

America's strategic position is stronger than it has been in decades. Despite reductions in personnel due to the decreasing threat from eastern Europe and Russia, the number of active duty personnel is expected to remain roughly constant through 2010. The U.S. Armed Forces' current goal is to maintain a sufficient force to fight and win two major regional conflicts occurring at the same time. Political events, however, could cause these plans to change.

Educational requirements will continue to rise as military jobs become more technical and complex. High school graduates and applicants with a college background will be sought to fill the ranks of enlisted personnel, while virtually all officers will need at least a bachelor's degree and, in some cases, an advanced degree as well.

Earnings

The earnings structure for military personnel is shown in Table 4. Most enlisted personnel started as recruits at Grade E-1 in 2000; however, those with special skills or above-average education started as high as Grade E-4. Most warrant officers started at Grade W-1 or W-2, depending upon their occupational and academic qualifications and the branch of service, but warrant officer is not an entry-level occupation and, consequently, these individuals all had previous military service. Most commissioned officers started at Grade O-1, while some with advanced education started as Grade O-2 and some highly trained officers—for example, physicians and dentists—started as high as Grade O-3. Pay varies by total years of service as well as rank. Because it usually takes many years to reach the higher ranks, most personnel in higher ranks receive the higher pay rates awarded to those with many years of service.

TABLE 4

Military Basic Monthly Pay by Grade for Active Duty Personnel, July 1, 2001

Grade	Years of Service: Less than 2	Over 4	Over 8	Over 12	Over 16	Over 20
O-10	$8,518.80	—	$9,156.90	$9,664.20	$10,356.00	$11,049.30
O-9	7,550.10	—	8,114.10	8,451.60	9,156.90	9,664.20
O-8	6,838.20	$7,252.20	7,747.80	8,114.10	8,451.60	9,156.90
O-7	5,682.30	6,112.50	6,514.50	6,915.90	7,747.80	—
O-6	4,211.40	—	5,160.90	—	6,005.40	6,617.40
O-5	3,368.70	4,280.40	—	4,831.80	5,481.60	5,790.30
O-4	2,839.20	3,739.50	4,127.70	4,629.30	4,935.00	—
O-3	2,638.20	3,489.30	3,839.70	4,189.80	—	—

Grade	Years of Service					
	Less than 2	**Over 4**	**Over 8**	**Over 12**	**Over 16**	**Over 20**
O-2	2,301.00	3,120.30	—	—	—	—
O-1	1,997.70	—	—	—	—	—
W-5	—	—	—	—	—	4,640.70
W-4	2,688.00	3,056.70	3,336.30	3,614.10	3,892.50	4,168.20
W-3	2,443.20	2,684.10	2,919.00	3,184.80	3,420.30	3,669.90
W-2	2,139.60	2,391.00	2,649.90	2,851.50	3,058.20	3,280.80
W-1	1,782.60	2,214.60	2,419.20	2,626.80	2,835.90	3,018.60
E-9	—	—	—	3,197.40	3,392.40	3,601.80
E-8	—	—	2,622.00	2,768.40	2,945.10	3,138.00
E-7	1,831.20	2,149.80	2,362.20	2,512.80	2,666.10	2,817.90
E-6	1,575.00	1,891.80	2,097.30	2,248.80	2,379.60	—
E-5	1,381.80	1,701.00	1,888.50	1,811.10	2,040.30	—
E-4	1,288.80	1,576.20	—	—	—	—
E-3	1,214.70	1,385.40	—	—	—	—
E-2	1,169.10	—	—	—	—	—
E-1 4mos+	1,042.80	—	—	—	—	—
E-1 <4mos	964.80	—	—	—	—	—

SOURCE: U.S. Department of Defense—Defense Finance and Accounting Service

In addition to basic pay, military personnel receive free room and board (or a tax-free housing and subsistence allowance), medical and dental care, a military clothing allowance, military supermarket and department store shopping privileges, 30 days of paid vacation a year (referred to as leave), and travel opportunities. In many duty stations, military personnel may receive a housing allowance that can be used for off-base housing. This allowance can be substantial, but varies greatly by rank and duty station. For example, in July 2001, the housing allowance for an E-4 with dependents was $462.90 per month; for a comparable individual without dependents, it was $323.40. The allowance for an O-4 with dependents was $881.70 per month; for a person without dependents, it was $766.50. Other allowances are paid for foreign duty, hazardous duty, submarine and flight duty, and employment as a medical officer. Athletic and other facilities—such as gymnasiums, tennis courts, golf courses, bowling centers, libraries, and movie theaters—are available on many military installations. Military personnel are eligible for retirement benefits after 20 years of service.

The Veterans Administration (VA) provides numerous benefits to those who have served at least 2 years in the armed forces. Veterans are eligible for free care in VA hospitals for all service-related disabilities, regardless of time served; those with other medical problems are eligible for free VA care if they are unable to pay the cost of hospitalization elsewhere. Admission to a VA medical center depends on the availability of beds, however. Veterans also are eligible for certain loans, including home loans. Veterans, regardless of health, can convert a military life insurance policy to an individual policy with any participating company in the veteran's state of residence. In addition, job counseling, testing, and placement services are available.

Veterans who participate in the New Montgomery GI Bill Program receive educational benefits. Under this program, armed forces personnel may elect to deduct up to $100 a month from their pay during the first 12 months of active duty, putting this money toward their future education. Veterans who serve on active duty for more than 2 years, or 2 years active duty plus 4 years in the Selected Reserve, will receive $528 a month in basic benefits for 36 months. Those who enlist and serve for 2 years will receive $429 a month for 36 months. In addition, each service provides its own additional contributions for future education. This sum becomes the service member's educational fund. Upon separation from active duty, the fund can be used to finance educational costs at any VA-approved institution. VA-approved schools include many vocational, correspondence, certification, business, technical, and flight training schools; community and junior colleges; and colleges and universities.

Sources of Additional Information

Each of the military services publishes handbooks, fact sheets, and pamphlets describing entrance requirements, training and advancement opportunities, and other aspects of military careers. These publications are widely available at all recruiting stations, at most state employment service offices, and in high schools, colleges, and public libraries. Information on educational and other veterans' benefits is available from VA offices located throughout the country.

In addition, the Defense Manpower Data Center, an agency of the U.S. Department of Defense, publishes *Military Career Guide Online*, a compendium of military occupational, training, and career information designed for use by students and job seekers.

Landscape Architects

O*NET 17-1012.00

Significant Points

- Almost 26 percent are self-employed—nearly 4 times the proportion for all professionals.
- A bachelor's degree in landscape architecture is the minimum requirement for entry-level jobs; many employers prefer to hire landscape architects who have completed at least one internship.
- Because many landscape architects work for small firms or are self-employed, benefits tend to be less generous than those provided to workers in large organizations.

Nature of the Work

Everyone enjoys attractively designed residential areas, public parks and playgrounds, college campuses, shopping centers, golf courses, parkways, and industrial parks. Landscape architects design these areas so that they are not only functional, but also beautiful, and compatible with the natural environment. They plan the location of buildings, roads, and walkways, and the arrangement of flowers, shrubs, and trees.

Increasingly, landscape architects are becoming involved with projects in environmental remediation, such as preservation and restoration of wetlands. Historic preservation is another important objective to which landscape architects may apply their knowledge of the environment, as well as their design and artistic talents.

Many types of organizations—from real estate development firms starting new projects to municipalities constructing airports or parks—hire landscape architects, who often are involved with the development of a site from its conception. Working with architects, surveyors, and engineers, landscape architects help determine the best arrangement of roads and buildings. They also collaborate with environmental scientists, foresters, and other professionals to find the best way to conserve or restore natural resources. Once these decisions are made, landscape architects create detailed plans indicating new topography, vegetation, walkways, and other landscaping details, such as fountains and decorative features.

In planning a site, landscape architects first consider the nature and purpose of the project and the funds available. They analyze the natural elements of the site, such as the climate, soil, slope of the land, drainage, and vegetation; observe where sunlight falls on the site at different times of the day and examine the site from various angles; and assess the effect of existing buildings, roads, walkways, and utilities on the project.

After studying and analyzing the site, landscape architects prepare a preliminary design. To account for the needs of the client as well as the conditions at the site, they frequently make changes before a final design is approved. They also take into account any local, state, or federal regulations, such as those protecting wetlands or historic resources. Computer-aided design (CAD) has become an essential tool for most landscape architects in preparing designs. Many landscape architects also use video simulation to help clients envision the proposed ideas and plans. For larger scale site planning, landscape architects also use geographic information systems technology, a computer mapping system.

Throughout all phases of the planning and design, landscape architects consult with other professionals involved in the project. Once the design is complete, they prepare a proposal for the client. They produce detailed plans of the site, including written reports, sketches, models, photographs, land-use studies, and cost estimates, and submit them for approval by the client and by regulatory agencies. When the plans are approved, landscape architects prepare working drawings showing all existing and proposed features. They also outline in detail the methods of construction and draw up a list of necessary materials.

Although many landscape architects supervise the installation of their design, some are involved in the construction of the site. However, the developer or landscape contractor usually does this.

Some landscape architects work on a variety of projects. Others specialize in a particular area, such as residential development, historic landscape restoration, waterfront improvement projects, parks and playgrounds, or shopping centers. Still others work in regional planning and resource management; feasibility, environmental impact, and cost studies; or site construction.

Most landscape architects do at least some residential work, but relatively few limit their practice to individual homeowners. Residential landscape design projects usually are too small to provide suitable income as compared with larger commercial or multiunit residential projects. Some nurseries offer residential landscape design services, but lesser-qualified landscape designers or others with training and experience in related areas usually perform these services.

Landscape architects who work for government agencies do site and landscape design for government buildings, parks, and other public lands, as well as park and recreation planning in national parks and forests. In addition, they prepare environmental impact statements and studies on environmental issues such as public land-use planning. Some restore degraded land, such as mines or landfills. Others architects use their skills in traffic-calming, the "art" of slowing traffic down through use of traffic design, enhancement of the physical environment, and greater attention to aesthetics.

Working Conditions

Landscape architects spend most of their time in offices creating plans and designs, preparing models and cost estimates, doing research, or attending meetings with clients and other professionals involved in a design or planning project. The remainder of their time is spent at the site. During the design and planning stage, landscape architects visit and analyze the site to verify that the design can be incorporated into the landscape. After the plans and specifications are completed, they may spend additional time at the site observing or supervising the construction.

Those who work in large firms may spend considerably more time out of the office because of travel to sites outside the local area.

Salaried employees in both government and landscape architectural firms usually work regular hours; however, they may work overtime to meet a project deadline. Hours of self-employed landscape architects vary.

Employment

Landscape architects held about 22,000 jobs in 2000. About 1 out of 3 salaried workers were employed in firms that provide landscape architecture services. Architectural and engineering firms employed most of the rest. The federal government also employs these workers, primarily in the U.S. Departments of Agriculture, Defense, and Interior. About 1 of every 4 landscape architects were self-employed.

Employment of landscape architects is concentrated in urban and suburban areas throughout the country; some landscape architects work in rural areas, particularly those employed by the federal government who plan and design parks and recreation areas.

Training, Other Qualifications, and Advancement

A bachelor's or master's degree in landscape architecture usually is necessary for entry into the profession. The bachelor's degree in landscape architecture takes 4 or 5 years to complete. There are two types of accredited master's degree programs. The master's degree as a first professional degree is a 3-year program designed for students with an undergraduate degree in another discipline; this is the most common type. The master's degree as the second professional degree is a 2-year program for students who have a bachelor's degree in landscape architecture and wish to teach or specialize in some aspect of landscape architecture, such as regional planning or golf course design.

In 2000, 58 colleges and universities offered 75 undergraduate and graduate programs in landscape architecture that were accredited by the Landscape Architecture Accreditation Board of the American Society of Landscape Architects.

College courses required in this field usually include technical subjects such as surveying, landscape design and construction, landscape ecology, site design, and urban and regional planning. Other courses include history of landscape architecture, plant and soil science, geology, professional practice, and general management. Many landscape architecture programs are adding courses that address environmental issues. In addition, most students at the undergraduate level take a year of prerequisite courses such as English, mathematics, and social and physical science. The design studio is an important aspect of many landscape architecture curriculums. Whenever possible, students are assigned real projects, providing them with valuable hands-on experience. While working on these projects, students become more proficient in the use of computer-aided design, geographic information systems, and video simulation.

In 2000, 46 states required landscape architects to be licensed or registered. Licensing is based on the Landscape Architect Registration Examination (L.A.R.E.), sponsored by the Council of Landscape Architectural Registration Boards and administered over a 3-day period. Admission to the exam usually requires a degree from an accredited school plus 1 to 4 years of work experience, although standards vary from state to state. Currently, 16 states require the passage of a state examination in addition to the L.A.R.E. to satisfy registration requirements. State examinations, which usually are 1 hour in length and completed at the end of the L.A.R.E., focus on laws, environmental regulations, plants, soils, climate, and any other characteristics unique to the state.

Because state requirements for licensure are not uniform, landscape architects may not find it easy to transfer their registration from one state to another. However, those who meet the national standards of graduating from an accredited program, serving 3 years of internship under the supervision of a registered landscape architect, and passing the L.A.R.E. can satisfy requirements in most states. Through this means, a landscape architect can obtain certification from the Council of Landscape Architectural Registration Boards, and so gain reciprocity (the right to work) in other states.

In the federal government, candidates for entry positions should have a bachelor's or master's degree in landscape architecture. The federal government does not require its landscape architects to be licensed.

Persons planning a career in landscape architecture should appreciate nature, enjoy working with their hands, and possess strong analytical skills. Creative vision and artistic talent also are desirable qualities. Good oral communication skills are essential; landscape architects must be able to convey their ideas to other professionals and clients and to make presentations before large groups. Strong writing skills also are valuable, as is knowledge of computer applications of all kinds, including word processing, desktop publishing, and spreadsheets. Landscape architects use these tools to develop presentations, proposals, reports, and land impact studies for clients, colleagues, and superiors. The ability to draft and design using CAD software is essential. Many employers recommend that prospective landscape architects complete at least one summer internship with a landscape architecture firm in order to gain an understanding of the day-to-day operations of a small business, including how to win clients, generate fees, and work within a budget.

In states where licensure is required, new hires may be called "apprentices" or "intern landscape architects" until they become licensed. Their duties vary depending on the type and size of the employing firm. They may do project research or prepare working drawings, construction documents, or base maps of the area to be landscaped. Some are allowed to participate in the actual design of a project. However, interns must perform all work under the supervision of a licensed landscape architect. Additionally, all drawings and specifications must be signed and sealed by the licensed landscape architect, who takes legal responsibility for the work. After gaining experience and becoming licensed, landscape architects usually can carry a design through all stages of development. After several years, they may become project managers, taking on the responsibility for meeting schedules and budgets, in addition to overseeing the project design; and later, associates or partners, with a proprietary interest in the business.

Many landscape architects are self-employed because start-up costs, after an initial investment in CAD software, are relatively low. Self-discipline, business acumen, and good marketing skills are important qualities for those who choose to open their own business. Even with these qualities, however, some may struggle while building a client base.

Those with landscape architecture training also qualify for jobs closely related to landscape architecture, and may, after gaining some experience, become construction supervisors, land or environmental planners, or landscape consultants.

Job Outlook

Employment of landscape architects is expected to increase faster than the average for all occupations through the year 2010. Overall, several factors are expected to increase demand for landscape architectural services over the long run: Anticipated growth in residential, commercial, and heavy construction; continued emphasis on preservation and restoration of wetlands; and growth in landscape ecology, the use of techniques from landscape architecture to address environmental problems.

Implementation of the Transportation Equity Act for the Twenty-First Century is expected to spur employment for landscape architects, particularly in state and local governments. This act, known as TEA-21, provides funds for surface transportation and transit programs, such as interstate highway maintenance and environment-friendly pedestrian and bicycle trails. Also, growth in construction of residential and commercial building is expected to contribute to demand for landscape architects. However, opportunities will vary from year to year, and by geographic region, depending on local economic conditions. During a recession, when real estate sales and construction slow down, landscape architects may face layoffs and greater competition for jobs. The need to replace landscape architects who retire or leave the labor force for other reasons is expected to produce nearly as many job openings as employment growth.

As the cost of land rises, the importance of good site planning and landscape design grows. Increasingly, new development is contingent upon compliance with environmental regulations and land use zoning, spurring demand for landscape architects to help plan sites and integrate man-made structures with the natural environment in the least disruptive way.

Budget tightening in the federal government might restrict hiring in the U.S. Forest Service and the National Park Service, agencies that traditionally employ the most landscape architects in the federal government. Instead, such agencies may increasingly contract out for landscape architecture services, providing additional employment opportunities in private landscape architecture firms.

In addition to the work related to new development and construction, landscape architects are expected to be involved in historic preservation, land reclamation, and refurbishment of existing sites. Because landscape architects can work on many different types of projects, they may have an easier time finding employment when traditional construction slows down than other design professionals may.

New graduates can expect to face competition for jobs in the largest and most prestigious landscape architecture firms. The number of professional degrees awarded in landscape architecture has remained steady over the years, even during times of fluctuating demand due to economic conditions. Opportunities will be best for landscape architects who develop strong technical skills—such as computer design—and communication skills, as well as knowledge of environmental codes and regulations. Those with additional training or experience in urban planning increase their opportunities for employment in landscape architecture firms that specialize in site planning as well as landscape design. Many employers prefer to hire entry-level landscape architects who have internship experience, which significantly reduces the amount of on-the-job training required.

Earnings

In 2000, median annual earnings for landscape architects were $43,540. The middle 50 percent earned between $32,990 and $59,490. The lowest 10 percent earned less than $26,300 and the highest 10 percent earned over $74,100. Landscape and horticultural services employed more landscape architects than any other industry, and their median annual earnings were $37,820 in 2000.

In 2001, the average annual salary for all landscape architects in the federal government in nonsupervisory, supervisory, and managerial positions was $62,824.

Because many landscape architects work for small firms or are self-employed, benefits tend to be less generous than those provided to workers in large organizations.

Related Occupations

Landscape architects use their knowledge of design, construction, land-use planning, and environmental issues to develop a landscape project. Others whose work requires similar skills are architects, except landscape and naval; surveyors, cartographers, photogrammetrists, and surveying technicians; civil engineers; and urban and regional planners. Landscape architects also know how to grow and use plants in the landscape. Some conservation scientists and foresters and biological and medical scientists study plants in general and do related work, while environmental scientists and geoscientists work in the area of environmental remediation.

Sources of Additional Information

Additional information, including a list of colleges and universities offering accredited programs in landscape architecture, is available from:

- American Society of Landscape Architects, Career Information, 636 Eye St. N.W., Washington, DC 20001. Internet: http://www.asla.org

General information on registration or licensing requirements is available from:

- Council of Landscape Architectural Registration Boards, 12700 Fair Lakes Circle, Suite 110, Fairfax, VA 22033. Internet: http://www.clarb.org

Librarians

O*NET 25-4021.00

Significant Points

- A master's degree in library science usually is required; special librarians often need an additional graduate or professional degree.
- Applicants for librarian jobs in large cities or suburban areas will face competition, whereas those willing to work in rural areas should have better job prospects.

Nature of the Work

The traditional concept of a library is being redefined from a place to access paper records or books, to one which also houses the most advanced media, including CD-ROM, the Internet, virtual libraries, and remote access to a wide range of resources. Consequently, librarians increasingly are combining traditional duties with tasks involving quickly changing technology. Librarians assist people in finding information and using it effectively for personal and professional purposes. Librarians must have knowledge of a wide variety of scholarly and public information sources, and follow trends related to publishing, computers, and the media to effectively oversee the selection and organization of library materials. They manage staff and develop and direct information programs and systems for the public to ensure that information is organized to meet users' needs.

Most librarian positions incorporate three aspects of library work: user services, technical services, and administrative services. Even librarians specializing in one of these areas perform other responsibilities. Librarians in user services, such as reference and children's librarians, work with the public to help them find the information they need. This involves analyzing users' needs to determine what information is appropriate, and searching for, acquiring, and providing information. It also includes an instructional role, such as showing users how to access information. For example, librarians commonly help users navigate the Internet, showing them how to most efficiently search for relevant information. Librarians in technical services, such as acquisitions and cataloguing, acquire and prepare materials for use and often do not deal directly with the public. Librarians in administrative services oversee the management and planning of libraries; negotiate contracts for services, materials, and equipment; supervise library employees; perform public relations and fundraising duties; prepare budgets; and direct activities to ensure that everything functions properly.

In small libraries or information centers, librarians usually handle all aspects of the work. They read book reviews, publishers' announcements, and catalogues to keep up with current literature and other available resources, and select and purchase materials from publishers, wholesalers, and distributors. Librarians prepare new materials by classifying them by subject matter, and describe books and other library materials so they are easy to find. They supervise assistants who prepare cards, computer records, or other access tools that direct users to resources. In large libraries, librarians often specialize in a single area, such as acquisitions, cataloguing, bibliography, reference, special collections, or administration. Teamwork is increasingly important to ensure quality service to the public.

Librarians also compile lists of books, periodicals, articles, and audiovisual materials on particular subjects; analyze collections; and recommend materials. They collect and organize books, pamphlets, manuscripts, and other materials in a specific field, such as rare books, genealogy, or music. In addition, they coordinate programs such as storytelling for children or literacy skills and book talks for adults; conduct classes; publicize services; provide reference help; write grants; and oversee other administrative matters.

Librarians are classified according to the type of library in which they work—public libraries, school library media centers, academic libraries, and special libraries. Some librarians work with specific groups, such as children, young adults, adults, or the disadvantaged. In school library media centers, librarians help teachers develop curricula, acquire materials for classroom instruction, and sometimes team-teach.

Librarians also work in information centers or libraries maintained by government agencies, corporations, law firms, advertising agencies, museums, professional associations, medical centers, hospitals, religious organizations, and research laboratories. They build and arrange an organization's information resources, which usually are limited to subjects of special interest to the organization. These special librarians can provide vital information services by preparing abstracts and indexes of current periodicals, organizing bibliographies, or analyzing background information and preparing reports on areas of particular interest. For example, a special librarian working for a corporation could provide the sales department with information on competitors or new developments affecting their field.

Many libraries have access to remote databases and maintain their own computerized databases. The widespread use of automation in libraries makes database searching skills important to librarians. Librarians develop and index databases and help train users to develop searching skills for the information they need. Some libraries are forming consortiums with other libraries through electronic mail. This allows patrons to simultaneously submit information requests to several libraries. The Internet also is expanding the amount of available reference information. Librarians must be aware of how to use these resources in order to locate information.

Librarians with computer and information systems skills can work as automated systems librarians (planning and operating computer systems) and information science librarians (designing information storage and retrieval systems and developing procedures for collecting, organizing, interpreting, and classifying information). These librarians analyze and plan for future information needs. The increased use of automated information systems enables librarians to focus on administrative and budgeting responsibilities, grant writing, and specialized research requests, while delegating more technical and user services responsibilities to technicians.

Increasingly, librarians apply their information management and research skills to arenas outside of libraries—for example, database development, reference tool development, information systems, publishing, Internet coordination, marketing, and training of database users. Entrepreneurial librarians sometimes start their own consulting practices, acting as freelance librarians or information brokers and providing services to other libraries, businesses, or government agencies.

Working Conditions

Librarians spend a significant portion of time at their desks or in front of computer terminals; extended work at video display terminals can cause eyestrain and headaches. Assisting users in obtaining information for their jobs, recreational purposes, and other tasks can be challenging and satisfying. At the same time, working with users under deadlines can be demanding and stressful. Some librarians lift and carry books, and some climb ladders to reach high stacks. Librarians in small organizations sometimes shelve books themselves.

More than 2 out of 10 librarians work part-time. Public and college librarians often work weekends and evenings, and have to work some holidays. School librarians usually have the same workday schedule as classroom teachers and similar vacation schedules. Special librarians usually work normal business hours, but, in fast-paced industries—such as advertising or legal services—they might work longer hours during peak times.

Employment

Librarians held about 149,000 jobs in 2000. Most were in school and academic libraries; others were in public and special libraries. A small number of librarians worked for hospitals and religious organizations. Others worked for governments.

Training, Other Qualifications, and Advancement

A master's degree in Library Science (MLS) is necessary for librarian positions in most public, academic, and special libraries, and in some school libraries. The federal government requires an MLS or the equivalent in education and experience of its librarians. Many colleges and universities offer MLS programs, but employers often prefer graduates of the approximately 56 schools accredited by the American Library Association. Most MLS programs require a bachelor's degree; any liberal arts major is appropriate.

Most MLS programs take one year to complete; others take two. A typical graduate program includes courses in the foundations of library and information science, including the history of books and printing, intellectual freedom and censorship, and the role of libraries and information in society. Other basic courses cover material selection and processing, the organization of information, reference tools and strategies, and user services. Courses are adapted to educate librarians to use new resources brought about by advancing technology, such as on-line reference systems, Internet search methods, and automated circulation systems. Course options can include resources for children or young adults; classification, cataloguing, indexing, and abstracting; library administration; and library automation. Computer-related coursework is an increasingly important part of an MLS degree. Some programs offer interdisciplinary degrees combining technical coursework in information science with traditional training in library science.

An MLS provides general preparation for library work, but some individuals specialize in a particular area, such as reference, technical services, or children's services. A PhD degree in library and information science is advantageous for a college teaching position, or a top administrative job in a college or university library or large library system.

In special libraries, an MLS usually is also required. In addition, most special librarians supplement their education with knowledge of the subject specialization, sometimes earning a master's, doctoral, or professional degree in the subject. Subject specializations include medicine, law, business, engineering, and the natural and social sciences. For example, a librarian working for a law firm may also be a licensed attorney, holding both library science and law degrees. In some jobs, knowledge of a foreign language is needed.

State certification requirements for public school librarians vary widely. Most states require school librarians, often called library media specialists, to be certified as teachers and to have taken courses in library science. An MLS is needed in some cases, perhaps with a library media specialization, or a master's in education with a specialty in school library media or educational media. Some states require certification of public librarians employed in municipal, county, or regional library systems.

Librarians participate in continuing training once they are on the job to keep abreast of new information systems brought about by changing technology.

Experienced librarians can advance to administrative positions, such as department head, library director, or chief information officer.

Job Outlook

Employment of librarians is expected to grow more slowly than the average for all occupations over the 2000–2010 period. The increasing use of computerized information storage and retrieval systems continues to contribute to slow growth in the demand for librarians. Computerized systems make cataloguing easier, which library technicians now handle. In addition, many libraries are equipped for users to access library computers directly from their homes or offices. These systems allow users to bypass librarians and conduct research on their own. However, librarians are needed to manage staff, help users develop database searching techniques, address complicated reference requests, and define users' needs. Despite expectations of slower-than-average employment growth, the need to replace librarians as they retire will result in numerous additional job openings.

Applicants for librarian jobs in large metropolitan areas, where most graduates prefer to work, usually face competition; those willing to work in rural areas should have better job prospects. Opportunities will be best for librarians outside traditional settings. Nontraditional library settings include information brokers, private corporations, and consulting firms. Many companies are turning to librarians because of their research and organizational skills, and knowledge of computer databases and library automation systems. Librarians can review vast amounts of information and analyze, evaluate, and organize it according to a company's specific needs. Librarians also are hired by organizations to set up information on the Internet. Librarians working in these settings may be classified as systems analysts, database specialists and trainers, webmasters or as developers, or LAN (local area network) coordinators.

Earnings

Salaries of librarians vary according to the individual's qualifications and the type, size, and location of the library. Librarians with primarily administrative duties often have greater earnings. Median annual earnings of librarians in 2000 were $41,700. The middle 50 percent earned between $32,840 and $52,110. The lowest 10 percent earned less than $25,030, and the highest 10 percent earned more than $62,990. Median annual earnings in the industries employing the largest numbers of librarians in 2000 were as follows:

Elementary and secondary schools	$43,320
Colleges and universities	43,050
Local government, except education and hospitals	38,370

The average annual salary for all librarians in the federal government in nonsupervisory, supervisory, and managerial positions was $63,651 in 2001.

Related Occupations

Librarians play an important role in the transfer of knowledge and ideas by providing people with access to the information they need and want. Jobs requiring similar analytical, organizational, and communicative skills include archivists, curators, and museum technicians; and computer and information scientists, research. The management aspect of a librarian's work is similar to the work of managers in a variety of business and government settings. School librarians have many duties similar to those of school teachers. Other jobs requiring the computer skills of some librarians include database administrators and computer systems analysts.

Sources of Additional Information

Information on librarianship, including information on scholarships or loans, is available from the American Library Association. For a listing of accredited library education programs, check their homepage:

- American Library Association, Office for Human Resource Development and Recruitment, 50 East Huron St., Chicago, IL 60611. Internet: http://www.ala.org

For information on a career as a special librarian, write to:

- Special Libraries Association, 1700 18th St. N.W., Washington, DC 20009

Information on graduate schools of library and information science can be obtained from:

- Association for Library and Information Science Education, P.O. Box 7640, Arlington, VA 22207. Internet: http://www.alise.org

For information on a career as a law librarian, scholarship information, and a list of ALA-accredited schools offering programs in law librarianship, contact:

- American Association of Law Libraries, 53 West Jackson Blvd., Suite 940, Chicago, IL 60604. Internet: http://www.aallnet.org

For information on employment opportunities as a health sciences librarian, scholarship information, credentialing information, and a list of MLA-accredited schools offering programs in health sciences librarianship, contact:

- Medical Library Association, 6 N. Michigan Ave., Suite 300, Chicago, IL 60602. Internet: http://www.mlanet.org

Information on acquiring a job as a librarian with the federal government may be obtained from the Office of Personnel Management through a telephone-based system. Consult your telephone directory under U.S. Government for a local number, or call (912) 757-3000; Federal Relay Service (800) 877-8339. The first number is not toll free and charges may result. Information also is available on the Internet http://www.usajobs.opm.gov

Information concerning requirements and application procedures for positions in the Library of Congress can be obtained directly from:

- Human Resources Office, Library of Congress, 101 Independence Ave. S.E., Washington, DC 20540-2231

State library agencies can furnish information on scholarships available through their offices, requirements for certification, and general information about career prospects in the state. Several of these agencies maintain job hotlines reporting openings for librarians.

State departments of education can furnish information on certification requirements and job opportunities for school librarians.

Many library science schools offer career placement services to their alumni and current students. Some allow nonaffiliated students and job seekers to use their services.

Library Technicians

O*NET 25-4031.00

Significant Points

- Training requirements range from a high school diploma to an associate or bachelor's degree, but computer skills are needed for many jobs.
- Increasing use of computerized circulation and information systems should spur job growth, but budget constraints of many libraries should moderate growth.
- Employment should grow rapidly in special libraries as growing numbers of professionals and other workers use those libraries.

Nature of the Work

Library technicians help Librarians acquire, prepare, and organize material, and assist users in finding information. Library technicians usually work under the supervision of a librarian, although they work independently in certain situations. Technicians in small libraries handle a range of duties; those in large libraries usually specialize. As libraries increasingly use new technologies—such as CD-ROM, the Internet, virtual libraries, and automated databases—the duties of library technicians will expand and evolve accordingly. Library technicians are assuming greater responsibilities, in some cases taking on tasks previously performed by Librarians.

Depending on the employer, library technicians can have other titles, such as library technical assistant or media aide. Library technicians direct library users to standard references, organize and maintain periodicals, prepare volumes for binding, handle interlibrary loan requests, prepare invoices, perform routine cataloguing and coding of library materials, retrieve information from computer databases, and supervise support staff.

The widespread use of computerized information storage and retrieval systems has resulted in technicians handling more technical and user services—such as entering catalogue information into the library's computer—that were once performed by Librarians. Technicians assist with customizing databases. In addition, technicians instruct patrons how to use computer systems to access data. The increased automation of recordkeeping has reduced the amount of clerical work performed by library technicians. Many libraries now offer self-service registration and circulations with computers, decreasing the time library technicians spend manually recording and inputting records.

Some library technicians operate and maintain audiovisual equipment, such as projectors, tape recorders, and videocassette recorders, and assist users with microfilm or microfiche readers. They also design posters, bulletin boards, or displays.

Library technicians in school libraries encourage and teach students to use the library and media center. They also help teachers obtain instructional materials and assist students with special assignments. Some work in special libraries maintained by government agencies, corporations, law firms, advertising agencies, museums, professional societies, medical centers, and research laboratories, where they conduct literature searches, compile bibliographies, and prepare abstracts, usually on subjects of particular interest to the organization.

To extend library services to more patrons, many libraries operate bookmobiles. Bookmobile drivers take trucks stocked with books to designated sites on a regular schedule. Bookmobiles serve community organizations such as shopping centers, apartment complexes, schools, and nursing homes. They also may be used to extend library service to patrons living in remote areas. Depending on local conditions, drivers may operate a bookmobile alone or may be accompanied by another library employee.

When working alone, the drivers answer patrons' questions, receive and check out books, collect fines, maintain the book collection, shelve materials, and occasionally operate audiovisual equipment to show slides or films. They participate and may assist in planning programs sponsored by the library such as reader advisory programs, used book sales, or outreach programs. Bookmobile drivers keep track of their mileage, the materials lent out, and the amount of fines collected. In some areas, they are responsible for maintenance of the vehicle and any photocopiers or other equipment in it. They record statistics on circulation and the number of people visiting the bookmobile. Drivers also may record requests for special items from the main library and arrange for the materials to be mailed or delivered to a patron during the next scheduled visit. Many bookmobiles are equipped with personal computers and CD-ROM systems linked to the main library system; this allows bookmobile drivers to reserve or locate books immediately. Some bookmobiles now offer Internet access to users.

Working Conditions

Technicians answer questions and provide assistance to library users. Those who prepare library materials sit at desks or computer terminals for long periods and can develop headaches or eyestrain from working with video display terminals. Some duties, like calculating circulation statistics, can be repetitive and boring. Others, such as performing computer searches using local and regional library networks and cooperatives, can be interesting and challenging. Library technicians may lift and carry books, and climb ladders to reach high stacks.

Library technicians in school libraries work regular school hours. Those in public libraries and college and university (academic) libraries also work weekends, evenings and some holidays. Library technicians in special libraries usually work normal business hours, although they often work overtime as well.

The schedules of bookmobile drivers depend on the size of the area being served. Some of these workers go out on their routes every day, while others go only on certain days. On these other days, they work at the library. Some also work evenings and weekends to give patrons as much access to the library as possible. Because bookmobile drivers may be the only link some people have to the library, much of their work is helping the public. They may assist handicapped or elderly patrons to the bookmobile, or shovel snow to assure their safety. They may enter hospitals or nursing homes to deliver books to patrons who are bedridden.

Employment

Library technicians held about 109,000 jobs in 2000. Most worked in school, academic, or public libraries. Some worked in hospitals and religious organizations. The federal government, primarily the U.S. Department of Defense and the U.S. Library of Congress, and state and local governments also employed library technicians.

Training, Other Qualifications, and Advancement

Training requirements for library technicians vary widely, ranging from a high school diploma to specialized postsecondary training. Some employers hire individuals with work experience or other training; others train inexperienced workers on the job. Other employers require that technicians have an associate or bachelor's degree. Given the rapid spread of automation in libraries, computer skills are needed for many jobs. Knowledge of databases, library automation systems, online library systems, online public access systems, and circulation systems is valuable.

Some 2-year colleges offer an associate of arts degree in library technology. Programs include both liberal arts and library-related study. Students learn about library and media organization and operation, and how to order, process, catalogue, locate, and circulate library materials and work with library automation. Libraries and associations offer continuing education courses to keep technicians abreast of new developments in the field.

Library technicians usually advance by assuming added responsibilities. For example, technicians often start at the circulation desk, checking books in and out. After gaining experience, they may become responsible for storing and verifying information. As they advance, they may become involved in budget and personnel matters in their department. Some library technicians advance to supervisory positions and are in charge of the day-to-day operation of their department.

Many bookmobile drivers are required to have a commercial driver's license.

Job Outlook

Employment of library technicians is expected to grow about as fast as the average for all occupations through 2010. In addition to employment growth, some job openings will result from the need to replace library technicians who transfer to other fields or leave the labor force.

The increasing use of library automation is expected to spur job growth among library technicians. Computerized information systems have simplified certain tasks, such as descriptive cataloguing, which can now be handled by technicians instead of Librarians. For example, technicians can now easily retrieve information from a central database and store it in the library's computer. Although efforts to contain costs could dampen employment growth of library technicians in school, public, and college and university libraries, cost containment efforts could also result in more hiring of library technicians than Librarians. Growth in the number of professionals and other workers who use special libraries should result in good job opportunities for library technicians in those settings.

Earnings

Median annual earnings of library technicians in 2000 were $23,170. The middle 50 percent earned between $17,820 and $29,840. The lowest 10 percent earned less than $13,810, and the highest 10 percent earned more than $35,660. Median annual earnings in the industries employing the largest numbers of library technicians in 2000 were as follows:

Colleges and universities	$25,320
Local government	22,910
Elementary and secondary schools	21,120

Salaries of library technicians in the federal government averaged $33,224 in 2001.

Related Occupations

Library technicians perform organizational and administrative duties. Workers in other occupations with similar duties include library assistants, clerical; information and record clerks; and medical records and health information technicians.

Sources of Additional Information

For information on training programs for library/media technical assistants, write to:

- American Library Association, Office for Human Resource Development and Recruitment, 50 East Huron St., Chicago, IL 60611. Internet: http://www.ala.org

Information on acquiring a job as a library technician with the federal government may be obtained from the Office of Personnel Management through a telephone-based system. Consult your telephone directory under U.S. Government for a local number, or call (912) 757-3000; Federal Relay Service (800) 877-8339. The first number is not toll free and charges may result. Information also is available on the Internet: http://www.usajobs.opm.gov

Information concerning requirements and application procedures for positions in the Library of Congress can be obtained directly from:

- Human Resources Office, Library of Congress, 101 Independence Ave. SE, Washington, DC 20540-2231

State library agencies can furnish information on requirements for technicians, and general information about career prospects in the state. Several of these agencies maintain job hotlines reporting openings for library technicians.

State departments of education can furnish information on requirements and job opportunities for school library technicians.

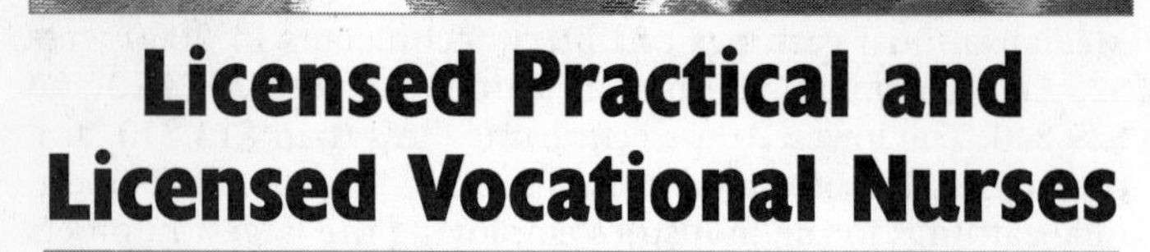

Licensed Practical and Licensed Vocational Nurses

O*NET 29-2061.00

Significant Points

- Training lasting about 1 year is available in about 1,100 state-approved programs, mostly in vocational or technical schools.
- Nursing homes will offer the most new jobs.
- Job seekers in hospitals may face competition as the number of hospital jobs for LPNs declines.

Nature of the Work

Licensed practical nurses (LPNs), or licensed vocational nurses (LVNs) as they are called in Texas and California, care for the sick, injured, convalescent, and disabled under the direction of physicians and registered nurses.

Most LPNs provide basic bedside care. They take vital signs such as temperature, blood pressure, pulse, and respiration. They also treat bedsores, prepare and give injections and enemas, apply dressings, give alcohol rubs and massages, apply ice packs and hot water bottles, and monitor catheters. LPNs observe patients and report adverse reactions to medications or treatments. They collect samples for testing, perform routine laboratory tests, feed patients, and record food and fluid intake and output. They help patients with bathing, dressing, and personal hygiene, keep them comfortable, and care for their emotional needs. In states where the law allows, they may administer prescribed medicines or start intravenous fluids. Some LPNs help deliver, care for, and feed infants. Experienced LPNs may supervise nursing assistants and aides.

LPNs in nursing homes provide routine bedside care, help evaluate residents' needs, develop care plans, and supervise the care provided by nursing aides. In doctors' offices and clinics, they also may make appointments, keep records, and perform other clerical duties. LPNs who work in private homes also may prepare meals and teach family members simple nursing tasks.

Working Conditions

Most licensed practical nurses in hospitals and nursing homes work a 40-hour week, but because patients need around-the-clock care, some work nights, weekends, and holidays. They often stand for long periods and help patients move in bed, stand, or walk.

LPNs may face hazards from caustic chemicals, radiation, and infectious diseases such as hepatitis. They are subject to back injuries when moving patients and shock from electrical equipment. They often must deal with the stress of heavy workloads. In addition, the patients they care for may be confused, irrational, agitated, or uncooperative.

Employment

Licensed practical nurses held about 700,000 jobs in 2000. Twenty-nine percent of LPNs worked in nursing homes, 28 percent worked in hospitals, and 14 percent in physicians' offices and clinics. Others worked for home healthcare services, residential care facilities, schools, temporary help agencies, or government agencies; about 1 in 5 worked part-time.

Training, Other Qualifications, and Advancement

All states and the District of Columbia require LPNs to pass a licensing examination after completing a state-approved practical nursing program. A high school diploma, or equivalent, usually is required for entry, although some programs accept candidates without a diploma or are designed as part of a high school curriculum.

In 2000, approximately 1,100 state-approved programs provided practical nursing training. Almost 6 out of 10 students were enrolled in technical or vocational schools, while 3 out of 10 were in community and junior colleges. Others were in high schools, hospitals, and colleges and universities.

Most practical nursing programs last about 1 year and include both classroom study and supervised clinical practice (patient care). Classroom study covers basic nursing concepts and patient-care related subjects, including anatomy, physiology, medical-surgical nursing, pediatrics, obstetrics, psychiatric nursing, administration of drugs, nutrition, and first aid. Clinical practice usually is in a hospital, but sometimes includes other settings.

LPNs should have a caring, sympathetic nature. They should be emotionally stable because work with the sick and injured can be stressful. They also should have keen observational, decision making, and communication skills. As part of a healthcare team, they must be able to follow orders and work under close supervision.

Job Outlook

Employment of LPNs is expected to grow about as fast as the average for all occupations through 2010 in response to the long-term care needs of a rapidly growing elderly population and the general growth of healthcare. Replacement needs will be a major source of job openings, as many workers leave the occupation permanently.

Employment of LPNs in nursing homes is expected to grow faster than the average. Nursing homes will offer the most new jobs for LPNs as the number of aged and disabled persons in need of long-term care rises. In addition to caring for the aged and disabled, nursing homes will be called on to care for the increasing number of patients who have been discharged from the hospital but who have not recovered enough to return home.

LPNs seeking positions in hospitals may face competition, as the number of hospital jobs for LPNs declines. An increasing proportion of sophisticated procedures, which once were performed only in hospitals, are being performed in physicians' offices and clinics, including ambulatory surgicenters and emergency medical centers, due largely to advances in technology. As a result, employment of LPNs is projected to grow much faster than average in these places as healthcare expands outside the traditional hospital setting.

Employment of LPNs is expected to grow much faster than average in home healthcare services. This is in response to a growing number of older persons with functional disabilities, consumer preference for care in the home, and technological advances, which make it possible to bring increasingly complex treatments into the home.

Earnings

Median annual earnings of licensed practical nurses were $29,440 in 2000. The middle 50 percent earned between $24,920 and $34,800. The lowest 10 percent earned less than $21,520, and the highest 10 percent earned more than $41,800. Median annual earnings in the industries employing the largest numbers of licensed practical nurses in 2000 were as follows:

Personnel supply services	$35,750
Home healthcare services	31,220
Nursing and personal care facilities	29,980
Hospitals	28,450
Offices and clinics of medical doctors	27,520

Related Occupations

LPNs work closely with people while helping them. So do emergency medical technicians and paramedics, social and human service assistants, surgical technologists, and teacher assistants.

Sources of Additional Information

For information about practical nursing, contact:

- National League for Nursing, 61 Broadway, New York, NY 10006. Internet: http://www.nln.org
- National Association for Practical Nurse Education and Service, Inc., 1400 Spring St., Suite 330, Silver Spring, MD 20910
- National Federation of Licensed Practical Nurses, Inc., 893 US Highway 70 West, Suite 202, Garner, NC 27529-2597

Machinists

O*NET 51-4041.00

Significant Points

- Machinists learn in apprenticeship programs, informally on the job, and in high schools, vocational schools, or community or technical colleges.
- Many entrants previously have worked as machine setters, operators, or tenders.
- Job opportunities are expected to be excellent.

Nature of the Work

Machinists use machine tools, such as lathes, milling machines, and spindles, to produce precision metal parts. Although they may produce large quantities of one part, precision machinists often produce small batches or one-of-a-kind items. They use their knowledge of the working properties of metals and their skill with machine tools to plan and carry out the operations needed to make machined products that meet precise specifications.

Before they machine a part, machinists must carefully plan and prepare the operation. These workers first review blueprints or written specifications for a job. Next, they calculate where to cut or bore into the workpiece (the piece of metal that is being shaped), how fast to feed the metal into the machine, and how much metal to remove. They then select tools and materials for the job, plan the sequence of cutting and finishing operations, and mark the metal stock to show where cuts should be made.

After this layout work is completed, machinists perform the necessary machining operations. They position the metal stock on the machine tool—spindle, drill press, lathe, milling machine, or other—set the controls, and make the cuts. During the machining process, they must constantly monitor the feed and speed of the machine. Machinists also ensure that the workpiece is being properly lubricated and cooled, because the machining of metal products generates a significant amount of heat. The temperature of the workpiece is a key concern because most metals expand when heated; machinists must adjust the size of their cuts relative to the temperature. Some rarer, but increasingly popular, metals, such as titanium, are machined at extremely high temperatures. Machinists also adjust cutting speeds to compensate for harmonic vibrations, which can decrease the accuracy of cuts, particularly on newer high-speed spindles and lathes.

Some machinists, often called production machinists, may produce large quantities of one part, especially parts requiring the use of complex operations and great precision. Production machinists work with complex computer numerically controlled (CNC) cutting machines. Frequently, machinists work with computer-control programmers to determine how the automated equipment will cut a part. The programmer determines the path of the cut, and the machinist determines the type of cutting tool, the speed of the cutting tool, and the feed rate. After the production process is designed, relatively simple and repetitive operations normally are performed by machine setters, operators, and tenders. Other machinists do maintenance work—repairing or making new parts for existing machinery. To repair a broken part, maintenance machinists may refer to blueprints and perform the same machining operations that were needed to create the original part.

Working Conditions

Most machine shops of today are relatively clean, well lit, and ventilated. Many computer-controlled machines are totally enclosed, minimizing the exposure of workers to noise, dust, and the lubricants used to cool workpieces during machining. Nevertheless, working around machine tools presents certain dangers, and workers must follow safety precautions. Machinists wear protective equipment such as safety glasses to shield against bits of flying metal and earplugs to dampen machinery noise. They also must exercise caution when handling hazardous coolants and lubricants. The job requires stamina, because machinists stand most of the day and, at times, may need to lift moderately heavy workpieces.

Most machinists work a 40-hour week. Evening and weekend shifts are becoming more common as companies justify investments in more expensive machinery by extending hours of operation. Overtime is common during peak production periods.

Employment

Machinists held about 430,000 jobs in 2000. Most machinists work in small machining shops or in manufacturing firms that produce durable goods, such as metalworking and industrial machinery, aircraft, or motor vehicles. Maintenance machinists work in most industries that use production machinery. Although machinists work in all parts of the country, jobs are most plentiful in the northeast, midwest, and west, where manufacturing is concentrated.

Training, Other Qualifications, and Advancement

Machinists train in apprenticeship programs, informally on the job, and in high schools, vocational schools, or community or technical colleges. Experience with machine tools is helpful. In fact, many entrants previously have worked as machine setters, operators, or tenders. Persons interested in becoming machinists should be mechanically inclined, able to work independently, and able to do highly accurate work that requires concentration and physical effort.

High school or vocational school courses in mathematics, blueprint reading, metalworking, and drafting are highly recommended. Apprenticeship programs consist of shop training and related classroom instruction. In shop training, apprentices work almost full time, and are supervised by an experienced machinist while learning to operate various machine tools. Classroom instruction includes math, physics, blueprint reading, mechanical drawing, and quality and safety practices. In addition, as machine shops have increased their use of computer-controlled equipment, training in the operation and programming of CNC machine tools has become essential. Apprenticeship classes are taught in cooperation with local community or vocational colleges. A growing number of machinists learn the trade through 2-year associate degree programs at community or technical colleges. Graduates of these programs still need significant on-the-job experience before they are fully qualified.

To boost the skill level of machinists and to create a more uniform standard of competency, a number of training facilities and colleges have recently begun implementing curriculums incorporating national skills standards developed by the National Institute of Metalworking Skills (NIMS). After completing such a curriculum and passing a performance requirement and written exam, a NIMS credential is granted to trainees, providing formal recognition of competency in a metalworking field. Completing a recognized certification program provides a machinist with better career opportunities.

As new automation is introduced, machinists normally receive additional training to update their skills. This training usually is provided by a representative of the equipment manufacturer or a local technical school. Some employers offer tuition reimbursement for job-related courses.

Machinists can advance in several ways. Experienced machinists may become computer-control programmers and operators, and some are promoted to supervisory or administrative positions in their firms. A few open their own shops.

Job Outlook

Despite projected slower-than-average employment growth, job opportunities for machinists should continue to be excellent. Many young people with the necessary educational and personal qualifications needed to obtain machining skills may prefer to attend college or may not wish to enter production-related occupations. Therefore, the number of workers obtaining the skills and knowledge necessary to fill machinist jobs is expected to be less than the number of job openings arising each year from employment growth and from the need to replace experienced machinists who transfer to other occupations or retire.

Employment of machinists is expected to grow more slowly than the average for all occupations over the 2000–2010 period because of rising productivity among machinists. Productivity gains are resulting from the expanded use of computer-controlled machine tools and new technologies, such as high-speed machining, which reduce the time required for machining operations. This allows fewer machinists to accomplish the same amount of work previously performed by more workers. Technology is not expected to affect the employment of machinists as significantly as that of most other production occupations, however, because many of the unique operations performed by machinists cannot be efficiently automated. Due to modern production techniques, employers prefer workers, such as machinists, who have a wide range of skills and are capable of performing almost any task in a machine shop. In addition, firms are likely to retain their most skilled workers to operate and maintain expensive new machinery.

Employment levels in this occupation are influenced by economic cycles—as the demand for machined goods falls, machinists involved in production may be laid off or forced to work fewer hours. Employment of machinists involved in plant maintenance, however, often is more stable because proper maintenance and repair of costly equipment remain vital concerns, even when production levels fall.

Earnings

Median hourly earnings of machinists were $14.78 in 2000. The middle 50 percent earned between $11.43 and $18.39. The lowest 10 percent earned less than $9.01, while the top 10 percent earned more than $21.84. Median hourly earnings in the manufacturing industries employing the largest number of machinists in 2000 were as follows:

Aircraft and parts	$16.86
Metalworking machinery	15.89
Industrial machinery, not elsewhere classified	14.66
Motor vehicles and equipment	14.24
Personnel supply services	8.80

Related Occupations

Occupations most closely related to that of machinist are other machining occupations. These include tool and die makers; machine setters, operators, and tenders—metal and plastic; and computer-control programmers and operators. Another occupation that requires precision and skill in working with metal is welding, soldering, and brazing workers.

Sources of Additional Information

For general information about machinists, contact:

- Precision Machine Products Association, 6700 West Snowville Rd., Brecksville, OH 44141-3292. Internet: http://www.pmpa.org

For a list of training centers and apprenticeship programs, contact:

- National Tooling and Metalworking Association, 9300 Livingston Rd., Fort Washington, MD 20744. Internet: http://www.ntma.org

For general occupational information and a list of training programs, contact:

- PMA Educational Foundation, 6363 Oak Tree Blvd., Independence, OH 44131-2500. Internet: http://www.pmaef.org

Management Analysts

O*NET 13-1111.00

Significant Points

- Thirty-three percent are self-employed, about twice the average for other management, business, and financial occupations.
- Most positions in private industry require a master's degree and 5 years of specialized experience; a bachelor's degree is sufficient for entry-level government jobs.
- Despite projected faster-than-average employment growth, intense competition is expected for jobs.

Nature of the Work

As business becomes more complex, the nation's firms are continually faced with new challenges. Firms increasingly rely on management analysts to help them remain competitive amidst these changes. Management analysts, often referred to as management consultants in private industry, analyze and propose ways to improve an organization's structure, efficiency, or profits. For example, a small but rapidly growing company that needs help improving the system of control over inventories and expenses may decide to employ a consultant who is an expert in just-in-time inventory management. In another case, a large company that has recently acquired a new division may hire management analysts to help reorganize the corporate structure and eliminate duplicate or nonessential jobs. In recent years, information technology and electronic commerce have provided new opportunities for management analysts. Companies hire consultants to develop strategies for entering and remaining competitive in the new electronic marketplace.

Firms providing management analysis range in size from a single practitioner to large international organizations employing thousands of consultants. Some analysts and consultants specialize in a specific industry while others specialize by type of business function, such as human resources or information systems. In government, management analysts tend to specialize by type of agency. The work of management analysts and consultants varies with each client or employer, and from project to project. Some projects require a team of consultants, each specializing in one area. In other projects, consultants work independently with the organization's managers. In all cases, analysts and consultants collect, review, and analyze information in order to make recommendations to managers.

Both public and private organizations use consultants for a variety of reasons. Some lack the internal resources needed to handle a project, while others need a consultant's expertise to determine what resources will be required and what problems may be encountered if they pursue a particular opportunity. To retain a consultant, a company first solicits proposals from a number of consulting firms specializing in the area in which it needs assistance. These proposals include the estimated cost and scope of the project, staffing requirements, references from a number of previous clients, and a completion deadline. The company then selects the proposal that best suits its needs.

After obtaining an assignment or contract, management analysts first define the nature and extent of the problem. During this phase, they analyze relevant data, which may include annual revenues, employment, or expenditures, and interview managers and employees while observing their operations. The analyst or consultant then develops solutions to the problem. In the course of preparing their recommendations, they take into account the nature of the organization, the relationship it has with others in the industry, and its internal organization and culture. Insight into the problem often is gained by building and solving mathematical models.

Once they have decided on a course of action, consultants report their findings and recommendations to the client. These suggestions usually are submitted in writing, but oral presentations regarding findings also are common. For some projects, management analysts are retained to help implement the suggestions they have made.

Management analysts in government agencies use the same skills as their private-sector colleagues do to advise managers on many types of issues. Most of these issues are similar to the problems faced by private firms. For example, if an agency is planning to purchase personal computers, it must first determine which type to buy, given its budget and data processing needs. In this case, management analysts would assess the prices and characteristics of various machines and determine which best meets the agency's needs.

Working Conditions

Management analysts usually divide their time between their offices and the client's site. In either situation, much of an analyst's time is spent indoors in clean, well-lit offices. Because they must spend a significant portion of their time with clients, analysts travel frequently.

Analysts and consultants generally work at least 40 hours a week. Uncompensated overtime is common, especially when project deadlines are approaching. Analysts may experience a great deal of stress as a result of trying to meet a client's demands, often on a tight schedule.

Self-employed consultants can set their workload and hours and work at home. On the other hand, their livelihood depends on their ability to maintain and expand their client base. Salaried consultants also must impress potential clients to get and keep clients for their company.

Employment

Management analysts held about 501,000 jobs in 2000. Thirty-three percent of these workers were self-employed, about twice the average for other management, business, and financial occupations. Management analysts are found throughout the country, but employment is concentrated in large metropolitan areas. Most work in management consulting and computer and data processing firms, and for federal, state, and local governments. The majority of those working for the federal government are in the U.S. Department of Defense.

Training, Other Qualifications, and Advancement

Educational requirements for entry-level jobs in this field vary widely between private industry and government. Most employers in private industry generally seek individuals with a master's degree in business administration or a related discipline. Some employers also require at least 5 years of experience in the field in which they plan to consult in addition to a master's degree. Most government agencies hire people with a bachelor's degree and no pertinent work experience for entry-level management analyst positions.

Many fields of study provide a suitable educational background for this occupation because of the wide range of areas addressed by management analysts. These include most academic programs in business and management, as well as computer and information sciences and engineering. In addition to the appropriate formal education, most entrants to this occupation have years of experience in management, human resources, information technology, or other specialties. Analysts also routinely attend conferences to keep abreast of current developments in their field.

Management analysts often work with minimal supervision, so they need to be self-motivated and disciplined. Analytical skills, the ability to get along with a wide range of people, strong oral and written communication skills, good judgment, time management skills, and creativity are other desirable qualities. The ability to work in teams also is an important attribute as consulting teams become more common.

As consultants gain experience, they often become solely responsible for a specific project, taking on more responsibility and managing their own hours. At the senior level, consultants may supervise lower level workers and become more involved in seeking out new business. Those with exceptional skills may eventually become a partner in the firm. Others with entrepreneurial ambition may open their own firm.

A high percentage of management consultants are self-employed, partly because business startup costs are low. Self-employed consultants also can share office space, administrative help, and other resources with other self-employed consultants or small consulting firms, thus reducing overhead costs. Because many small consulting firms fail each year for lack of managerial expertise and clients, those interested in opening their own firm must have good organizational and marketing skills and several years of consulting experience.

The Institute of Management Consultants USA, Inc. (IMC USA) offers a wide range of professional development programs and resources, such as meetings and workshops, that can be helpful for management consultants. The IMC USA also offers the Certified Management Consultant (CMC) designation to those who pass an examination and meet minimum levels of education and experience. Certification is not mandatory for management consultants, but it may give a job seeker a competitive advantage.

Job Outlook

Despite projected rapid employment growth, keen competition is expected for jobs as management analysts. Because analysts can come from such diverse educational backgrounds, the pool of applicants from which employers can draw is quite large. Furthermore, the independent and challenging nature of the work, combined with high earnings potential, makes this occupation attractive to many. Job opportunities are expected to be best for those with a graduate degree, industry expertise, and a talent for salesmanship and public relations.

Employment of management analysts is expected to grow faster than the average for all occupations through 2010, as industry and government increasingly rely on outside expertise to improve the performance of their organizations. Job growth is projected in very large consulting firms with international expertise and in smaller consulting firms that specialize in specific areas, such as biotechnology, healthcare, information technology, human resources, engineering, and telecommunications. Growth in the number of individual practitioners may be hindered, however, by increasing use of consulting teams, which permits examination of a variety of different issues and problems within an organization.

Employment growth of management analysts and consultants has been driven by a number of changes in the business environment that have forced American firms to take a closer look at their operations. These changes include developments in information technology and the growth of electronic commerce. Traditional companies hire analysts to help design intranets or company Web sites, or establish online businesses. New Internet start-up companies hire analysts not only to design Web sites, but also to advise them in more traditional business practices, such as pricing strategies, marketing, and inventory and human resource management. In order to offer clients better quality and a wider variety of services, consulting firms are partnering with traditional computer software and technology firms. Also, many computer firms are developing consulting practices of their own in order to take advantage of this expanding market. Although information technology consulting should remain one of the fastest growing consulting areas, the volatility of the computer and data processing services industry necessitates that the most successful management analysts have knowledge of traditional business practices in addition to computer applications, systems integration, and Web design and management skills.

The growth of international business also has contributed to an increase in demand for management analysts. As U.S. firms expand their business abroad, many will hire management analysts to help them form the right strategy for entering the market. Such managers also advise on legal matters pertaining to a specific countries or help with organizational, administrative, and other issues, especially if the U.S. company is involved in a partnership or merger with a local firm. These trends provide management analysts with more opportunities to travel or work abroad, but also require that they have a more comprehensive knowledge of international business and foreign cultures and languages.

Furthermore, as international and domestic markets have become more competitive, firms have needed to use resources more efficiently. Management analysts increasingly are sought to help reduce costs, streamline operations, and develop marketing strategies. As this process continues and businesses downsize, even more opportunities will be created for analysts to perform duties that previously were handled internally. Finally, management analysts also will be in greater demand in the public sector, as federal, state, and local government agencies seek ways to become more efficient.

Earnings

Salaries for management analysts vary widely by experience, education, and employer. Median annual earnings of management analysts in 2000 were $55,040. The middle 50 percent earned between $41,970 and $72,630. The lowest 10 percent earned less than $32,860, and the highest 10 percent earned more than $98,210. Median annual earnings in the industries employing the largest numbers of management analysts and consultants in 2000 were as follows:

Accounting, auditing, and bookkeeping	$62,230
Management and public relations	61,290
Federal government	59,780
Computer and data processing services	56,070
State government	43,470

According to a 2000 survey by the Association of Management Consulting Firms, earnings—including bonuses and profit sharing—for research associates in member firms averaged $39,200; for entry-level consultants, $58,000; for management consultants, $76,300; for senior consultants, $100,300; for junior partners, $133,500; and for senior partners, $259,500.

Salaried management analysts usually receive common benefits such as health and life insurance, a retirement plan, vacation, and sick leave, as well as less common benefits such as profit sharing and bonuses for outstanding work. In addition, the employer usually reimburses all travel expenses. Self-employed consultants have to maintain their own office and provide their own benefits.

Related Occupations

Management analysts collect, review, and analyze data; make recommendations; and implement their ideas. Others who use similar skills include systems analysts, computer scientists, and database administrators; operations research analysts; economists and market and survey researchers; and financial analysts and personal financial advisors.

Sources of Additional Information

Information about career opportunities in management consulting is available from:

- The Association of Management Consulting Firms, 380 Lexington Ave., Suite 1700, New York, NY 10168. Internet: http://www.amcf.org

Information about the Certified Management Consultant designation can be obtained from:

- The Institute of Management Consultants USA, Inc., 2025 M St. N.W., Suite 800, Washington DC 20036. Internet: http://www.imcusa.org

Information on obtaining a management analyst position with the federal government is available from the Office of Personnel Management (OPM) through a telephone-based system. Consult your telephone directory under U.S. Government for a local number or call (912) 757-3000; Federal Relay Service: (800) 877-8339. The first number is not toll free, and charges may result. Information also is available from the OPM Internet site: http://www.usajobs.opm.gov

Materials Engineers

O*NET 17-2131.00

Significant Points

- Overall job opportunities in engineering are expected to be good.
- A bachelor's degree is required for most entry-level jobs.
- Starting salaries are significantly higher than those of college graduates in other fields.
- Continuing education is critical to keep abreast of the latest technology.

Nature of the Work

Materials engineers are involved in the extraction, development, processing, and testing of the materials used to create a diversity of products, from computer chips and television screens to golf clubs and snow skis. They work with metals, ceramics, plastics, semiconductors, and combinations of materials called composites to create new materials that meet certain mechanical, electrical, and chemical requirements. They also are involved in selecting materials for new applications.

There are numerous new developments within materials engineering that make it possible to manipulate and use materials in various ways. For example, materials engineers have developed the ability to create and then study materials at an atomic level using advanced processes, electrons, neutrons, or X rays and to replicate the characteristics of materials and their components with computers.

Materials engineers specializing in metals can be considered *metallurgical engineers*, while those specializing in ceramics can be considered *ceramic engineers*. Most metallurgical engineers work in one of the three main branches of metallurgy—extractive or chemical, physical, and process. Extractive metallurgists are concerned with removing metals from ores and refining and alloying them to obtain useful metal. Physical metallurgists study the nature, structure, and physical properties of metals and their alloys, and relate them to the methods of processing them into final products. Process metallurgists develop and improve metalworking processes such as casting, forging, rolling, and drawing. Ceramic engineers develop ceramic materials and the processes for making ceramic materials into useful products. Ceramics include all nonmetallic, inorganic materials that generally require high temperatures in their processing. Ceramic engineers work on products as diverse as glassware, automobile and aircraft engine components, fiber-optic communication lines, tile, and electric insulators.

Working Conditions

Most engineers work in office buildings, laboratories, or industrial plants. Others may spend time outdoors at construction sites, mines, and oil and gas exploration and production sites, where they monitor or direct operations or solve onsite problems. Some engineers travel extensively to plants or work sites.

Many engineers work a standard 40-hour week. At times, deadlines or design standards may bring extra pressure to a job. When this happens, engineers may work longer hours and experience considerable stress.

Employment

Materials engineers held about 33,000 jobs in 2000. Because materials are building blocks for other goods, materials engineers are widely distributed among manufacturing industries. In fact, 84 percent of materials engineers worked in manufacturing industries, primarily metal production and processing, electronic and other electrical equipment, transportation equipment, and industrial machinery and equipment. They also worked in services industries such as engineering and management and research and testing services. Most remaining materials engineers worked for federal and state governments.

Training, Other Qualifications, and Advancement

A bachelor's degree in engineering is required for almost all entry-level engineering jobs. College graduates with a degree in a physical science or mathematics occasionally may qualify for some engineering jobs, especially in specialties in high demand. Most engineering degrees are granted in electrical, electronics, mechanical, or civil engineering. However, engineers trained in one branch may work in related branches. This flexibility allows employers to meet staffing needs in new technologies and specialties in which engineers are in short supply. It also allows engineers to shift to fields with better employment prospects or to those that more closely match their interests.

Most engineering programs involve a concentration of study in an engineering specialty, along with courses in both mathematics and science. Most programs include a design course, sometimes accompanied by a computer or laboratory class or both.

In addition to the standard engineering degree, many colleges offer 2- or 4-year degree programs in engineering technology. These programs, which usually include various hands-on laboratory classes that focus on current issues, prepare students for practical design and production work, rather than for jobs which require more theoretical and scientific knowledge. Graduates of 4-year technology programs may get jobs similar to those obtained by graduates with a bachelor's degree in engineering. Engineering technology graduates, however, are not qualified to register as professional engineers under the same terms as graduates with degrees in engineering. Some employers regard technology program graduates as having skills between those of a technician and an engineer.

Graduate training is essential for engineering faculty positions and many research and development programs, but is not required for the majority of entry-level engineering jobs. Many engineers obtain graduate degrees in engineering or business administration to learn new technology and broaden their education. Many high-level executives in government and industry began their careers as engineers.

About 330 colleges and universities offer bachelor's degree programs in engineering that are accredited by the Accreditation Board for Engineering and Technology (ABET), and about 250 colleges offer accredited bachelor's degree programs in engineering technology. ABET accreditation is based on an examination of an engineering program's student achievement, program improvement, faculty, curricular content, facilities, and institutional commitment. Although most institutions offer programs in the major branches of engineering, only a few offer programs in the smaller specialties. Also, programs of the same title may vary in content. For example, some programs emphasize industrial practices, preparing students for a job in industry, whereas others are more theoretical and are designed to prepare students for graduate work. Therefore, students should investigate curricula and check accreditations carefully before selecting a college.

Admissions requirements for undergraduate engineering schools include a solid background in mathematics (algebra, geometry, trigonometry, and calculus) and sciences (biology, chemistry, and physics), and courses in English, social studies, humanities, and computers. Bachelor's degree programs in engineering typically are designed to last 4 years, but many students find that it takes between 4 and 5 years to complete their studies. In a typical 4-year college curriculum, the first 2 years are spent studying mathematics, basic sciences, introductory engineering, humanities, and social sciences. In the last 2 years, most courses are in engineering, usually with a concentration in one branch. Some programs offer a general engineering curriculum; students then specialize in graduate school or on the job.

Some engineering schools and 2-year colleges have agreements whereby the 2-year college provides the initial engineering education, and the engineering school automatically admits students for their last 2 years. In addition, a few engineering schools have arrangements whereby a student spends 3 years in a liberal arts college studying pre-engineering subjects and 2 years in an engineering school studying core subjects, and then receives a bachelor's degree from each school. Some colleges and universities offer 5-year master's degree programs. Some 5- or even 6-year cooperative plans combine classroom study and practical work, permitting students to gain valuable experience and finance part of their education. All 50 states and the District of Columbia usually require licensure for engineers who offer their services directly to the public. Engineers who are licensed are called Professional Engineers (PE). This licensure generally requires a degree from an ABET-accredited engineering program, 4 years of relevant work experience, and successful completion of a state examination. Recent graduates can start the licensing process by taking the examination in two stages. The initial Fundamentals of Engineering (FE) examination can be taken upon graduation. Engineers who pass this examination commonly are called Engineers in Training (EIT) or Engineer Interns (EI). The EIT certification usually is valid for 10 years. After acquiring suitable work experience, EITs can take the second examination, the Principles and Practice of Engineering Exam. Several states have imposed mandatory continuing education requirements for relicensure. Most states recognize licensure from other states.

Engineers should be creative, inquisitive, analytical, and detail oriented. They should be able to work as part of a team and to communicate well, both orally and in writing. Communication abilities are becoming more important because much of their work is becoming more diversified, meaning that engineers interact with specialists in a wide range of fields outside engineering.

Beginning engineering graduates usually work under the supervision of experienced engineers and, in large companies, also may receive formal classroom or seminar-type training. As new engineers gain knowledge and experience, they are assigned more difficult projects with greater independence to develop designs, solve problems, and make decisions. Engineers may advance to become technical specialists or to supervise a staff or team of engineers and technicians. Some may eventually become engineering managers or enter other managerial or sales jobs.

Job Outlook

Employment of materials engineers is expected to grow more slowly than the average for all occupations through 2010. More materials engineers will be needed to develop new materials for electronics and plastics products. However, many of the manufacturing industries in which materials engineers are concentrated—such as primary metals, industrial machinery and equipment, and stone, clay, and glass products—are expected to experience declines in employment, reducing employment opportunities for materials engineers. As firms contract out to meet their materials engineering needs, however, employment growth is expected in many services industries, including research and testing, personnel supply, health, and engineering and architectural services. In addition to growth, job openings will result from the need to replace materials engineers who transfer to other occupations or leave the labor force.

Earnings

Median annual earnings of materials engineers were $59,100 in 2000. The middle 50 percent earned between $47,320 and $72,900. The lowest 10 percent earned less than $37,680, and the highest 10 percent earned more than $87,630.

According to a 2001 salary survey by the National Association of Colleges and Employers, bachelor's degree candidates in materials engineering received starting offers averaging $49,936 a year.

Related Occupations

Materials engineers apply the principles of physical science and mathematics in their work. Other workers who use scientific and mathematical principles include architects, except landscape and naval; engineering and natural sciences managers; computer and information systems managers; mathematicians; drafters; engineering technicians; sales engineers; science technicians; and physical and life scientists, including agricultural and food scientists, biological and medical scientists, conservation scientists and foresters, atmospheric scientists, chemists and materials scientists, environmental scientists and geoscientists, and physicists and astronomers.

Sources of Additional Information

For further information about materials engineers, contact:

- Minerals, Metals, & Materials Society, 184 Thorn Hill Rd., Warrendale, PA 15086. Internet: http://www.tms.org
- ASM International Foundation, Materials Park, OH 44073-0002. Internet: http://www.asm-intl.org

High school students interested in obtaining a full package of guidance materials and information (product number SP-01) on a variety of engineering disciplines should contact the Junior Engineering Technical Society by sending $3.50 to:

- JETS-Guidance, 1420 King St., Suite 405, Alexandria, VA 22314-2794. Internet: http://www.jets.org

High school students interested in obtaining information on ABET-accredited engineering programs should contact:

- The Accreditation Board for Engineering and Technology, Inc., 111 Market Place, Suite 1050, Baltimore, MD 21202-4012. Internet: http://www.abet.org

Non-licensed engineers and college students interested in obtaining information on Professional Engineer licensure should contact:

- The National Society of Professional Engineers, 1420 King St., Alexandria, VA 22314-2794. Internet: http://www.nspe.org
- National Council of Examiners for Engineers and Surveying, P.O. Box 1686, Clemson, SC 29633-1686. Internet: http://www.ncees.org

Information on general engineering education and career resources is available from:

- American Society for Engineering Education, 1818 N St. N.W., Suite 600, Washington, DC 20036-2479. Internet: http://www.asee.org

Information on obtaining an engineering position with the federal government is available from the Office of Personnel Management (OPM) through a telephone-based system. Consult your telephone directory under U.S. Government for a local number or call (912) 757-3000; Federal Relay Service: (800) 877-8339. The first number is not toll free, and charges may result. Information also is available from the OPM Internet site: http://www.usajobs.opm.gov

Non-high school students and those wanting more detailed information should contact societies representing the individual branches of engineering. Each can provide information about careers in the particular branch. The individual statements elsewhere in this book also provide other information in detail on aerospace; agricultural; biomedical; chemical; civil; computer hardware; electrical and electronics; environmental; industrial, including health and safety; mechanical; mining and geological, including mining safety; nuclear; and petroleum engineering.

Mathematicians

O*NET 15-2021.00

Significant Points

- A doctoral degree in mathematics usually is the minimum education needed, except in the federal government.
- Employment is expected to decline because very few jobs with the title mathematician are available.
- Master's and PhD degree holders with a strong background in mathematics and a related discipline, such as computer science or engineering, should have good employment opportunities in related occupations.

Nature of the Work

Mathematics is one of the oldest and most fundamental sciences. Mathematicians use mathematical theory, computational techniques, algorithms, and the latest computer technology to solve economic, scientific, engineering, physics, and business problems. The work of mathematicians falls into two broad classes—theoretical (pure) mathematics and applied mathematics. These classes, however, are not sharply defined, and often overlap.

Theoretical mathematicians advance mathematical knowledge by developing new principles and recognizing previously unknown relationships between existing principles of mathematics. Although they seek to increase basic knowledge without necessarily considering its practical use, such pure and abstract knowledge has been instrumental in producing or furthering many scientific and engineering achievements. Many theoretical mathematicians are employed as university faculty and divide their time between teaching and conducting research.

Applied mathematicians, on the other hand, use theories and techniques, such as mathematical modeling and computational methods, to formulate and solve practical problems in business, government, engineering, and in the physical, life, and social sciences. For example, they may analyze the most efficient way to schedule airline routes between cities, the effect and safety of new drugs, the aerodynamic characteristics of an experimental automobile, or the cost-effectiveness of alternate manufacturing processes. Applied mathematicians working in industrial research and development may develop or enhance mathematical methods when solving a difficult problem. Some mathematicians, called cryptanalysts, analyze and decipher encryption systems designed to transmit military, political, financial, or law enforcement-related information in code.

Applied mathematicians start with a practical problem, envision the separate elements of the process under consideration, and then reduce the elements into mathematical variables. They often use computers to analyze relationships among the variables and solve complex problems by developing models with alternate solutions.

Much of the work in applied mathematics is done by individuals with titles other than mathematician. In fact, because mathematics is the foundation upon which so many other academic disciplines are built, the number of workers using mathematical techniques is much greater than the number formally designated as mathematicians. For example, engineers, computer scientists, physicists, and economists are among those who use mathematics extensively. Some professionals, including statisticians, actuaries, and operations research analysts, actually are specialists in a particular branch of mathematics. Frequently, applied mathematicians are required to collaborate with other workers in their organizations to achieve common solutions to problems.

Working Conditions

Mathematicians usually work in comfortable offices. They often are part of an interdisciplinary team that may include economists, engineers, computer scientists, physicists, technicians, and others. Deadlines, overtime work, special requests for information or analysis, and prolonged travel to attend seminars or conferences may be part of their jobs. Mathematicians who work in academia usually have a mix of teaching and research responsibilities. These mathematicians often conduct research alone, or are aided by graduate students interested in the topic being researched.

Employment

Mathematicians held about 3,600 jobs in 2000. In addition, about 20,000 persons held full-time mathematics faculty positions in colleges and universities in 2000, according to the American Mathematical Society.

Many nonfaculty mathematicians work for federal or state governments. The U.S. Department of Defense is the primary federal employer, accounting for about three-fourths of the mathematicians employed by the federal government. In the private sector, major employers include research and testing services, educational services, security and commodity exchanges, and management and public relations services. Within manufacturing, the aerospace and drug industries are the key employers. Some mathematicians also work for banks and insurance companies.

Training, Other Qualifications, and Advancement

A doctoral degree in mathematics usually is the minimum education needed for prospective mathematicians, except in the federal government. In the federal government, entry-level job candidates usually must have a 4-year degree with a major in mathematics or a 4-year degree with the equivalent of a mathematics major—24 semester hours of mathematics courses.

In private industry, candidates for mathematician jobs typically need a master's or PhD degree. Most of the positions designated for mathematicians are in research and development laboratories as part of technical teams. Research scientists in such positions engage either in basic research on pure mathematical principles or in applied research on developing or improving specific products or processes. The majority of those with a bachelor's or master's degree in mathematics who work in private industry do so not as mathematicians, but in related fields such as computer science, where they have titles such as computer programmer, systems analyst, or systems engineer.

Most colleges and universities offer a bachelor's degree in mathematics. Mathematics courses usually required for this degree include calculus, differential equations, and linear and abstract algebra. Additional courses might include probability theory and statistics, mathematical analysis, numerical analysis, topology, discrete mathematics, and mathematical logic. Many colleges and universities urge or require students majoring in mathematics to take courses in a field that is closely related to mathematics, such as computer science, engineering, life science, physical science, or economics. A double major in mathematics and another discipline such as computer science, economics, or another one of the sciences is particularly desirable to many employers. A prospective college mathematics major should take as many mathematics courses as possible while in high school.

In 2001, about 200 colleges and universities offered a master's degree as the highest degree in either pure or applied mathematics; about 200 offered a PhD degree in pure or applied mathematics. In graduate school, students conduct research and take advanced courses, usually specializing in a subfield of mathematics.

For jobs in applied mathematics, training in the field in which the mathematics will be used is very important. Mathematics is used extensively in physics, actuarial science, statistics, engineering, and operations research. Computer science, business and industrial management, economics, finance, chemistry, geology, life sciences, and behavioral sciences are likewise dependent on applied mathematics. Mathematicians also should have substantial knowledge of computer programming because most complex mathematical computation and much mathematical modeling is done on a computer.

Mathematicians need good reasoning ability and persistence in order to identify, analyze, and apply basic principles to technical problems. Communication skills are important, as mathematicians must be able to interact and discuss proposed solutions with people who may not have an extensive knowledge of mathematics.

Job Outlook

Employment of mathematicians is expected to decline through 2010, because very few jobs with the title mathematician are available. However, master's and PhD degree holders with a strong background in mathematics and a related discipline, such as engineering or computer science, should have good job opportunities. However, many of these workers have job titles that reflect their occupation, rather than the title mathematician.

Advancements in technology usually lead to expanding applications of mathematics, and more workers with knowledge of mathematics will be required in the future. However, jobs in industry and government often require advanced knowledge of related scientific disciplines in addition to mathematics. The most common fields in which mathematicians study and find work are computer science and software development, physics, engineering, and operations research. More mathematicians also are becoming involved in financial analysis. Mathematicians must compete for jobs, however, with people who have degrees in these other disciplines. The most successful job seekers will be able to apply mathematical theory to real-world problems, and possess good communication, teamwork, and computer skills.

Private industry jobs require at least a master's degree in mathematics or in one of the related fields. Bachelor's degree holders in mathematics usually are not qualified for most jobs, and many seek advanced degrees in mathematics or a related discipline. However, bachelor's degree holders who meet state certification requirement may become primary or secondary school mathematics teachers.

Holders of a master's degree in mathematics will face very strong competition for jobs in theoretical research. Because the number of PhD degrees awarded in mathematics continues to exceed the number of university positions available, many of these graduates will need to find employment in industry and government.

Earnings

Median annual earnings of mathematicians were $68,640 in 2000. The middle 50 percent earned between $50,740 and $85,520. The lowest 10 percent had earnings of less than $35,390, while the highest 10 percent earned over $101,900.

According to a 2001 survey by the National Association of Colleges and Employers, starting salary offers averaged $46,466 a year for mathematics graduates with a bachelor's degree, and $55,938 for those with a master's degree. Doctoral degree candidates averaged $53,440.

In early 2001, the average annual salary for mathematicians employed by the federal government in supervisory, nonsupervisory, and managerial positions was $76,460; for mathematical statisticians, it was $76,530, and for cryptanalysts, $70,840.

Related Occupations

Other occupations that require extensive knowledge of mathematics or, in some cases, a degree in mathematics include actuaries; statisticians; computer programmers; systems analysts, computer scientists, and database administrators; computer software engineers; and operations research analysts. A strong background in mathematics also facilitates employment as teachers—postsecondary, engineers, economists and survey and market researchers, financial analysts and personal financial advisors, and physicists and astronomers.

Sources of Additional Information

For more information about careers and training in mathematics, especially for doctoral-level employment, contact:

- American Mathematical Society, 201 Charles St., Providence, RI 02940. Internet: http://www.ams.org

For specific information on careers in applied mathematics, contact:

- Society for Industrial and Applied Mathematics, 3600 University City Science Center, Philadelphia, PA 19104-2688. Internet: http://www.siam.org/alterindex.htm

Information on obtaining a mathematician position with the federal government is available from the Office of Personnel Management (OPM) through a telephone-based system. Consult your telephone directory under U.S. Government for a local number or call (912) 757-3000; Federal Relay Service: (800) 877-8339. The first number is not toll free, and charges may result. Information also is available from the OPM Internet site: http://www.usajobs.opm.gov

Mechanical Engineers

O*NET 17-2141.00

Significant Points

- Overall job opportunities in engineering are expected to be good.
- A bachelor's degree is required for most entry-level jobs.
- Starting salaries are significantly higher than those of college graduates in other fields.
- Continuing education is critical to keep abreast of the latest technology.

Nature of the Work

Mechanical engineers research, develop, design, manufacture, and test tools, engines, machines, and other mechanical devices. They work on power-producing machines such as electric generators, internal combustion engines, and steam and gas turbines. They also develop power-using machines such as refrigeration and air-conditioning equipment, machine tools, material handling systems, elevators and escalators, industrial production equipment, and robots used in manufacturing. Mechanical engineers also design tools needed by other engineers for their work. The field of nanotechnology, which involves the creation of high-performance materials and components by integrating atoms and molecules, is introducing entirely new principles to the design process.

Computers assist mechanical engineers by accurately and efficiently performing computations and by aiding the design process by permitting the modeling and simulation of new designs. Computer-Aided Design (CAD) and Computer-Aided Manufacturing (CAM) are used for design data processing and for developing alternative designs.

Mechanical engineers work in many industries, and their work varies by industry and function. Some specialties include applied mechanics; computer-aided design and manufacturing; energy systems; pressure vessels and piping; and heating, refrigeration, and air-conditioning systems. Mechanical engineering is one of the broadest engineering disciplines. Mechanical engineers may work in production operations in manufacturing or agriculture, maintenance, or technical sales; many are administrators or managers.

Working Conditions

Most engineers work in office buildings, laboratories, or industrial plants. Others may spend time outdoors at construction sites, mines, and oil and gas exploration and production sites, where they monitor or direct operations or solve onsite problems. Some engineers travel extensively to plants or work sites.

Many engineers work a standard 40-hour week. At times, deadlines or design standards may bring extra pressure to a job. When this happens, engineers may work longer hours and experience considerable stress.

Employment

Mechanical engineers held about 221,000 jobs in 2000. More than 1 out of 2 jobs were in manufacturing—mostly in machinery, transportation equipment, electrical equipment, instruments, and fabricated metal products industries. Engineering and management services, business services, and the federal government provided most of the remaining jobs.

Training, Other Qualifications, and Advancement

A bachelor's degree in engineering is required for almost all entry-level engineering jobs. College graduates with a degree in a physical science or mathematics occasionally may qualify for some engineering jobs, especially in specialties in high demand. Most engineering degrees are granted in electrical, electronics, mechanical, or civil engineering. However, engineers trained in one branch may work in related branches. This flexibility allows employers to meet staffing needs in new technologies and specialties in which engineers are in short supply. It also allows engineers to shift to fields with better employment prospects or to those that more closely match their interests.

Most engineering programs involve a concentration of study in an engineering specialty, along with courses in both mathematics and science. Most programs include a design course, sometimes accompanied by a computer or laboratory class or both.

In addition to the standard engineering degree, many colleges offer 2- or 4-year degree programs in engineering technology. These programs, which usually include various hands-on laboratory classes that focus on current issues, prepare students for practical design and production work, rather than for jobs which require more theoretical and scientific knowledge. Graduates of 4-year technology programs may get jobs similar to those obtained by graduates with a bachelor's degree in engineering. Engineering technology graduates, however, are not qualified to register as professional engineers under the same terms as graduates with degrees in engineering. Some employers regard technology program graduates as having skills between those of a technician and an engineer.

Graduate training is essential for engineering faculty positions and many research and development programs, but is not required for the majority of entry-level engineering jobs. Many engineers obtain graduate degrees in engineering or business administration to learn new technology and broaden their education. Many high-level executives in government and industry began their careers as engineers.

About 330 colleges and universities offer bachelor's degree programs in engineering that are accredited by the Accreditation Board for Engineering and Technology (ABET), and about 250 colleges offer accredited bachelor's degree programs in engineering technology. ABET accreditation is based on an examination of an engineering program's student achievement, program improvement, faculty, curricular content, facilities, and institutional commitment. Although most institutions offer programs in the major branches of engineering, only a few offer programs in the smaller specialties. Also, programs of the same title may vary in content. For example, some programs emphasize industrial practices, preparing students for a job in industry, whereas others are more theoretical and are designed to prepare students for graduate work. Therefore, students should investigate curricula and check accreditations carefully before selecting a college.

Admissions requirements for undergraduate engineering schools include a solid background in mathematics (algebra, geometry, trigonometry, and calculus) and sciences (biology, chemistry, and physics), and courses in English, social studies, humanities, and computers. Bachelor's degree programs in engineering typically are designed to last 4 years, but many students find that it takes between 4 and 5 years to complete their studies. In a typical 4-year college curriculum, the first 2 years are spent studying mathematics, basic sciences, introductory engineering, humanities, and social sciences. In the last 2 years, most courses are in engineering, usually with a concentration in one branch. Some programs offer a general engineering curriculum; students then specialize in graduate school or on the job.

Some engineering schools and 2-year colleges have agreements whereby the 2-year college provides the initial engineering education, and the engineering school automatically admits students for their last 2 years. In addition, a few engineering schools have arrangements whereby a student spends 3 years in a liberal arts college studying pre-engineering subjects and 2 years in an engineering school studying core subjects, and then receives a bachelor's degree from each school. Some colleges and universities offer 5-year master's degree programs. Some 5- or even 6-year cooperative plans combine classroom study and practical work, permitting students to gain valuable experience and finance part of their education. All 50 states and the District of Columbia usually require licensure for engineers who offer their services directly to the public. Engineers who are licensed are called Professional Engineers (PE). This licensure generally requires a degree from an ABET-accredited engineering program, 4 years of relevant work experience, and successful completion of a state examination. Recent graduates can start the licensing process by taking the examination in two stages. The initial Fundamentals of Engineering (FE) examination can be taken upon graduation. Engineers who pass this examination commonly are called Engineers in Training (EIT) or Engineer Interns (EI). The EIT certification usually is valid for 10 years. After acquiring suitable work experience, EITs can take the second examination, the Principles and Practice of Engineering Exam. Several states have imposed mandatory continuing education requirements for relicensure. Most states recognize licensure from other states. Many mechanical engineers are licensed as PEs.

Engineers should be creative, inquisitive, analytical, and detail oriented. They should be able to work as part of a team and to communicate well, both orally and in writing. Communication abilities are becoming more important because much of their work is becoming more diversified, meaning that engineers interact with specialists in a wide range of fields outside engineering.

Beginning engineering graduates usually work under the supervision of experienced engineers and, in large companies, also may receive formal classroom or seminar-type training. As new engineers gain knowledge and experience, they are assigned more difficult projects with greater independence to develop designs, solve problems, and make decisions. Engineers may advance to become technical specialists or to supervise a staff or team of engineers and technicians. Some may eventually become engineering managers or enter other managerial or sales jobs.

Job Outlook

Employment of mechanical engineers is projected to grow about as fast as the average for all occupations through 2010. Although overall manufacturing employment is expected to grow slowly, employment of mechanical engineers in manufacturing should increase more rapidly as the demand for improved machinery and machine tools grows and industrial machinery and processes become increasingly complex. Also, emerging technologies in information technology, biotechnology, and nanotechnology will create new job opportunities for mechanical engineers.

Employment of mechanical engineers in business and engineering services firms is expected to grow faster than average as other industries in the economy increasingly contract out to these firms to solve engineering problems. In addition to job openings from growth, many openings should result from the need to replace workers who transfer to other occupations or leave the labor force.

Earnings

Median annual earnings of mechanical engineers were $58,710 in 2000. The middle 50 percent earned between $47,600 and $72,850. The lowest 10 percent earned less than $38,770, and the highest 10 percent earned more than $88,610. Median annual earnings in the industries employing the largest numbers of mechanical engineers in 2000 were:

Personnel supply services	$81,080
Federal government	66,320
Engineering and architectural services	59,800
Motor vehicles and equipment	59,400
Construction and related machinery	54,480

According to a 2001 salary survey by the National Association of Colleges and Employers, bachelor's degree candidates in mechanical engineering received starting offers averaging $48,426 a year, master's degree candidates had offers averaging $55,994, and PhD candidates were initially offered $72,096.

Related Occupations

Mechanical engineers apply the principles of physical science and mathematics in their work. Other workers who use scientific and mathematical principles include architects, except landscape and naval; engineering and natural sciences managers; computer and information systems managers; mathematicians; drafters; engineering technicians; sales engineers; science technicians; and physical and life scientists, including agricultural and food scientists, biological and medical scientists, conservation scientists and foresters, atmospheric scientists, chemists and materials scientists, environmental scientists and geoscientists, and physicists and astronomers.

Sources of Additional Information

Further information about mechanical engineers is available from:

- The American Society of Mechanical Engineers, Three Park Ave., New York, NY 10016. Internet: http://www.asme.org

High school students interested in obtaining a full package of guidance materials and information (product number SP-01) on a variety of engineering disciplines should contact the Junior Engineering Technical Society by sending $3.50 to:

- JETS-Guidance, 1420 King St., Suite 405, Alexandria, VA 22314-2794. Internet: http://www.jets.org

High school students interested in obtaining information on ABET-accredited engineering programs should contact:

- The Accreditation Board for Engineering and Technology, Inc., 111 Market Place, Suite 1050, Baltimore, MD 21202-4012. Internet: http://www.abet.org

Non-licensed engineers and college students interested in obtaining information on Professional Engineer licensure should contact:

- The National Society of Professional Engineers, 1420 King St., Alexandria, VA 22314-2794. Internet: http://www.nspe.org
- National Council of Examiners for Engineers and Surveying, P.O. Box 1686, Clemson, SC 29633-1686. Internet: http://www.ncees.org

Information on general engineering education and career resources is available from:

- American Society for Engineering Education, 1818 N St. N.W., Suite 600, Washington, DC 20036-2479. Internet: http://www.asee.org

Information on obtaining an engineering position with the federal government is available from the Office of Personnel Management (OPM) through a telephone-based system. Consult your telephone directory under U.S. Government for a local number or call (912) 757-3000; Federal Relay Service: (800) 877-8339. The first number is not toll free, and charges may result. Information also is available from the OPM Internet site: http://www.usajobs.opm.gov

Non-high school students and those wanting more detailed information should contact societies representing the individual branches of engineering. Each can provide information about careers in the particular branch. The individual statements elsewhere in this book also provide other information in detail on aerospace; agricultural; biomedical; chemical; civil; computer hardware; electrical and electronics; environmental; industrial, including health and safety; materials; mining and geological, including mining safety; nuclear; and petroleum engineering.

Medical Assistants

O*NET 31-9092.00

Significant Points

- The job of medical assistant is expected to be one of the fastest growing occupations through the year 2010.
- Job prospects should be best for medical assistants with formal training or experience.

Nature of the Work

Medical assistants perform routine administrative and clinical tasks to keep the offices and clinics of physicians, podiatrists, chiropractors, and optometrists running smoothly. They should not be confused with physician assistants who examine, diagnose, and treat patients under the direct supervision of a physician.

The duties of medical assistants vary from office to office, depending on office location, size, and specialty. In small practices, medical assistants usually are "generalists," handling both administrative and clinical duties and reporting directly to an office manager, physician, or other health practitioner. Those in large practices tend to specialize in a particular area under the supervision of department administrators.

Medical assistants perform many administrative duties. They answer telephones, greet patients, update and file patient medical records, fill out insurance forms, handle correspondence, schedule appointments, arrange for hospital admission and laboratory services, and handle billing and bookkeeping.

Clinical duties vary according to state law and include taking medical histories and recording vital signs, explaining treatment procedures to patients, preparing patients for examination, and assisting the physician during the examination. Medical assistants collect and prepare laboratory specimens or perform basic laboratory tests on the premises, dispose of contaminated supplies, and sterilize medical instruments. They instruct patients about medication and special diets, prepare and administer medications as directed by a physician, authorize drug refills as directed, telephone prescriptions to a pharmacy, draw blood, prepare patients for X rays, take electrocardiograms, remove sutures, and change dressings.

Medical assistants also may arrange examining room instruments and equipment, purchase and maintain supplies and equipment, and keep waiting and examining rooms neat and clean.

Assistants who specialize have additional duties. *Podiatric medical assistants* make castings of feet, expose and develop X rays, and assist podiatrists in surgery. *Ophthalmic medical assistants* help ophthalmologists provide medical eye care. They conduct diagnostic tests, measure and record vision, and test eye muscle function. They also show patients how to insert, remove, and care for contact lenses; and they apply eye dressings. Under the direction of the physician, they may administer eye medications. They also maintain optical and surgical instruments and may assist the ophthalmologist in surgery.

Working Conditions

Medical assistants work in well-lighted, clean environments. They constantly interact with other people, and may have to handle several responsibilities at once.

Most full-time medical assistants work a regular 40-hour week. Some work part-time, evenings, or weekends.

Employment

Medical assistants held about 329,000 jobs in 2000. Sixty percent were in physicians' offices, and about 15 percent were in hospitals, including inpatient and outpatient facilities. The rest were in nursing homes, offices of other health practitioners, and other healthcare facilities.

Training, Other Qualifications, and Advancement

Most employers prefer graduates of formal programs in medical assisting. Such programs are offered in vocational-technical high schools, postsecondary vocational schools, community and junior colleges, and in colleges and universities. Postsecondary programs usually last either 1 year, resulting in a certificate or diploma, or 2 years, resulting in an associate degree. Courses cover anatomy, physiology, and medical terminology as well as typing, transcription, recordkeeping, accounting, and insurance processing. Students learn laboratory techniques, clinical and diagnostic procedures, pharmaceutical principles, medication administration, and first aid. They study office practices, patient relations, medical law, and ethics. Accredited programs include an internship that provides practical experience in physicians' offices, hospitals, or other healthcare facilities.

Two agencies recognized by the U.S. Department of Education accredit programs in medical assisting: the Commission on Accreditation of Allied Health Education Programs (CAAHEP) and the Accrediting Bureau of Health Education Schools (ABHES). In 2001, there were about 500 medical assisting programs accredited by CAAHEP and about 170 accredited by ABHES. The Committee on Accreditation for Ophthalmic Medical Personnel approved 14 programs in ophthalmic medical assisting.

Formal training in medical assisting, while generally preferred, is not always required. Some medical assistants are trained on the job, although this is less common than in the past. Applicants usually need a high school diploma or the equivalent. Recommended high school courses include mathematics, health, biology, typing, bookkeeping, computers, and office skills. Volunteer experience in the healthcare field also is helpful.

Although there is no licensing for medical assistants, some states require them to take a test or a course before they can perform certain tasks, such as taking X rays. Employers prefer to hire experienced workers or certified applicants who have passed a national examination, indicating that the medical assistant meets certain standards of competence. The American Association of Medical Assistants awards the Certified Medical Assistant credential. The American Medical Technologists awards the Registered Medical Assistant credential. The American Society of Podiatric Medical Assistants awards the Podiatric Medical Assistant Certified credential, and the Joint Commission on Allied Health Personnel in Ophthalmology awards credentials at three levels—Certified Ophthalmic Assistant, Certified Ophthalmic Technician, and Certified Ophthalmic Medical Technologist.

Medical assistants may be able to advance to office manager. They may qualify for a variety of administrative support occupations, or may teach medical assisting. Some, with additional education, enter other health occupations such as nursing and medical technology.

Medical assistants deal with the public; therefore, they must be neat and well-groomed and have a courteous, pleasant manner. Medical assistants must be able to put patients at ease and explain physicians' instructions. They must respect the confidential nature of medical information. Clinical duties require a reasonable level of manual dexterity and visual acuity.

Job Outlook

Employment of medical assistants is expected to grow much faster than the average for all occupations through the year 2010 as the health services industry expands because of technological advances in medicine, and a growing and aging population. It is one of the fastest growing occupations.

Employment growth will be driven by the increase in the number of group practices, clinics, and other healthcare facilities that need a high proportion of support personnel, particularly the flexible medical assistant who can handle both administrative and clinical duties. Medical assistants primarily work in outpatient settings, where much faster-than-average growth is expected.

In view of the preference of many healthcare employers for trained personnel, job prospects should be best for medical assistants with formal training or experience, particularly those with certification.

Earnings

The earnings of medical assistants vary, depending on experience, skill level, and location. Median annual earnings of medical assistants were $23,000 in 2000. The middle 50 percent earned between $19,460 and $27,460 a year. The lowest 10 percent earned less than $16,700, and the highest 10 percent earned more than $32,850 a year. Median annual earnings in the industries employing the largest number of medical assistants in 2000 were as follows:

Offices and clinics of medical doctors	$23,610
Hospitals	22,950
Health and allied services, not elsewhere classified	22,860
Offices of osteopathic physicians	21,420
Offices of other health practitioners	20,860

Related Occupations

Workers in other medical support occupations include dental assistants, medical records and health information technicians, medical secretaries, occupational therapist assistants and aides, pharmacy aides, and physical therapist assistants and aides.

Sources of Additional Information

Information about career opportunities, CAAHEP-accredited educational programs in medical assisting, and the Certified Medical Assistant exam is available from:

- The American Association of Medical Assistants, 20 North Wacker Dr., Suite 1575, Chicago, IL 60606. Internet: http://www.aama-ntl.org

Information about career opportunities and the Registered Medical Assistant certification exam is available from:

- Registered Medical Assistants of American Medical Technologists, 710 Higgins Rd., Park Ridge, IL 60068-5765

For a list of ABHES-accredited educational programs in medical assisting, contact:

- Accrediting Bureau of Health Education Schools, 803 West Broad St., Suite 730, Falls Church, VA 22046. Internet: http://www.abhes.org

Information about career opportunities, training programs, and the Certified Ophthalmic Assistant exam is available from:

- Joint Commission on Allied Health Personnel in Ophthalmology, 2025 Woodlane Dr., St. Paul, MN 55125-2995. Internet: http://www.jcahpo.org

Information about careers for podiatric assistants is available from:

- American Society of Podiatric Medical Assistants, 2124 S. Austin Blvd., Cicero, IL 60650

Medical Records and Health Information Technicians

O*NET 29-2071.00

Significant Points

- Medical records and health information technicians are projected to be one of the fastest growing occupations.
- High school students can improve chances of acceptance into a medical record and health information education program by taking anatomy, physiology, medical terminology, and computer courses.
- Most technicians will be employed by hospitals, but job growth will be faster in offices and clinics of physicians, nursing homes, and home health agencies.

Nature of the Work

Every time healthcare personnel treat a patient, they record what they observed, and how the patient was treated medically. This record includes information the patient provides concerning their symptoms and medical history, the results of examinations, reports of X rays and laboratory tests, diagnoses, and treatment plans. Medical records and health information technicians organize and evaluate these records for completeness and accuracy.

Medical records and health information technicians begin to assemble patients' health information by first making sure their initial medical charts are complete. They ensure all forms are completed and properly identified and signed, and all necessary information is in the computer. Sometimes, they communicate with physicians or others to clarify diagnoses or get additional information.

Technicians assign a code to each diagnosis and procedure. They consult classification manuals and rely, also, on their knowledge of disease processes. Technicians then use a software program to assign the patient to one of several hundred "diagnosis-related groups," or DRGs. The DRG determines the amount the hospital will be reimbursed if the patient is covered by Medicare or other insurance programs using the DRG system. Technicians who specialize in coding are called health information coders, medical record coders, coder/abstractors, or coding specialists. In addition to the DRG system, coders use other coding systems, such as those geared towards ambulatory settings.

Technicians also use computer programs to tabulate and analyze data to help improve patient care, control costs, for use in legal actions, in response to surveys, or for use in research studies. *Tumor registrars* compile and maintain records of patients who have cancer to provide information to physicians and for research studies.

Medical records and health information technicians' duties vary with the size of the facility. In large to medium facilities, technicians may specialize in one aspect of health information, or supervise health information clerks and transcriptionists while a *medical records and health information administrator* manages the department. In small facilities, a credentialed medical records and health information technician sometimes manages the department.

Working Conditions

Medical records and health information technicians usually work a 40-hour week. Some overtime may be required. In hospitals—where health information departments often are open 24 hours a day, 7 days a week—technicians may work day, evening, and night shifts.

Medical records and health information technicians work in pleasant and comfortable offices. This is one of the few health occupations in which there is little or no physical contact with patients. Because accuracy is essential, technicians must pay close attention to detail. Technicians who work at computer monitors for prolonged periods must guard against eyestrain and muscle pain.

Employment

Medical records and health information technicians held about 136,000 jobs in 2000. About 4 out of 10 jobs were in hospitals. The rest were mostly in nursing homes, medical group practices, clinics, and home health agencies. Insurance firms that deal in health matters employ a small number of health information technicians to tabulate and analyze health information. Public health departments also hire technicians to supervise data collection from healthcare institutions and to assist in research.

Training, Other Qualifications, and Advancement

Medical records and health information technicians entering the field usually have an associate degree from a community or junior college. In addition to general education, coursework includes medical terminology, anatomy and physiology, legal aspects of health information, coding and abstraction of data, statistics, database management, quality improvement methods, and computer training. Applicants can improve their chances of admission into a program by taking biology, chemistry, health, and computer courses in high school.

Hospitals sometimes advance promising health information clerks to jobs as medical records and health information technicians, although this practice may be less common in the future. Advancement usually requires 2 to 4 years of job experience and completion of a hospital's in-house training program.

Most employers prefer to hire registered health information technicians (RHIT), who must pass a written examination offered by AHIMA. To take the examination, a person must graduate from a 2-year associate degree program accredited by the Commission on Accreditation of Allied Health Education Programs (CAAHEP) of the American Medical Association. Technicians trained in non-CAAHEP accredited programs, or on the job, are not eligible to take the examination. In 2001, CAAHEP accredited 177 programs for health information technicians. Technicians who specialize in coding may also obtain voluntary certification.

Experienced medical records and health information technicians usually advance in one of two ways—by specializing or managing. Many senior technicians specialize in coding, particularly Medicare coding, or in tumor registry.

In large medical records and health information departments, experienced technicians may advance to section supervisor, overseeing the work of the coding, correspondence, or discharge sections, for example. Senior technicians with RHIT credentials may become director or assistant director of a medical records and health information department in a small facility. However, in larger institutions, the director is usually an administrator, with a bachelor's degree in medical records and health information administration.

Job Outlook

Job prospects for formally trained technicians should be very good. Employment of medical records and health information technicians is expected to grow much faster than the average for all occupations through 2010, due to rapid growth in the number of medical tests, treatments, and procedures which will be increasingly scrutinized by third-party payers, regulators, courts, and consumers.

Hospitals will continue to employ a large percentage of health information technicians, but growth will not be as fast as in other areas. Increasing demand for detailed records in offices and clinics of physicians should result in fast employment growth, especially in large group practices. Rapid growth is also expected in nursing homes and home health agencies.

Earnings

Median annual earnings of medical records and health information technicians were $22,750 in 2000. The middle 50 percent earned between $18,700 and $28,590. The lowest 10 percent earned less than $15,710, and the highest 10 percent earned more than $35,170. Median annual earnings in the industries employing the largest numbers of medical records and health information technicians in 2000 were as follows:

Nursing and personal care facilities	$23,760
Hospitals	23,540
Offices and clinics of medical doctors	21,090

Related Occupations

Medical records and health information technicians need a strong clinical background to analyze the contents of medical records. Workers in other occupations requiring knowledge of medical terminology, anatomy, and physiology without physical contact with the patient are medical secretaries and medical transcriptionists.

Sources of Additional Information

Information on careers in medical records and health information technology, including a list of CAAHEP-accredited programs is available from:

- American Health Information Management Association, 233 N. Michigan Ave., Suite 2150, Chicago, IL 60601-5800. Internet: http://www.ahima.org

Medical Transcriptionists

O*NET 31-9094.00

Significant Points

- Employers prefer medical transcriptionists who have completed a vocational school or community college program.
- Employment is projected to grow faster than average due to increasing demand for medical transcription services.
- Some medical transcriptionists enjoy the flexibility of working at home, especially those with previous experience in a hospital or clinic setting.

Nature of the Work

Medical transcriptionists, also called *medical transcribers* and *medical stenographers*, listen to dictated recordings made by physicians and other healthcare professionals and transcribe them into medical reports, correspondence, and other administrative material. They generally listen to recordings on a special headset, using a foot pedal to pause the recording when necessary, and key the text into a personal computer or word processor, editing as necessary for grammar and clarity. The documents they produce include discharge summaries, history and physical examination reports, operating room reports, consultation reports, autopsy reports, diagnostic imaging studies, and referral letters. Medical transcriptionists return transcribed documents to the dictator for review and signature, or correction. These documents eventually become part of patients' permanent files.

To understand and accurately transcribe dictated reports into a format that is clear and comprehensible for the reader, medical transcriptionists must understand medical terminology, anatomy and physiology, diagnostic procedures, and treatment. They also must be able to translate medical jargon and abbreviations into their expanded forms. To help identify terms appropriately, transcriptionists refer to standard medical reference materials—both printed and electronic; some of these are available over the Internet. Medical transcriptionists must comply with specific standards that apply to the style of medical records, in addition to the legal and ethical requirements involved with keeping patient records confidential.

Experienced transcriptionists spot mistakes or inconsistencies in a medical report and check back with the dictator to correct the information. Their ability to understand and correctly transcribe patient assessments and treatments reduces the chance of patients receiving ineffective or even harmful treatments and ensures high quality patient care.

Currently, most healthcare providers transmit dictation to medical transcriptionists using either digital or analog dictating equipment. With the emergence of the Internet, some transcriptionists receive dictation over the Internet and are able to quickly return transcribed documents to clients for approval. As confidentiality concerns are resolved, this practice will become more prevalent. Another emerging trend is the implementation of speech recognition technology, which electronically translates sound into text and creates drafts of reports. Reports are then formatted; edited for mistakes in translation, punctuation, or grammar; and checked for consistency and possible medical errors. Transcriptionists working in specialized areas with more standard terminology, such as radiology or pathology, are more likely to encounter speech recognition technology. However, use of speech recognition technology will become more widespread as the technology becomes more sophisticated.

Medical transcriptionists who work in physicians' offices and clinics may have other office duties, such as receiving patients, scheduling appointments, answering the telephone, and handling incoming and outgoing mail. Medical secretaries may also transcribe as part of their jobs. Court reporters have similar duties, but with a different focus. They take verbatim reports of speeches, conversations, legal proceedings, meetings, and other events when written accounts of spoken words are necessary for correspondence, records, or legal proof.

Working Conditions

The majority of these workers are employed in comfortable settings, such as hospitals, physicians' offices, clinics, laboratories, medical libraries, government medical facilities, or at home. An increasing number of medical transcriptionists telecommute from home-based offices as employees or subcontractors for hospitals and transcription services or as self-employed independent contractors.

Work in this occupation presents few hazards, although sitting in the same position for long periods can be tiring, and workers can suffer wrist, back, neck, or eye problems due to strain and risk repetitive motion injuries such as carpal tunnel syndrome. The pressure to be accurate and fast also can be stressful.

Many medical transcriptionists work a standard 40-hour week. Self-employed medical transcriptionists are more likely to work irregular hours—including part-time, evenings, weekends, or on an on-call basis.

Employment

Medical transcriptionists held about 102,000 jobs in 2000. About 2 out of 5 worked in hospitals and about another 2 out of 5 in physicians' offices and clinics. Others worked for laboratories, colleges and universities, transcription services, and temporary help agencies.

Training, Other Qualifications, and Advancement

Employers prefer to hire transcriptionists who have completed postsecondary training in medical transcription, offered by many vocational schools, community colleges, and distance-learning programs. Completion of a 2-year associate degree or 1-year certificate program—including coursework in anatomy, medical terminology, medicolegal issues, and English grammar and punctuation—is highly recommended, but not always required. Many of these programs include supervised on-the-job experience. Some transcriptionists, especially those already familiar with medical terminology due to previous experience as a nurse or medical secretary, become proficient through on-the-job training.

The American Association for Medical Transcription (AAMT) awards the voluntary designation, Certified Medical Transcriptionist (CMT), to those who earn passing scores on written and practical examinations. As in many other fields, certification is recognized as a sign of competence. Because medical terminology is constantly evolving, medical transcriptionists are encouraged to regularly update their skills. Every 3 years, CMTs must earn continuing education credits to be recertified.

In addition to understanding medical terminology, transcriptionists must have good English grammar and punctuation skills, as well as familiarity with personal computers and word processing software. Normal hearing acuity and good listening skills also are necessary. Employers often require applicants to take pre-employment tests.

With experience, medical transcriptionists can advance to supervisory positions, home-based work, consulting, or teaching. With additional education or training, some become medical records and health information technicians, medical coders, or medical records and health information administrators.

Job Outlook

Employment of medical transcriptionists is projected to grow faster than the average for all occupations through 2010. Demand for medical transcription services will be spurred by a growing and aging population. Older age groups receive proportionately greater numbers of medical tests, treatments, and procedures that require documentation. A high level of demand for transcription services also will be sustained by the continued need for electronic documentation that can be easily shared among providers, third-party payers, regulators, and consumers. Growing numbers of medical transcriptionists will be needed to amend patients' records, edit for grammar, and discover discrepancies in medical records.

Advancements in speech recognition technology are not projected to significantly reduce the need for medical transcriptionists because these workers will continue to be needed to review and edit drafts for accuracy. In spite of the advances in this technology, it has been difficult for the software to grasp and analyze the human voice and the English language with all its diversity. There will continue to be a need for skilled medical transcriptionists to identify and appropriately edit the inevitable errors created by speech recognition systems, and create a final document.

Hospitals will continue to employ a large percentage of medical transcriptionists, but job growth will not be as fast as in other areas. Increasing demand for standardized records in offices and clinics of physicians should result in rapid employment growth, especially in large group practices. Job opportunities should be the best for those who earn an associate degree or certification from the American Association for Medical Transcription.

Earnings

Medical transcriptionists had median hourly earnings of $12.15 in 2000. The middle 50 percent earned between $10.07 and $14.41. The lowest 10 percent earned less than $8.66, and the highest 10 percent earned more than $16.70. Median hourly earnings in the industries employing the largest numbers of medical transcriptionists in 2000 were as follows:

Offices and clinics of medical doctors	$12.25
Hospitals	12.14
Mailing, reproduction, and stenographic services	11.47

Compensation methods for medical transcriptionists vary. Some are paid based on the number of hours they work or on the number of lines they transcribe. Others receive a base pay per hour with incentives for extra production. Large hospitals and healthcare organizations usually prefer to pay for the time an employee works. Independent contractors and employees of transcription services almost always receive production-based pay.

According to a 1999 study conducted by Hay Management Consultants for the American Association for Medical Transcription, entry-level medical transcriptionists had median hourly earnings of $10.32 and the most experienced transcriptionists had median hourly earnings of $13.00. Earnings were highest in organizations employing 1,000 or more workers. Transcriptionists receiving production-based pay earned about 7 to 8.5 cents per standardized line (based on a 65-character line, counting all keystrokes). However, independent contractors—who have higher expenses than their corporate counterparts, receive no benefits, and face higher risk of termination than employed transcriptionists—typically charge about 12 to 13 cents per standardized line.

Related Occupations

A number of other workers type, record information, and process paperwork. Among these are court reporter, secretaries and administrative assistants, receptionists and information clerks, and human resources assistants, except payroll and timekeeping. Other workers who provide medical support include medical assistants and medical records and health information technicians.

Sources of Additional Information

For information on a career as a medical transcriptionist, send a self-addressed, stamped envelope to:

- American Association for Medical Transcription, 3460 Oakdale Rd., Suite M, Modesto, CA 95355-9690. Internet: http://www.aamt.org

State employment service offices can provide information about job openings for medical transcriptionists.

Mining and Geological Engineers, Including Mining Safety Engineers

O*NET 17-2151.00

Significant Points

- Overall job opportunities in engineering are expected to be good.
- A bachelor's degree is required for most entry-level jobs.
- Starting salaries are significantly higher than those of college graduates in other fields.
- Continuing education is critical to keep abreast of the latest technology.

Nature of the Work

Mining and geological engineers find, extract, and prepare coal, metals, and minerals for use by manufacturing industries and utilities. They design open pit and underground mines, supervise the construction of mine shafts and tunnels in underground operations, and devise methods for transporting minerals to processing plants. Mining engineers are responsible for the safe, economical, and environmentally sound operation of mines. Some mining engineers work with geologists and metallurgical engineers to locate and appraise new ore deposits. Others develop new mining equipment or direct mineral processing operations to separate minerals from the dirt, rock, and other materials with which they are mixed. Mining engineers frequently specialize in the mining of one mineral or metal, such as coal or gold. With increased emphasis on protecting the environment, many mining engineers work to solve problems related to land reclamation and water and air pollution.

Mining safety engineers use their knowledge of mine design and practices to ensure the safety of workers and to comply with state and federal safety regulations. They inspect walls and roof surfaces, test air samples, and examine mining equipment for compliance with safety practices.

Working Conditions

Most engineers work in office buildings, laboratories, or industrial plants. Others may spend time outdoors at construction sites, mines, and oil and gas exploration and production sites, where they monitor or direct operations or solve onsite problems. Some engineers travel extensively to plants or work sites.

Many engineers work a standard 40-hour week. At times, deadlines or design standards may bring extra pressure to a job. When this happens, engineers may work longer hours and experience considerable stress.

Employment

Mining and geological engineers, including mining safety engineers, held about 6,500 jobs in 2000. While one-half worked in the mining industry, other mining engineers worked in government agencies or engineering consulting firms.

Mining engineers usually are employed at the location of natural deposits, often near small communities, and sometimes outside the United States. Those in research and development, management, consulting, or sales, however, often are located in metropolitan areas.

Training, Other Qualifications, and Advancement

A bachelor's degree in engineering is required for almost all entry-level engineering jobs. College graduates with a degree in a physical science or mathematics occasionally may qualify for some engineering jobs, especially in specialties in high demand. Most engineering degrees are granted in electrical, electronics, mechanical, or civil engineering. However, engineers trained in one branch may work in related branches. This flexibility allows employers to meet staffing needs in new technologies and specialties in which engineers are in short supply. It also allows engineers to shift to fields with better employment prospects or to those that more closely match their interests.

Most engineering programs involve a concentration of study in an engineering specialty, along with courses in both mathematics and science. Most programs include a design course, sometimes accompanied by a computer or laboratory class or both.

In addition to the standard engineering degree, many colleges offer 2- or 4-year degree programs in engineering technology. These programs, which usually include various hands-on laboratory classes that focus on current issues, prepare students for practical design and production work, rather than for jobs which require more theoretical and scientific knowledge. Graduates of 4-year technology programs may get jobs similar to those obtained by graduates with a bachelor's degree in engineering. Engineering technology graduates, however, are not qualified to register as professional engineers under the same terms as graduates with degrees in engineering. Some employers regard technology program graduates as having skills between those of a technician and an engineer.

Graduate training is essential for engineering faculty positions and many research and development programs, but is not required for the majority of entry-level engineering jobs. Many engineers obtain graduate degrees in engineering or business administration to learn new technology and broaden their education. Many high-level executives in government and industry began their careers as engineers.

About 330 colleges and universities offer bachelor's degree programs in engineering that are accredited by the Accreditation Board for Engineering and Technology (ABET), and about 250 colleges offer accredited bachelor's degree programs in engineering technology. ABET accreditation is based on an examination of an engineering program's student achievement, program improvement, faculty, curricular content, facilities, and institutional commitment. Although most institutions offer programs in the major branches of engineering, only a few offer programs in the smaller specialties. Also, programs of the same title may vary in content. For example, some programs emphasize industrial practices, preparing students for a job in industry, whereas others are more theoretical and are designed to prepare students for graduate work. Therefore, students should investigate curricula and check accreditations carefully before selecting a college.

Admissions requirements for undergraduate engineering schools include a solid background in mathematics (algebra, geometry, trigonometry, and calculus) and sciences (biology, chemistry, and physics), and courses in English, social studies, humanities, and computers. Bachelor's degree programs in engineering typically are designed to last 4 years, but many students find that it takes between 4 and 5 years to complete their studies. In a typical 4-year college curriculum, the first 2 years are spent studying mathematics, basic sciences, introductory engineering, humanities, and social sciences. In the last 2 years, most courses are in engineering, usually with a concentration in one branch. Some programs offer a general engineering curriculum; students then specialize in graduate school or on the job.

Some engineering schools and 2-year colleges have agreements whereby the 2-year college provides the initial engineering education, and the engineering school automatically admits students for their last 2 years. In addition, a few engineering schools have arrangements whereby a student spends 3 years in a liberal arts college studying pre-engineering subjects and 2 years in an engineering school studying core subjects, and then receives a bachelor's degree from each school. Some colleges and universities offer 5-year master's degree programs. Some 5- or even 6-year cooperative plans combine classroom study and practical work, permitting students to gain valuable experience and finance part of their education. All 50 states and the District of Columbia usually require licensure for engineers who offer their services directly to the public. Engineers who are licensed are called Professional Engineers (PE). This licensure generally requires a degree from an ABET-accredited engineering program, 4 years of relevant work experience, and successful completion of a state examination. Recent graduates can start the licensing process by taking the examination in two stages. The initial Fundamentals of Engineering (FE) examination can be taken upon graduation. Engineers who pass this examination commonly are called Engineers in Training (EIT) or Engineer Interns (EI). The EIT certification usually is valid for 10 years. After acquiring suitable work experience, EITs can take the second examination, the Principles and Practice of Engineering Exam. Several states have imposed mandatory continuing education requirements for relicensure. Most states recognize licensure from other states.

Engineers should be creative, inquisitive, analytical, and detail oriented. They should be able to work as part of a team and to communicate well, both orally and in writing. Communication abilities are becoming more important because much of their work is becoming more diversified, meaning that engineers interact with specialists in a wide range of fields outside engineering.

Beginning engineering graduates usually work under the supervision of experienced engineers and, in large companies, also may receive formal classroom or seminar-type training. As new engineers gain knowledge and experience, they are assigned more difficult projects with greater independence to develop designs, solve problems, and make decisions. Engineers may advance to become technical specialists or to supervise a staff or team of engineers and technicians. Some may eventually become engineering managers or enter other managerial or sales jobs.

Job Outlook

Employment of mining and geological engineers, including mining safety engineers, is expected to decline through 2010. Most of the industries in which mining engineers are concentrated—such as coal, metal, and mineral mining, as well as stone, clay, and glass products manufacturing—are expected to experience declines in employment.

Although no job openings are expected to result from employment growth, there should be openings resulting from the need to replace mining engineers who transfer to other occupations or leave the labor force. A large number of mining engineers currently employed are approaching retirement age. In addition, relatively few schools offer mining engineering programs, and the small number of graduates is not expected to increase.

Mining operations around the world recruit graduates of U.S. mining engineering programs. Consequently, job opportunities may be better worldwide than within the United States. As a result, graduates should be prepared for the possibility of frequent travel or even living abroad.

Earnings

Median annual earnings of mining and geological engineers, including mining safety engineers, were $60,820 in 2000. The middle 50 percent earned between $47,320 and $78,720. The lowest 10 percent earned less than $36,070, and the highest 10 percent earned more than $100,050.

According to a 2001 salary survey by the National Association of Colleges and Employers, bachelor's degree candidates in mining engineering received starting offers averaging $42,507 a year, and master's degree candidates, on average, were offered $54,038.

Related Occupations

Mining and geological engineers apply the principles of physical science and mathematics in their work. Other workers who use scientific and mathematical principles include architects, except landscape and naval; engineering and natural sciences managers; computer and information systems managers; mathematicians; drafters; engineering technicians; sales engineers; science technicians; and physical and life scientists, including agricultural and food scientists, biological and medical scientists, conservation scientists and foresters, atmospheric scientists, chemists and materials scientists, environmental scientists and geoscientists, and physicists and astronomers.

Sources of Additional Information

For general information about mining engineers, contact:

- The Society for Mining, Metallurgy, and Exploration, Inc., P.O. Box 625002, Littleton, CO 80162-5002. Internet: http://www.smenet.org

High school students interested in obtaining a full package of guidance materials and information (product number SP-01) on a variety of engineering disciplines should contact the Junior Engineering Technical Society by sending $3.50 to:

- JETS-Guidance, 1420 King St., Suite 405, Alexandria, VA 22314-2794. Internet: http://www.jets.org

High school students interested in obtaining information on ABET-accredited engineering programs should contact:

- The Accreditation Board for Engineering and Technology, Inc., 111 Market Place, Suite 1050, Baltimore, MD 21202-4012. Internet: http://www.abet.org

Non-licensed engineers and college students interested in obtaining information on Professional Engineer licensure should contact:

- The National Society of Professional Engineers, 1420 King St., Alexandria, VA 22314-2794. Internet: http://www.nspe.org
- National Council of Examiners for Engineers and Surveying, P.O. Box 1686, Clemson, SC 29633-1686. Internet: http://www.ncees.org

Information on general engineering education and career resources is available from:

- American Society for Engineering Education, 1818 N St. N.W., Suite 600, Washington, DC 20036-2479. Internet: http://www.asee.org

Information on obtaining an engineering position with the federal government is available from the Office of Personnel Management (OPM) through a telephone-based system. Consult your telephone directory under U.S. Government for a local number or call (912) 757-3000; Federal Relay Service: (800) 877-8339. The first number is not toll free, and charges may result. Information also is available from the OPM Internet site: http://www.usajobs.opm.gov

Non-high school students and those wanting more detailed information should contact societies representing the individual branches of engineering. Each can provide information about careers in the particular branch. The individual statements elsewhere in this book also provide other information in detail on aerospace; agricultural; biomedical; chemical; civil; computer hardware; electrical and electronics; environmental; industrial, including health and safety; materials; mechanical; nuclear; and petroleum engineering.

News Analysts, Reporters, and Correspondents

O*NET 27-3021.00, 27-3022.00

Significant Points

- Most employers prefer individuals with a bachelor's degree in journalism and experience.
- Competition will be keen for jobs on large metropolitan newspapers and broadcast stations and on national magazines; most entry-level openings arise on small publications.
- Jobs often are stressful because of irregular hours, frequent night and weekend work, and pressure to meet deadlines.

Nature of the Work

News analysts, reporters, and correspondents play a key role in our society. They gather information, prepare stories, and make broadcasts that inform us about local, state, national, and international events; present points of view on current issues; and report on the actions of public officials, corporate executives, special-interest groups, and others who exercise power.

News analysts examine, interpret, and broadcast news received from various sources, and also are called *newscasters* or *news anchors*. News anchors present news stories and introduce videotaped news or live transmissions from on-the-scene reporters. Some newscasters at large stations and networks usually specialize in a particular type of news, such as sports or weather. *Weathercasters*, also called weather reporters, report current and forecasted weather conditions. They gather information from national satellite weather services, wire services, and local and regional weather bureaus. Some weathercasters are trained meteorologists and can develop their own weather forecasts. *Sportscasters* select, write, and deliver sports news. This may include interviews with sports personalities and coverage of games and other sporting events.

In covering a story, *reporters* investigate leads and news tips, look at documents, observe events at the scene, and interview people. Reporters take notes and also may take photographs or shoot videos. At their office, they organize the material, determine the focus or emphasis, write their stories, and edit accompanying video material. Many reporters enter information or write stories on laptop computers, and electronically submit them to their offices from remote locations. In some cases, *newswriters* write a story from information collected and submitted by reporters. Radio and television reporters often compose stories and report "live" from the scene. At times, they later tape an introduction or commentary to their story in the studio. Some journalists also interpret the news or offer opinions to readers, viewers, or listeners. In this role, they are called *commentators* or *columnists*.

General assignment reporters write news, such as an accident, a political rally, the visit of a celebrity, or a company going out of business, as assigned. Large newspapers and radio and television stations assign reporters to gather news about specific categories or beats, such as crime or education. Some reporters specialize in fields such as health, politics, foreign affairs, sports, theater, consumer affairs, social events, science, business, and religion. Investigative reporters cover stories that take many days or weeks of information gathering. Some publications use teams of reporters instead of assigning specific beats, allowing reporters to cover a greater variety of stories. News teams may include reporters, editors, graphic artists, and photographers, working together to complete a story.

News correspondents report on news occurring in the large U.S. and foreign cities where they are stationed. Reporters on small publications cover all aspects of the news. They take photographs, write headlines, lay out pages, edit wire service copy, and write editorials. Some also solicit advertisements, sell subscriptions, and perform general office work.

Working Conditions

The work of news analysts, reporters, and correspondents usually is hectic. They are under great pressure to meet deadlines and broadcasts sometimes are made with little time for preparation. Some work in comfortable, private offices; others work in large rooms filled with the sound of keyboards and computer printers, as well as the voices of other reporters. Curious onlookers, police, or other emergency workers can distract those reporting from the scene for radio and television. Covering wars, political uprisings, fires, floods, and similar events often is dangerous.

Working hours vary. Reporters on morning papers often work from late afternoon until midnight. Those on afternoon or evening papers generally work from early morning until early afternoon or mid afternoon. Radio and television reporters usually are assigned to a day or evening shift. Magazine reporters usually work during the day.

Reporters sometimes have to change their work hours to meet a deadline, or to follow late-breaking developments. Their work demands long hours, irregular schedules, and some travel. Many stations and networks are on the air 24 hours a day, so newscasters can expect to work unusual hours.

Employment

News analysts, reporters, and correspondents held about 78,000 jobs in 2000. Nearly half worked for newspapers—either large city dailies or suburban and small town dailies or weeklies. About 28 percent worked in radio and television broadcasting, and others worked for magazines and wire services. About 12,000 news analysts, reporters, and correspondents were self-employed.

Training, Other Qualifications, and Advancement

Most employers prefer individuals with a bachelor's degree in journalism, but some hire graduates with other majors. They look for experience on school newspapers or broadcasting stations and internships with news organizations. Large city newspapers and stations also may prefer candidates with a degree in a subject-matter specialty such as economics, political science, or business. Large newspapers and broadcasters also require a minimum of 3 to 5 years of experience as a reporter.

Bachelor's degree programs in journalism are available in over 400 colleges or universities. About three-fourths of the courses in a typical curriculum are in liberal arts; the remainder are in journalism. Journalism courses include introductory mass media, basic reporting and copy editing, history of journalism, and press law and ethics. Students planning a career in broadcasting take courses in radio and television newscasting and production. Those planning newspaper or magazine careers usually specialize in news-editorial journalism. Those planning careers in new media, such as online newspapers or magazines, require a merging of traditional and new journalism skills. To create a story for online presentation, they need to know how to use computer software to combine online story text with audio and video elements and graphics.

Many community and junior colleges offer journalism courses or programs; credits may be transferable to 4-year journalism programs.

About 120 schools offered a master's degree in journalism in 2000; about 35 schools offered a PhD degree. Some graduate programs are intended primarily as preparation for news careers, while others prepare journalism teachers, researchers and theorists, and advertising and public relations workers.

High school courses in English, journalism, and social studies provide a good foundation for college programs. Useful college liberal arts courses include English with an emphasis on writing, sociology, political science, economics, history, and psychology. Courses in computer science, business, and speech are useful, as well. Fluency in a foreign language is necessary in some jobs.

Although reporters need good word-processing skills, computer graphics and desktop publishing skills also are useful. Computer-assisted reporting involves the use of computers to analyze data in search of a story. This technique and the interpretation of the results require strong math skills and familiarity with databases. Knowledge of news photography also is valuable for entry-level positions, which sometimes combine reporter/camera operator or reporter/photographer responsibilities.

Experience in a part-time or summer job or an internship with a news organization is very important. (Most newspapers, magazines, and broadcast news organizations offer reporting and editing internships.) Work on high school and college newspapers, at broadcasting stations, or on community papers or U.S. Armed Forces publications also helps. In addition, journalism scholarships, fellowships, and assistantships awarded to college journalism students by universities, newspapers, foundations, and professional organizations are helpful. Experience as a stringer or freelancer, a part-time reporter who is paid only for stories printed, also is advantageous.

Reporters should be dedicated to providing accurate and impartial news. Accuracy is important, both to serve the public and because untrue or libelous statements can lead to costly lawsuits. A nose for news, persistence, initiative, poise, resourcefulness, a good memory, and physical stamina are important, as well as the emotional stability to deal with pressing deadlines, irregular hours, and dangerous assignments. Broadcast reporters and news analysts must be comfortable on camera. All reporters must be at ease in unfamiliar places and with a variety of people. Positions involving on-air work require a pleasant voice and appearance.

Most reporters start at small publications or broadcast stations as general assignment reporters or copy editors. Large publications and stations hire few recent graduates; as a rule, they require new reporters to have several years of experience.

Beginning reporters cover court proceedings and civic and club meetings, summarize speeches, and write obituaries. With experience, they report more difficult assignments, cover an assigned beat, or specialize in a particular field.

Some news analysts and reporters can advance by moving to large newspapers or stations. A few experienced reporters become columnists, correspondents, writers, announcers, or public relations specialists. Others become editors in print journalism or program managers in broadcast journalism, who supervise reporters. Some eventually become broadcasting or publications industry managers.

Job Outlook

Employment of news analysts, reporters, and correspondents is expected to grow more slowly than the average for all occupations through the year 2010—the result of mergers, consolidations, and closures of newspapers; decreased circulation; increased expenses; and a decline in advertising profits. Despite little change in overall employment, some job growth is expected in radio and television stations, and even more rapid growth is expected in new media areas, such as online newspapers and magazines. Job openings also will result from the need to replace workers who leave these occupations permanently. Some news analysts, reporters, and correspondents find the work too stressful and hectic or do not like the lifestyle, and transfer to other occupations.

Competition will continue to be keen for jobs on large metropolitan newspapers and broadcast stations and on national magazines. Talented writers who can handle highly specialized scientific or technical subjects have an advantage. Also, newspapers increasingly are hiring stringers and freelancers.

Most entry-level openings arise on small publications, as reporters and correspondents become editors or reporters on larger publications or leave the field. Small town and suburban newspapers will continue to offer most opportunities for persons seeking to enter this field.

Journalism graduates have the background for work in closely related fields such as advertising and public relations, and many take jobs in these fields. Other graduates accept sales, managerial, or other nonmedia positions, because of the difficulty in finding media jobs.

The newspaper and broadcasting industries are sensitive to economic ups and downs, because these industries depend on advertising revenue. During recessions, few new reporters are hired, and some reporters lose their jobs.

Earnings

Salaries for news analysts, reporters, and correspondents vary widely but, in general, are relatively high, except at small stations and small publications, where salaries often are very low. Median annual earnings of news analysts, reporters, and correspondents were $29,110 in 2000. The middle 50 percent earned between $21,320 and $45,540. The lowest 10 percent earned less than $16,540, and the highest 10 percent earned more than $69,300. Median annual earnings of news analysts, reporters, and correspondents were $33,550 in radio and television broadcasting and $26,900 in newspapers in 2000.

According to a 1999 survey conducted by the National Association of Broadcasters and the Broadcast Cable Financial Management Association, the annual average salary, including bonuses, was $83,400 for weekday anchors and $44,200 for those working on weekends. Television news reporters earned on average $33,700. Weekday sportscasters typically earned $68,900, while weekend sportscasters earned $37,200. Weathercasters averaged $68,500 during the week and $36,500 on weekends. According to the 2001 survey, the annual average salary, including bonuses, was $55,100 for radio news reporters and $53,300 for sportscasters in radio broadcasting.

Related Occupations

News analysts, reporters, and correspondents must write clearly and effectively to succeed in their profession. Others for whom good writing ability is essential include writers and editors, and public relations specialists. Many news analysts, reporters, and correspondents also must communicate information orally. Others for whom oral communication skills are vital are announcers, interpreters and translators, sales and related occupations, and teachers.

Sources of Additional Information

For information on careers in broadcast news and related scholarships and internships, contact:

- Radio and Television News Directors Foundation, 1000 Connecticut Ave. N.W., Suite 615, Washington, DC 20036. Internet: http://www.rtndf.org

General information on the broadcasting industry is available from:

- National Association of Broadcasters, 1771 N St. N.W., Washington, DC 20036. Internet: http://www.nab.org

Career information, including the pamphlets *Newspaper Career Guide* and *Newspaper: What's In It For Me?* is available from:

- Newspaper Association of America, 1921 Gallows Rd., Suite 600, Vienna, VA 22182. Internet: http://www.naa.org

Information on careers in journalism, colleges and universities offering degree programs in journalism or communications, and journalism scholarships and internships may be obtained from:

- Dow Jones Newspaper Fund, Inc., P.O. Box 300, Princeton, NJ 08543-0300

Information on union wage rates for newspaper and magazine reporters is available from:

- Newspaper Guild, Research and Information Department, 501 3rd St. N.W., Suite 250, Washington, DC 20001. Internet: http://www.newsguild.org

For a list of schools with accredited programs in journalism, send a stamped, self-addressed envelope to:

- Accrediting Council on Education in Journalism and Mass Communications, University of Kansas School of Journalism and Mass Communications, Stauffer-Flint Hall, Lawrence, KS 66045. Internet: http://www.ukans.edu/~acejmc

Information on newspaper careers and community newspapers is available from:

- National Newspaper Association, 1010 North Glebe Rd., Suite 450, Arlington, VA 22201. Internet: http://www.nnafoundation.org

Names and locations of newspapers and a list of schools and departments of journalism are published in the *Editor and Publisher International Year Book*, available in most public libraries and newspaper offices.

Nuclear Engineers

O*NET 17-2161.00

Significant Points

- Overall job opportunities in engineering are expected to be good.
- A bachelor's degree is required for most entry-level jobs.
- Starting salaries are significantly higher than those of college graduates in other fields.
- Continuing education is critical to keep abreast of the latest technology.

Nature of the Work

Nuclear engineers research and develop the processes, instruments, and systems used to derive benefits from nuclear energy and radiation. They design, develop, monitor, and operate nuclear plants used to generate power. They may work on the nuclear fuel cycle—the production, handling, and use of nuclear fuel and the safe disposal of waste produced by nuclear energy—or on fusion energy. Some specialize in the development of nuclear power sources for spacecraft; others find industrial and medical uses for radioactive materials, such as equipment to diagnose and treat medical problems.

Working Conditions

Most engineers work in office buildings, laboratories, or industrial plants. Others may spend time outdoors at construction sites, mines, and oil and gas exploration and production sites, where they monitor or direct operations or solve onsite problems. Some engineers travel extensively to plants or work sites.

Many engineers work a standard 40-hour week. At times, deadlines or design standards may bring extra pressure to a job. When this happens, engineers may work longer hours and experience considerable stress.

Employment

Nuclear engineers held about 14,000 jobs in 2000. About 58 percent were in utilities, 26 percent in engineering consulting firms, and 14 percent in the federal government. More than half of all federally employed nuclear engineers were civilian employees of the navy, and most of the rest worked for the Department of

Energy. Most nonfederally employed nuclear engineers worked for public utilities or engineering consulting companies. Some worked for defense manufacturers or manufacturers of nuclear power equipment.

Training, Other Qualifications, and Advancement

A bachelor's degree in engineering is required for almost all entry-level engineering jobs. College graduates with a degree in a physical science or mathematics occasionally may qualify for some engineering jobs, especially in specialties in high demand. Most engineering degrees are granted in electrical, electronics, mechanical, or civil engineering. However, engineers trained in one branch may work in related branches. This flexibility allows employers to meet staffing needs in new technologies and specialties in which engineers are in short supply. It also allows engineers to shift to fields with better employment prospects or to those that more closely match their interests.

Most engineering programs involve a concentration of study in an engineering specialty, along with courses in both mathematics and science. Most programs include a design course, sometimes accompanied by a computer or laboratory class or both.

In addition to the standard engineering degree, many colleges offer 2- or 4-year degree programs in engineering technology. These programs, which usually include various hands-on laboratory classes that focus on current issues, prepare students for practical design and production work, rather than for jobs which require more theoretical and scientific knowledge. Graduates of 4-year technology programs may get jobs similar to those obtained by graduates with a bachelor's degree in engineering. Engineering technology graduates, however, are not qualified to register as professional engineers under the same terms as graduates with degrees in engineering. Some employers regard technology program graduates as having skills between those of a technician and an engineer.

Graduate training is essential for engineering faculty positions and many research and development programs, but is not required for the majority of entry-level engineering jobs. Many engineers obtain graduate degrees in engineering or business administration to learn new technology and broaden their education. Many high-level executives in government and industry began their careers as engineers.

About 330 colleges and universities offer bachelor's degree programs in engineering that are accredited by the Accreditation Board for Engineering and Technology (ABET), and about 250 colleges offer accredited bachelor's degree programs in engineering technology. ABET accreditation is based on an examination of an engineering program's student achievement, program improvement, faculty, curricular content, facilities, and institutional commitment. Although most institutions offer programs in the major branches of engineering, only a few offer programs in the smaller specialties. Also, programs of the same title may vary in content. For example, some programs emphasize industrial practices, preparing students for a job in industry, whereas others are more theoretical and are designed to prepare students for graduate work. Therefore, students should investigate curricula and check accreditations carefully before selecting a college.

Admissions requirements for undergraduate engineering schools include a solid background in mathematics (algebra, geometry, trigonometry, and calculus) and sciences (biology, chemistry, and physics), and courses in English, social studies, humanities, and computers. Bachelor's degree programs in engineering typically are designed to last 4 years, but many students find that it takes between 4 and 5 years to complete their studies. In a typical 4-year college curriculum, the first 2 years are spent studying mathematics, basic sciences, introductory engineering, humanities, and social sciences. In the last 2 years, most courses are in engineering, usually with a concentration in one branch. Some programs offer a general engineering curriculum; students then specialize in graduate school or on the job.

Some engineering schools and 2-year colleges have agreements whereby the 2-year college provides the initial engineering education, and the engineering school automatically admits students for their last 2 years. In addition, a few engineering schools have arrangements whereby a student spends 3 years in a liberal arts college studying pre-engineering subjects and 2 years in an engineering school studying core subjects, and then receives a bachelor's degree from each school. Some colleges and universities offer 5-year master's degree programs. Some 5- or even 6-year cooperative plans combine classroom study and practical work, permitting students to gain valuable experience and finance part of their education. All 50 states and the District of Columbia usually require licensure for engineers who offer their services directly to the public. Engineers who are licensed are called Professional Engineers (PE). This licensure generally requires a degree from an ABET-accredited engineering program, 4 years of relevant work experience, and successful completion of a state examination. Recent graduates can start the licensing process by taking the examination in two stages. The initial Fundamentals of Engineering (FE) examination can be taken upon graduation. Engineers who pass this examination commonly are called Engineers in Training (EIT) or Engineer Interns (EI). The EIT certification usually is valid for 10 years. After acquiring suitable work experience, EITs can take the second examination, the Principles and Practice of Engineering Exam. Several states have imposed mandatory continuing education requirements for relicensure. Most states recognize licensure from other states.

Engineers should be creative, inquisitive, analytical, and detail oriented. They should be able to work as part of a team and to communicate well, both orally and in writing. Communication abilities are becoming more important because much of their work is becoming more diversified, meaning that engineers interact with specialists in a wide range of fields outside engineering.

Beginning engineering graduates usually work under the supervision of experienced engineers and, in large companies, also may receive formal classroom or seminar-type training. As new engineers gain knowledge and experience, they are assigned more difficult projects with greater independence to develop designs, solve problems, and make decisions. Engineers may advance to become technical specialists or to supervise a staff or team of engineers and technicians. Some may eventually become engineering managers or enter other managerial or sales jobs.

Job Outlook

Good opportunities should exist for nuclear engineers because the small number of nuclear engineering graduates is likely to be in rough balance with the number of job openings. Because this is a small occupation, projected job growth will generate few openings; consequently, most openings will result from the need to replace nuclear engineers who transfer to other occupations or leave the labor force.

Little or no change in employment of nuclear engineers is expected through 2010. Due to public concerns over the cost and safety of nuclear power, no commercial nuclear power plants are under construction in the United States. Nevertheless, nuclear engineers will be needed to operate existing plants. In addition, nuclear engineers will be needed to work in defense-related areas, to develop nuclear medical technology, and to improve and enforce waste management and safety standards.

Earnings

Median annual earnings of nuclear engineers were $79,360 in 2000. The middle 50 percent earned between $67,590 and $89,310. The lowest 10 percent earned less than $58,030, and the highest 10 percent earned more than $105,930. In 2000, the median annual earnings of nuclear engineers in electric services were $77,890. In the federal government, nuclear engineers in supervisory, nonsupervisory, and management positions earned an average of $71,700 a year in 2001.

According to a 2001 salary survey by the National Association of Colleges and Employers, bachelor's degree candidates in nuclear engineering received starting offers averaging $49,609 a year, and master's degree candidates, on average, were offered $56,299.

Related Occupations

Nuclear engineers apply the principles of physical science and mathematics in their work. Other workers who use scientific and mathematical principles include architects, except landscape and naval; engineering and natural sciences managers; computer and information systems managers; mathematicians; drafters; engineering technicians; sales engineers; science technicians; and physical and life scientists, including agricultural and food scientists, biological and medical scientists, conservation scientists and foresters, atmospheric scientists, chemists and materials scientists, environmental scientists and geoscientists, and physicists and astronomers.

Sources of Additional Information

General information about nuclear engineers is available from:

- American Nuclear Society, 555 North Kensington Ave., LaGrange Park, IL 60525. Internet: http://www.ans.org

High school students interested in obtaining a full package of guidance materials and information (product number SP-01) on a variety of engineering disciplines should contact the Junior Engineering Technical Society by sending $3.50 to:

- JETS-Guidance, 1420 King St., Suite 405, Alexandria, VA 22314-2794. Internet: http://www.jets.org

High school students interested in obtaining information on ABET-accredited engineering programs should contact:

- The Accreditation Board for Engineering and Technology, Inc., 111 Market Place, Suite 1050, Baltimore, MD 21202-4012. Internet: http://www.abet.org

Non-licensed engineers and college students interested in obtaining information on Professional Engineer licensure should contact:

- The National Society of Professional Engineers, 1420 King St., Alexandria, VA 22314-2794. Internet: http://www.nspe.org
- National Council of Examiners for Engineers and Surveying, P.O. Box 1686, Clemson, SC 29633-1686. Internet: http://www.ncees.org

Information on general engineering education and career resources is available from:

- American Society for Engineering Education, 1818 N St. N.W., Suite 600, Washington, DC 20036-2479. Internet: http://www.asee.org

Information on obtaining an engineering position with the federal government is available from the Office of Personnel Management (OPM) through a telephone-based system. Consult your telephone directory under U.S. Government for a local number or call (912) 757-3000; Federal Relay Service: (800) 877-8339. The first number is not toll free, and charges may result. Information also is available from the OPM Internet site: http://www.usajobs.opm.gov

Non-high school students and those wanting more detailed information should contact societies representing the individual branches of engineering. Each can provide information about careers in the particular branch. The individual statements elsewhere in this book also provide other information in detail on aerospace; agricultural; biomedical; chemical; civil; computer hardware; electrical and electronics, except computer; environmental; industrial, including health and safety; materials; mechanical; mining and geological, including mining safety; and petroleum engineering.

Nuclear Medicine Technologists

O*NET 29-2033.00

Significant Points

- Faster-than-average growth will arise from an increase in the number of middle-aged and elderly persons, who are the primary users of diagnostic procedures.
- Technologists with cross training in radiologic technology or other modalities will have the best prospects.

Nature of the Work

In nuclear medicine, radionuclides—unstable atoms that emit radiation spontaneously—are used to diagnose and treat disease. Radionuclides are purified and compounded like other drugs to form radiopharmaceuticals. Nuclear medicine technologists administer these radiopharmaceuticals to patients, then monitor the characteristics and functions of tissues or organs in which they localize. Abnormal areas show higher or lower concentrations of radioactivity than normal.

Nuclear medicine technologists operate cameras that detect and map the radioactive drug in the patient's body to create an image on photographic film or a computer monitor. Radiologic technologists and technicians also operate diagnostic imaging equipment, but their equipment creates an image by projecting an X ray through the patient.

Nuclear medicine technologists explain test procedures to patients. They prepare a dosage of the radiopharmaceutical and administer it by mouth, injection, or other means. When preparing radiopharmaceuticals, technologists adhere to safety standards that keep the radiation dose to workers and patients as low as possible.

Technologists position patients and start a gamma scintillation camera, or "scanner," which creates images of the distribution of a radiopharmaceutical as it localizes in and emits signals from the patient's body. Technologists produce the images on a computer screen or on film for a physician to interpret. Some nuclear medicine studies, such as cardiac function studies, are processed with the aid of a computer.

Nuclear medicine technologists also perform radioimmunoassay studies that assess the behavior of a radioactive substance inside the body. For example, technologists may add radioactive substances to blood or serum to determine levels of hormones or therapeutic drug content.

Technologists keep patient records and record the amount and type of radionuclides received, used, and disposed of.

Working Conditions

Nuclear medicine technologists generally work a 40-hour week. This may include evening or weekend hours in departments that operate on an extended schedule. Opportunities for part-time and shift work are also available. In addition, technologists in hospitals may have on-call duty on a rotational basis.

Because technologists are on their feet much of the day, and may lift or turn disabled patients, physical stamina is important.

Although there is potential for radiation exposure in this field, it is kept to a minimum by the use of shielded syringes, gloves, and other protective devices and adherence to strict radiation safety guidelines. Technologists also wear badges that measure radiation levels. Because of safety programs, however, badge measurements rarely exceed established safety levels.

Employment

Nuclear medicine technologists held about 18,000 jobs in 2000. About two-thirds of all jobs were in hospitals. The rest were in physicians' offices and clinics, including diagnostic imaging centers.

Training, Other Qualifications, and Advancement

Nuclear medicine technology programs range in length from 1 to 4 years and lead to a certificate, associate degree, or bachelor's degree. Generally, certificate programs are offered in hospitals, associate programs in community colleges, and bachelor's programs in 4-year colleges and in universities. Courses cover physical sciences, the biological effects of radiation exposure, radiation protection and procedures, the use of radiopharmaceuticals, imaging techniques, and computer applications.

One-year certificate programs are for health professionals, especially radiologic technologists and diagnostic medical sonographers, who wish to specialize in nuclear medicine. They also attract medical technologists, registered nurses, and others who wish to change fields or specialize. Others interested in the nuclear medicine technology field have three options: A 2-year certificate program, a 2-year associate program, or a 4-year bachelor's program.

The Joint Review Committee on Education Programs in Nuclear Medicine Technology accredits most formal training programs in nuclear medicine technology. In 2000, there were 95 accredited programs in the continental United States and Puerto Rico.

All nuclear medicine technologists must meet the minimum federal standards on the administration of radioactive drugs and the operation of radiation detection equipment. In addition, about half of all states require technologists to be licensed. Technologists also may obtain voluntary professional certification or registration. Registration or certification is available from the American Registry of Radiologic Technologists and from the Nuclear Medicine Technology Certification Board. Most employers prefer to hire certified or registered technologists.

Nuclear medicine technologists should be sensitive to patients' physical and psychological needs. They must pay attention to detail, follow instructions, and work as part of a team. In addition, operating complicated equipment requires mechanical ability and manual dexterity.

Technologists may advance to supervisor, then to chief technologist, and to department administrator or director. Some technologists specialize in a clinical area such as nuclear cardiology or computer analysis or leave patient care to take positions in research laboratories. Some become instructors or directors in nuclear medicine technology programs, a step that usually requires a bachelor's degree or a master's in nuclear medicine technology. Others leave the occupation to work as sales or training representatives for medical equipment and radiopharmaceutical manufacturing firms, or as radiation safety officers in regulatory agencies or hospitals.

Job Outlook

Employment of nuclear medicine technologists is expected to grow faster than the average for all occupations through the year 2010. The number of openings each year will be very low because the occupation is small. Growth will arise from an increase in the number of middle-aged and older persons who are the primary users of diagnostic procedures, including nuclear medicine tests.

Technological innovations may increase the diagnostic uses of nuclear medicine. One example is the use of radiopharmaceuticals in combination with monoclonal antibodies to detect cancer at far earlier stages than is customary today, and without resorting to surgery. Another is the use of radionuclides to examine the heart's ability to pump blood. Wider use of nuclear medical imaging to observe metabolic and biochemical changes for neurology, cardiology, and oncology procedures, also will spur some demand for nuclear medicine technologists.

On the other hand, cost considerations will affect the speed with which new applications of nuclear medicine grow. Some promising nuclear medicine procedures, such as positron emission tomography (PET), are extremely costly, and hospitals contemplating them will have to consider equipment costs, reimbursement policies, and the number of potential users.

Earnings

Median annual earnings of nuclear medicine technologists were $44,130 in 2000. The middle 50 percent earned between $38,150 and $52,190. The lowest 10 percent earned less than $31,910, and the highest 10 percent earned more than $58,500. Median annual earnings of nuclear medicine technologists in 2000 were $44,000 in hospitals.

Related Occupations

Nuclear medical technologists operate sophisticated equipment to help physicians and other health practitioners diagnose and treat patients. Cardiovascular technologists and technicians, clinical laboratory technologists and technicians, diagnostic medical sonographers, radiation therapists, radiologic technologists and technicians, and respiratory therapists also perform similar functions.

Sources of Additional Information

Additional information on a career as a nuclear medicine technologist is available from:

- The Society of Nuclear Medicine-Technologist Section, 1850 Samuel Morse Dr., Reston, VA 22090. Internet: http://www.snm.org

For career information, send a stamped, self-addressed business size envelope with your request to:

- American Society of Radiologic Technologists, Customer Service Department, 15000 Central Ave. S.E., Albuquerque, NM 87123-3917, or call (800) 444-2778. Internet: http://www.asrt.org/asrt.htm

For a list of accredited programs in nuclear medicine technology, write to:

- Joint Review Committee on Educational Programs in Nuclear Medicine Technology, PMB 418, 1 2nd Avenue East, Suite C, Polson, MT 59860-2107. Internet: http://www.jrcnmt.org

Information on certification is available from:

- American Registry of Radiologic Technologists, 1255 Northland Dr., St. Paul, MN 55120-1155. Internet: http://www.arrt.org
- Nuclear Medicine Technology Certification Board, 2970 Clairmont Rd., Suite 610, Atlanta, GA 30329. Internet: http://www.nmtcb.org

Nursing, Psychiatric, and Home Health Aides

O*NET 31-1011.00, 31-1012.00, 31-1013.00

Significant Points

- Job prospects for nursing and home health aides will be very good because of fast growth and high replacement needs in these large occupations.
- Minimum education or training is generally required for entry-level jobs, but earnings are low.

Nature of the Work

Nursing and psychiatric aides help care for physically or mentally ill, injured, disabled, or infirm individuals confined to hospitals, nursing and personal care facilities, and mental health settings. Home health aides duties are similar, but they work in patients' homes or residential care facilities.

Nursing aides, also known as nursing assistants, geriatric aides, unlicensed assistive personnel, or hospital attendants, perform routine tasks under the supervision of nursing and medical staff. They answer patients' call bells, deliver messages, serve meals, make beds, and help patients eat, dress, and bathe. Aides also may provide skin care to patients; take temperatures, pulse, respiration, and blood pressure; and help patients get in and out of bed and walk. They also may escort patients to operating and examining rooms, keep patients' rooms neat, set up equipment, store and move supplies, or assist with some procedures. Aides observe patients' physical, mental, and emotional conditions and report any change to the nursing or medical staff.

Nursing aides employed in nursing homes often are the principal caregivers, having far more contact with residents than other members of the staff. Because some residents may stay in a nursing home for months or even years, aides develop ongoing relationships with them and interact with them in a positive, caring way.

Psychiatric aides, also known as mental health assistants or psychiatric nursing assistants, care for mentally impaired or emotionally disturbed individuals. They work under a team that may include psychiatrists, psychologists, psychiatric nurses, social workers, and therapists. In addition to helping patients dress, bathe, groom, and eat, psychiatric aides socialize with them and lead them in educational and recreational activities. Psychiatric aides may play games such as cards with the patients, watch television with them, or participate in group activities such as sports or field trips. They observe patients and report any physical or behavioral signs that might be important for the professional staff to know. They accompany patients to and from examinations and treatments. Because they have such close contact with patients, psychiatric aides can have a great deal of influence on their outlook and treatment.

Home health aides help elderly, convalescent, or disabled persons live in their own homes instead of in a health facility. Under the direction of nursing or medical staff, they provide health-related services, such as administering oral medications. Like nursing aides, home health aides may check pulse, temperature, and respiration; help with simple prescribed exercises; keep patients' rooms neat; and help patients move from bed, bathe, dress, and groom. Occasionally, they change nonsterile dressings, give massages and alcohol rubs, or assist with braces and artificial limbs. Experienced home health aides also may assist with medical equipment such as ventilators, which help patients breathe.

Most home health aides work with elderly or disabled persons who need more extensive care than family or friends can provide. Some help discharged hospital patients who have relatively short-term needs.

In home healthcare agencies, a registered nurse, physical therapist, or social worker usually assigns specific duties and supervises home health aides. Aides keep records of services performed and patients' condition and progress. They report changes in patients' conditions to the supervisor or case manager.

Working Conditions

Most full-time aides work about 40 hours a week, but because patients need care 24 hours a day, some aides work evenings, nights, weekends, and holidays. Many work part-time. Aides spend many hours standing and walking, and they often face heavy workloads. Because they may have to move patients in and out of bed or help them stand or walk, aides must guard against back injury. Aides also may face hazards from minor infections and major diseases, such as hepatitis, but can avoid infections by following proper procedures.

Aides often have unpleasant duties, such as emptying bedpans and changing soiled bed linens. The patients they care for may be disoriented, irritable, or uncooperative. Psychiatric aides must be prepared to care for patients whose illness may cause violent behavior. While their work can be emotionally demanding, many aides gain satisfaction from assisting those in need.

Home health aides may go to the same patient's home for months or even years. However, most aides work with a number of different patients, each job lasting a few hours, days, or weeks. Home health aides often visit multiple patients on the same day.

Home health aides generally work alone, with periodic visits by their supervisor. They receive detailed instructions explaining when to visit patients and what services to perform. Aides are individually responsible for getting to patients' homes, and they may spend a good portion of the working day traveling from one patient to another. Because mechanical lifting devices available in institutional settings are seldom available in patients' homes, home health aides are particularly susceptible to injuries resulting from overexertion when assisting patients.

Employment

Nursing, psychiatric, and home health aides held about 2.1 million jobs in 2000. Nursing aides held about 1.4 million jobs, home health aides held roughly 615,000 jobs, and psychiatric aides held about 65,000 jobs. About one-half of nursing aides worked in nursing homes, and about one-fourth worked in hospitals. Most home health aides were employed by home health agencies, visiting nurse associations, social services agencies, residential care facilities, and temporary-help firms. Others worked for home health departments of hospitals and nursing facilities, public health agencies, and community volunteer agencies. Most psychiatric aides worked in psychiatric units of general hospitals, psychiatric hospitals, state and county mental institutions, homes for mentally retarded and psychiatric patients, and community mental health centers.

Training, Other Qualifications, and Advancement

In many cases, neither a high school diploma nor previous work experience is necessary for a job as a nursing, psychiatric, or home health aide. A few employers, however, require some training or experience. Hospitals may require experience as a nursing aide or home health aide. Nursing homes often hire inexperienced workers who must complete a minimum of 75 hours of mandatory training and pass a competency evaluation program within 4 months of employment. Aides who complete the program are certified and placed on the state registry of nursing aides. Some states require psychiatric aides to complete a formal training program.

The federal government has enacted guidelines for home health aides whose employers receive reimbursement from Medicare. Federal law requires home health aides to pass a competency test covering 12 areas: communication skills; documentation of patient status and care provided; reading and recording vital signs; basic infection control procedures; basic body functions; maintenance of a healthy environment; emergency procedures; physical, emotional, and developmental characteristics of patients; personal hygiene and grooming; safe transfer techniques; normal range of motion and positioning; and basic nutrition.

A home health aide may take training before taking the competency test. Federal law suggests at least 75 hours of classroom and practical training supervised by a registered nurse. Training and testing programs may be offered by the employing agency, but must meet the standards of the Healthcare Financing Administration. Training programs vary depending upon state regulations.

The National Association for Home Care offers national certification for home health aides. The certification is a voluntary demonstration that the individual has met industry standards.

Nursing aide training is offered in high schools, vocational-technical centers, some nursing homes, and some community colleges. Courses cover body mechanics, nutrition, anatomy and physiology, infection control, communication skills, and resident rights. Personal care skills such as how to help patients bathe, eat, and groom also are taught.

Some facilities, other than nursing homes, provide classroom instruction for newly hired aides, while others rely exclusively on informal on-the-job instruction from a licensed nurse or an experienced aide. Such training may last several days to a few months. From time to time, aides may also attend lectures, workshops, and in-service training.

These occupations can offer individuals an entry into the world of work. The flexibility of night and weekend hours also provides high school and college students a chance to work during the school year.

Applicants should be tactful, patient, understanding, healthy, emotionally stable, dependable, and have a desire to help people. They should also be able to work as part of a team, have good communication skills, and be willing to perform repetitive, routine tasks. Home health aides should be honest, and discreet because they work in private homes.

Aides must be in good health. A physical examination, including state regulated tests such as those for tuberculosis, may be required.

Opportunities for advancement within these occupations are limited. To enter other health occupations, aides generally need additional formal training. Some employers and unions provide opportunities by simplifying the educational paths to advancement. Experience as an aide can also help individuals decide whether to pursue a career in the healthcare field.

Job Outlook

Overall employment of nursing, psychiatric, and home health aides is projected to grow faster than the average through the year 2010, although individual occupational growth rates vary. Home health aides are expected to grow the fastest, as a result of growing demand for home healthcare from an aging population and efforts to contain healthcare costs by moving patients out of hospitals and nursing facilities as quickly as possible. Consumer preference for care in the home and improvements in medical technologies for in-home treatment also will contribute to much faster-than-average employment growth for home health aides.

Nursing aide employment will not grow as fast as home health aide employment, largely because nursing aides are concentrated in the relatively slower-growing nursing home sector. Nevertheless, employment of nursing aides is expected to grow faster than the average for all occupations in response to increasing emphasis on rehabilitation and the long-term care needs of a rapidly growing elderly population. Financial pressure on hospitals to discharge patients as soon as possible should produce more nursing home admissions. Modern medical technology will also increase the employment of nursing aides. This technology, while saving and extending more lives, increases the need for long-term care provided by aides.

Employment of psychiatric aides—the smallest of the three occupations—is expected to grow as fast as the average. The number of jobs for psychiatric aides in hospitals, where one-half of psychiatric aides work, will decline due to attempts to contain costs by limiting inpatient psychiatric treatment. Employment in other sectors will rise in response to growth in the number of older persons. Many elderly will require mental health services, increasing public acceptance of formal treatment for drug abuse and alcoholism, and a lessening of the stigma attached to those receiving mental healthcare.

Numerous openings for nursing and home health aides will arise from a combination of fast growth and high replacement needs for these large occupations. Turnover is high, a reflection of modest entry requirements, low pay, high physical and emotional demands, and lack of advancement opportunities. For these same reasons, many people are unwilling to perform this kind of work. Therefore, persons who are interested in this work and suited for it should have excellent job opportunities.

Earnings

Median hourly earnings of nursing aides, orderlies, and attendants were $8.89 in 2000. The middle 50 percent earned between $7.51 and $10.59 an hour. The lowest 10 percent earned less than $6.48, and the highest 10 percent earned more than $12.69 an hour. Median hourly earnings in the industries employing the largest numbers of nursing aides, orderlies, and attendants in 2000 were as follows:

Personnel supply services	$9.82
Local government	9.66
Hospitals	9.42
Nursing and personal care facilities	8.61
Residential care	7.96

Median hourly earnings of psychiatric aides were $10.45 in 2000. The middle 50 percent earned between $8.38 and $13.02 an hour. The lowest 10 percent earned less than $7.10, and the highest 10 percent earned more than $15.50 an hour. Median hourly earnings of psychiatric aides in 2000 were $12.61 in state government and $10.50 in hospitals.

Nursing and psychiatric aides in hospitals generally receive at least 1 week's paid vacation after 1 year of service. Paid holidays and sick leave, hospital and medical benefits, extra pay for late-shift work, and pension plans also are available to many hospital and some nursing home employees.

Median hourly earnings of home health aides were $8.23 in 2000. The middle 50 percent earned between $7.13 and $9.88 an hour. The lowest 10 percent earned less than $6.14, and the highest 10 percent earned more than $11.93 an hour. Median hourly earnings in the industries employing the largest numbers of home health aides in 2000 were as follows:

Nursing and personal care facilities	$8.65
Personnel supply services	8.60
Residential care	8.16
Home healthcare services	7.91
Individual and family services	7.89

Home health aides receive slight pay increases with experience and added responsibility. They usually are paid only for the time worked in the home; they normally are not paid for travel time between jobs. Most employers hire only on-call hourly workers and provide no benefits.

Related Occupations

Nursing, psychiatric, and home health aides help people who need routine care or treatment. So do childcare workers, medical assistants, occupational therapist assistants and aides, personal and home care aides, and physical therapist assistants and aides.

Sources of Additional Information

Information about employment opportunities may be obtained from local hospitals, nursing homes, home healthcare agencies, psychiatric facilities, state boards of nursing, and local offices of the state employment service.

General information about training and referrals to state and local agencies about opportunities for home health aides, a list of relevant publications, and information on certification are available from:

- National Association for Home Care, 228 7th St. S.E., Washington, DC 20003. Internet: http://www.nahc.org

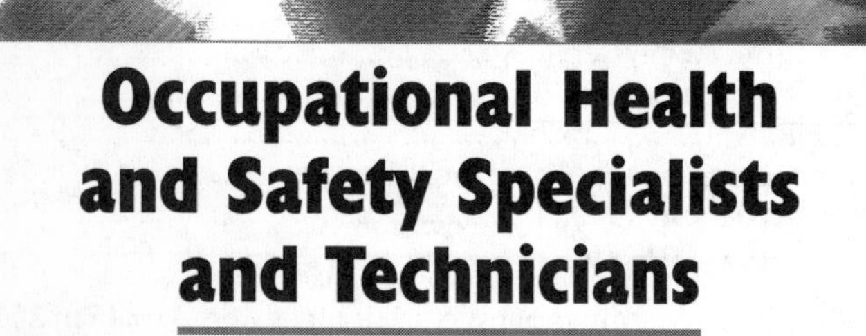

Occupational Health and Safety Specialists and Technicians

O*NET 29-9011.00, 29-9012.00

Significant Points

- Almost half of occupational health and safety specialists and technicians work in federal, state, and local government agencies that enforce rules on health and safety.
- For positions as specialists, many employers, including the federal government, require 4-year college degrees in safety or a related field.

Nature of the Work

Occupational health and safety specialists and technicians, also known as *occupational health and safety inspectors* and *industrial hygienists*, help keep workplaces safe and workers unscathed. They promote occupational health and safety within organizations by developing safer, healthier, and more efficient ways of working. *Occupational health and safety specialists* analyze work environments and design programs to control, eliminate, and prevent disease or injury caused by chemical, physical, and biological agents or ergonomic factors. They may conduct inspections and enforce adherence to laws, regulations, or employer policies governing worker health and safety. *Occupational health and safety technicians* collect data on work environments for analysis by occupational health and safety specialists. Usually working under the supervision of specialists, they help implement and evaluate programs designed to limit risks to workers.

Occupational health and safety specialists and technicians identify hazardous conditions and practices. Sometimes, they develop methods to predict hazards from experience, historical data, and other information sources. Then they identify potential hazards in existing or future systems, equipment, products, facilities, or processes. After reviewing the causes or effects of hazards, they evaluate the probability and severity of accidents that may result. For example, they might uncover patterns in injury data that implicate a specific cause such as system failure, human error, incomplete or faulty decision making, or a weakness in existing policies or practices. Then they develop and help enforce a plan to eliminate hazards, conducting training sessions for management, supervisors, and workers on health and safety practices and regulations, as necessary. Lastly, they may check on the progress of the safety plan after its implementation. If improvements are not satisfactory, a new plan might be designed and put into practice.

Many occupational health and safety specialists inspect and test machinery and equipment, such as lifting devices, machine shields, or scaffolding, to ensure they meet appropriate safety regulations. They may check that personal protective equipment, such as masks, respirators, safety glasses, or safety helmets, is being used in workplaces according to regulations. They also check that dangerous materials are stored correctly. They test and identify work areas for potential accident and health hazards, such as toxic fumes and explosive gas-air mixtures, and may implement appropriate control measures, such as adjustments to ventilation systems. Their investigations might involve talking with workers and observing their work, as well as inspecting elements in their work environment, such as lighting, tools, and equipment.

To measure and control hazardous substances, such as the noise or radiation levels, occupational health and safety specialists and technicians prepare and calibrate scientific equipment. Samples of dust, gases, vapors, and other potentially toxic materials must be collected and handled properly to ensure safety and accurate test results.

If an accident occurs, occupational health and safety specialists help investigate unsafe working conditions, study possible causes, and recommend remedial action. Some occupational health and safety specialists and technicians assist with the rehabilitation of workers after accidents and injuries, and make sure they return to work successfully.

Frequent communication with management may be necessary to report on the status of occupational health and safety programs. Consultation with engineers or physicians also may be required.

Occupational health and safety specialists prepare reports including observations, analysis of contaminants, and recommendation for control and correction of hazards. Those who develop expertise in certain areas may develop occupational health and safety systems, including policies, procedures, and manuals.

Working Conditions

Occupational health and safety specialists and technicians work with many different people in a variety of environments. Their jobs often involve considerable fieldwork, and some travel frequently. Many occupational health and safety specialists and technicians work long and often irregular hours.

Occupational health and safety specialists and technicians may experience unpleasant, stressful, and dangerous working conditions. For example, health and safety inspectors are exposed to many of the same physically strenuous conditions and hazards as industrial employees, and the work may be performed in unpleasant, stressful, and dangerous working conditions. Health and safety inspectors may find themselves in adversarial roles when the organization or individual being inspected objects to the process or its consequences.

Employment

Occupational health and safety specialists and technicians held about 35,000 jobs in 2000. The federal government—chiefly the Department of Labor—employed 8 percent, state governments employed 17 percent, and local governments employed 19 percent. The remainder were employed throughout the private sector in schools, hospitals, management consulting firms, public utilities, and manufacturing firms.

Within the federal government, most jobs are as Occupational Health and Safety Administration (OSHA) inspectors, who enforce U.S. Department of Labor regulations that ensure adequate safety principles, practices, and techniques are applied in workplaces. Employers may be fined for violation of OSHA standards. Within the U.S. Department of Health and Human Services, occupational health and safety specialists working for the National Institute of Occupational Safety and Health (NIOSH) provide private companies with an avenue to evaluate the health and safety of their employees without the risk of being fined. Most large government agencies also employ occupational health and safety specialists and technicians who work to protect agency employees.

Most private companies either employ their own safety personnel or contract safety professionals to ensure OSHA compliance, as needed.

Training, Other Qualifications, and Advancement

Requirements include a combination of education, experience, and passing scores on written examinations. Many employers, including the federal government, require a 4-year college degree in safety or a related field for some positions. Experience as a safety professional is also a prerequisite for many positions.

All occupational health and safety specialists and technicians are trained in the applicable laws or inspection procedures through some combination of classroom and on-the-job training. In general, people who want to enter this occupation should be responsible and like detailed work. Occupational health and safety specialists and technicians should be able to communicate well. Recommended high school courses include English, chemistry, biology, and physics.

Certification is available through the Board of Certified Safety Professionals (BCSP) and the American Board of Industrial Hygiene (ABIH). The BCSP offers the Certified Safety Professional (CSP) credential, while the ABIH proffers the Certified Industrial Hygienist (CIH) credential. Also, the Council on Certification of Health, Environmental, and Safety Technologists, a joint effort between the BCSP and ABIH, awards the Occupational Health and Safety Technologist (OHST) credential. Requirements for the OHST credential are less stringent than those for the CSP or CIH credentials. Once education and experience requirements have been met, certification may be obtained through an examination. Continuing education is required for recertification. Although voluntary, many employers encourage certification.

Federal government occupational health and safety specialists and technicians whose job performance is satisfactory advance through their career ladder to a specified full-performance level. For positions above this level, usually supervisory positions, advancement is competitive and based on agency needs and individual merit. Advancement opportunities in state and local governments and the private sector are often similar to those in the federal government.

With additional experience or education, promotion to a managerial position is possible. Research or related teaching positions at the college level require advanced education.

Job Outlook

Employment of occupational health and safety specialists and technicians is expected to grow about as fast as the average for all occupations through 2010, reflecting a balance of continuing public demand for a safe and healthy work environment against the desire for smaller government and fewer regulations. Additional job openings will arise from the need to replace those who transfer to other occupations, retire, or leave the labor force for other reasons. In private industry, employment growth will reflect industry growth and the continuing self-enforcement of government and company regulations and policies.

Employment of occupational health and safety specialists and technicians is seldom affected by general economic fluctuations. Federal, state, and local governments—which employ almost half of all specialists and technicians—provide considerable job security.

Earnings

Median annual earnings of occupational health and safety specialists and technicians were $42,750 in 2000. The middle 50 percent earned between $32,060 and $54,880. The lowest 10 percent earned less than $23,780, while the highest 10 percent earned over $67,760. Median annual earnings of occupational health and safety specialists and technicians in 2000 were $41,330 in local government and $41,110 in state government.

Most occupational health and safety specialists and technicians work for federal, state, and local governments or in large private firms. The latter generally offer more generous benefits than smaller firms offer.

Related Occupations

Occupational health and safety specialists and technicians ensure that laws and regulations are obeyed. Others who enforce laws and regulations include agricultural inspectors, construction and building inspectors, correctional officers, financial examiners, fire inspectors, police and detectives, and transportation inspectors.

Sources of Additional Information

Information about jobs in federal, state, and local government as well as in private industry is available from the states' employment service offices.

For information on a career as an industrial hygienist and a list of colleges and universities offering programs in industrial hygiene, contact:

- American Industrial Hygiene Association, 2700 Prosperity Ave., Suite 250, Fairfax, VA 22031. Internet: http://www.aiha.org

For a list of colleges and universities offering safety and related degrees, including correspondence courses, contact:

- American Society of Safety Engineers, 1800 E Oakton St., Des Plaines, IL 60018. Internet: http://www.asse.org

For information on the Certified Safety Professional credential, contact:

- Board of Certified Safety Professionals, 208 Burwash Ave., Savoy, IL 61874. Internet: http://www.bcsp.org

For information on the Certified Industrial Hygiene credential, contact:

- American Board of Industrial Hygiene, 6015 West St. Joseph, Suite 102, Lansing, MI 48917. Internet: http://www.abih.org

For information on the Occupational Health and Safety Technologist credential, contact:

- Council on Certification of Health, Environmental, and Safety Technologists, 208 Burwash Ave., Savoy, IL 61874. Internet: http://www.cchest.org

For additional career information, contact:

- U.S. Department of Health and Human Services, Center for Disease Control and Prevention, National Institute of Occupational Safety and Health, Hubert H. Humphrey Bldg., 200 Independence Ave. SW, Room 715H, Washington, DC 20201. Internet: http://www.cdc.gov/niosh/homepage.html
- U.S. Department of Labor, Occupational Safety and Health Administration, 200 Constitution Ave. NW, Washington, DC 20210. Internet: http://www.osha.gov

Information on obtaining positions as occupational health and safety specialists and technicians with the federal government is available from the Office of Personnel Management through a telephone-based system. Consult your telephone directory under U.S. Government for a local number or call (912) 757-3000; Federal Relay Service: (800) 877-8339. The first number is not toll free, and charges may result. Information also is available from the Internet site: http://www.usajobs.opm.gov

Occupational Therapist Assistants and Aides

O*NET 31-2011.00, 31-2012.00

Significant Points

- Certified occupational therapist assistants must complete an associate degree or certificate program. In contrast, occupational therapist aides usually receive most of their training on the job.
- Aides are not licensed, so by law they are not allowed to perform as wide a range of tasks as occupational therapist assistants do.
- Employment is projected to increase much faster than the average, as rapid growth in the number of middle-aged and elderly individuals increases the demand for therapeutic services.

Nature of the Work

Occupational therapist assistants and aides work under the direction of occupational therapists to provide rehabilitative services to persons with mental, physical, emotional, or developmental impairments. The ultimate goal is to improve clients' quality of life by helping them compensate for limitations. For example, occupational therapist assistants help injured workers reenter the labor force by helping them improve their motor skills or help persons with learning disabilities increase their independence, by teaching them to prepare meals or use public transportation.

Occupational therapist assistants help clients with rehabilitative activities and exercises outlined in a treatment plan developed in collaboration with an occupational therapist. Activities range from teaching the proper method of moving from a bed into a wheelchair, to the best way to stretch and limber the muscles of the hand. Assistants monitor an individual's activities to make sure they are performed correctly and to provide encouragement. They also record their client's progress for use by the occupational therapist. If the treatment is not having the intended effect, or the client is not improving as expected, the therapist may alter the treatment program in hopes of obtaining better results. In addition, occupational therapist assistants document billing of the client's health insurance provider.

Occupational therapist aides typically prepare materials and assemble equipment used during treatment and are responsible for a range of clerical tasks. Duties can include scheduling appointments, answering the telephone, restocking or ordering depleted supplies, and filling out insurance forms or other paperwork. Aides are not licensed, so by law they are not allowed to perform as wide a range of tasks as occupational therapist assistants.

Working Conditions

The hours and days that occupational therapist assistants and aides work vary, depending on the facility and whether they are full or part-time employees. Many outpatient therapy offices and clinics have evening and weekend hours, to help coincide with patients' personal schedules.

Occupational therapist assistants and aides need to have a moderate degree of strength, due to the physical exertion required in assisting patients with their treatment. For example, in some cases, assistants and aides need to help lift patients. Additionally, constant kneeling, stooping, and standing for long periods all are part of the job.

Employment

Occupational therapist assistants and aides held 25,000 jobs in 2000. Occupational therapist assistants held about 17,000 jobs, and occupational therapist aides held about 8,500. About 30 percent of assistants and aides worked in hospitals, 25 percent worked in offices of occupational therapists, and 20 percent worked in nursing and personal care facilities. The remainder primarily worked in offices and clinics of physicians, social services agencies, outpatient rehabilitation centers, and home health agencies.

Training, Other Qualifications, and Advancement

Persons must complete an associate degree or certificate program from an accredited community college or technical school to qualify for occupational therapist assistant jobs. In contrast, occupational therapist aides usually receive most of their training on the job.

There were 185 accredited occupational therapist assistant programs in the United States in 2000. The first year of study typically involves an introduction to healthcare, basic medical terminology, anatomy, and physiology. In the second year, courses are more rigorous and usually include occupational therapist courses in areas such as mental health, gerontology, and pediatrics. Students also must complete supervised fieldwork in a clinic or community setting. Applicants to occupational therapist assistant programs can improve their chances of admission by taking high school courses in biology and health and by performing volunteer work in nursing homes, occupational or physical therapist's offices, or elsewhere in the healthcare field.

Occupational therapist assistants are regulated in most states, and must pass a national certification examination after they graduate. Those who pass the test are awarded the title of certified occupational therapist assistant.

Occupational therapist aides usually receive most of their training on the job. Qualified applicants must have a high school diploma, strong interpersonal skills, and a desire to help people in need. Applicants may increase their chances of getting a job by volunteering their services, thus displaying initiative and aptitude to the employer.

Assistants and aides must be responsible, patient, and willing to take directions and work as part of a team. Furthermore, they should be caring and want to help people who are not able to help themselves.

Job Outlook

Employment of occupational therapist assistants and aides is expected to grow much faster than the average for all occupations through 2010. Federal legislation imposing limits on reimbursement for therapy services may adversely affect the job market for occupational therapist assistants and aides in the near term. However, over the long run, demand for occupational therapist assistants and aides will continue to rise, with growth in the number of individuals with disabilities or limited function. Growth will result from an increasing population in older age groups, including the baby-boom generation, which increasingly needs occupational therapy services as they become older. Demand also will result from advances in medicine that allow more people with critical problems to survive and then need rehabilitative therapy. Third-party payers, concerned with rising healthcare costs may begin to encourage occupational therapists to delegate more of the hands-on therapy work to occupational therapist assistants and aides. By having assistants and aides work more closely with clients under the guidance of a therapist, the cost of therapy should be more modest than otherwise.

Earnings

Median annual earnings of occupational therapist assistants were $34,340 in 2000. The middle 50 percent earned between $29,280 and $40,690. The lowest 10 percent earned less than $23,970, and the highest 10 percent earned more than $45,370. Median annual earnings of occupational therapist assistants in 2000 were $33,390 in hospitals.

Median annual earnings of occupational therapist aides were $20,710 in 2000. The middle 50 percent earned between $16,510 and $28,470. The lowest 10 percent earned less than $14,370, and the highest 10 percent earned more than $35,900.

Related Occupations

Occupational therapist assistants and aides work under the direction of occupational therapists. Other occupations in the healthcare field that work under the supervision of professionals include dental assistants, medical assistants, pharmacy technicians, and physical therapist assistants and aides.

Sources of Additional Information

For information on a career as an occupational therapist assistant and a list of accredited programs, contact:

- The American Occupational Therapy Association, 4720 Montgomery Ln., P.O. Box 31220, Bethesda, MD 20824-1220. Internet: http://www.aota.org

Occupational Therapists

O*NET 29-1122.00

Significant Points

- Employment is projected to increase faster than the average, as rapid growth in the number of middle-aged and elderly individuals increases the demand for therapeutic services.
- Occupational therapists are increasingly taking on supervisory roles.
- More than one-third of occupational therapists work part-time.

Nature of the Work

Occupational therapists (OTs) help people improve their ability to perform tasks in their daily living and working environments. They work with individuals who have conditions that are mentally, physically, developmentally, or emotionally disabling. They also help them to develop, recover, or maintain daily living and work skills. Occupational therapists not only help clients improve basic motor functions and reasoning abilities, but also compensate for permanent loss of function. Their goal is to help clients have independent, productive, and satisfying lives.

Occupational therapists assist clients in performing activities of all types, ranging from using a computer, to caring for daily needs such as dressing, cooking, and eating. Physical exercises may be used to increase strength and dexterity, while paper and pencil exercises may be chosen to improve visual acuity and the ability to discern patterns. A client with short-term memory loss, for instance, might be encouraged to make lists to aid recall. A person with coordination problems might be assigned exercises to improve hand-eye coordination. Occupational therapists also use computer programs to help clients improve decision making, abstract reasoning, problem solving, and perceptual skills, as well as memory, sequencing, and coordination—all of which are important for independent living.

For those with permanent functional disabilities, such as spinal cord injuries, cerebral palsy, or muscular dystrophy, therapists instruct in the use of adaptive equipment such as wheelchairs, splints, and aids for eating and dressing. They also design or make special equipment needed at home or at work. Therapists develop computer-aided adaptive equipment and teach clients with severe limitations how to use it. This equipment enables clients to communicate better and to control other aspects of their environment.

Some occupational therapists, called *industrial therapists*, treat individuals whose ability to function in a work environment has been impaired. They arrange employment, plan work activities, and evaluate the client's progress.

Occupational therapists may work exclusively with individuals in a particular age group, or with particular disabilities. In schools, for example, they evaluate children's abilities, recommend and provide therapy, modify classroom equipment, and in general, help children participate as fully as possible in school programs and activities. Occupational therapy is also beneficial to the elderly population. Therapists help senior citizens lead more productive, active and independent lives through a variety of methods, including the use of adaptive equipment.

Occupational therapists in mental health settings treat individuals who are mentally ill, mentally retarded, or emotionally disturbed. To treat these problems, therapists choose activities that help people learn to cope with daily life. Activities include time management skills, budgeting, shopping, homemaking, and use of public transportation. They may also work with individuals who are dealing with alcoholism, drug abuse, depression, eating disorders, or stress related disorders.

Recording a client's activities and progress is an important part of an occupational therapist's job. Accurate records are essential for evaluating clients, billing, and reporting to physicians and others.

Working Conditions

Occupational therapists in hospitals and other healthcare and community settings usually work a 40-hour week. Those in schools may also participate in meetings and other activities, during and after the school day. More than one-third of occupational therapists work part-time.

In large rehabilitation centers, therapists may work in spacious rooms equipped with machines, tools, and other devices generating noise. The job can be tiring, because therapists are on their feet much of the time. Those providing home healthcare may spend time driving from appointment to appointment. Therapists also face hazards such as back strain from lifting and moving clients and equipment.

Therapists are increasingly taking on supervisory roles. Due to rising healthcare costs, third party payers are beginning to encourage occupational therapist assistants and aides to take more hands-on responsibility. By having assistants and aides work more closely with clients under the guidance of a therapist, the cost of therapy should be more modest.

Employment

Occupational therapists held about 78,000 jobs in 2000. About 1 in 6 occupational therapists held more than one job in 2000. The largest number of jobs was in hospitals, including many in rehabilitation and psychiatric hospitals. Other major employers include offices and clinics of occupational therapists and other health practitioners, school systems, home health agencies, nursing homes, community mental health centers, adult daycare programs, job training services, and residential care facilities.

Some occupational therapists are self-employed in private practice. They see clients referred by physicians or other health professionals, or provide contract or consulting services to nursing homes, schools, adult daycare programs, and home health agencies.

Training, Other Qualifications, and Advancement

A bachelor's degree in occupational therapy is the minimum requirement for entry into this field. All states, Puerto Rico, and the District of Columbia regulate occupational therapy. To obtain a license, applicants must graduate from an accredited educational program, and pass a national certification examination. Those who pass the test are awarded the title of registered occupational therapist.

In 1999, entry-level education was offered in 88 bachelor's degree programs; 11 postbachelor's certificate programs for students with a degree other than occupational therapy; and 53 entry-level master's degree programs. Nineteen programs offered a combined bachelor's and master's degree and 2 offered an entry-level doctoral degree. Most schools have full-time programs, although a growing number also offer weekend or part-time programs.

Occupational therapy coursework includes physical, biological, and behavioral sciences, and the application of occupational therapy theory and skills. Completion of 6 months of supervised fieldwork also is required.

Persons considering this profession should take high school courses in biology, chemistry, physics, health, art, and the social sciences. College admissions offices also look favorably at paid or volunteer experience in the healthcare field.

Occupational therapists need patience and strong interpersonal skills to inspire trust and respect in their clients. Ingenuity and imagination in adapting activities to individual needs are assets. Those working in home healthcare must be able to successfully adapt to a variety of settings.

Job Outlook

Employment of occupational therapists is expected to increase faster than the average for all occupations through 2010. Federal legislation imposing limits on reimbursement for therapy services may adversely affect the job market for occupational therapists in the near term. However, over the long run, the demand for occupational therapists should continue to rise as a result of growth in the number of individuals with disabilities or limited function requiring therapy services. The baby-boom generation's movement into middle age, a period when the incidence of heart attack and stroke increases, will increase the demand for therapeutic services. The rapidly growing population 75 years of age and above (an age that suffers from a high incidence of disabling conditions), also will demand additional services. Medical advances now enable more patients with critical problems to survive. These patients may need extensive therapy.

Hospitals will continue to employ a large number of occupational therapists to provide therapy services to acutely ill inpatients. Hospitals will also need occupational therapists to staff their outpatient rehabilitation programs.

Employment growth in schools will result from expansion of the school-age population and extended services for disabled students. Therapists will be needed to help children with disabilities prepare to enter special education programs.

Earnings

Median annual earnings of occupational therapists were $49,450 in 2000. The middle 50 percent earned between $40,460 and $57,890. The lowest 10 percent earned less than $32,040, and the highest 10 percent earned more than $70,810. Median annual earnings in the industries employing the largest numbers of occupational therapists in 2000 were as follows:

Nursing and personal care facilities	$51,220
Hospitals	50,430
Offices of other health practitioners	49,520
Elementary and secondary schools	45,340

Related Occupations

Occupational therapists use specialized knowledge to help individuals perform daily living skills and achieve maximum independence. Other workers performing similar duties include chiropractors, physical therapists, recreational therapists, rehabilitation counselors, respiratory therapists, and speech-language pathologists and audiologists.

Sources of Additional Information

For more information on occupational therapy as a career and a list of education programs, send a self-addressed label and $5.00 to:

- The American Occupational Therapy Association, 4720 Montgomery Ln., P.O. Box 31220, Bethesda, MD 20824-1220. Internet: http://www.aota.org

Operations Research Analysts

O*NET 15-2031.00

Significant Points

- Individuals with a master's or PhD degree in management science, operations research, or a closely related field should have good job prospects.
- Employment growth is projected to be slower than average because few job openings are expected to have the title operations research analyst.

Nature of the Work

Operations research and management science are terms that are used interchangeably to describe the discipline of applying advanced analytical techniques to help make better decisions and to solve problems. The procedures of operations research gave effective assistance during World War II in missions such as deploying radar, searching for enemy submarines, and getting supplies where they were most needed. Following the war, new analytical methods were developed and numerous peacetime applications emerged, leading to the use of operations research in many industries and occupations.

The prevalence of operations research in the nation's economy reflects the growing complexity of managing large organizations that require the effective use of money, materials, equipment, and people. Operations research analysts help determine better ways to coordinate these elements by applying analytical methods from mathematics, science, and engineering. They solve problems in different ways and propose alternative solutions to management, which then chooses the course of action that best meets the organization's goals. In general, operations research analysts may be concerned with diverse issues such as top-level strategy, planning, forecasting, resource allocation, performance measurement, scheduling, design of production facilities and systems, supply chain management, pricing, transportation and distribution, and analysis of data in large databases.

The duties of the operations research analyst vary according to the structure and management philosophy of the employer or client. Some firms centralize operations research in one department; others use operations research in each division. Operations research analysts also may work closely with senior managers to identify and solve a variety of problems. Some organizations contract operations research services with a consulting firm. Economists, systems analysts, mathematicians, industrial engineers, and others also may apply operations research techniques to address problems in their respective fields.

Regardless of the type or structure of the client organization, operations research in its classical role entails a similar set of procedures in carrying out analysis to support management's quest for performance improvement. Managers begin the process by describing the symptoms of a problem to the analyst, who then formally defines the problem. For example, an operations research analyst for an auto manufacturer may be asked to determine the best inventory level for each of the parts needed on a production line and to determine the number of windshields to be kept in inventory. Too many windshields would be wasteful and expensive, while too few could result in an unintended halt in production.

Operations research analysts study such problems, then break them into their component parts. Analysts then gather information about each of these parts from a variety of sources. To determine the most efficient amount of inventory to be kept on hand, for example, operations research analysts might talk with engineers about production levels, discuss purchasing arrangements with buyers, and examine data on storage costs provided by the accounting department.

With this information in hand, the analyst is ready to select the most appropriate analytical technique. Analysts could use several techniques—including simulation, linear and nonlinear programming, dynamic programming, queuing and other stochastic-process models, Markov decision processes, econometric methods, data envelopment analysis, neural networks, expert systems, decision analysis, and the analytic hierarchy process. Nearly all of these techniques, however, involve the construction of a mathematical model that attempts to describe the system being studied. The use of models enables the analyst to assign values to the different components, and clarify the relationships between components. These values can be altered to examine what may happen to the system under different circumstances.

In most cases, the computer program developed to solve the model must be modified and run repeatedly to obtain different solutions. A model for airline flight scheduling, for example, might include variables for the cities to be connected, amount of fuel required to fly the routes, projected levels of passenger demand, varying ticket and fuel prices, pilot scheduling, and maintenance costs. By locating the right combination of variable values, the analyst is able to produce the best flight schedule consistent with particular assumptions.

Upon concluding the analysis, the operations research analyst presents to management recommendations based on the results. Additional computer runs to consider different assumptions may be needed before deciding on the final recommendation. Once management reaches a decision, the analyst usually works with others in the organization to ensure the plan's successful implementation.

Working Conditions

Operations research analysts generally work regular hours in an office environment. Because they work on projects that are of immediate interest to top management, operations research analysts often are under pressure to meet deadlines and work more than a 40-hour week.

Employment

Operations research analysts held about 47,000 jobs in 2000. Major employers include telecommunication companies, air carriers, computer and data processing services firms, financial institutions, insurance carriers, engineering and management services firms, and the federal government. Most operations research analysts in the federal government work for the U.S. Armed Forces, and many operations research analysts in private industry work directly or indirectly on national defense. About 1 out of 5 analysts work for engineering, management and public relations, and research and testing, agencies that do operations research consulting.

Training, Other Qualifications, and Advancement

Employers generally prefer applicants with at least a master's degree in operations research, engineering, business, mathematics, information systems, or management science, coupled with a bachelor's degree in computer science or a quantitative discipline such as economics, mathematics, or statistics. Dual graduate degrees in operations research and computer science are especially attractive to employers. Operations research analysts also must be able to think logically and work well with people, and employers prefer workers with good oral and written communication skills.

In addition to formal education, employers often sponsor training for experienced workers, helping them keep up with new developments in operations research techniques and computer science. Some analysts attend advanced university classes on these subjects at their employer's expense.

Because computers are the most important tools for in-depth analysis, training and experience in programming are required. Operations research analysts typically need to be proficient in database collection and management, programming, and in the development and use of sophisticated software programs.

Beginning analysts usually perform routine work under the supervision of more experienced analysts. As they gain knowledge and experience, they are assigned more complex tasks and given greater autonomy to design models and solve problems. Operations research analysts advance by assuming positions as technical specialists or supervisors. The skills acquired by operations research analysts are useful for higher level management jobs, so experienced analysts may leave the field to assume nontechnical managerial or administrative positions. Operations research analysts with significant experience might become consultants and some may even open their own consulting practice.

Job Outlook

Employment of operations research analysts is expected to grow more slowly than the average for all occupations through 2010, because few job openings in this field are expected to have the title *operations research analyst.* Job opportunities in operations research should be good, however, because of interest in improving productivity, effectiveness, and competitiveness, and because of the extensive availability of data, computers, and software. Many jobs in operations research have other titles such as *operations analyst, management analyst, systems analyst, or policy analyst.* Individuals who hold a master's or PhD degree in operations research, management science, or a closely related field should find good job opportunities as the number of openings generated by employment growth and the need to replace those leaving the occupation is expected to exceed the number of persons graduating with these credentials.

Organizations today face pressure from growing domestic and international competition and must work to make their operations as effective as possible. As a result, businesses will increasingly rely on operations research analysts to optimize profits by improving productivity and reducing costs. As new technology is introduced into the marketplace, operations research analysts will be needed to determine how to utilize the technology in the best way.

Opportunities for operations research analysts exist in almost every industry because of the diversity of applications for their work. However, opportunities should be especially good in highly competitive industries, such as manufacturing, transportation, and telecommunications, and finance. As businesses and government agencies continue to contract out jobs to cut costs, many operations research analysts also will find opportunities as consultants, either working for a consulting firm or setting up their own practice. Opportunities in the military also exist, but will depend on the size of future military budgets. As the military develops new weapons systems and strategies, military leaders will rely on operations research analysts to test and evaluate their accuracy and effectiveness.

Earnings

Median annual earnings of operations research analysts were $53,420 in 2000. The middle 50 percent earned between $40,530 and $70,790. The lowest 10 percent had earnings of less than $31,860, while the highest 10 percent earned more than $88,870. In 2000, median annual earnings in the industries employing the largest numbers of operations research analysts were:

Computer and data processing services	$65,420
Federal government	62,990
Aircraft and parts	52,960
Engineering and architectural services	47,480

The average annual salary for operations research analysts in the federal government in nonsupervisory, supervisory, and managerial positions was $77,730 in 2001.

Related Occupations

Operations research analysts apply advanced analytical methods to large, complicated problems. Workers in other occupations that stress advanced analysis include systems analysts, computer scientists, and database administrators; computer programmers; engineers; mathematicians; statisticians; and economists and market and survey researchers. Because its goal is improved organizational effectiveness, operations research also is closely allied to managerial occupations, such as computer and information systems managers, and management analysts.

Sources of Additional Information

Information on career opportunities for operations research analysts is available from:

- Institute for Operations Research and the Management Sciences, 901 Elkridge Landing Rd., Suite 400, Linthicum, MD 21090. Internet: http://www.informs.org

For information on operations research careers in the armed forces and U.S. Department of Defense, contact:

- Military Operations Research Society, 101 South Whiting St., Suite 202, Alexandria, VA 22304. Internet: http://www.mors.org

Ophthalmic Laboratory Technicians

O*NET 51-9083.01, 51-9083.02

Significant Points

- Nearly all ophthalmic laboratory technicians learn their skills on the job.
- Increasing use of automated equipment will result in relatively slow job growth.
- Only a small number of job openings will be created each year because the occupation is small and slower-than-average growth is expected.

Nature of the Work

Ophthalmic laboratory technicians—also known as *manufacturing opticians*, *optical mechanics*, or *optical goods workers*—make prescription eyeglass or contact lenses. Prescription lenses are curved in such a way that light is correctly focused onto the retina of the patient's eye, improving vision. Some ophthalmic laboratory technicians manufacture lenses for other optical instruments, such as telescopes and binoculars. Ophthalmic laboratory technicians cut, grind, edge, and finish lenses according to specifications provided by dispensing opticians, optometrists, or ophthalmologists. They also may insert lenses into frames to produce finished glasses. Although some lenses still are produced by hand, technicians increasingly use automated equipment to make lenses.

Ophthalmic laboratory technicians should not be confused with workers in other vision care occupations. Ophthalmologists and optometrists are "eye doctors" who examine eyes, diagnose and treat vision problems, and prescribe corrective lenses. Ophthalmologists are physicians who perform eye surgery. Dispensing opticians, who also may do work described here, help patients select frames and lenses, and adjust finished eyeglasses.

Ophthalmic laboratory technicians read prescription specifications, then select standard glass or plastic lens blanks and mark them to indicate where the curves specified on the prescription should be ground. They place the lens in the lens grinder, set the dials for the prescribed curvature, and start the machine. After a minute or so, the lens is ready to be "finished" by a machine that rotates it against a fine abrasive to grind it and smooth out rough edges. The lens is then placed in a polishing machine with an even finer abrasive, to polish it to a smooth, bright finish.

Next, the technician examines the lens through a lensometer, an instrument similar in shape to a microscope, to make sure the degree and placement of the curve is correct. The technician then cuts the lenses and bevels the edges to fit the frame, dips each lens into dye if the prescription calls for tinted or coated lenses, polishes the edges, and assembles the lenses and frame parts into a finished pair of glasses.

In small laboratories, technicians usually handle every phase of the operation. In large ones, technicians may be responsible for operating computerized equipment where virtually every phase of the operation is automated. Technicians also inspect the final product for quality and accuracy.

Working Conditions

Ophthalmic laboratory technicians work in relatively clean and well-lighted laboratories and have limited contact with the public. Surroundings are relatively quiet despite the humming of machines. At times, technicians wear goggles to protect their eyes, and may spend a great deal of time standing.

Most ophthalmic laboratory technicians work a 5-day, 40-hour week, which may include weekends, evenings, or occasionally some overtime. Some work part-time.

Ophthalmic laboratory technicians need to take precautions against the hazards associated with cutting glass, handling chemicals, and working near machinery.

Employment

Ophthalmic laboratory technicians held about 32,000 jobs in 2000. Thirty-one percent were in retail optical stores that manufacture and sell prescription glasses and contact lenses, and 23 percent were in optical laboratories. These laboratories manufacture eyewear and contact lenses for sale by retail stores, as well as by ophthalmologists and optometrists. Most of the rest were in wholesalers or in optical laboratories that manufacture lenses for other optical instruments, such as telescopes and binoculars.

Training, Other Qualifications, and Advancement

Nearly all ophthalmic laboratory technicians learn their skills on the job. Employers filling trainee jobs prefer applicants who are high school graduates. Courses in science, mathematics, and computers are valuable; manual dexterity and the ability to do precision work are essential.

Technician trainees producing lenses by hand start on simple tasks such as marking or blocking lenses for grinding, then progress to lens grinding, lens cutting, edging, beveling, and eyeglass assembly. Depending on individual aptitude, it may take up to 6 months to become proficient in all phases of the work.

Technicians using automated systems will find computer skills valuable. Training is completed on the job and varies in duration depending on the type of machinery and individual aptitude.

A very small number of ophthalmic laboratory technicians learn their trade in the armed forces or in the few programs in optical technology offered by vocational-technical institutes or trade schools. These programs have classes in optical theory, surfacing and lens finishing, and the reading and applying of prescriptions. Programs vary in length from 6 months to 1 year and award certificates or diplomas.

Ophthalmic laboratory technicians can become supervisors and managers. Some technicians become dispensing opticians, although further education or training generally is required.

Job Outlook

Overall employment of ophthalmic laboratory technicians is expected to grow more slowly than the average for all occupations through the year 2010. Employment is expected to increase slowly in manufacturing as firms invest in automated machinery.

Demographic trends make it likely that many more Americans will need vision care in the years ahead. Not only will the population grow, but also the proportion of middle-aged and older adults is projected to increase rapidly. Middle age is a time when many people use corrective lenses for the first time, and elderly persons usually require more vision care than others.

Fashion, too, influences demand. Frames come in a variety of styles and colors—encouraging people to buy more than one pair. Demand also is expected to grow in response to the availability of new technologies that improve the quality and look of corrective lenses, such as antireflective coatings and bifocal lenses without the line visible in traditional bifocals.

Most job openings will arise from the need to replace technicians who transfer to other occupations or leave the labor force. However, only a small number of job openings will be created each year because the occupation is small.

Earnings

Median hourly earnings of ophthalmic laboratory technicians were $9.88 in 2000. The middle 50 percent earned between $8.25 and $12.07 an hour. The lowest 10 percent earned less than $7.19, and the highest 10 percent earned more than $14.71 an hour. In 2000, median hourly earnings of ophthalmic laboratory technicians were $10.25 in ophthalmic goods manufacturing and $9.79 in retail stores, not elsewhere classified, including optical goods stores.

Related Occupations

Workers in other precision production occupations include dental laboratory technicians, orthotists and prosthetists, and precision instrument and equipment repairers.

Sources of Additional Information

For a list of accredited programs in ophthalmic laboratory technology, contact:

- Commission on Opticianry Accreditation, 7023 Little River Turnpike, Suite 207, Annandale, VA 22003

State employment service offices can provide information about job openings for ophthalmic laboratory technicians.

Opticians, Dispensing

O*NET 29-2081.00

Significant Points

- Most dispensing opticians receive training on-the-job or through apprenticeships lasting 2 or more years; 22 states require a license.
- Projected employment growth reflects steadfast demand for corrective lenses and trends in fashion.
- The number of job openings will be relatively small because the occupation is small.

Nature of the Work

Dispensing opticians fit eyeglasses and contact lenses, following prescriptions written by ophthalmologists or optometrists.

Dispensing opticians examine written prescriptions to determine lens specifications. They recommend eyeglass frames, lenses, and lens coatings after considering the prescription and the customer's occupation, habits, and facial features. Dispensing opticians measure clients' eyes, including the distance between the centers of the pupils and the distance between the eye surface and the lens. For customers without prescriptions, dispensing opticians may use a lensometer to record the present eyeglass prescription. They also may obtain a customer's previous record, or verify a prescription with the examining optometrist or ophthalmologist.

Dispensing opticians prepare work orders that give ophthalmic laboratory technicians information needed to grind and insert lenses into a frame. The work order includes lens prescriptions and information on lens size, material, color, and style. Some dispensing opticians grind and insert lenses themselves. After the glasses are made, dispensing opticians verify that the lenses have been ground to specifications. Then they may reshape or bend the frame, by hand or using pliers, so that the eyeglasses fit the customer properly and comfortably. Some also fix, adjust, and refit broken frames. They instruct clients about adapting to, wearing, or caring for eyeglasses.

Some dispensing opticians specialize in fitting contacts, artificial eyes, or cosmetic shells to cover blemished eyes. To fit contact lenses, dispensing opticians measure eye shape and size, select the type of contact lens material, and prepare work orders specifying the prescription and lens size. Fitting contact lenses requires considerable skill, care, and patience. Dispensing opticians observe customers' eyes, corneas, lids, and contact lenses with special instruments and microscopes. During several visits, opticians show customers how to insert, remove, and care for their contacts, and ensure the fit is correct.

Dispensing opticians keep records on customer prescriptions, work orders, and payments; track inventory and sales; and perform other administrative duties.

Working Conditions

Dispensing opticians work indoors in attractive, well-lighted, and well-ventilated surroundings. They may work in medical offices or small stores where customers are served one at a time, or in large stores where several dispensing opticians serve a number of customers at once. Opticians spend a lot of time on their feet. If they prepare lenses, they need to take precautions against the hazards associated with glass cutting, chemicals, and machinery.

Most dispensing opticians work a 40-hour week, although some work longer hours. Those in retail stores may work evenings and weekends. Some work part-time.

Employment

Dispensing opticians held about 68,000 jobs in 2000. Almost half worked for ophthalmologists or optometrists who sell glasses directly to patients. Many also work in retail optical stores that offer one-stop shopping. Customers may have their eyes examined, choose frames, and have glasses made on the spot. Some work in optical departments of drug and department stores.

Training, Other Qualifications, and Advancement

Employers usually hire individuals with no background in opticianry or those who have worked as Ophthalmic Laboratory Technicians and then provide the required training. Most dispensing opticians receive training on-the-job or through apprenticeships lasting 2 or more years. Some employers, however, seek people with postsecondary training in opticianry.

Knowledge of physics, basic anatomy, algebra, geometry, and mechanical drawing is particularly valuable because training usually includes instruction in optical mathematics, optical physics, and the use of precision measuring instruments and other machinery and tools. Dispensing opticians deal directly with the public, so they should be tactful, pleasant, and communicate well. Manual dexterity and the ability to do precision work are essential.

Large employers usually offer structured apprenticeship programs, and small employers provide more informal on-the-job training. In the 22 states that require dispensing opticians to be licensed, individuals without postsecondary training work from 2 to 4 years as apprentices. Apprenticeship or formal training is offered in most states as well.

Apprentices receive technical training and learn office management and sales. Under the supervision of an experienced optician, optometrist, or ophthalmologist, apprentices work directly with patients, fitting eyeglasses and contact lenses. In the 21 states requiring licensure, information about apprenticeships and licensing procedures is available from the state board of occupational licensing.

Formal opticianry training is offered in community colleges and a few colleges and universities. In 2000, the Commission on Opticianry Accreditation accredited 25 programs that awarded 2-year associate degrees in opticianry. There also are shorter programs of 1 year or less. Some states that offer a license to dispensing opticians allow graduates to take the licensure exam immediately upon graduation; others require a few months to a year of experience.

Dispensing opticians may apply to the American Board of Opticianry (ABO) and the National Contact Lens Examiners (NCLE) for certification of their skills. Certification must be renewed every 3 years through continuing education. Those licensed in states where licensing renewal requirements include continuing education credits may use proof of their renewed state license to meet the recertification requirements of the ABO. Likewise, the NCLE will accept proof of license renewal from any state that has contact lens requirements.

Many experienced dispensing opticians open their own optical stores. Others become managers of optical stores or sales representatives for wholesalers or manufacturers of eyeglasses or lenses.

Job Outlook

Employment of dispensing opticians is expected to increase about as fast as the average for all occupations through 2010 as demand grows for corrective lenses. The number of middle-aged and elderly persons is projected to increase rapidly. Middle age is a time when many individuals use corrective lenses for the first time, and elderly persons generally require more vision care than others.

Fashion, too, influences demand. Frames come in a growing variety of styles and colors—encouraging people to buy more than one pair. Demand also is expected to grow in response to the availability of new technologies that improve the quality and look of corrective lenses, such as anti-reflective coatings and bifocal lenses without the line visible in old-style bifocals. Improvements in bifocal, extended wear, and disposable contact lenses also will spur demand.

The need to replace those who leave the occupation will result in additional job openings. Nevertheless, the total number of job openings will be relatively small because the occupation is small. This occupation is vulnerable to changes in the business cycle because eyewear purchases often can be deferred for a time. Employment of opticians can fall somewhat during economic downturns.

Earnings

Median annual earnings of dispensing opticians were $24,430 in 2000. The middle 50 percent earned between $19,200 and $31,770. The lowest 10 percent earned less than $15,900, and the highest 10 percent earned more than $39,660.

Related Occupations

Other workers who deal with customers and perform delicate work include camera and photographic equipment repairers, dental laboratory technicians, jewelers and precious stone and metal workers, locksmiths and safe repairers, ophthalmic laboratory technicians, orthotists and prosthetists, and watch repairers.

Sources of Additional Information

For general information about a career as a dispensing optician and about continuing education, as well as a list of state licensing boards for opticianry, contact:

- Opticians Association of America, 7023 Little River Turnpike, Suite 207, Annandale, VA 22003. Internet: http://www.opticians.org

For general information about a career as a dispensing optician and a list of accredited training programs, contact:

- Commission on Opticianry Accreditation, 7023 Little River Turnpike, Suite 207, Annandale, VA 22003

For general information on opticianry and a list of home-study programs, seminars, and review materials, contact:

- National Academy of Opticianry, 8401 Corporate Dr., Suite 605, Landover, MD 20785. Internet: http://www.nao.org

To learn about voluntary certification for opticians who fit spectacles, as well as state licensing boards of opticianry, contact:

- American Board of Opticianry, 6506 Loisdale Rd., Suite 209, Springfield, VA 22150. Internet: http://www.abo.org

For information on voluntary certification for dispensing opticians who fit contact lenses, contact:

- National Contact Lens Examiners, 6506 Loisdale Rd., Suite 209, Springfield, VA 22150. Internet: http://www.abo.org

Paralegals and Legal Assistants

O*NET 23-2011.00

Significant Points

- While some paralegals train on the job, employers increasingly prefer graduates of postsecondary paralegal education programs, especially graduates of 4-year paralegal programs or college graduates who have completed paralegal certificate programs.
- Paralegals are projected to rank among the fastest growing occupations in the economy, as they increasingly perform many legal tasks formerly carried out by lawyers.
- Stiff competition is expected, as the number of graduates of paralegal training programs and others seeking to enter the profession outpaces job growth.

Nature of the Work

While lawyers assume ultimate responsibility for legal work, they often delegate many of their tasks to paralegals. In fact, paralegals—also called legal assistants—continue to assume a growing range of tasks in the nation's legal offices and perform many of the same tasks as lawyers. Nevertheless, they are still explicitly prohibited from carrying out duties that are considered to be the practice of law, such as setting legal fees, giving legal advice, and presenting cases in court.

One of a paralegal's most important tasks is helping lawyers prepare for closings, hearings, trials, and corporate meetings. Paralegals investigate the facts of cases and ensure that all relevant information is considered. They also identify appropriate laws, judicial decisions, legal articles, and other materials that are relevant to assigned cases. After they analyze and organize the information, paralegals may prepare written reports that attorneys use in determining how cases should be handled. Should attorneys decide to file lawsuits on behalf of clients, paralegals may help prepare the legal arguments, draft pleadings and motions to be filed with the court, obtain affidavits, and assist attorneys during trials. Paralegals also organize and track files of important case documents and make them available and easily accessible to attorneys.

In addition to this preparatory work, paralegals also perform a number of other vital functions. For example, they help draft contracts, mortgages, separation agreements, and trust instruments. They also may assist in preparing tax returns and planning estates. Some paralegals coordinate the activities of other law office employees and maintain financial office records. Various additional tasks may differ, depending on the employer.

Paralegals are found in all types of organizations, but most are employed by law firms, corporate legal departments, and various government offices. In these organizations, they may work in all areas of the law, including litigation, personal injury, corporate law, criminal law, employee benefits, intellectual property, labor law, bankruptcy, immigration, family law, and real estate. Within specialties, functions often are broken down further so that paralegals may deal with a specific area. For example, paralegals specializing in labor law may deal exclusively with employee benefits.

The duties of paralegals also differ widely based on the type of organization in which they are employed. Paralegals who work for corporations often assist attorneys with employee contracts, shareholder agreements, stock-option plans, and employee benefit plans. They also may help prepare and file annual financial reports, maintain corporate minute books and resolutions, and secure loans for the corporation. Paralegals often monitor and review government regulations to ensure that the corporation operates within the law.

The duties of paralegals who work in the public sector usually vary within each agency. In general, they analyze legal material for internal use, maintain reference files, conduct research for attorneys, and collect and analyze evidence for agency hearings. They may then prepare informative or explanatory material on laws, agency regulations, and agency policy for general use by the agency and the public. Paralegals employed in community legal-service projects help the poor, the aged, and others in need of legal assistance. They file forms, conduct research, prepare documents, and when authorized by law, may represent clients at administrative hearings.

Paralegals in small and medium-sized law firms usually perform a variety of duties that require a general knowledge of the law. For example, they may research judicial decisions on improper police arrests or help prepare a mortgage contract. Paralegals employed by large law firms, government agencies, and corporations, however, are more likely to specialize in one aspect of the law.

Computer use and technical knowledge has become essential to paralegal work. Computer software packages and the Internet are increasingly used to search legal literature stored in computer databases and on CD-ROM. In litigation involving many supporting documents, paralegals may use computer databases to retrieve, organize, and index various materials. Imaging software allows paralegals to scan documents directly into a database, while billing programs help them to track hours billed to clients. Computer software packages also may be used to perform tax computations and explore the consequences of possible tax strategies for clients.

Working Conditions

Paralegals employed by corporations and government usually work a standard 40-hour week. Although most paralegals work year round, some are temporarily employed during busy times of the year, then released when the workload diminishes. Paralegals who work for law firms sometimes work very long hours when they are under pressure to meet deadlines. Some law firms reward such loyalty with bonuses and additional time off.

These workers handle many routine assignments, particularly when they are inexperienced. As they gain experience, paralegals usually assume more varied tasks with additional responsibility. Paralegals do most of their work at desks in offices and law libraries. Occasionally, they travel to gather information and perform other duties.

Employment

Paralegals and legal assistants held about 188,000 jobs in 2000. Private law firms employed the vast majority; most of the remainder worked for corporate legal departments and various levels of government. Within the federal government, the U.S. Department of Justice is the largest employer, followed by the U.S. Departments of Treasury and Defense, and the Federal Deposit Insurance Corporation. Other employers include state and local governments, publicly funded legal-service centers, banks, real estate development companies, and insurance companies. A small number of paralegals own their own businesses and work as freelance legal assistants, contracting their services to attorneys or corporate legal departments.

Training, Other Qualifications, and Advancement

There are several ways to become a paralegal. Employers usually require formal paralegal training obtained through associate or bachelor's degree programs or through a certification program. Increasingly, employers prefer graduates of 4-year paralegal programs or college graduates who have completed paralegal certificate programs. Some employers prefer to train paralegals on the job, hiring college graduates with no legal experience or promoting experienced legal secretaries. Other entrants have experience in a technical field that is useful to law firms, such as a background in tax preparation for tax and estate practice, or nursing or health administration for personal injury practice.

More than 800 formal paralegal training programs are offered by 4-year colleges and universities, law schools, community and junior colleges, business schools, and proprietary schools. There are currently 247 programs approved by the American Bar Association (ABA). Although this approval is neither required nor sought by many programs, graduation from an ABA-approved program can enhance one's employment opportunities. The requirements for admission to these programs vary. Some require certain college courses or a bachelor's degree; others accept high school graduates or those with legal experience; and a few schools require standardized tests and personal interviews.

Paralegal programs include 2-year associate degree programs, 4-year bachelor's degree programs, and certificate programs that take only a few months to complete. Many certificate programs only require a high school diploma or GED for admission, but they usually are designed for students who already hold an associate or baccalaureate degree. Programs typically include courses on law and legal research techniques, in addition to courses covering specialized areas of law, such as real estate, estate planning and probate, litigation, family law, contracts, and criminal law. Many employers prefer applicants with specialized training.

The quality of paralegal training programs varies; the better programs usually include job placement. Programs increasingly include courses introducing students to the legal applications of computers. Many paralegal training programs include an internship in which students gain practical experience by working for several months in a private law firm, office of a public defender or attorney general, bank, corporate legal department, legal-aid organization, or government agency. Experience gained in internships is an asset when seeking a job after graduation. Prospective students should examine the experiences of recent graduates before enrolling in those programs.

Although most employers do not require certification, earning a voluntary certificate from a professional society may offer advantages in the labor market. The National Association of Legal Assistants, for example, has established standards for certification requiring various combinations of education and experience. Paralegals who meet these standards are eligible to take a 2-day examination, given three times each year at several regional testing centers. Those who pass this examination may use the designation Certified Legal Assistant (CLA). In addition, the Paralegal Advanced Competency Exam, established in 1996 and administered through the National Federation of Paralegal Associations, offers professional recognition to paralegals with a bachelor's degree and at least 2 years of experience. Those who pass this examination may use the designation Registered Paralegal (RP).

Paralegals must be able to document and present their findings and opinions to their supervising attorney. They need to understand legal terminology and have good research and investigative skills. Familiarity with the operation and applications of computers in legal research and litigation support also is increasingly important. Paralegals should stay informed of new developments in the laws that affect their area of practice. Participation in continuing legal education seminars allows paralegals to maintain and expand their legal knowledge.

Because paralegals frequently deal with the public, they should be courteous and uphold the ethical standards of the legal profession. The National Association of Legal Assistants, the National Federation of Paralegal Associations, and a few states have established ethical guidelines for paralegals to follow.

Paralegals usually are given more responsibilities and less supervision as they gain work experience. Experienced paralegals who work in large law firms, corporate legal departments, and government agencies may supervise and delegate assignments to other paralegals and clerical staff. Advancement opportunities also include promotion to managerial and other law-related positions within the firm or corporate legal department. However, some paralegals find it easier to move to another law firm when seeking increased responsibility or advancement.

Job Outlook

Paralegals and legal assistants are projected to rank among the fastest growing occupations in the economy through 2010. Employment growth stems from law firms and other employers with legal staffs increasingly hiring paralegals to lower the cost and increase the availability and efficiency of legal services. The majority of job openings for paralegals in the future will be new jobs created by rapid employment growth, but additional job openings will arise as people leave the occupation. Despite projections of fast employment growth, stiff competition for jobs should continue as the number of graduates of paralegal training programs and others seeking to enter the profession outpaces job growth.

Private law firms will continue to be the largest employers of paralegals, but a growing array of other organizations, such as corporate legal departments, insurance companies, real estate and title insurance firms, and banks will also continue to hire paralegals. Demand for paralegals is expected to grow as an increasing population requires additional legal services, especially in areas such as intellectual property, healthcare, international, elder, sexual harassment, and environmental law. The growth of prepaid legal plans also should contribute to the demand for legal services. Paralegal employment is expected to increase as organizations presently employing paralegals assign them a growing range of tasks, and as paralegals are increasingly employed in small and medium-sized establishments. A growing number of experienced paralegals are expected to establish their own businesses.

Job opportunities for paralegals will expand in the public sector as well. Community legal-service programs, which provide assistance to the poor, aged, minorities, and middle-income families, will employ additional paralegals to minimize expenses and serve the most people. Federal, state, and local government agencies, consumer organizations, and the courts also should continue to hire paralegals in increasing numbers.

To a limited extent, paralegal jobs are affected by the business cycle. During recessions, demand declines for some discretionary legal services, such as planning estates, drafting wills, and handling real estate transactions. Corporations are less inclined to initiate litigation when falling sales and profits lead to fiscal belt tightening. As a result, full-time paralegals employed in offices adversely affected by a recession may be laid off or have their work hours reduced. On the other hand, during recessions, corporations and individuals are more likely to face other problems that require legal assistance, such as bankruptcies, foreclosures, and divorces. Paralegals, who provide many of the same legal services as lawyers at a lower cost, tend to fare relatively better in difficult economic conditions.

Earnings

Earnings of paralegals and legal assistants vary greatly. Salaries depend on education, training, experience, type and size of employer, and geographic location of the job. In general, paralegals who work for large law firms or in large metropolitan areas earn more than those who work for smaller firms or in less populated regions. In 2000, full-time, wage and salary paralegals and legal assistants had median annual earnings of $35,360. The middle 50 percent earned between $28,700 and $45,010. The top 10 percent earned more than $56,060, while the bottom 10 percent earned less than $23,350. Median annual earnings in the industries employing the largest numbers of paralegals in 2000 were as follows:

Federal government	$48,560
Legal services	34,230
Local government	34,120
State government	32,680

According to the National Association of Legal Assistants, paralegals had an average salary of $38,000 in 2000. In addition to a salary, many paralegals received a bonus, which averaged about $2,400. According to the National Federation of Paralegal Associations, starting salaries of paralegals with 1 year or less experience averaged $38,100 in 1999.

Related Occupations

Several other occupations call for a specialized understanding of the law and the legal system, but do not require the extensive training of a lawyer. These include law clerks; title examiners, abstractors, and searchers; claims adjusters, appraisers, examiners, and investigators; and occupational health and safety specialists and technicians.

Sources of Additional Information

General information on a career as a paralegal can be obtained from:

- Standing Committee on Legal Assistants, American Bar Association, 541 North Fairbanks Court, Chicago, IL 60611. Internet: http://www.abanet.org

For information on the Certified Legal Assistant exam, schools that offer training programs in a specific state, and standards and guidelines for paralegals, contact:

- National Association of Legal Assistants, Inc., 1516 South Boston St., Suite 200, Tulsa, OK 74119. Internet: http://www.nala.org

Information on a career as a paralegal, schools that offer training programs, job postings for paralegals, the Paralegal Advanced

Competency Exam, and local paralegal associations can be obtained from:

- National Federation of Paralegal Associations, P.O. Box 33108, Kansas City, MO 64114. Internet: http://www.paralegals.org

Information on paralegal training programs, including the pamphlet "How to Choose a Paralegal Education Program," may be obtained from:

- American Association for Paralegal Education, 2965 Flowers Road South, Atlanta, GA 30341. Internet: http://www.aafpe.org

Information on obtaining a position as a paralegal specialist with the federal government is available from the Office of Personnel Management (OPM) through a telephone-based system. Consult your telephone directory under U.S. Government for a local number or call (912) 757-3000; Federal Relay Service: (800) 877-8339. The first number is not toll free, and charges may result. Information also is available from the OPM Internet site: http://www.usajobs.opm.gov

Petroleum Engineers

O*NET 17-2171.00

Significant Points

- Overall job opportunities in engineering are expected to be good.
- A bachelor's degree is required for most entry-level jobs.
- Starting salaries are significantly higher than those of college graduates in other fields.
- Continuing education is critical to keep abreast of the latest technology.

Nature of the Work

Petroleum engineers search the world for reservoirs containing oil or natural gas. Once these are discovered, petroleum engineers work with geologists and other specialists to understand the geologic formation and properties of the rock containing the reservoir, determine the drilling methods to be used, and monitor drilling and production operations. They design equipment and processes to achieve the maximum profitable recovery of oil and gas. Petroleum engineers rely heavily on computer models to simulate reservoir performance using different recovery techniques. They also use computer models for simulations of the effects of various drilling options.

Because only a small proportion of oil and gas in a reservoir will flow out under natural forces, petroleum engineers develop and use various enhanced recovery methods. These include injecting water, chemicals, gases, or steam into an oil reservoir to force out more of the oil, and computer-controlled drilling or fracturing to connect a larger area of a reservoir to a single well. Because even the best techniques in use today recover only a portion of the oil and gas in a reservoir, petroleum engineers research and develop technology and methods to increase recovery and lower the cost of drilling and production operations.

Working Conditions

Most engineers work in office buildings, laboratories, or industrial plants. Others may spend time outdoors at construction sites, mines, and oil and gas exploration and production sites, where they monitor or direct operations or solve onsite problems. Some engineers travel extensively to plants or work sites.

Many engineers work a standard 40-hour week. At times, deadlines or design standards may bring extra pressure to a job. When this happens, engineers may work longer hours and experience considerable stress.

Employment

Petroleum engineers held about 9,000 jobs in 2000, mostly in the oil and gas extraction, petroleum refining, and engineering and architectural services. Employers include major oil companies and hundreds of smaller, independent oil exploration, production, and service companies. Engineering consulting firms and government agencies also employ many petroleum engineers.

Most petroleum engineers work where oil and gas are found. Large numbers are employed in Texas, Louisiana, Oklahoma, and California, including offshore sites. Many American petroleum engineers also work overseas in oil-producing countries.

Training, Other Qualifications, and Advancement

A bachelor's degree in engineering is required for almost all entry-level engineering jobs. College graduates with a degree in a physical science or mathematics occasionally may qualify for some engineering jobs, especially in specialties in high demand. Most engineering degrees are granted in electrical, electronics, mechanical, or civil engineering. However, engineers trained in one branch may work in related branches. This flexibility allows employers to meet staffing needs in new technologies and specialties in which engineers are in short supply. It also allows engineers to shift to fields with better employment prospects or to those that more closely match their interests.

Most engineering programs involve a concentration of study in an engineering specialty, along with courses in both mathematics and science. Most programs include a design course, sometimes accompanied by a computer or laboratory class or both.

In addition to the standard engineering degree, many colleges offer 2- or 4-year degree programs in engineering technology. These programs, which usually include various hands-on laboratory classes that focus on current issues, prepare students for practical design and production work, rather than for jobs which require more theoretical and scientific knowledge. Graduates of 4-year technology programs may get jobs similar to those obtained by graduates with a bachelor's degree in engineering. Engineering technology graduates, however, are not qualified to register as professional engineers under the same terms as graduates with degrees in engineering. Some employers regard technology program graduates as having skills between those of a technician and an engineer.

Graduate training is essential for engineering faculty positions and many research and development programs, but is not required for the majority of entry-level engineering jobs. Many engineers obtain graduate degrees in engineering or business administration to learn new technology and broaden their education. Many high-level executives in government and industry began their careers as engineers.

About 330 colleges and universities offer bachelor's degree programs in engineering that are accredited by the Accreditation Board for Engineering and Technology (ABET), and about 250 colleges offer accredited bachelor's degree programs in engineering technology. ABET accreditation is based on an examination of an engineering program's student achievement, program improvement, faculty, curricular content, facilities, and institutional commitment. Although most institutions offer programs in the major branches of engineering, only a few offer programs in the smaller specialties. Also, programs of the same title may vary in content. For example, some programs emphasize industrial practices, preparing students for a job in industry, whereas others are more theoretical and are designed to prepare students for graduate work. Therefore, students should investigate curricula and check accreditations carefully before selecting a college.

Admissions requirements for undergraduate engineering schools include a solid background in mathematics (algebra, geometry, trigonometry, and calculus) and sciences (biology, chemistry, and physics), and courses in English, social studies, humanities, and computers. Bachelor's degree programs in engineering typically are designed to last 4 years, but many students find that it takes between 4 and 5 years to complete their studies. In a typical 4-year college curriculum, the first 2 years are spent studying mathematics, basic sciences, introductory engineering, humanities, and social sciences. In the last 2 years, most courses are in engineering, usually with a concentration in one branch. Some programs offer a general engineering curriculum; students then specialize in graduate school or on the job.

Some engineering schools and 2-year colleges have agreements whereby the 2-year college provides the initial engineering education, and the engineering school automatically admits students for their last 2 years. In addition, a few engineering schools have arrangements whereby a student spends 3 years in a liberal arts college studying pre-engineering subjects and 2 years in an engineering school studying core subjects, and then receives a bachelor's degree from each school. Some colleges and universities offer 5-year master's degree programs. Some 5- or even 6-year cooperative plans combine classroom study and practical work, permitting students to gain valuable experience and finance part of their education. All 50 states and the District of Columbia usually require licensure for engineers who offer their services directly to the public. Engineers who are licensed are called Professional Engineers (PE). This licensure generally requires a degree from an ABET-accredited engineering program, 4 years of relevant work experience, and successful completion of a state examination. Recent graduates can start the licensing process by taking the examination in two stages. The initial Fundamentals of Engineering (FE) examination can be taken upon graduation. Engineers who pass this examination commonly are called Engineers in Training (EIT) or Engineer Interns (EI). The EIT certification usually is valid for 10 years. After acquiring suitable work experience, EITs can take the second examination, the Principles and Practice of Engineering Exam. Several states have imposed mandatory continuing education requirements for relicensure. Most states recognize licensure from other states.

Engineers should be creative, inquisitive, analytical, and detail oriented. They should be able to work as part of a team and to communicate well, both orally and in writing. Communication abilities are becoming more important because much of their work is becoming more diversified, meaning that engineers interact with specialists in a wide range of fields outside engineering.

Beginning engineering graduates usually work under the supervision of experienced engineers and, in large companies, also may receive formal classroom or seminar-type training. As new engineers gain knowledge and experience, they are assigned more difficult projects with greater independence to develop designs, solve problems, and make decisions. Engineers may advance to become technical specialists or to supervise a staff or team of engineers and technicians. Some may eventually become engineering managers or enter other managerial or sales jobs.

Job Outlook

Employment of petroleum engineers is expected to decline through 2010 because most of the potential petroleum-producing areas in the United States already have been explored. Even so, favorable opportunities are expected for petroleum engineers because the number of job openings is likely to exceed the relatively small number of graduates. All job openings should result from the need to replace petroleum engineers who transfer to other occupations or leave the labor force.

Also, petroleum engineers work around the world, and many foreign employers seek U.S.-trained petroleum engineers. In fact, the best employment opportunities may be in other countries.

Earnings

Median annual earnings of petroleum engineers were $78,910 in 2000. The middle 50 percent earned between $60,610 and $100,210. The lowest 10 percent earned less than $48,120, and the highest 10 percent earned more than $118,630.

According to a 2001 salary survey by the National Association of Colleges and Employers, bachelor's degree candidates in petroleum engineering received starting offers averaging $53,878 year and master's degree candidates, on average, were offered $58,500.

Related Occupations

Petroleum engineers apply the principles of physical science and mathematics in their work. Other workers who use scientific and mathematical principles include architects, except landscape and naval; engineering and natural sciences managers; computer and information systems managers; mathematicians; drafters; engineering technicians; sales engineers; science technicians; and physical and life scientists, including agricultural and food scientists, biological and medical scientists, conservation scientists and foresters, atmospheric scientists, chemists and materials scientists, environmental scientists and geoscientists, and physicists and astronomers.

Sources of Additional Information

For further information about petroleum engineers, contact:

- Society of Petroleum Engineers, P.O. Box 833836, Richardson, TX 75083-3836. Internet: http://www.spe.org

High school students interested in obtaining a full package of guidance materials and information (product number SP-01) on a variety of engineering disciplines should contact the Junior Engineering Technical Society by sending $3.50 to:

- JETS-Guidance, 1420 King St., Suite 405, Alexandria, VA 22314-2794. Internet: http://www.jets.org

High school students interested in obtaining information on ABET-accredited engineering programs should contact:

- The Accreditation Board for Engineering and Technology, Inc., 111 Market Place, Suite 1050, Baltimore, MD 21202-4012. Internet: http://www.abet.org

Non-licensed engineers and college students interested in obtaining information on Professional Engineer licensure should contact:

- The National Society of Professional Engineers, 1420 King St., Alexandria, VA 22314-2794. Internet: http://www.nspe.org
- National Council of Examiners for Engineers and Surveying, P.O. Box 1686, Clemson, SC 29633-1686. Internet: http://www.ncees.org

Information on general engineering education and career resources is available from:

- American Society for Engineering Education, 1818 N St. N.W., Suite 600, Washington, DC 20036-2479. Internet: http://www.asee.org

Information on obtaining an engineering position with the federal government is available from the Office of Personnel Management (OPM) through a telephone-based system. Consult your telephone directory under U.S. Government for a local number or call (912) 757-3000; Federal Relay Service: (800) 877-8339. The first number is not toll free, and charges may result. Information also is available from the OPM Internet site: http://www.usajobs.opm.gov

Non-high school students and those wanting more detailed information should contact societies representing the individual branches of engineering. Each can provide information about careers in the particular branch. The individual statements elsewhere in this book also provide other information in detail on aerospace; agricultural; biomedical; chemical; civil; computer hardware; electrical and electronics, except computer; environmental; industrial, including health and safety; materials; mechanical; mining and geological, including mining safety; and nuclear engineering.

Pharmacists

O*NET 29-1051.00

Significant Points

- Pharmacists are becoming more involved in drug therapy decision-making and patient counseling.
- A license is required; one must serve an internship under a licensed pharmacist, graduate from an accredited college of pharmacy, and pass a state examination.
- Very good employment opportunities are expected.
- Earnings are very high, but some pharmacists work long hours, nights, weekends, and holidays.

Nature of the Work

Pharmacists dispense drugs prescribed by physicians and other health practitioners and provide information to patients about medications and their use. They advise physicians and other health practitioners on the selection, dosages, interactions, and side effects of medications. Pharmacists must understand the use; clinical effects; and composition of drugs, including their chemical, biological, and physical properties. Compounding—the actual mixing of ingredients to form powders, tablets, capsules, ointments, and solutions—is only a small part of a pharmacist's practice, because most medicines are produced by pharmaceutical companies in a standard dosage and drug delivery form. Most pharmacists work either in a community setting, such as a retail drug store, or in a hospital or clinic.

Pharmacists in community or retail pharmacies counsel patients and answer questions about prescription drugs, such as those about possible adverse reactions or interactions. They provide information about over-the-counter drugs and make recommendations after asking a series of health questions, such as whether the customer is taking any other medications. They also give advice about durable medical equipment and home healthcare supplies. They also may complete third-party insurance forms and other paperwork. Those who own or manage community pharmacies may sell nonhealth-related merchandise, hire and supervise personnel, and oversee the general operation of the pharmacy. Some community pharmacists provide specialized services to help patients manage conditions such as diabetes, asthma, smoking cessation, or high blood pressure.

Pharmacists in hospitals and clinics dispense medications and advise the medical staff on the selection and effects of drugs. They may make sterile solutions and buy medical supplies. They also assess, plan, and monitor drug programs or regimens. They counsel patients on the use of drugs while in the hospital, and on their use at home when the patients are discharged. Pharmacists also may evaluate drug use patterns and outcomes for patients in hospitals or managed care organizations.

Pharmacists who work in home healthcare monitor drug therapy and prepare infusions—solutions that are injected into patients—and other medications for use in the home.

Most pharmacists keep confidential computerized records of patients' drug therapies to ensure that harmful drug interactions do not occur. They frequently teach pharmacy students serving as interns in preparation for graduation and licensure.

Some pharmacists specialize in specific drug therapy areas, such as intravenous nutrition support, oncology (cancer), nuclear pharmacy (used for chemotherapy), and pharmacotherapy (the treatment of mental disorders with drugs).

Pharmacists are responsible for the accuracy of every prescription that is filled, but they often rely upon pharmacy technicians and pharmacy aides to assist them. Thus, the pharmacist may delegate prescription-filling and administrative tasks and supervise their completion.

Working Conditions

Pharmacists usually work in clean, well lighted, and well-ventilated areas. Many pharmacists spend most of their workday on their feet. When working with sterile or potentially dangerous pharmaceutical products, pharmacists wear gloves and masks and work with other special protective equipment. Many community and hospital pharmacies are open for extended hours or around the clock, so pharmacists may work evenings, nights, weekends, and holidays. Consultant pharmacists may travel to nursing homes or other facilities to monitor patient's drug therapy.

About 1 out of 7 pharmacists worked part-time in 2000. Most full-time salaried pharmacists worked about 40 hours a week. Some, including many self-employed pharmacists, worked more than 50 hours a week.

Employment

Pharmacists held about 217,000 jobs in 2000. About 6 out of 10 worked in community pharmacies, either independently owned or part of a drug store chain, grocery store, department store, or mass merchandiser. Most community pharmacists were salaried employees, but some were self-employed owners. About 21 percent of salaried pharmacists worked in hospitals, and others worked in clinics, mail-order pharmacies, pharmaceutical wholesalers, home healthcare agencies, or the federal government.

Training, Other Qualifications, and Advancement

A license to practice pharmacy is required in all states, the District of Columbia, and U.S. territories. To obtain a license, one must serve an internship under a licensed pharmacist, graduate from an accredited college of pharmacy, and pass a state examination. All states, except California and Florida, currently grant a license without extensive re-examination to qualified pharmacists already licensed by another state; one should check with state boards of pharmacy for details. Many pharmacists are licensed to practice in more than one state. States may require continuing education for license renewal.

In 2000, the American Council on Pharmaceutical Education accredited 82 colleges of pharmacy to confer degrees. Pharmacy programs grant the degree of Doctor of Pharmacy (PharmD), which requires at least 6 years of postsecondary study and the passing of the licensure examination of a state board of pharmacy. The PharmD is a 4-year program that requires at least 2 years of college study prior to admittance. This degree has replaced the Bachelor of Science (BS) degree, which will cease to be awarded after 2005.

Colleges of pharmacy require at least 2 years of college-level prepharmacy education. Entry requirements usually include mathematics and natural sciences, such as chemistry, biology, and physics, as well as courses in the humanities and social sciences. Some colleges require the applicant to take the Pharmacy College Admissions Test.

All colleges of pharmacy offer courses in pharmacy practice, designed to teach students to dispense prescriptions and to communicate with patients and other health professionals. Such courses also strengthen students' understanding of professional ethics and allow them to practice management responsibilities. Pharmacists' training increasingly emphasizes direct patient care, as well as consultative services to other health professionals.

In the 2000–2001 academic year, 64 colleges of pharmacy awarded the master of science degree or the PhD degree. Both the master's and PhD degrees are awarded after completion of a PharmD degree. These degrees are designed for those who want more laboratory and research experience. Many master's and PhD holders work in research for a drug company or teach at a university. Other options for pharmacy graduates who are interested in further training include 1- or 2-year residency programs or fellowships. Pharmacy residencies are postgraduate training programs in pharmacy practice. Pharmacy fellowships are highly individualized programs designed to prepare participants to work in research laboratories. Some pharmacists who run their own pharmacy obtain a master's degree in business administration (MBA).

Areas of graduate study include pharmaceutics and pharmaceutical chemistry (physical and chemical properties of drugs and dosage forms), pharmacology (effects of drugs on the body), and pharmacy administration.

Prospective pharmacists should have scientific aptitude, good communication skills, and a desire to help others. They also must be conscientious and pay close attention to detail, because the decisions they make affect human lives.

In community pharmacies, pharmacists usually begin at the staff level. After they gain experience and secure the necessary capital, some become owners or part owners of pharmacies. Pharmacists in chain drug stores may be promoted to pharmacy supervisor or manager at the store level, then to manager at the district or regional level, and later to an executive position within the chain's headquarters.

Hospital pharmacists may advance to supervisory or administrative positions. Pharmacists in the pharmaceutical industry may advance in marketing, sales, research, quality control, production, packaging, or other areas.

Job Outlook

Very good employment opportunities are expected for pharmacists over the 2000–2010 period because the number of degrees granted in pharmacy are not expected to be as numerous as the number of job openings created by employment growth and the need to replace pharmacists who retire or otherwise leave the occupation.

Employment of pharmacists is expected to grow faster than the average for all occupations through the year 2010, due to the increased pharmaceutical needs of a larger and older population and greater use of medication. The growing numbers of middle-aged and elderly people—who, on average, use more prescription drugs than do younger people—will continue to spur demand for pharmacists in all practice settings. Other factors likely to increase the demand for pharmacists include scientific advances that will make more drug products available, new developments in genome research and medication distribution systems, and increasingly sophisticated consumers seeking more information about drugs.

Retail pharmacies are taking steps to increase their prescription volume to make up for declining dispensing fees. Automation of drug dispensing and greater use of pharmacy technicians and pharmacy aides will help them to dispense more prescriptions. The number of community pharmacists needed in the future will depend on the rate of expansion of chain drug stores and the willingness of insurers to reimburse pharmacists for providing clinical services to patients taking prescription medications. With its emphasis on cost control, managed care encourages growth of lower-cost prescription drug distributors, such as mail-order firms, for certain medications. Faster-than-average employment growth is expected in retail pharmacies.

Employment in hospitals is expected to grow about as fast as average, as hospitals reduce inpatient stays, downsize, and consolidate departments. Pharmacy services are shifting to long-term, ambulatory, and home care settings, where opportunities for pharmacists will be best. New opportunities are emerging for pharmacists in managed-care organizations, where they may analyze trends and patterns in medication use for their populations of patients, and for pharmacists trained in research, disease management, and pharmacoeconomics—determining the costs and benefits of different drug therapies.

Cost-conscious insurers and health systems may continue to emphasize the role of pharmacists in primary and preventive health services. They realize that the expense of using medication to treat diseases and conditions often is considerably less than the potential costs for patients whose conditions go untreated. Pharmacists also can reduce the expenses resulting from unexpected complications due to allergic reactions or medication interactions.

Earnings

Median annual earnings of pharmacists in 2000 were $70,950. The middle 50 percent earned between $61,860 and $81,690 a year. The lowest 10 percent earned less than $51,570, and the highest 10 percent, more than $89,010 a year. Median annual earnings in the industries employing the largest numbers of pharmacists in 2000 were as follows:

Department stores	$73,730
Grocery stores	72,440
Drug stores and proprietary stores	72,110
Hospitals	68,760

According to a survey by *Drug Topics* magazine, published by Medical Economics Co., average starting base salaries of full-time, salaried pharmacists were $67,824 in 2000. Pharmacists working in chain drug stores had an average annual base salary of $71,486 while pharmacists working in independent drug stores averaged $62,040 and hospital pharmacists averaged $61,250. Many pharmacists also receive compensation in the form of bonuses, overtime, and profit-sharing.

Related Occupations

Pharmacy technicians and pharmacy aides also work in pharmacies. Persons in other professions who may work with pharmaceutical compounds include biological and medical scientists and chemists and materials scientists.

Sources of Additional Information

For information on pharmacy as a career, preprofessional and professional requirements, programs offered by colleges of pharmacy, and student financial aid, contact:

- American Association of Colleges of Pharmacy, 1426 Prince St., Alexandria, VA 22314. Internet: http://www.aacp.org
- National Association of Boards of Pharmacy, 700 Busse Highway, Park Ridge, IL 60068. Internet: http://www.nabp.net

General information on careers in pharmacy is available from:

- National Association of Chain Drug Stores, 413 N. Lee St., P.O. Box 1417-D49, Alexandria, VA 22313-1480. Internet: http://www.nacds.org

State licensure requirements are available from each state's Board of Pharmacy.

Information on specific college entrance requirements, curriculums, and financial aid is available from any college of pharmacy.

Pharmacy Technicians

O*NET 29-2052.00

Significant Points

- Job opportunities are expected to be good, especially for those with certification or previous work experience.

- Many technicians work evenings, weekends, and some holidays.
- Two-thirds of all jobs are in retail pharmacies.

Nature of the Work

Pharmacy technicians help licensed pharmacists provide medication and other healthcare products to patients. Technicians usually perform routine tasks to help prepare prescribed medication for patients, such as counting tablets and labeling bottles. Technicians refer any questions regarding prescriptions, drug information, or health matters to a pharmacist.

Pharmacy aides work closely with pharmacy technicians. They are often clerks or cashiers who primarily answer telephones, handle money, stock shelves, and perform other clerical duties. Pharmacy technicians usually perform more complex tasks than do pharmacy aides, although, in some states, their duties and job titles overlap.

Pharmacy technicians who work in retail pharmacies have varying responsibilities, depending on state rules and regulations. Technicians receive written prescriptions or requests for prescription refills from patients. They also may receive prescriptions sent electronically from the doctor's office. They must verify that the information on the prescription is complete and accurate. To prepare the prescription, technicians must retrieve, count, pour, weigh, measure, and sometimes mix the medication. Then, they prepare the prescription labels, select the type of prescription container, and affix the prescription and auxiliary labels to the container. Once the prescription is filled, technicians price and file the prescription, which must be checked by a pharmacist before it is given to a patient. Technicians may establish and maintain patient profiles, prepare insurance claim forms, and stock and take inventory of prescription and over-the-counter medications.

In hospitals, technicians have added responsibilities. They read patient charts and prepare and deliver the medicine to patients. The pharmacist must check the order before it is delivered to the patient. The technician then copies the information about the prescribed medication onto the patient's profile. Technicians also may assemble a 24-hour supply of medicine for every patient. They package and label each dose separately. The package is then placed in the medicine cabinet of each patient until the supervising pharmacist checks it for accuracy. It is then given to the patient.

Working Conditions

Pharmacy technicians work in clean, organized, well-lighted, and well-ventilated areas. Most of their workday is spent on their feet. They may be required to lift heavy boxes or to use stepladders to retrieve supplies from high shelves.

Technicians work the same hours as pharmacists. This may include evenings, nights, weekends, and holidays. Because some hospital and retail pharmacies are open 24 hours a day, technicians may work varying shifts. As their seniority increases, technicians often have increased control over the hours they work. There are many opportunities for part-time work in both retail and hospital settings.

Employment

Pharmacy technicians held about 190,000 jobs in 2000. Two-thirds of all jobs were in retail pharmacies, either independently owned or part of a drug store chain, grocery store, department store, or mass retailer. More than 2 out of 10 jobs were in hospitals and a small number were in mail-order and Internet pharmacies, clinics, pharmaceutical wholesalers, and the federal government.

Training, Other Qualifications, and Advancement

Although most pharmacy technicians receive informal on-the-job training, employers favor those who have completed formal training and certification. However, there are currently few state and no federal requirements for formal training or certification of pharmacy technicians. Employers who can neither afford, nor have the time to give, on-the-job training often seek formally educated pharmacy technicians. Formal education programs and certification emphasize the technicians' interest in and dedication to the work to potential employers. In addition to the military, some hospitals, proprietary schools, vocational or technical colleges, and community colleges offer formal education programs.

Formal pharmacy-technician education programs require classroom and laboratory work in a variety of areas, including medical and pharmaceutical terminology, pharmaceutical calculations, pharmacy record keeping, pharmaceutical techniques, and pharmacy law and ethics. Technicians also are required to learn medication names, actions, uses, and doses. Many training programs include internships, in which students gain hands-on experience in actual pharmacies. Students receive a diploma, certificate, or an associate degree, depending on the program.

Prospective pharmacy technicians with experience working as an aide in a community pharmacy or volunteering in a hospital may have an advantage. Employers also prefer applicants with strong customer service and communication skills and with experience managing inventories, counting, measuring, and using computers. Technicians entering the field need strong mathematics, spelling, and reading skills. A background in chemistry, English, and health education also may be beneficial. Some technicians are hired without formal training, but under the condition that they obtain certification within a specified period to retain employment.

The Pharmacy Technician Certification Board administers the National Pharmacy Technician Certification Examination. This exam is voluntary and displays the competency of the individual to act as a pharmacy technician. Eligible candidates must have a high school diploma or GED, and those who pass the exam earn the title of Certified Pharmacy Technician (CPhT). The exam is offered several times per year at various locations nationally. Employers, often Pharmacists, know that individuals who pass the exam have a standardized body of knowledge and skills.

Certified technicians must be recertified every 2 years. Technicians must complete 20 contact hours of pharmacy-related topics within the 2-year certification period to become eligible for recertification. Contact hours are awarded for on-the-job training, attending lectures, and college coursework. At least 1 contact hour must be in pharmacy law. Contact hours can be earned from several different sources, including pharmacy associations, pharmacy colleges, and pharmacy technician training programs. Up to 10 contact hours can be earned when the technician is employed under the direct supervision and instruction of a pharmacist.

Successful pharmacy technicians are alert, observant, organized, dedicated, and responsible. They should be willing and able to take directions. They must enjoy precise work; details are sometimes a matter of life and death. Although a pharmacist must check and approve all their work, they should be able to work on their own without constant instruction from the pharmacist. Candidates interested in becoming pharmacy technicians cannot have prior records of drug or substance abuse.

Strong interpersonal and communication skills are needed because there is a lot of interaction with patients, coworkers, and healthcare professionals. Teamwork is very important because technicians are often required to work with pharmacists, aides, and other technicians.

Job Outlook

Good job opportunities are expected for full-time and part-time work, especially for technicians with formal training or previous experience. Job openings for pharmacy technicians will result from the expansion of retail pharmacies and other employment settings, and from the need to replace workers who transfer to other occupations or leave the labor force.

Employment of pharmacy technicians is expected to grow much faster than the average for all occupations through 2010 due to the increased pharmaceutical needs of a larger and older population, and to the greater use of medication. The increased number of middle-aged and elderly people—who, on average, use more prescription drugs than do younger people—will spur demand for technicians in all practice settings. With advances in science, more medications are becoming available to treat more conditions.

Cost-conscious insurers, pharmacies, and health systems will continue to emphasize the role of technicians. As a result, pharmacy technicians will assume responsibility for more routine tasks previously performed by pharmacists. Pharmacy technicians also will need to learn and master new pharmacy technology as it surfaces. For example, robotic machines are used to dispense medicine into containers; technicians must oversee the machines, stock the bins, and label the containers. Thus, while automation is increasingly incorporated into the job, it will not necessarily reduce the need for technicians.

Almost all states have legislated the maximum number of technicians who can safely work under a pharmacist at a time. In some states, increased demand for technicians has encouraged an expanded ratio of technicians to pharmacists. Changes in these laws could directly affect employment.

Earnings

Median hourly earnings of pharmacy technicians in 2000 were $9.93. The middle 50 percent earned between $8.12 and $12.26; the lowest 10 percent earned less than $7.00, and the highest 10 percent earned more than $14.56. Median hourly earnings in the industries employing the largest numbers of pharmacy technicians in 2000 were as follows:

Hospitals	$11.44
Grocery stores	10.57
Drugs, proprietaries, and sundries	10.09
Drug stores and proprietary stores	9.00
Department stores	8.75

Certified technicians may earn more. Shift differentials for working evenings or weekends also can increase earnings. Some technicians belong to unions representing hospital or grocery store workers.

Related Occupations

This occupation is most closely related to pharmacists and pharmacy aides. Workers in other medical support occupations include dental assistants, licensed practical and licensed vocational nurses, medical transcriptionists, medical records and health information technicians, occupational therapist assistants and aides, physical therapist assistants and aides, secretaries and administrative assistants, and surgical technologists.

Sources of Additional Information

For information on certification and a National Pharmacy Technician Certification Examination Candidate Handbook, contact:

- Pharmacy Technician Certification Board, 2215 Constitution Ave. NW, Washington DC 20037. Internet: http://www.ptcb.org

Photographers

O*NET 27-4021.01, 27-4021.02

Significant Points

- Technical expertise, a "good eye," imagination, and creativity are essential.
- Only the most skilled and talented who have good business sense maintain long-term careers.
- More than half of all photographers are self-employed, a much higher proportion than the average for all occupations.

Nature of the Work

Photographers produce and preserve images that paint a picture, tell a story, or record an event. To create commercial quality photographs, photographers need both technical expertise and creativity. Producing a successful picture requires choosing and presenting a subject to achieve a particular effect and selecting the appropriate equipment. For example, photographers may enhance the subject's appearance with lighting or draw attention to a particular aspect of the subject by blurring the background.

Today, many cameras adjust settings like shutter speed and aperture automatically. They also let the photographer adjust these settings manually, allowing greater creative and technical control over the picture-taking process. In addition to automatic and manual cameras, photographers use an array of film, lenses, and equipment—from filters, tripods, and flash attachments to specially constructed lighting equipment.

Photographers use either a traditional camera or a newer digital camera that electronically records images. A traditional camera records images on silver halide film that is developed into prints. Some photographers send their film to laboratories for processing. Color film requires expensive equipment and exacting conditions for correct processing and printing. Other photographers, especially those who use black-and-white film or require special effects, prefer to develop and print their own photographs. Photographers who do their own film developing must have the technical skill to operate a fully equipped darkroom or the appropriate computer software to process prints digitally.

Recent advances in electronic technology now make it possible for the professional photographer to develop and scan standard 35mm or other types of film, and use flatbed scanners and photofinishing laboratories to produce computer-readable, digital images from film. After converting the film to a digital image, photographers can edit and electronically transmit images, making it easier and faster to shoot, develop, and transmit pictures from remote locations.

Using computers and specialized software, photographers also can manipulate and enhance the scanned or digital image to create a desired effect. Images can be stored on compact disc (CD) the same way as music. Digital technology also allows the production of larger, more colorful, and more accurate prints or images for use in advertising, photographic art, and scientific research. Some photographers use this technology to create electronic portfolios, as well. Because much photography now involves the use of computer technology, photographers must have hands-on knowledge of computer editing software.

Some photographers specialize in areas such as portrait, commercial and industrial, scientific, news, or fine arts photography. *Portrait photographers* take pictures of individuals or groups of people and often work in their own studios. Some specialize in weddings or school photographs and may work on location. Portrait photographers who are business owners arrange for advertising, schedule appointments, set and adjust equipment, develop and retouch negatives, and mount and frame pictures. They also purchase supplies, keep records, bill customers, and may hire and train employees.

Commercial and industrial photographers take pictures of various subjects, such as buildings, models, merchandise, artifacts, and landscapes. These photographs are used in a variety of media, including books, reports, advertisements, and catalogs. Industrial photographers often take pictures of equipment, machinery, products, workers, and company officials. The pictures then are used for analyzing engineering projects, publicity, or as records of equipment development or deployment, such as placement of an offshore rig. This photography frequently is done on location.

Scientific photographers photograph a variety of subjects to illustrate or record scientific or medical data or phenomena, using knowledge of scientific procedures. They typically possess additional knowledge in areas such as engineering, medicine, biology, or chemistry.

News photographers, also called *photojournalists*, photograph newsworthy people, places, sports, political activities, and community events for newspapers, journals, magazines, or television. Some news photographers are salaried staff; others are self-employed and are known as freelance photographers.

Fine arts photographers sell their photographs as fine artwork. In addition to technical proficiency, fine arts photographers need artistic talent and creativity.

Self-employed, or freelance, photographers may license the use of their photographs through stock photo agencies or contract with clients or agencies to provide photographs as necessary. Stock agencies grant magazines and other customers the right to purchase the use of photographs, and, in turn, pay the photographer on a commission basis. Stock photo agencies require an application from the photographer and a sizable portfolio. Once accepted, a large number of new submissions usually are required from the photographer each year.

Working Conditions

Working conditions for photographers vary considerably. Photographers employed in government and advertising agencies usually work a 5-day, 40-hour week. On the other hand, news photographers often work long, irregular hours and must be available to work on short notice. Many photographers work part-time or variable schedules.

Portrait photographers usually work in their own studios but also may travel to take photographs at the client's location, such as a school, a company office, or a private home. News and commercial photographers frequently travel locally, stay overnight on assignments, or travel to distant places for long periods.

Some photographers work in uncomfortable, or even dangerous surroundings, especially news photographers covering accidents, natural disasters, civil unrest, or military conflicts. Many photographers must wait long hours in all kinds of weather for an event to take place and stand or walk for long periods while carrying heavy equipment. News photographers often work under strict deadlines.

Self-employment allows for greater autonomy, freedom of expression, and flexible scheduling. However, income can be uncertain and the continuous, time-consuming search for new clients can be stressful. Some self-employed photographers hire assistants who help seek out new business.

Employment

Photographers held about 131,000 jobs in 2000. More than half were self-employed, a much higher proportion than the average for all occupations. Some self-employed photographers contracted with advertising agencies, magazines, or others to do individual projects at a predetermined fee, while others operated portrait studios or provided photographs to stock photo agencies.

Most salaried photographers worked in portrait or commercial photography studios. Newspapers, magazines, television broadcasters, advertising agencies, and government agencies employed most of the others. Most photographers worked in metropolitan areas.

Training, Other Qualifications, and Advancement

Employers usually seek applicants with a "good eye," imagination, and creativity, as well as a good technical understanding of photography. Entry-level positions in photojournalism, industrial, or scientific photography generally require a college degree in journalism or photography. Freelance and portrait photographers need technical proficiency, whether gained through a degree program, vocational training, or extensive work experience.

Many universities, community and junior colleges, vocational-technical institutes, and private trade and technical schools offer photography courses. Basic courses in photography cover equipment, processes, and techniques. Bachelor's degree programs, especially those including business courses, provide a well-rounded education. Art schools offer useful training in design and composition.

Individuals interested in photography should subscribe to photographic newsletters and magazines, join camera clubs, and seek summer or part-time employment in camera stores, newspapers, or photo studios.

Photographers may start out as assistants to experienced photographers. Assistants learn to mix chemicals, develop film, print photographs, and the other skills necessary to run a portrait or commercial photography business. Freelance photographers also should develop an individual style of photography in order to differentiate themselves from the competition. Some photographers enter the field by submitting unsolicited photographs to magazines and art directors at advertising agencies. For freelance photographers, a good portfolio of their work is critical.

Photographers need good eyesight, artistic ability, and hand-eye coordination. They should be patient, accurate, and detail oriented. Photographers should be able to work well with others, as they frequently deal with clients, graphic designers, or advertising and publishing specialists. Increasingly, photographers need to know computer software programs and applications that allow them to prepare and edit images.

Portrait photographers need the ability to help people relax in front of the camera. Commercial and fine arts photographers must be imaginative and original. News photographers not only must be good with a camera, but also must understand the story behind an event so their pictures match the story. They must be decisive in recognizing a potentially good photograph and act quickly to capture it.

Photographers who operate their own businesses, or freelance, need business skills as well as talent. These individuals must know how to prepare a business plan; submit bids; write contracts; hire models, if needed; get permission to shoot on locations that normally are not open to the public; obtain releases to use photographs of people; license and price photographs; secure copyright protection for their work; and keep financial records.

After several years of experience, magazine and news photographers may advance to photography or picture editor positions. Some photographers teach at technical schools, film schools, or universities.

Job Outlook

Photographers can expect keen competition for job openings because the work is attractive to many people. The number of individuals interested in positions as commercial and news photographers usually is much greater than the number of openings. Those who succeed in landing a salaried job or attracting enough work to earn a living by freelancing are likely to be the most creative, able to adapt to rapidly changing technologies, and adept at operating a business. Related work experience, job-related training, or some unique skill or talent—such as a background in computers or electronics—also are beneficial to prospective photographers.

Employment of photographers is expected to increase about as fast as the average for all occupations through 2010. Demand for portrait photographers should increase as the population grows. And, as the number of electronic versions of magazines, journals, and newspapers grows on the Internet, photographers will be needed to provide digital images.

Employment growth of photographers will be constrained somewhat by the widespread use of digital photography. Besides increasing photographers' productivity, improvements in digital technology will allow individual consumers and businesses to produce, store, and access photographic images on their own. Declines in the newspaper industry will reduce demand for photographers to provide still images for print.

Earnings

Median annual earnings of salaried photographers were $22,300 in 2000. The middle 50 percent earned between $16,790 and $33,020. The lowest 10 percent earned less than $13,760, and the highest 10 percent earned more than $46,890.

Salaried photographers (more of whom work full time) tend to earn more than those who are self-employed. Because most freelance and portrait photographers purchase their own equipment, they incur considerable expense acquiring and maintaining cameras and accessories. Unlike news and commercial photographers, few fine arts photographers are successful enough to support themselves solely through their art.

Related Occupations

Other occupations requiring artistic talent include architects, except landscape and naval; artists and related workers; designers; and television, video, and motion picture camera operators and editors.

Sources of Additional Information

Career information on photography is available from:

- Professional Photographers of America, Inc., 229 Peachtree St. N.E., Suite 2200, Atlanta, GA 30303
- National Press Photographers Association, Inc., 3200 Croasdaile Dr., Suite 306, Durham, NC 27705. Internet: http://www.nppa.org/default.cfm

Photographic Process Workers and Processing Machine Operators

O*NET 51-9131.01, 51-9131.02, 51-9131.03, 51-9131.04, 51-9132.00

Significant Points

- Employment is expected to show little or no change as digital photography becomes commonplace.
- Most receive on-the-job training from their companies, manufacturers' representatives, and experienced workers.

Nature of the Work

Both amateur and professional photographers rely heavily on photographic process workers and processing machine operators to develop film, make prints or slides, and do related tasks, such as enlarging or retouching photographs. *Photographic processing machine operators* operate various machines, such as mounting presses and motion picture film printing, photographic printing, and film developing machines. *Photographic process workers* perform more delicate tasks, such as retouching photographic negatives and prints to emphasize or correct specific features.

Photographic processing machine operators often have specialized jobs. *Film process technicians* operate machines that develop exposed photographic film or sensitized paper in a series of chemical and water baths to produce negative or positive images. First, technicians mix developing and fixing solutions, following a formula. They then load the film in the machine, which immerses the exposed film in a developer solution. This brings out the latent image. The next steps include immersing the negative in a stop-bath to halt the developer action, transferring it to a hyposolution to fix the image, and then immersing it in water to remove the chemicals. The technician then dries the film. In some cases, these steps are performed by hand.

Color printer operators control equipment that produces color prints from negatives. These workers read customer instructions to determine processing requirements. They load film into color printing equipment, examine negatives to determine equipment control settings, set controls, and produce a specified number of prints. Finally, they inspect the finished prints for defects, remove any that are found, and insert the processed negatives and prints into an envelope for return to the customer.

Photographic process workers, sometimes known as *digital imaging technicians*, use computer images of conventional negatives and specialized computer software to vary the contrast of images, remove unwanted background, or combine features from different photographs. Although computers and digital technology are replacing much manual work, some photographic process workers, especially those who work in portrait studios, still perform many specialized tasks by hand directly on the photo or negative. *Airbrush artists* restore damaged and faded photographs, and may color or shade drawings to create photographic likenesses using an airbrush. *Photographic retouchers* alter photographic negatives, prints, or images to accentuate the subject. *Colorists* apply oil colors to portrait photographs to create natural, lifelike appearances. *Photographic spotters* remove imperfections on photographic prints and images.

Working Conditions

Photographic process workers and processing machine operators generally spend their work hours in clean, appropriately lighted, well-ventilated, and air-conditioned offices, photofinishing laboratories, or 1-hour minilabs. In recent years, more commercial photographic processing has been done on computers than in darkrooms; and this trend is expected to continue.

Some photographic process workers and processing machine operators are exposed to the chemicals and fumes associated with developing and printing. These workers must wear rubber gloves and aprons and take precautions against these hazards. Those who use computers for extended periods may experience back pain, eye strain, or fatigue.

Photographic processing machine operators must do repetitive work at a rapid pace without any loss of accuracy. Photographic process workers do detailed tasks, such as airbrushing and spotting, which can contribute to eye fatigue.

Many photo laboratory employees work a 40-hour week, including evenings and weekends, and may work overtime during peak seasons.

Employment

Photographic process workers held about 26,000 jobs in 2000. Nearly one-third of photographic process workers were employed in photofinishing laboratories and 1-hour minilabs. About one-fourth worked for portrait studios or commercial laboratories that specialize in processing the work of professional photographers for advertising and other industries.

Photographic processing machine operators held about 50,000 jobs in 2000. One-half worked in retail establishments, such as department stores and drug stores. About one-quarter worked in photofinishing laboratories and 1-hour minilabs. A small number were self-employed.

Employment fluctuates somewhat over the course of the year. Typically, employment peaks during school graduation and summer vacation periods, and again during the winter holiday season.

Training, Other Qualifications, and Advancement

Most photographic process workers and processing machine operators receive on-the-job training from their companies, manufacturers' representatives, and experienced workers. New employees gradually learn to use the machines and chemicals that develop and print film.

Employers prefer applicants who are high school graduates or those who have some experience in the field. Familiarity with computers is essential for photographic processing machine operators. The ability to perform simple mathematical calculations also is helpful. Photography courses that include instruction in film processing are valuable preparation. Such courses are available through high schools, vocational-technical institutes, private trade schools, and colleges and universities.

On-the-job training in photographic processing occupations can range from just a few hours for print machine operators to several months for photographic processing workers like airbrush artists and colorists. Some workers attend periodic training seminars to maintain a high level of skill. Manual dexterity, good hand-eye coordination, and good vision, including normal color perception, are important qualifications for photographic process workers.

Photographic process machine workers can sometimes advance from jobs as machine operators to supervisory positions in laboratories or to management positions within retail stores.

Job Outlook

Overall employment of photographic process workers and processing machine operators is expected to show little or no change through the year 2010. Employment of processing machine operators will grow slower than average, while employment of photographic process workers will decline. Most openings will result from replacement needs, which are higher for machine operators than for photographic process workers.

In recent years, the use of digital cameras, which use electronic memory rather than film to record images, has grown rapidly among professional photographers and advanced amateurs. As the cost of digital photography drops, the use of such cameras will become more widespread among amateur photographers, reducing the demand for traditional photographic processing machine operators. However, conventional cameras, which use film to record images, are expected to continue to be the camera of choice among most casual photographers. Population growth and the popularity of amateur and family photography will contribute to an ongoing need for photographic processing machine operators to process the film used in conventional cameras, including increasingly sophisticated disposable cameras. This need will prevent what otherwise would be even slower growth in the numbers of these workers.

Employment of photographic process workers is expected to decline, as digital cameras and imaging become more commonplace. Using digital cameras and technology, consumers who have a personal computer and the proper software will be able to download and view pictures on their computer, as well as manipulate, correct, and retouch their own photographs. No matter what improvements occur in camera technology though, there will be some photographic processing tasks that require skillful manual treatment.

Earnings

Earnings of photographic process workers vary greatly depending on skill level, experience, and geographic location. Median hourly earnings for photographic process workers were $9.44 in 2000. The middle 50 percent earned between $7.56 and $12.54. The lowest 10 percent earned less than $6.44, and the highest 10 percent earned more than $16.61. Median hourly earnings were $9.55 in miscellaneous business services, including photofinishing laboratories.

Median hourly earning for photographic processing machine operators were $8.39 in 2000. The middle 50 percent earned between $7.06 and $10.56. The lowest 10 percent earned less than $6.06, and the highest 10 percent earned more than $14.48. Median hourly earnings in the industries employing the largest numbers of photographic processing machine operators were as follows:

Miscellaneous business services	$9.03
Department stores	7.97
Drug stores and proprietary stores	6.89

Related Occupations

Photographic process workers and processing machine operators need specialized knowledge of the photodeveloping process. Other workers who apply specialized technical knowledge include clinical laboratory technologists and technicians, computer operators, jewelers and precious stone and metal workers, prepress technicians and workers; printing machine operators; and science technicians.

Sources of Additional Information

For information about employment opportunities in photographic laboratories and schools that offer degrees in photographic technology, contact:

- Photo Marketing Association International, 3000 Picture Place, Jackson, MI 49201

Physical Therapist Assistants and Aides

O*NET 31-2021.00, 31-2022.00

Significant Points

- Employment is projected to increase much faster than the average, as rapid growth in the number of middle-aged and elderly individuals increases the demand for therapeutic services.
- Licensed physical therapist assistants have an associate degree, but physical therapist aides usually learn skills on the job.
- More than two-thirds of jobs for physical therapist assistants and aides were in hospitals or offices of physical therapists.

Nature of the Work

Physical therapist assistants and aides perform components of physical therapy procedures and related tasks selected by a supervising physical therapist. These workers assist physical therapists in providing services that help improve mobility, relieve pain, and prevent or limit permanent physical disabilities of patients suffering from injuries or disease. Patients include accident victims and individuals with disabling conditions, such as low back pain, arthritis, heart disease, fractures, head injuries, and cerebral palsy.

Physical therapist assistants perform a variety of tasks. Components of treatment procedures performed by these workers, under the direction and supervision of physical therapists, involve exercises, massages, electrical stimulation, paraffin baths, hot and cold packs, traction, and ultrasound. Physical therapist assistants record the patient's responses to treatment and report to the physical therapist the outcome of each treatment.

Physical therapist aides help make therapy sessions productive, under the direct supervision of a physical therapist or physical therapist assistant. They usually are responsible for keeping the treatment area clean and organized and preparing for each patient's therapy. When patients need assistance moving to or from a treatment area, aides push them in a wheelchair, or provide them with a shoulder to lean on. Because they are not licensed, aides do not perform the clinical tasks of a physical therapist assistant.

The duties of aides include some clerical tasks, such as ordering depleted supplies, answering the phone, and filling out insurance forms and other paperwork. The extent to which an aide or an assistant performs clerical tasks depends on the size and location of the facility.

Working Conditions

The hours and days that physical therapist assistants and aides work vary, depending on the facility and on whether they are full or part-time employees. Many outpatient physical therapy offices and clinics have evening and weekend hours, to help coincide with patients' personal schedules.

Physical therapist assistants and aides need to have a moderate degree of strength, due to the physical exertion required in assisting patients with their treatment. For example, in some cases, assistants and aides need to help lift patients. Additionally, constant kneeling, stooping, and standing for long periods are all part of the job.

Employment

Physical therapist assistants and aides held 80,000 jobs in 2000. Physical therapist assistants held about 44,000 jobs; and physical therapist aides held about 36,000. They work alongside physical therapists in a variety of settings. More than two-thirds of jobs for assistants and aides were in hospitals or offices of physical therapists. Others work in nursing and personal care facilities, outpatient rehabilitation centers, offices and clinics of physicians, and home health agencies.

Training, Other Qualifications, and Advancement

Physical therapist aides are trained on the job, but physical therapist assistants typically earn an associate degree from an accredited physical therapist assistant program. Licensure or registration is not required in all states for the physical therapist assistant to practice. The states that require licensure stipulate specific educational and examination criteria. Complete information on practice acts and regulations can be obtained from the state licensing boards. Additional requirements may include certification in CPR and other first aid and a minimum number of hours of clinical experience.

According to the American Physical Therapy Association, there were 268 accredited physical therapist assistant programs in the United States as of 2001. Accredited physical therapist assistant programs are designed to last 2 years, or 4 semesters, and culminate in an associate degree. Programs are divided into academic study and hands on clinical experience. Academic coursework includes algebra, anatomy and physiology, biology, chemistry, and psychology. Before students begin their clinical field experience, many programs require that they complete a semester of anatomy and physiology and have certifications in CPR and other first aid. Both educators and prospective employers view clinical experience as an integral part of ensuring that students understand the responsibilities of a physical therapist assistant.

Employers typically require physical therapist aides to have a high school diploma, strong interpersonal skills, and a desire to assist people in need. Most employers provide clinical on-the-job training.

Job Outlook

Employment of physical therapist assistants and aides is expected to grow much faster than the average through the year 2010. Federal legislation imposing limits on reimbursement for therapy services may adversely affect the job market for physical therapist assistants and aides in the near term. However, over the long run, demand for physical therapist assistants and aides will continue to rise, with growth in the number of individuals with disabilities or limited function. The rapidly growing elderly population is particularly vulnerable to chronic and debilitating conditions that require therapeutic services. These patients often need additional assistance in their treatment, making the roles of assistants and aides vital. The large baby-boom generation is entering the prime age for heart attacks and strokes, further increasing the demand for cardiac and physical rehabilitation. Additionally, future medical developments should permit an increased percentage of trauma victims to survive, creating added demand for therapy services.

Licensed physical therapist assistants can enhance the cost-effective provision of physical therapy services. Once a patient is evaluated, and a treatment plan is designed by the physical therapist, the physical therapist assistant can provide many aspects of treatment, as prescribed by the therapist.

Earnings

Median annual earnings of physical therapist assistants were $33,870 in 2000. The middle 50 percent earned between $28,830 and $40,440. The lowest 10 percent earned less than $23,150, and the highest 10 percent earned more than $45,610. Median annual earnings of physical therapist assistants in 2000 were $33,660 in offices of other healthcare practitioners and $33,820 in hospitals.

Median annual earnings of physical therapist aides were $19,670 in 2000. The middle 50 percent earned between $16,460 and $23,390. The lowest 10 percent earned less than $14,590, and the highest 10 percent earned more than $28,800. Median annual earnings of physical therapist aides in 2000 were $18,320 in offices of other healthcare practitioners and $19,840 in hospitals.

Related Occupations

Physical therapist assistants and aides work under the supervision of physical therapists. Other occupations in the healthcare field that work under the supervision of professionals include dental assistants, medical assistants, occupational therapist assistants and aides, and pharmacy technicians.

Sources of Additional Information

Information on a career as a physical therapist assistant and a list of schools offering accredited programs can be obtained from:

- The American Physical Therapy Association, 1111 North Fairfax St., Alexandria, VA 22314-1488. Internet: http://www.apta.org

Physical Therapists

O*NET 29-1123.00

Significant Points

- Employment is expected to increase faster than the average, as rapid growth in the number of middle-aged and elderly individuals increases the demand for therapeutic services.
- After graduating from an accredited physical therapist educational program, therapists must pass a licensure exam before they can practice.

Nature of the Work

Physical therapists (PTs) provide services that help restore function, improve mobility, relieve pain, and prevent or limit permanent physical disabilities of patients suffering from injuries or disease. They restore, maintain, and promote overall fitness and health. Their patients include accident victims and individuals with disabling conditions such as low back pain, arthritis, heart disease, fractures, head injuries, and cerebral palsy.

Therapists examine patients' medical histories, then test and measure their strength, range of motion, balance and coordination, posture, muscle performance, respiration, and motor function. They also determine patients' ability to be independent and reintegrate into the community or workplace after injury or illness. Next, they develop treatment plans describing a treatment strategy, its purpose, and anticipated outcome. Physical therapist assistants, under the direction and supervision of a physical therapist, may be involved in implementing treatment plans with patients. Physical therapist aides perform routine support tasks, as directed by the therapist.

Treatment often includes exercise for patients who have been immobilized and lack flexibility, strength, or endurance. They encourage patients to use their own muscles to further increase flexibility and range of motion before finally advancing to other exercises improving strength, balance, coordination, and endurance. Their goal is to improve how an individual functions at work and home.

Physical therapists also use electrical stimulation, hot packs or cold compresses, and ultrasound to relieve pain and reduce swelling. They may use traction or deep-tissue massage to relieve pain. Therapists also teach patients to use assistive and adaptive devices such as crutches, prostheses, and wheelchairs. They also may show patients exercises to do at home to expedite their recovery.

As treatment continues, physical therapists document progress, conduct periodic examinations, and modify treatments when necessary. Such documentation is used to track the patient's progress, and identify areas requiring more or less attention.

Physical therapists often consult and practice with a variety of other professionals, such as physicians, dentists, nurses, educators, social workers, occupational therapists, speech-language pathologists, and audiologists.

Some physical therapists treat a wide range of ailments; others specialize in areas such as pediatrics, geriatrics, orthopedics, sports medicine, neurology, and cardiopulmonary physical therapy.

Working Conditions

Physical therapists practice in hospitals, clinics, and private offices that have specially equipped facilities, or they treat patients in hospital rooms, homes, or schools.

Most full-time physical therapists work a 40-hour week, which may include some evenings and weekends. The job can be physically demanding because therapists often have to stoop, kneel, crouch, lift, and stand for long periods. In addition, physical therapists move heavy equipment and lift patients or help them turn, stand, or walk.

Employment

Physical therapists held about 132,000 jobs in 2000; about 1 in 4 worked part-time. The number of jobs is greater than the number of practicing physical therapists because some physical therapists hold two or more jobs. For example, some may work in a private practice, but also work part-time in another health facility.

About two-thirds of physical therapists were employed in either hospitals or offices of physical therapists. Other jobs were in home health agencies, outpatient rehabilitation centers, offices and clinics of physicians, and nursing homes. Some physical therapists are self-employed in private practices. They may provide services to individual patients or contract to provide services in hospitals, rehabilitation centers, nursing homes, home health agencies, adult daycare programs, and schools. They may be in solo practice or be part of a consulting group. Physical therapists also teach in academic institutions and conduct research.

Training, Other Qualifications, and Advancement

All states require physical therapists to pass a licensure exam before they can practice, after graduating from an accredited physical therapist educational program.

According to the American Physical Therapy Association, there were 199 accredited physical therapist programs in 2001. Of the accredited programs, 165 offered master's degrees, and 33 offered doctoral degrees. By 2002, all physical therapist programs seeking accreditation are required to offer degrees at the master's degree level and above, in accordance with the Commission on Accreditation in Physical Therapy Education.

Physical therapist programs start with basic science courses such as biology, chemistry, and physics, and then introduce specialized courses such as biomechanics, neuroanatomy, human growth and development, manifestations of disease, examination techniques, and therapeutic procedures. Besides classroom and laboratory instruction, students receive supervised clinical experience. Courses useful when applying to physical therapist educational programs include anatomy, biology, chemistry, social science, mathematics, and physics. Before granting admission, many professional education programs require experience as a volunteer in a physical therapy department of a hospital or clinic.

Physical therapists should have strong interpersonal skills to successfully educate patients about their physical therapy treatments. They should also be compassionate and possess a desire to help patients. Similar traits also are needed to interact with the patient's family.

Physical therapists are expected to continue professional development by participating in continuing education courses and workshops. A number of states require continuing education to maintain licensure.

Job Outlook

Employment of physical therapists is expected to grow faster than the average for all occupations through 2010. Federal legislation imposing limits on reimbursement for therapy services may adversely affect the job market for physical therapists in the near term. However, over the long run, the demand for physical therapists should continue to rise as a result of growth in the number of individuals with disabilities or limited function requiring therapy services. The rapidly growing elderly population is particularly vulnerable to chronic and debilitating conditions that require therapeutic services. Also, the baby-boom generation is entering the prime age for heart attacks and strokes, increasing the demand for cardiac and physical rehabilitation. More young people will need physical therapy as technological advances save the lives of a larger proportion of newborns with severe birth defects.

Future medical developments should also permit a higher percentage of trauma victims to survive, creating additional demand for rehabilitative care. Growth also may result from advances in medical technology which permit treatment of more disabling conditions.

Widespread interest in health promotion also should increase demand for physical therapy services. A growing number of employers are using physical therapists to evaluate work sites, develop exercise programs, and teach safe work habits to employees in the hope of reducing injuries.

Earnings

Median annual earnings of physical therapists were $54,810 in 2000. The middle 50 percent earned between $46,660 and

$67,390. The lowest 10 percent earned less than $38,510, and the highest 10 percent earned more than $83,370.

Related Occupations

Physical therapists rehabilitate persons with physical disabilities. Others who work in the rehabilitation field include occupational therapists, recreational therapists, rehabilitation counselors, respiratory therapists, and speech-language pathologists and audiologists.

Sources of Additional Information

Additional information on a career as a physical therapist and a list of accredited educational programs in physical therapy are available from:

- American Physical Therapy Association, 1111 North Fairfax St., Alexandria, VA 22314-1488. Internet: http://www.apta.org

Physician Assistants

O*NET 29-1071.00

Significant Points

- The typical physician assistant program lasts about 2 years and usually requires at least 2 years of college and some healthcare experience for admission.
- Earnings are high and job opportunities should be good.

Nature of the Work

Physician assistants (PAs) provide healthcare services under the supervision of physicians. They should not be confused with medical assistants, who perform routine clinical and clerical tasks. PAs are formally trained to provide diagnostic, therapeutic, and preventive healthcare services, as delegated by a physician. Working as members of the healthcare team, they take medical histories, examine and treat patients, order and interpret laboratory tests and X rays, make diagnoses, and prescribe medications. They also treat minor injuries by suturing, splinting, and casting. PAs record progress notes, instruct and counsel patients, and order or carry out therapy. In 47 states and the District of Columbia, physician assistants may prescribe medications. PAs also may have managerial duties. Some order medical and laboratory supplies and equipment and may supervise technicians and assistants.

Physician assistants work under the supervision of a physician. However, PAs may be the principal care providers in rural or inner city clinics, where a physician is present for only 1 or 2 days each week. In such cases, the PA confers with the supervising physician and other medical professionals as needed or as required by law. PAs also may make house calls or go to hospitals and nursing homes to check on patients and report back to the physician.

The duties of physician assistants are determined by the supervising physician and by state law. Aspiring PAs should investigate the laws and regulations in the states in which they wish to practice.

Many PAs work in primary care areas such as general internal medicine, pediatrics, and family medicine. Others work in specialty areas, such as general and thoracic surgery, emergency medicine, orthopedics, and geriatrics. PAs specializing in surgery provide pre- and postoperative care, and may work as first or second assistants during major surgery.

Working Conditions

Although PAs usually work in a comfortable, well-lighted environment, those in surgery often stand for long periods, and others do considerable walking. Schedules vary according to practice setting, and often depend on the hours of the supervising physician. The workweek of PAs in physicians' offices may include weekends, night hours, or early morning hospital rounds to visit patients. These workers also may be on call. PAs in clinics usually work a 40-hour week.

Employment

Physician assistants held about 58,000 jobs in 2000. The number of jobs is greater than the number of practicing PAs because some hold two or more jobs. For example, some PAs work with a supervising physician, but also work in another practice, clinic, or hospital. According to the American Academy of Physician Assistants, there were about 40,469 certified PAs in clinical practice as of January 2000.

Almost 56 percent of jobs for PAs were in the offices and clinics of physicians, dentists, or other health practitioners. About 32 percent were in hospitals. The rest were mostly in public health clinics, temporary help agencies, schools, prisons, home healthcare agencies, and the U.S. Department of Veterans Affairs.

According to the American Academy of Physician Assistants, about one-third of all PAs provide healthcare to communities with fewer than 50,000 residents, in which physicians may be in limited supply.

Training, Other Qualifications, and Advancement

All states require that new PAs complete an accredited, formal education program. As of July 2001, there were 129 accredited or provisionally accredited educational programs for physician assistants; 64 of these programs offered a master's degree. The rest offered either a bachelor's degree or an associate degree. Most PA graduates have at least a bachelor's degree.

Admission requirements vary, but many programs require 2 years of college and some work experience in the healthcare field. Students should take courses in biology, English, chemistry, math, psychology, and social sciences. More than two-thirds of all applicants hold bachelor's or master's degrees. Many applicants are former emergency medical technicians, other allied health professionals, or nurses.

PA programs usually last at least 2 years. Most programs are in schools of allied health, academic health centers, medical schools, or 4-year colleges; a few are in community colleges, the military, or hospitals. Many accredited PA programs have clinical teaching affiliations with medical schools.

PA education includes classroom instruction in biochemistry, pathology, human anatomy, physiology, microbiology, clinical pharmacology, clinical medicine, geriatric and home healthcare, disease prevention, and medical ethics. Students obtain supervised clinical training in several areas, including primary care medicine, inpatient medicine, surgery, obstetrics and gynecology, geriatrics, emergency medicine, psychiatry, and pediatrics. Sometimes, PA students serve one or more of these "rotations" under the supervision of a physician who is seeking to hire a PA. These rotations often lead to permanent employment.

All states and the District of Columbia have legislation governing the qualifications or practice of physician assistants. All jurisdictions require physician assistants to pass the Physician Assistants National Certifying Examination, administered by the National Commission on Certification of Physician Assistants (NCCPA)—open to graduates of accredited PA educational programs. Only those successfully completing the examination may use the credential "Physician Assistant-Certified (PA-C)." In order to remain certified, PAs must complete 100 hours of continuing medical education every 2 years. Every 6 years, they must pass a recertification examination or complete an alternate program combining learning experiences and a take-home examination.

Some PAs pursue additional education in a specialty area such as surgery, neonatology, or emergency medicine. PA postgraduate residency training programs are available in areas such as internal medicine, rural primary care, emergency medicine, surgery, pediatrics, neonatology, and occupational medicine. Candidates must be graduates of an accredited program and be certified by the NCCPA.

Physician assistants need leadership skills, self-confidence, and emotional stability. They must be willing to continue studying throughout their career to keep up with medical advances.

As they attain greater clinical knowledge and experience, PAs can advance to added responsibilities and higher earnings. However, by the very nature of the profession, clinically practicing PAs always are supervised by physicians.

Job Outlook

Employment opportunities are expected to be good for physician assistants, particularly in areas or settings that have difficulty attracting physicians, such as rural and inner city clinics. Employment of PAs is expected to grow much faster than the average for all occupations through the year 2010 due to anticipated expansion of the health services industry and an emphasis on cost containment.

Physicians and institutions are expected to employ more PAs to provide primary care and to assist with medical and surgical procedures because PAs are cost-effective and productive members of the healthcare team. Physician assistants can relieve physicians of routine duties and procedures. Telemedicine—using technology to facilitate interactive consultations between physicians and physician assistants—also will expand the use of physician assistants.

Besides the traditional office-based setting, PAs should find a growing number of jobs in institutional settings such as hospitals, academic medical centers, public clinics, and prisons. Additional PAs may be needed to augment medical staffing in inpatient teaching hospital settings if the number of physician residents is reduced. In addition, state-imposed legal limitations on the numbers of hours worked by physician residents are increasingly common and encourage hospitals to use PAs to supply some physician resident services. Opportunities will be best in states that allow PAs a wider scope of practice.

Earnings

Median annual earnings of physician assistants were $61,910 in 2000. The middle 50 percent earned between $47,970 and $73,890. The lowest 10 percent earned less than $32,690, and the highest 10 percent earned more than $88,100. Median annual earnings of physician assistants in 2000 were $64,430 in offices and clinics of medical doctors and $61,460 in hospitals.

According to the American Academy of Physician Assistants, median income for physician assistants in full-time clinical practice in 2000 was about $65,177; median income for first-year graduates was about $56,977. Income varies by specialty, practice setting, geographical location, and years of experience.

Related Occupations

Other health workers who provide direct patient care that requires a similar level of skill and training include occupational therapists, physical therapists, and speech-language pathologists and audiologists.

Sources of Additional Information

For information on a career as a physician assistant, contact:

- American Academy of Physician Assistants Information Center, 950 North Washington St., Alexandria, VA 22314-1552. Internet: http://www.aapa.org

For a list of accredited programs and a catalog of individual PA training programs, contact:

- Association of Physician Assistant Programs, 950 North Washington St., Alexandria, VA 22314-1552. Internet: http://www.apap.org

For eligibility requirements and a description of the Physician Assistant National Certifying Examination, contact:

- National Commission on Certification of Physician Assistants, Inc., 157 Technology Pkwy., Suite 800, Norcross, GA 30092-2913. Internet: http://www.nccpa.net

Physicists and Astronomers

O*NET 19-2011.00, 19-2012.00

Significant Points

- A doctoral degree is the usual educational requirement because most jobs are in basic research and development; a bachelor's or master's degree is sufficient for some jobs in applied research and development.
- Because funding for research grows slowly, new PhD graduates will face competition for basic research jobs.

Nature of the Work

Physicists explore and identify basic principles governing the structure and behavior of matter, the generation and transfer of energy, and the interaction of matter and energy. Some physicists use these principles in theoretical areas, such as the nature of time and the origin of the universe; others apply their physics knowledge to practical areas, such as the development of advanced materials, electronic and optical devices, and medical equipment.

Physicists design and perform experiments with lasers, cyclotrons, telescopes, mass spectrometers, and other equipment. Based on observations and analysis, they attempt to discover and explain laws describing the forces of nature, such as gravity, electromagnetism, and nuclear interactions. Physicists also find ways to apply physical laws and theories to problems in nuclear energy, electronics, optics, materials, communications, aerospace technology, navigation equipment, and medical instrumentation.

Astronomy is sometimes considered a subfield of physics. *Astronomers* use the principles of physics and mathematics to learn about the fundamental nature of the universe, including the sun, moon, planets, stars, and galaxies. They also apply their knowledge to solve problems in navigation, space flight, and satellite communications, and to develop the instrumentation and techniques used to observe and collect astronomical data.

Most physicists work in research and development. Some do basic research to increase scientific knowledge. Physicists who conduct applied research build upon the discoveries made through basic research and work to develop new devices, products, and processes. For example, basic research in solid-state physics led to the development of transistors and, then, of integrated circuits used in computers.

Physicists also design research equipment. This equipment often has additional unanticipated uses. For example, lasers are used in surgery, microwave devices are used in ovens, and measuring instruments can analyze blood or the chemical content of foods. A small number of physicists work in inspection, testing, quality control, and other production-related jobs in industry.

Much physics research is done in small or medium-size laboratories. However, experiments in plasma, nuclear, and high energy and in some other areas of physics require extremely large, expensive equipment, such as particle accelerators. Physicists in these subfields often work in large teams. Although physics research may require extensive experimentation in laboratories, research physicists still spend time in offices planning, recording, analyzing, and reporting on research.

Almost all astronomers do research. Some are theoreticians, working on the laws governing the structure and evolution of astronomical objects. Others analyze large quantities of data gathered by observatories and satellites, and write scientific papers or reports on their findings. Some astronomers actually operate, usually as part of a team, large space- or ground-based telescopes. However, astronomers may spend only a few weeks each year making observations with optical telescopes, radio telescopes, and other instruments. For many years, satellites and other space-based instruments have provided tremendous amounts of astronomical data. New technology resulting in improvements in analytical techniques and instruments, such as computers and optical telescopes and mounts, is leading to a resurgence in ground-based research. A small number of astronomers work in museums housing planetariums. These astronomers develop and revise programs presented to the public and may direct planetarium operations.

Physicists generally specialize in one of many subfields—elementary particle physics, nuclear physics, atomic and molecular physics, physics of condensed matter (solid-state physics), optics, acoustics, space physics, plasma physics, or the physics of fluids. Some specialize in a subdivision of one of these subfields. For example, within condensed matter physics, specialties include superconductivity, crystallography, and semiconductors. However, all physics involves the same fundamental principles, so specialties may overlap, and physicists may switch from one subfield to another. Also, growing numbers of physicists work in combined fields, such as biophysics, chemical physics, and geophysics.

Working Conditions

Physicists often work regular hours in laboratories and offices. At times, however, those who are deeply involved in research may work long or irregular hours. Most do not encounter unusual hazards in their work. Some physicists temporarily work away from home at national or international facilities with unique equipment, such as particle accelerators. Astronomers who make observations using ground-based telescopes may spend long periods in observatories; this work usually involves travel to remote locations. Long hours, including routine night work, may create temporarily stressful conditions.

Physicists and astronomers whose work is dependent on grant money often are under pressure to write grant proposals to keep their work funded.

Employment

Physicists and astronomers held about 10,000 jobs in 2000. Jobs for astronomers accounted for only a small number—10 percent—of the total. About 40 percent of all nonfaculty physicists and astronomers worked for commercial or noncommercial research, development, and testing laboratories. The federal government employed almost 35 percent, mostly in the U.S. Department of Defense, but also in the National Aeronautics and Space Administration (NASA), and in the U.S. Departments of Commerce, Health and Human Services, and Energy. Other physicists and astronomers worked in colleges and universities in nonfaculty positions, or for state governments, drug companies, or electronic equipment manufacturers.

Besides the jobs described above, many physicists and astronomers held faculty positions in colleges and universities.

Although physicists and astronomers are employed in all parts of the country, most work in areas in which universities, large research and development laboratories, or observatories are located.

Training, Other Qualifications, and Advancement

Because most jobs are in basic research and development, a doctoral degree is the usual educational requirement for physicists and astronomers. Additional experience and training in a postdoctoral research appointment, although not required, is important for physicists and astronomers aspiring to permanent positions in basic research in universities and government laboratories. Many physics and astronomy PhD holders ultimately teach at the college or university level.

Master's degree holders usually do not qualify for basic research positions but do qualify for many kinds of jobs requiring a physics background, including positions in manufacturing and applied research and development. Physics departments in some colleges and universities are creating professional master's degree programs to specifically prepare students for physics-related research and development in private industry that do not require a PhD degree. A master's degree may suffice for teaching jobs in 2-year colleges. Those with bachelor's degrees in physics are rarely qualified to fill positions as research or teaching physicists. They are, however, usually qualified to work in engineering-related areas, in software development and other scientific fields, as technicians, or to assist in setting up computer networks and sophisticated laboratory equipment. Some may qualify for applied research jobs in private industry or nonresearch positions in the federal government. Some become science teachers in secondary schools. Astronomy bachelor's or master's degree holders often enter a field unrelated to astronomy, and they are qualified to work in planetariums running science shows, to assist astronomers doing research, and to operate space- and ground-based telescopes and other astronomical instrumentation.

About 507 colleges and universities offer a bachelor's degree in physics. Undergraduate programs provide a broad background in the natural sciences and mathematics. Typical physics courses include electromagnetism, optics, thermo-dynamics, atomic physics, and quantum mechanics.

In 2000, 183 colleges and universities had departments offering PhD degrees in physics. Another 72 departments offered a master's as their highest degree. Graduate students usually concentrate in a subfield of physics, such as elementary particles or condensed matter. Many begin studying for their doctorate immediately after receiving their bachelor's degree.

About 69 universities grant degrees in astronomy, either through an astronomy department, a physics department, or a combined physics/astronomy department. Applicants to astronomy doctoral programs face competition for available slots. Those planning a career in astronomy should have a very strong physics background. In fact, an undergraduate degree in either physics or astronomy is excellent preparation, followed by a PhD in astronomy.

Mathematical ability, problem-solving and analytical skills, an inquisitive mind, imagination, and initiative are important traits for anyone planning a career in physics or astronomy. Prospective physicists who hope to work in industrial laboratories applying physics knowledge to practical problems should broaden their educational background to include courses outside of physics, such as economics, computer technology, and business management. Good oral and written communication skills also are important because many physicists work as part of a team, write research papers or proposals, or have contact with clients or customers with nonphysics backgrounds.

Many physics and astronomy PhDs begin their careers in a postdoctoral research position, where they may work with experienced physicists as they continue to learn about their specialty and develop ideas and results to be used in later work. Initial work may be under the close supervision of senior scientists. After some experience, physicists perform increasingly complex tasks and work more independently. Those who develop new products or processes sometimes form their own companies or join new firms to exploit their own ideas.

Job Outlook

Historically, many physicists and astronomers have been employed on research projects—often defense-related. Because defense expenditures are expected to increase over the next decade, employment of physicists and astronomers is projected to grow about as fast as the average for all occupations, through the year 2010. The need to replace physicists and astronomers who retire will, however, account for most expected job openings. The federal government funds numerous noncommercial research facilities. The Federally Funded Research and Development Centers (FFRDCs), whose missions include a significant physics component, are largely funded by the Department of Energy (DOE) or the Department of Defense (DOD), and their R&D budgets did not keep pace with inflation during much of the 1990s. However, federal budgets have recently increased for physics-related research at these centers, as well as other agencies such as NASA. If R&D funding continues to grow at these agencies, job opportunities for physicists and astronomers, especially those dependent on federal research grants, should be better than they have been in many years.

Although research and development budgets in private industry will continue to grow, many research laboratories in private industry are expected to continue to reduce basic research, which includes much physics research, in favor of applied or manufacturing research and product and software development. Nevertheless, many persons with a physics background continue to be in demand in the areas of information technology, semiconductor technology, and other applied sciences. This trend is expected to continue; however, many of these positions will be under job titles such as computer software engineer, computer programmer, engineer, and systems developer, rather than physicist.

For several years, the number of doctorates granted in physics has been much greater than the number of openings for physicists, resulting in keen competition, particularly for research positions in colleges and universities and in research and development centers. Competitive conditions are beginning to ease, because the number of doctorate degrees awarded has begun dropping, following recent declines in enrollment in graduate physics programs. However, new doctoral graduates should still expect to face competition for research jobs, not only from fellow graduates, but also from an existing supply of postdoctoral workers seeking to leave low-paying, temporary positions and non-U.S. citizen applicants.

Opportunities may be more numerous for those with a master's degree, particularly graduates from programs preparing students for applied research and development, product design, and manufacturing positions in industry. Many of these positions, however, will have titles other than physicist, such as engineer or computer scientist.

Persons with only a bachelor's degree in physics or astronomy are not qualified to enter most physicist or astronomer research jobs but may qualify for a wide range of positions in engineering, technician, mathematics, and computer- and environment-related occupations. Those who meet state certification requirements can become high school physics teachers, an occupation reportedly in strong demand in many school districts. Despite competition for traditional physics and astronomy research jobs, individuals with a physics degree at any level will find their skills useful for entry to many other occupations.

Earnings

Median annual earnings of physicists and astronomers in 2000 were $82,535. Median annual earnings of astronomers were $74,510, while physicists earned $83,310. The middle 50 percent of physicists earned between $65,820 and $102,270. The lowest 10 percent earned less than $51,680, and the highest 10 percent earned more than $116,290.

According to a 2001 National Association of Colleges and Employers survey, the average annual starting salary offer to physics doctoral degree candidates was $68,273.

The American Institute of Physics reported a median annual salary of $78,000 in 2000 for its members with PhDs (excluding those in postdoctoral positions); with master's degrees, $63,800; and with bachelor's degrees, $60,000. Those working in temporary postdoctoral positions earned significantly less.

The average annual salary for physicists employed by the federal government was $86,799 in 2001; for astronomy and space scientists, $89,734.

Related Occupations

The work of physicists and astronomers relates closely to that of engineers, chemists, atmospheric scientists, computer scientists, computer programmers, and mathematicians.

Sources of Additional Information

General information on career opportunities in physics is available from:

- American Institute of Physics, Career Services Division and Education and Employment Division, One Physics Ellipse, College Park, MD 20740-3843. Internet: http://www.aip.org
- The American Physical Society, One Physics Ellipse, College Park, MD 20740-3844. Internet: http://www.aps.org

Power Plant Operators, Distributors, and Dispatchers

O*NET 51-8011.00, 51-8012.00, 51-8013.01, 51-8013.02

Significant Points

- Overall employment of operators, distributors, and dispatchers is expected to change little due to increasing industry competition.
- Opportunities will be best for operators with training in automated systems.
- Little or no change in employment and low turnover will result in few job opportunities.

Nature of the Work

Electricity is vital for most everyday activities. From the moment you flip the first switch each morning, you are connecting to a huge network of people, electric lines, and generating equipment. Power plant operators control the machinery that generates electricity. Power distributors and dispatchers control the flow of electricity from the power plant over a network of transmission lines, to industrial plants and substations, and, finally, over distribution lines to residential users.

Power plant operators control and monitor boilers, turbines, generators, and auxiliary equipment in power generating plants. Operators distribute power demands among generators, combine the current from several generators, and monitor instruments to maintain voltage and regulate electricity flows from the plant. When power requirements change, these workers start or stop generators and connect or disconnect them from circuits. They often use computers to keep records of switching operations and loads on generators, lines, and transformers. Operators also may use computers to prepare reports of unusual incidents, malfunctioning equipment, or maintenance performed during their shift.

Operators in plants with automated control systems work mainly in a central control room and usually are called *control room operators* and *control room operator trainees* or *assistants*. In older plants, the controls for the equipment are not centralized, and *switchboard operators* control the flow of electricity from a central point, whereas *auxiliary equipment operators* work throughout the plant, operating and monitoring valves, switches, and gauges.

The Nuclear Regulatory Commission (NRC) licenses operators of nuclear power plants. *Reactor operators* are authorized to control equipment that affects the power of the reactor in a nuclear power plant. In addition, an NRC-licensed *senior reactor operator* must be on duty during each shift to act as the plant supervisor and supervise the operation of all controls in the control room.

Power distributors and dispatchers, also called *load dispatchers* or *systems operators*, control the flow of electricity through transmission lines to industrial plants and substations that supply residential electric needs. They operate current converters, voltage transformers, and circuit breakers. Dispatchers monitor equipment and record readings at a pilot board, which is a map of the transmission grid system showing the status of transmission circuits and connections with substations and industrial plants.

Dispatchers also anticipate power needs, such as those caused by changes in the weather. They call control room operators to start or stop boilers and generators, to bring production into balance with needs. They handle emergencies such as transformer or transmission line failures and route current around affected areas. They also operate and monitor equipment in substations, which step up or step down voltage, and operate switchboard levers to control the flow of electricity in and out of substations.

Working Conditions

Because electricity is provided around the clock, operators, distributors, and dispatchers usually work one of three daily 8-hour shifts or one of two 12-hour shifts on a rotating basis. Shift assignments may change periodically, so that all operators can share duty on less desirable shifts. Work on rotating shifts can be stressful and fatiguing, because of the constant change in living and sleeping patterns. Operators, distributors, and dispatchers who work in control rooms generally sit or stand at a control station. This work is not physically strenuous but requires constant attention. Operators who work outside the control room may be exposed to danger from electric shock, falls, and burns.

Nuclear power plant operators are subject to random drug and alcohol tests, as are most workers at nuclear power plants.

Employment

Power plant operators, distributors, and dispatchers held about 55,000 jobs in 2000. Jobs are located throughout the country. About 8 in 10 worked for utility companies and government agencies that produced electricity. Others worked for manufacturing establishments that produced electricity for their own use.

Training, Other Qualifications, and Advancement

Employers seek high school graduates for entry-level operator, distributor, and dispatcher positions. Candidates with strong math and science skills are preferred. College-level courses or prior experience in a mechanical or technical job may be helpful. Employers increasingly require computer proficiency, as computers are used to keep records, generate reports, and track maintenance. Most entry-level positions are helper or laborer jobs, such as in powerline construction. Depending on the results of aptitude tests, worker preferences, and availability of openings, workers may be assigned to train for one of many utility positions.

Workers selected for training as a fossil-fueled power plant operator or distributor undergo extensive on-the-job and classroom training. Several years of training and experience are required to become a fully qualified control room operator or power distributor. With further training and experience, workers may advance to shift supervisor. Utilities generally promote from within; therefore, opportunities to advance by moving to another employer are limited.

Extensive training and experience are necessary to pass the Nuclear Regulatory Commission (NRC) examinations for reactor operators and senior reactor operators. To maintain their license, licensed reactor operators must pass an annual practical plant operation exam and a biennial written exam administered by their employer. Training may include simulator and on-the-job training, classroom instruction, and individual study. Entrants to nuclear power plant operator trainee jobs must have strong math and science skills. Experience in other power plants or with Navy nuclear propulsion plants also is helpful. With further training and experience, reactor operators may advance to senior reactor operator positions.

In addition to preliminary training as a power plant operator, distributor, or dispatcher, most workers are given periodic refresher training. Nuclear power plant operators are given frequent refresher training. This training is usually taken on plant simulators designed specifically to replicate procedures and situations that might be encountered working at the trainee's plant.

Job Outlook

Little or no change in employment of power plant operators, distributors, and dispatchers is expected through the year 2010, as the industry continues to restructure in response to deregulation and increasing competition. The Energy Policy Act of 1992 has had a tremendous impact on the organization of the utilities industry. This legislation has increased competition in power generating utilities by allowing independent power producers to sell power directly to industrial and other wholesale customers. Utilities, which historically operated as regulated local monopolies, are restructuring operations to reduce costs and compete effectively and, as a result, the number of jobs is decreasing.

People who want to become power plant operators, distributors, and dispatchers are expected to encounter keen competition for these high-paying jobs. Little or no change in employment and low turnover in this occupation will result in few job opportunities. The slow pace of new plant construction also will limit opportunities for power plant operators, distributors, and dispatchers. Increasing use of automatic controls and more efficient equipment should increase productivity and decrease the demand for operators. Individuals with training in computers and automated equipment will have the best job prospects.

Earnings

Median annual earnings of power plant operators were $46,090 in 2000. The middle 50 percent earned between $37,320 and $54,200 a year. The lowest 10 percent earned less than $28,700, and the highest 10 percent earned more than $62,020 a year. Median annual earnings of power plant operators in 2000 were $48,350 in electric services and $40,160 in local government.

Median annual earnings of nuclear power reactor operators were $57,220 in 2000. The middle 50 percent earned between $50,720 and $67,320 a year. The lowest 10 percent earned less than $46,890, and the highest 10 percent earned more than $74,370 a year.

Median annual earnings of power distributors and dispatchers were $48,570 in 2000. The middle 50 percent earned between $39,880 and $58,290 a year. The lowest 10 percent earned less than $31,760, and the highest 10 percent earned more than $69,260 a year. Median annual earnings in electric services, the industry employing the largest numbers of power distributors and dispatchers, were $49,070.

Related Occupations

Other workers who monitor and operate plant and systems equipment include chemical plant and system operators; petroleum pump system operators, refinery operators, and gaugers; stationary engineers and boiler operators; and water and wastewater treatment plant and system operators.

Sources of Additional Information

For information about employment opportunities, contact local electric utility companies, locals of unions mentioned below, and state employment service offices.

For general information about power plant operators, nuclear power reactor operators, and power distributors and dispatchers, contact:

- International Brotherhood of Electrical Workers, 1125 15th St. N.W., Washington, DC 20005. Utility Workers Union of America, 815 16th St. N.W., Suite 605, Washington, DC 20006

Precision Instrument and Equipment Repairers

O*NET 49-9061.00, 49-9062.00, 49-9063.01, 49-9063.02, 49-9063.03, 49-9063.04, 49-9064.00, 49-9069.99

Significant Points

- Training requirements include a high school diploma and, in some cases, postsecondary education, coupled with significant on-the-job training.
- Good opportunities are expected for most types of jobs.
- Overall employment is expected to grow about as fast as average, but projected growth varies by detailed occupation.
- About 1 out of 4 is self-employed.

Nature of the Work

Repairing and maintaining watches, cameras, musical instruments, medical equipment, and other precision instruments requires a high level of skill and attention to detail. For example, some devices contain tiny gears that must be manufactured to within one one-hundredth of a millimeter of design specifications, and other devices contain sophisticated electronic controls.

Camera and photographic equipment repairers work through a series of steps in fixing a camera. The first step is determining whether a repair would be profitable. Many inexpensive cameras cost more to repair than to replace. The most complicated or expensive problems are referred back to the manufacturer. If the repairers decide to proceed, they diagnose the problem, often by disassembling numerous small parts in order to reach the source. They then make needed adjustments or replace a defective part. Many problems are caused by the electronic circuits used in many cameras, which require an understanding of electronics. Camera repairers also maintain cameras by removing and replacing broken or worn parts and cleaning and lubricating gears and springs. Many of the components and parts involved are extremely small, requiring a great deal of manual dexterity. Frequently, older camera parts are no longer available, requiring repairers to build replacement parts or strip junked cameras. When machining new parts, workers often use a small lathe, a grinding wheel, and other metalworking tools.

Camera repairers also repair the increasingly popular digital cameras. Repairs on such cameras are similar to those for most modern cameras, but, because digital cameras have no film to wind, they employ fewer moving parts.

Watch and clock repairers work almost exclusively on expensive timepieces, as moderately priced timepieces are cheaper to replace than to repair. Electrically powered quartz watches and clocks function with almost no moving parts, limiting necessary maintenance to replacing the battery. Many expensive timepieces still employ old-style mechanical movements and a manual winding mechanism or spring. This type of timepiece requires regular adjustment and maintenance. Any repair or maintenance work on a mechanical timepiece requires the disassembly of many fine gears and components. Each part is inspected for signs of significant wear. Some gears or springs may need to be replaced or machined. All of the parts are cleaned and oiled.

As for older cameras, replacement parts are frequently unavailable for antique watches or clocks. In such cases, watch repairers must machine their own parts. They employ small lathes and other machines in creating tiny parts.

Musical instrument repairers and tuners combine their love of music with a highly skilled craft. Musical instrument repairers and tuners, often referred to as technicians, work in four specialties: band instruments, pianos and organs, violins, and guitars.

Band instrument repairers, brass and wind instrument repairers, and percussion instrument repairers focus on woodwind, brass, reed, and percussion instruments damaged through deterioration or by accident. They move mechanical parts or play scales to find problems. They may unscrew and remove rod pins, keys, worn cork pads, and pistons and remove soldered parts using gas torches. They repair dents in metal and wood using filling techniques or a mallet. Drums often need new drumheads, which are cut from animal skin. These repairers use gas torches, grinding wheels, shears, mallets, and small hand tools.

Piano repairers use similar techniques, skills, and tools. Repairers often earn additional income by tuning pianos, which involves tightening and loosening different strings to achieve the proper tone or pitch. Pipe-organ repairers do work similar to that of piano repairers on a larger scale. Additionally, they assemble new organs. Because pipe organs are too large to transport, they must be assembled on site. Even with repairers working in teams or with assistants, the organ assembly process can take several weeks or even months, depending upon the size of the organ.

Violin repairers and guitar repairers adjust and repair string instruments. Initially, repairers play and inspect the instrument to find any defects. They replace or repair cracked or broken sections and damaged parts. They also restring the instruments and repair damage to their finish.

The work of *medical equipment repairers* differs significantly from other precision instrument and equipment repair work. Although medical equipment repairers work on fine mechanical systems, the larger scale of their tasks requires less precision. The machines that they repair include electric wheelchairs, mechanical lifts, hospital beds, and customized vehicles.

Medical equipment repairers use various tools, including ammeters, voltmeters, and other measuring devices to diagnose problems. They use handtools and machining equipment, such as small lathes and other metalworking equipment, to make repairs.

Other precision instrument and equipment repairers service, repair, and replace a wide range of equipment associated with automated or instrument-controlled manufacturing processes. A precision instrument repairer working at an electric power plant, for example, would repair and maintain instruments that monitor the operation of the plant, such as pressure and temperature gauges. These workers use many of the same tools that medical equipment repairers use. Malfunctioning parts are often replaced, but sometimes repair is necessary. Replacement parts are not always available, so repairers sometimes machine or fabricate a new part. Preventive maintenance involves regular lubrication, cleaning, and adjustment of many measuring devices.

Working Conditions

Camera, watch, and musical instrument repairers work under fairly similar solitary, low-stress conditions with minimal supervision. A quiet, well-lighted workshop or repair shop is typical, while a few of these repairers travel to the instrument being repaired, such as a piano, organ, or grandfather clock.

Medical equipment and precision instrument and equipment repairers normally work daytime hours. But, like other hospital and factory employees, some repairers work irregular hours. Precision instrument repairers work under a wide array of conditions, from hot, dirty, noisy factories to air-conditioned workshops to outdoor fieldwork. Attention to safety is essential, as the work sometimes involves dangerous machinery or toxic chemicals. Due to the individual nature of the work, supervision is fairly minimal.

Employment

Precision instrument and equipment repairers held 63,000 jobs in 2000. The overwhelming majority of medical equipment repairers and other precision instrument and equipment repairers were wage and salary workers. Medical equipment repairers often work for hospitals or wholesale equipment suppliers, while most precision instrument repairers work in manufacturing. On the other hand, about 1 out of 4 watch, camera and photographic equipment, and musical instrument repairers was self-employed. The following table presents employment by occupation:

Medical equipment repairers	28,000
Camera and photographic equipment repairers	7,200
Musical instrument repairers and tuners	7,100
Watch repairers	5,200
All other precision instrument and equipment repairers	15,000

Training, Other Qualifications, and Advancement

Most employers require at least a high school diploma for beginning precision instrument and equipment repairers. Many employers prefer applicants with some postsecondary education. Much training takes place on the job. The ability to read and understand technical manuals is important. Necessary physical qualities include good fine motor skills and vision. Also, precision equipment repairers must be able to pay close attention to details, enjoy problem solving, and have the desire to disassemble machines to see how they work. Most precision equipment repairers must be able to work alone with minimal supervision.

The educational background required for camera and photographic equipment repairers varies, but some background in electronics is necessary. Some workers complete postsecondary training, such as an associate degree, in this field. The job requires the ability to read an electronic schematic diagram and comprehend other technical information, in addition to good manual dexterity. New employees are trained on the job in two stages over about a year. First, they assist a senior repairer for about 6 months. Then, they refine their skills by performing repairs on their own for an additional 6 months. Finally, repairers continually hone and improve their skills by attending manufacturer-sponsored seminars on the specifics of particular models.

Medical equipment repairers are trained in a similar manner. A background in electronics is helpful, but not required. Like camera repairers, they often specialize in a model or brand. Medical equipment repair requires less training than other precision equipment repair specialties. There are no schools to train these repairers; instead, they learn through hands-on experience and observation. New repairers begin by observing and assisting an experienced worker over a period of 3 to 6 months. Gradually, they begin working independently, while still under close supervision.

Training also varies for watch and clock repairers. Several associations, including the American Watchmakers-Clockmakers Institute (AWI) and the National Association of Watch and Clock Collectors, offer certifications. Some certifications can be completed in a few months; some require simply passing an examination; and the most demanding certifications require 3,000 hours, over 2 years, of classroom time in technical institutes or colleges. (Clock repairers generally require less training than watch repairers because watches have smaller components and require greater precision.) Some repairers opt to learn through assisting a master watch repairer. Nevertheless, developing proficiency in watch or clock repair requires several years of education and experience.

For musical instrument repairers and tuners, employers prefer people with post-high school training in music repair technology. According to a 1997 Piano Technicians Guild membership survey, more than 85 percent of respondents had completed at least some college work; at least 50 percent had a bachelor's or higher degree, although not always in music repair technology. A few technical schools and colleges offer courses in instrument repair. Graduates of these programs normally receive additional training on the job, working with an experienced repairer. A few musical instrument repairers and tuners begin learning their trade on the job as assistants, but employers strongly prefer those with technical school training. Trainees perform a variety of tasks around the shop. Full qualification usually requires 2 to 5 years of training and practice.

Educational requirements for other precision instrument and equipment repair jobs also vary, but include a high school diploma, with a focus on mathematics and science courses. Most employers require postsecondary courses, as repairers need to understand blueprints, electrical schematic diagrams, and electrical, hydraulic, and electromechanical systems. In addition to formal education, a year or two of on-the-job training is required before a repairer is considered fully qualified. Some advancement opportunities exist, but many supervisory positions require more formal education.

Job Outlook

Good opportunities are expected for most types of precision instrument and equipment repairer jobs. Overall employment is projected to grow about as fast as the average for all occupations over the 2000–2010 period. However, projected growth varies by detailed occupation.

Job growth among medical equipment repairers should grow about as fast as the average for all occupations over the projected period. The expanding elderly population should spark strong demand for medical equipment and, in turn, create good employment opportunities in this occupation.

On the other hand, employment of musical instrument repairers is expected to increase more slowly than average. Replacement needs will provide the most job opportunities as many repairers and tuners near retirement. While the expected increase in the number of school-age children involved with music should spur demand for repairers, music must compete with other extracurricular activities and interests. Without new musicians, there will be a slump in instrument rentals, purchases, and repairs. Because training in the repair of musical instruments is difficult to obtain—there are only a few schools that offer training programs, and few experienced workers are willing to take on apprentices—opportunities should be good for those who receive training.

Employment of camera and photographic equipment repairers is expected to decline. The camera repair business is fairly immune to downturns in the business cycle, as consumers are more likely to repair an expensive camera than to buy a new one. However, the popularity of inexpensive cameras adversely affects employment in this occupation, as most point-and-shoot cameras are cheaper to replace than repair.

Employment of watch repairers is expected to grow more slowly than average. However, applicants should have very good opportunities because a large proportion of watch and clock repairers are approaching retirement age and because of trends in watch fashions. Over the past few decades, changes in technology, including the invention of digital and quartz watches that need few repairs, caused a significant decline in the demand for watch repairers. In recent years, there has been a rapidly growing demand for antique and expensive mechanical watches, resulting in increased need for watch repairers.

The projected slower-than-average employment growth of other precision instrument and equipment repairers reflects the expected lack of employment growth in manufacturing and other industries in which they are employed. Nevertheless, good employment opportunities are expected for other precision instrument and equipment repairers due to the relatively small number of people entering the occupation and the need to replace repairers who retire.

Earnings

The following table shows median hourly earnings for various precision instrument and equipment repairers in 2000. Earnings ranged from less than $6.48 for the lowest 10 percent of watch repairers, to more than $31.47 for the highest 10 percent of musical instrument repairers and tuners.

Occupation	Earnings
Medical equipment repairers	$16.99
Musical instrument repairers and tuners	15.10
Camera and photographic equipment repairers	13.94
Watch repairers	12.08
All other precision instrument and equipment repairers	19.87

Earnings within the different occupations vary significantly, depending upon skill levels. For example, a watch and clock repairer may simply change batteries and replace worn wrist straps, while highly skilled watch and clock repairers, with years of training and experience, may rebuild and replace worn parts. According to a survey by the American Watchmakers-Clockmakers Institute, the median annual earnings of highly skilled watch and clock repairers were about $40,000 in 2000.

Related Occupations

Many precision instrument and equipment repairers work with precision mechanical and electronic equipment. Other workers who repair precision mechanical and electronic equipment include computer, automated teller, and office machine repairers and coin, vending, and amusement machine servicers and repairers. Other workers who make precision items include dental laboratory technicians and ophthalmic laboratory technicians. Some precision instrument and equipment repairers work with a wide array of industrial equipment. Their work environment and responsibilities are similar to those of industrial machinery installation, maintenance, and repair workers. Much of the work of watch repairers is similar to that of jewelers and precious stone and metal workers. Camera repairers' work is similar to that of electronic home entertainment equipment installers and repairers. Both occupations work with consumer electronics that are based around a circuit board, but that also involve numerous moving mechanical parts.

Sources of Additional Information

For more information about camera repair careers, contact:

- The National Association of Photo Equipment Technicians (NAPET), 3000 Picture Pl., Jackson, MI 49201

For additional information on medical equipment repair, contact your local medical equipment repair shop or hospital.

For information on musical instrument repair, including schools offering training, contact:

- National Association of Professional Band Instrument Repair Technicians (NAPBIRT), P.O. Box 51, Normal, IL 61761. Internet: http://www.napbirt.org

For additional information on piano repair work, contact:

- Piano Technicians Guild, 3930 Washington St., Kansas City, MO 64111-2963. Internet: http://www.ptg.org

For information about training, mentoring programs, and schools with programs in precision instrument repair, contact:

- ISA-The Instrumentation, Systems, and Automation Society, 67 Alexander Dr., P.O. Box 12277, Research Triangle Park, NC 27709. Internet: http://www.isa.org

For information about watch and clock repair and a list of schools with related programs of study, contact:

- American Watchmakers-Clockmakers Institute (AWI), 701 Enterprise Dr., Harrison, OH 45030-1696. Internet: http://www.awi-net.org

Prepress Technicians and Workers

O*NET 51-5021.00, 51-5022.01, 51-5022.02, 51-5022.03, 51-5022.04, 51-5022.05, 51-5022.06, 51-5022.07, 51-5022.08, 51-5022.09, 51-5022.10, 51-5022.11, 51-5022.12, 51-5022.13

Significant Points

- Most workers train on the job; some complete formal graphics arts programs or other postsecondary programs in printing technology. Most employers prefer to hire experienced prepress technicians and workers.
- Employment is projected to decline as the increased use of computers in typesetting and page layout eliminates many prepress jobs.

Nature of the Work

The printing process has three stages—prepress, press, and binding or postpress. Prepress technicians and workers prepare material for printing presses. They perform a variety of tasks involved with transforming text and pictures into finished pages and making printing plates of the pages.

Advances in computer software and printing technology continue to change prepress work. Customers, as well as prepress technicians and workers, use their computers to produce material that looks like the desired finished product. Customers, using their own computers, increasingly do much of the typesetting and page layout work formerly done by prepress technicians and workers. This process, called "desktop publishing," poses new challenges for the printing industry. Instead of receiving simple typed text from customers, prepress technicians and workers get the material on a computer disk. Because of this, customers are increasingly likely to have already settled on a format on their own, rather than relying on suggestions from prepress technicians and workers. Furthermore, the printing industry is rapidly moving toward complete "digital imaging," by which customers' material received on computer disks is converted directly into printing plates. Other innovations in prepress work are digital color page makeup systems, electronic page layout systems, and off-press color proofing systems.

Typesetting and page layouts also have been affected by technological changes. The old "hot type" method of text composition—which used molten lead to create individual letters, paragraphs, and full pages of text—is nearly extinct. Today, composition work is done with computers and "cold type" technology. Cold type, which is any of a variety of methods creating type without molten lead, has traditionally used "photo typesetting" to ready text and pictures for printing. Although this method has many variations, all use photography to create images on paper. The images are assembled into page format and re-photographed to create film negatives from which the actual printing plates are made. However, newer cold type methods are becoming more common. These automate the photography or make printing plates directly from electronic files.

In one common form of phototypesetting, text is entered into a computer programmed to hyphenate, space, and create columns of text. Typesetters or data entry clerks may do keyboarding of text at the printing establishment. Increasingly, however, authors do this work before the job is sent out for composition. The coded text then is transferred to a typesetting machine, which uses photography, a cathode-ray tube, or a laser to create an image on typesetting paper or film. Once it has been developed the paper or film is sent to a lithographer who makes the actual printing plate.

New technologies have had a significant impact on the role of other composition workers. Sophisticated publishing software allows an entire newspaper, catalog, or book page, complete with artwork and graphics, to be made up on the computer screen exactly as it will appear in print. Although generally this is the work of desktop publishers, improvements in packaged software allow customers to do more of their own typesetting and layout work. Operators, however, still transmit the pages for production into film and then into plates or directly into plates. "Imagesetters" read text from computer memory and then "beam" it directly onto film, paper, or plates, bypassing the slower photographic process traditionally used. In small shops, *job printers* may be responsible for composition and page layout, reading proofs for errors and clarity, correcting mistakes, and printing.

With traditional photographic processes, the material is arranged and typeset, and then passed on to workers who further prepare it for the presses. *Camera operators* usually are classified as line camera operators, halftone operators, or color separation photographers. Line camera operators start the process of making a lithographic plate by photographing and developing film negatives or positives of the material to be printed. They adjust light and expose film for a specified length of time, and then develop film in a series of chemical baths. They may load exposed film in machines that automatically develop and fix the image. The use of film in printing will decline, as electronic imaging becomes more prevalent. With decreased costs and improved quality, electronic imaging has become the method of choice in the industry.

The lithographic printing process requires that images be made up of tiny dots coming together to form a picture. Photographs cannot be printed without them. When normal "continuous-tone" photographs need to be reproduced, halftone camera operators separate the photograph into pictures containing the dots. Color separation photography is more complex. In this process, camera operators produce four-color separation negatives from a continuous-tone color print or transparency.

Most of this separation work is done electronically on scanners. *Scanner operators* use computerized equipment to capture photographs or art as digital data, or to create film negatives or positives of photographs or art. The computer controls the color separation of the scanning process, and with the help of the operator, corrects for mistakes, or compensates for deficiencies in the original color print or transparency. Each scan produces a dotted image, or halftone, of the original in one of four primary printing colors—yellow, magenta, cyan, and black. The images are used to produce printing plates that print each of these colors, with transparent colored inks, one at a time. These produce "secondary" color combinations of red, green, blue, and black, which can be combined to produce the colors and hues of the original photograph.

Scanners that can perform color correction during the color separation procedure are rapidly replacing *lithographic dot etchers,* who retouch film negatives or positives by sharpening or reshaping images. They work by hand, using chemicals, dyes, and special tools. Dot etchers must know the characteristics of all types of paper and must produce fine shades of color. Like camera operators, they are usually assigned to only one phase of the work, and may have job titles such as dot etcher, retoucher, or letterer.

New technology also is lessening the need for *film strippers,* who cut the film to the required size and arrange and tape the negatives onto "flats"—or layout sheets used by platemakers to make press plates. When completed, flats resemble large film negatives of the text in its final form. In large printing establishments such as newspapers, arrangement is done automatically.

Platemakers use a photographic process to make printing plates. The film assembly or flat is placed on top of a thin metal plate coated with a light-sensitive resin. Exposure to ultraviolet light activates the chemical in parts not protected by the film's dark areas. The plate then is developed in a solution that removes the unexposed non-image area, exposing bare metal. The chemical on areas of the plate exposed to the light hardens and becomes water repellent. The hardened parts of the plate form the text.

A growing number of printing plants use lasers to directly convert electronic data to plates without any use of film. Entering, storing, and retrieving information from computer-aided equipment require technical skills. In addition to operating and maintaining the equipment, lithographic platemakers must make sure that plates meet quality standards.

During the printing process, the plate is first covered with a thin coat of water. The water adheres only to the bare metal non-image areas, and is repelled by the hardened areas that were exposed to light. Next, the plate comes in contact with a rubber roller covered with oil-based ink. Because oil and water do not mix, the ink is repelled by the water-coated area and sticks to the hardened areas. The ink covering the hardened text is transferred to paper.

Although computers perform a wider variety of tasks, printing still involves text composition, page layout, and plate making, so printing still will require prepress technicians and workers. As computer skills become increasingly important, these workers will need to demonstrate a desire and an ability to benefit from the frequent retraining required by rapidly changing technology.

Working Conditions

Prepress technicians and workers usually work in clean, air-conditioned areas with little noise. Some workers, such as typesetters and compositors, may develop eyestrain from working in front of a video display terminal, or musculoskeletal problems such as backaches. Lithographic artists and film strippers may find working with fine detail tiring to the eyes. Platemakers, who work with toxic chemicals, face the hazard of skin irritations. Workers often are subject to stress and the pressures of short deadlines and tight work schedules.

Prepress employees usually work an 8-hour day. Some workers—particularly those employed by newspapers—work night shifts, weekends, and holidays.

Employment

Prepress technicians and workers held about 162,000 jobs in 2000. Of these, almost 56,000 were job printers.

Most prepress jobs were found in firms that handle commercial or business printing, and in newspaper plants. Commercial printing firms print newspaper inserts, catalogs, pamphlets, and advertisements, while business form establishments print material such as sales receipts. A large number of jobs also are found in printing trade services firms and "in-plant" operations. Establishments in printing trade services typically perform custom compositing, platemaking, and related prepress services.

The printing and publishing industry is one of the most geographically dispersed in the United States, and prepress jobs are found throughout the country. However, job prospects may be best in large metropolitan cities such as New York, Chicago, Los Angeles, Philadelphia, Dallas, and Washington, DC.

Training, Other Qualifications, and Advancement

Most prepress technicians and workers train on the job; the length of training varies by occupation. Some skills, such as typesetting, can be learned in a few months, but they are the most likely to be automated in the future. Other skills, such as stripping (image assembly), require years of experience to master. However, these workers also should expect to receive intensive retraining.

Workers often start as helpers who are selected for on-the-job training programs after demonstrating their reliability and interest in the occupation. They begin with instruction from an experienced craft worker and advance based on their demonstrated mastery of skills at each level. All workers should expect to be retrained from time to time to handle new, improved equipment. As workers gain experience, they advance to positions with greater responsibility. Some move into supervisory positions.

Apprenticeship is another way to become a skilled prepress worker, although few apprenticeships have been offered in recent years. Apprenticeship programs emphasize a specific craft—such as camera operator, film stripper, lithographic etcher, scanner operator, or platemaker—but apprentices are introduced to all phases of printing.

Usually, most employers prefer to hire high school graduates who possess good communication skills, both oral and written. Prepress technicians and workers should be able to deal courteously with people, because in small shops they may take customer orders. They also may perform computations to estimate job costs. Persons interested in working for firms using advanced printing technology need to know the basics of electronics and computers. Mathematical skills also are essential for operating many of the software packages used to run modern, computerized prepress equipment.

Prepress technicians and workers need good manual dexterity, and they must be able to pay attention to detail and work independently. Good eyesight, including visual acuity, depth perception, field of view, color vision, and the ability to focus quickly, also are assets. Artistic ability is often a plus. Employers also seek persons who are even-tempered and adaptable, important qualities for workers who often must meet deadlines and learn how to operate new equipment.

Formal graphic arts programs, offered by community and junior colleges and some 4-year colleges, are a good way to learn about the industry. These programs provide job-related training, which will help when seeking full-time employment. Bachelor's degree programs in graphic arts usually are intended for students who may eventually move into management positions, and 2-year associate degree programs are designed to train skilled workers.

Courses in various aspects of printing also are available at vocational-technical institutes, industry-sponsored update and retraining programs, and private trade and technical schools.

As workers gain experience, they may advance to positions with greater responsibility. Some move into supervisory positions.

Job Outlook

Overall employment of prepress technicians and workers is expected to decline through 2010. Demand for printed material should continue to grow, spurred by rising levels of personal income, increasing school enrollments, higher levels of educational attainment, and expanding markets. However, increased use of computers in desktop publishing should eliminate many prepress jobs.

Technological advances will have a varying effect on employment among the prepress occupations. This reflects the increasing proportion of page layout and design that will be performed using computers. Most prepress technicians and workers such as paste-up, composition and typesetting, photoengraving, platemaking, film stripping, and camera operator occupations are expected to experience declines as handwork becomes automated. Computerized equipment allowing reporters and editors to specify type and style, and to format pages at a desktop computer terminal, already has eliminated many typesetting and composition jobs; more may disappear in the years ahead. In contrast, employment of job printers is expected to grow slightly because at certain times, it is more advantageous for a company to contract out this same service to a small shop. Contracting out is likely to benefit job printers because they usually are found in small shops.

Job prospects also will vary by industry. Changes in technology have shifted many employment opportunities away from the traditional printing plants into advertising agencies, public relations firms, and large corporations. Many companies are turning to in-house typesetting or desktop publishing, as personal computers with elaborate graphic capabilities have become common. Corporations are finding it more profitable to print their own newsletters and other reports than to send them out to trade shops.

Some new jobs for prepress technicians and workers are expected to emerge in commercial printing establishments. New equipment should reduce the time needed to complete a printing job, and allow commercial printers to make inroads into new markets that require fast turnaround. Because small establishments predominate, commercial printing should provide the best opportunities for inexperienced workers who want to gain a good background in all facets of printing.

Most employers prefer to hire experienced prepress technicians and workers. Among persons without experience, however, opportunities should be best for those with computer backgrounds who have completed postsecondary programs in printing technology. Many employers prefer graduates of these programs because the comprehensive training they receive helps them learn the printing process and adapt more rapidly to new processes and techniques.

Earnings

Median hourly earnings of prepress technicians and workers were $14.57 in 2000. The middle 50 percent earned between $10.70 and $19.12 an hour. The lowest 10 percent earned less than $8.20, and the highest 10 percent earned more than $23.57 an hour. Median hourly earnings in commercial printing, the industry employing the largest number of prepress technicians and workers, were $15.26 in 2000.

Median hourly earnings of job printers were $13.61 in 2000. The middle 50 percent earned between $10.00 and $17.67 an hour. The lowest 10 percent earned less than $7.81, and the highest 10 percent earned more than $21.88 an hour. Median hourly earnings in commercial printing, the industry employing the largest number of job printers, were $14.68 in 2000.

Wage rates for prepress technicians and workers vary according to occupation, level of experience, training, location, and size of the firm, and whether they are union members.

Related Occupations

Prepress technicians and workers use artistic skills in their work. These skills also are essential for artists and related workers, etchers and engravers, designers, and desktop publishers. In addition to typesetters, other workers who operate machines equipped with keyboards include data entry and information processing workers.

Sources of Additional Information

Details about apprenticeship and other training programs may be obtained from local employers such as newspapers and printing shops, or from local offices of the state employment service.

For information on careers and training in printing and the graphic arts, write to:

- Printing Industries of America, 100 Daingerfield Rd., Alexandria, VA 22314. Internet: http://www.gain.org/servlet/gateway/PIA_GATF/non_index.html
- Graphic Communications Council, 1899 Preston White Dr., Reston, VA 20191. Internet: http://www.npes.org/edcouncil/index.html
- Graphic Communications International Union, 1900 L St. N.W., Washington, DC 20036. Internet: http://www.gciu.org
- Graphic Arts Technical Foundation, 200 Deer Run Rd., Sewickley, PA 15143. Internet: http://www.gatf.org

Printing Machine Operators

O*NET 51-5023.01, 51-5023.02, 51-5023.03, 51-5023.04, 51-5023.05, 51-5023.06, 51-5023.07, 51-5023.08, 51-5023.09

Significant Points

- Most are trained informally on the job.
- Employment growth will be slowed by the increasing use of new, more efficient computerized printing presses.
- Job seekers are likely to face keen competition; opportunities should be best for persons who qualify for formal apprenticeship training or who complete postsecondary training programs.

Nature of the Work

Printing machine operators prepare, operate, and maintain the printing presses in a pressroom. Duties of printing machine operators vary according to the type of press they operate—offset lithography, gravure, flexography, screen printing, or letterpress. Offset lithography, which transfers an inked impression from a rubber-covered cylinder to paper or other material, is the dominant printing process. With gravure, the recesses on an etched plate or cylinder are inked and pressed to paper. Flexography is a form of rotary printing in which ink is applied to the surface by a flexible rubber printing plate with a raised image area. Gravure and flexography should increase in use, but letterpress, in which an inked, raised surface is pressed against paper, will be phased out. In addition to the major printing processes, plateless or nonimpact processes are coming into general use. Plateless processes—including electronic, electrostatic, and ink-jet printing—are used for copying, duplicating, and document and specialty printing, usually by quick and in-house printing shops.

To prepare presses for printing, machine operators install and adjust the printing plate, adjust pressure, ink the presses, load paper, and adjust the press to the paper size. Press operators ensure that paper and ink meet specifications, and adjust margins and the flow of ink to the inking rollers accordingly. They then feed paper through the press cylinders and adjust feed and tension controls.

While printing presses are running, press operators monitor their operation and keep the paper feeders well stocked. They make adjustments to correct uneven ink distribution, speed, and temperatures in the drying chamber, if the press has one. If paper jams or tears and the press stops, which can happen with some offset presses, operators quickly correct the problem to minimize downtime. Similarly, operators working with other high-speed presses constantly look for problems, making quick corrections to avoid expensive losses of paper and ink. Throughout the run, operators occasionally pull sheets to check for any printing imperfections.

In most shops, press operators also perform preventive maintenance. They oil and clean the presses and make minor repairs.

Machine operators' jobs differ from one shop to another because of differences in the kinds and sizes of presses. Small commercial shops are operated by one person and tend to have relatively small presses, which print only one or two colors at a time. Operators who work with large presses have assistants and helpers. Large newspaper, magazine, and book printers use giant "in-line web" presses that require a crew of several press operators and press assistants. These presses are fed paper in big rolls, called "webs," up to 50 inches or more in width. Presses print the paper on both sides; trim, assemble, score, and fold the pages; and count the finished sections as they come off the press.

Most plants have or will soon have installed printing presses with computers and sophisticated instruments to control press operations, making it possible to set up for jobs in less time. Computers allow press operators to perform many of their tasks electronically. With this equipment, press operators monitor the printing process on a control panel or computer monitor, which allows them to adjust the press electronically.

Working Conditions

Operating a press can be physically and mentally demanding, and sometimes tedious. Printing machine operators are on their feet most of the time. Often, operators work under pressure to meet deadlines. Most printing presses are capable of high printing speeds, and adjustments must be made quickly to avoid waste. Pressrooms are noisy, and workers in certain areas wear ear protectors. Working with press machinery can be hazardous, but accidents can be avoided when safe work practices are observed. The threat of accidents is less with newer computerized presses because operators make most adjustments from a control panel. Many press operators work evening, night, and overtime shifts.

Employment

Printing machine operators held about 222,000 jobs in 2000. Most press operator jobs were in newspaper plants or in firms handling commercial or business printing. Commercial printing firms print newspaper inserts, catalogs, pamphlets, and the advertisements found in mailboxes, and business form establishments print items such as business cards, sales receipts, and paper used in computers. Additional jobs were in the "in-plant" section of organizations and businesses that do their own printing—such as banks, insurance companies, and government agencies.

The printing and publishing industry is one of the most geographically dispersed in the United States, and press operators can find jobs throughout the country. However, jobs are concentrated in large printing centers such as New York, Los Angeles, Chicago, Philadelphia, Dallas, and Washington, DC.

Training, Other Qualifications, and Advancement

Although completion of a formal apprenticeship or a postsecondary program in printing equipment operation continue to be the best ways to learn the trade, most printing machine operators are trained informally on the job while working as assistants or helpers to experienced operators. Beginning press operators load, unload, and clean presses. With time, they move up to operating one-color sheet-fed presses and eventually advance to multicolor presses. Operators are likely to gain experience on many kinds of printing presses during the course of their career.

Apprenticeships for press operators in commercial shops take 4 years. In addition to on-the-job instruction, apprenticeships include related classroom or correspondence school courses. Once the dominant method for preparing for this occupation, apprenticeships are becoming less prevalent.

In contrast, formal postsecondary programs in printing equipment operation offered by technical and trade schools and community colleges are growing in importance. Some postsecondary school programs require 2 years of study and award an associate degree, but most programs can be completed in 1 year or less. Postsecondary courses in printing are increasingly important because they provide the theoretical knowledge needed to operate advanced equipment.

Persons who wish to become printing machine operators need mechanical aptitude to make press adjustments and repairs. Oral and writing skills also are required. Operators should possess the mathematical skills necessary to compute percentages, weights, and measures, and to calculate the amount of ink and paper needed to do a job. Because of technical developments in the printing industry, courses in chemistry, electronics, color theory, and physics are helpful.

Technological changes have had a tremendous effect on the skills needed by printing machine operators. New presses now require operators to possess basic computer skills. Even experienced operators periodically receive retraining and skill updating. For example, printing plants that change from sheet-fed offset presses to Web offset presses have to retrain the entire press crew because skill requirements for the two types of presses are different. Web offset presses, with their faster operating speeds, require faster decisions, monitoring of more variables, and greater physical effort. In the future, workers are expected to need to retrain several times during their career.

Printing machine operators may advance in pay and responsibility by working on a more complex printing press. Through experience and demonstrated ability, for example, a one-color sheet-fed press operator may become a four-color sheet-fed press operator. Others may advance to pressroom supervisor and become responsible for an entire press crew.

Job Outlook

Persons seeking jobs as printing machine operators are likely to face keen competition from experienced operators and prepress workers who have been displaced by new technology, particularly those who have completed retraining programs. Opportunities to become printing machine operators are likely to be best for persons who qualify for formal apprenticeship training or who complete postsecondary training programs.

Employment of printing machine operators is expected to grow more slowly than the average for all occupations through 2010. Although demand for printed materials will grow, employment growth will be slowed by the increased use of new, more efficient computerized printing presses. Most job openings will result from the need to replace operators who retire or leave the occupation.

Most new jobs will result from expansion of the printing industry as demand for printed material increases in response to demographic trends, U.S. expansion into foreign markets, and growing use of direct mail by advertisers. Demand for books and magazines will increase as school enrollments rise, and as substantial growth in the middle-aged and older population spurs adult education and leisure reading. Additional growth should stem from increased foreign demand for domestic trade publications, professional and scientific works, and mass-market books such as paperbacks.

Continued employment in commercial printing will be spurred by increased expenditures for print advertising materials to be mailed directly to prospective customers. New market research techniques are leading advertisers to increase spending on messages targeted to specific audiences, and should continue to require the printing of a wide variety of newspaper inserts, catalogs, direct mail enclosures, and other kinds of print advertising.

Other printing, such as newspapers, books, and greeting cards, also will provide jobs. Experienced press operators will fill most of these jobs because many employers are under severe pressure to meet deadlines and have limited time to train new employees.

Earnings

Median hourly earnings of printing machine operators were $13.57 in 2000. The middle 50 percent earned between $10.38 and $17.80 an hour. The lowest 10 percent earned less than $8.09, and the highest 10 percent earned more than $21.92 an hour. Median hourly earnings in the industries employing the largest numbers of printing machine operators in 2000, were as follows:

Commercial printing	$14.91
Newspapers	14.71
Paperboard containers and boxes	14.44
Miscellaneous converted paper products	13.78
Mailing, reproduction, and stenographic services	10.92

The basic wage rate for a printing machine operator depends on the type of press being run and the geographic area in which the work is located. Workers covered by union contracts usually had higher earnings.

Related Occupations

Other workers who set up and operate production machinery include machine setters, operators, and tenders—metal and plastic, bookbinders and bindery workers, and various precision machine operators.

Sources of Additional Information

Details about apprenticeships and other training opportunities may be obtained from local employers such as newspapers and printing shops, local offices of the Graphic Communications International Union, local affiliates of Printing Industries of America, or local offices of the state employment service.

For general information about press operators, write to:

- Graphic Communications International Union, 1900 L St. N.W., Washington, DC 20036. Internet: http://www.gciu.org

For information on careers and training in printing and the graphic arts, write to:

- Printing Industries of America, 100 Daingerfield Rd., Alexandria, VA 22314. Internet: http://www.gain.org/servlet/gateway/PIA_GATF/non_index.html
- Graphic Communications Council, 1899 Preston White Dr., Reston, VA 20191. Internet: http://www.npes.org/edcouncil/index.html
- Graphic Arts Technical Foundation, 200 Deer Run Rd., Sewickley, PA 15143. Internet: http://www.gatf.org

Radio and Telecommunications Equipment Installers and Repairers

O*NET 49-2021.00, 49-2022.01, 49-2022.02, 49-2022.03, 49-2022.04, 49-2022.05

Significant Points

- Employment is projected to decline.
- Applicants with electronics training and computer skills should have the best opportunities.
- Weekend and holiday hours are common; repairers may be on call around the clock in case of emergencies.

Nature of the Work

Telephones and radios depend on a variety of equipment to transmit communications signals. Electronic switches route telephone signals to their destinations. Switchboards direct telephone calls within a single location or organization. Radio transmitters and receivers relay signals from wireless phones and radios to their destinations. Newer telecommunications equipment is computerized and can communicate a variety of information, including data, graphics, and video. The workers who set up and maintain this sophisticated equipment are radio and telecommunications equipment installers and repairers.

Central office installers set up switches, cables, and other equipment in central offices. These locations are the hubs of a telecommunications network—they contain the switches and routers that direct packets of information to their destinations. *PBX installers and repairers* set up private branch exchange (PBX) switchboards, which relay incoming, outgoing, and interoffice calls within a single location or organization. To install switches and switchboards, installers first connect the equipment to power lines and communications cables and install frames and supports. They test the connections to ensure that adequate power is available and that the communication links function. They also install equipment such as power systems, alarms, and telephone sets. New switches and switchboards are computerized; workers install software or may program the equipment to provide specific features. For example, as a cost-cutting feature, an installer may program a PBX switchboard to route calls over different lines at different times of the day. However, other workers, such as computer support specialists, rather than installers, generally handle complex programming. Finally, the installer performs tests to verify that the newly installed equipment functions properly.

The increasing reliability of telephone switches and routers has simplified maintenance. New telephone switches are self-monitoring and alert repairers to malfunctions. Some switches allow repairers to diagnose and correct problems from remote locations. When faced with a malfunction, the repairer may refer to manufacturers' manuals that provide maintenance instructions. PBX repairers determine if the problem is located within the PBX system, or if it originates in the telephone lines maintained by the local phone company.

When problems with telecommunications equipment arise, telecommunications equipment repairers diagnose the source of the problem by testing each of the different parts of the equipment, which requires an understanding of how the software and hardware interact. Repairers often use spectrum and/or network analyzers to locate the problem. A network analyzer sends a signal through the equipment to detect any distortion in the signal. The nature of the signal distortion often directs the repairer to the source of the problem. To fix the equipment, repairers may use small hand tools, including pliers and screwdrivers, to remove and replace defective components such as circuit boards or wiring. Newer equipment is easier to repair, since whole boards and parts are designed to be quickly removed and replaced. Repairers also may install updated software or programs that maintain existing software.

Station installers and repairers, telephone—commonly known as *telephone installers and repairers*—install and repair telephone wiring and equipment on customers' premises. They install telephone service by connecting customers' telephone wires to outside service lines. These lines run on telephone poles or in underground conduits. The installer may climb poles or ladders to make the connections. Once the telephone is connected, the line is tested to insure that it receives a dial tone. When a maintenance problem occurs, repairers test the customers' lines to determine if the problem is located in the customers' premises or in the outside service lines. When onsite procedures fail to resolve installation or maintenance problems, repairers may request support from their technical service center. Line installers and repairers install the wires and cables that connect customers with central offices.

Radio mechanics install and maintain radio transmitting and receiving equipment. This includes stationary equipment mounted on transmission towers and mobile equipment, such as radio communications systems in service and emergency vehicles. Their work does not include cellular communications towers and equipment. Newer radio equipment is self-monitoring and may alert mechanics to potential malfunctions. When malfunctions occur, these mechanics examine equipment for damaged components and loose or broken wires. They use electrical measuring instruments to monitor signal strength, transmission capacity, interference, and signal delay, as well as hand tools to replace defective components and parts and to adjust equipment so it performs within required specifications.

Working Conditions

Radio and telecommunications equipment installers and repairers generally work in clean, well-lighted, air-conditioned surroundings, such as a telephone company's central office, a customer's PBX location, or an electronic repair shop or service center. Telephone installers and repairers work on rooftops, ladders, and telephone poles. Radio mechanics may maintain equipment located on the tops of transmissions towers. While working outdoors, these workers are subject to a variety of weather conditions.

Nearly all radio and telecommunications equipment installers and repairers work full time. Many work regular business hours to meet the demand for repair services during the workday. Schedules are more irregular at companies that need repair services 24 hours a day or where installation and maintenance must take place after business hours. At these locations, mechanics work a variety of shifts, including weekend and holiday hours. Repairers may be on call around the clock, in case of emergencies, and may have to work overtime.

The work of most repairers involves lifting, reaching, stooping, crouching, and crawling. Adherence to safety precautions is important to guard against work hazards. These hazards include falls, minor burns, electrical shock, and contact with hazardous materials.

Employment

Radio and telecommunications equipment installers and repairers held about 196,000 jobs in 2000. About 189,000 were telecommunications equipment installers and repairers, except line installers, and the rest were radio mechanics. Most worked for telephone communications companies but many radio mechanics worked in electrical repair shops.

Training, Other Qualifications, and Advancement

Most employers seek applicants with postsecondary training in electronics and a familiarity with computers. Training sources include 2- and 4-year college programs in electronics or communications, trade schools, and equipment and software manufacturers. Military experience with communications equipment is highly valued by many employers.

Newly hired repairers usually receive some training from their employers. This may include formal classroom training in electronics, communications systems, or software and informal, hands-on training with communications equipment. Large companies may send repairers to outside training sessions to keep these employees informed of new equipment and service procedures. As networks have become more sophisticated—often including equipment from a variety of companies—the knowledge needed for installation and maintenance also has increased.

Repairers must be able to distinguish colors, because wires are color-coded, and they must be able to hear distinctions in the various tones on a telephone system. For positions that require climbing poles and towers, workers must be in good physical shape. Repairers who handle assignments alone at a customer's site must be able to work without close supervision. For workers who frequently contact customers, a pleasant personality, neat appearance, and good communications skills also are important.

Experienced repairers with advanced training may become specialists or troubleshooters who help other repairers diagnose difficult problems, or may work with engineers in designing equipment and developing maintenance procedures. Because of their familiarity with equipment, repairers are particularly well qualified to become manufacturers' sales workers. Workers with leadership ability also may become maintenance supervisors or service managers. Some experienced workers open their own repair services or shops or become wholesalers or retailers of electronic equipment.

Job Outlook

Employment of radio and telecommunications equipment installers and repairers is expected to decline through 2010. Although the need for installation work will grow as companies seek to upgrade their telecommunications networks, there will be a declining need for maintenance work—performed by telecommunications equipment installers and repairers, except line installers—because of increasingly reliable self-monitoring and self-diagnosing equipment. The replacement of two-way radio systems by wireless systems, especially in service vehicles, has eliminated the need in many companies for onsite radio mechanics. The increased reliability of wireless equipment and the use of self-monitoring systems also will continue to lessen the need for radio mechanics. Applicants with electronics training and computer skills should have the best opportunities for radio and telecommunications equipment installer and repairer jobs.

Job opportunities will vary by specialty. For example, opportunities should be available for central office and PBX installers and repairers as the growing popularity of the Internet, expanded multimedia offerings such as video on demand, and other telecommunications services continue to place additional demand on telecommunications networks. These new services require high data transfer rates, which can only be achieved by installing new optical switching and routing equipment. Extending high speed communications from central offices to customers also will require the installation of more advanced switching and routing equipment. Whereas increased reliability and automation of switching equipment will limit opportunities, these effects will be offset by the strong demand for installation and upgrading of switching equipment.

Station installers and repairers, on the other hand, can expect keen competition. Pre-wired buildings and the increasing reliability of telephone equipment will reduce the need for installation and maintenance of customers' telephones. The number of pay phones is declining as cellular telephones have increased in popularity, which also will adversely affect employment in this specialty as pay phone installation and maintenance is one of their major functions.

Earnings

In 2000, median hourly earnings of telecommunications equipment installers and repairers, except line installers were $21.17. The middle 50 percent earned between $16.55 and $24.99. The bottom 10 percent earned less than $12.04, whereas the top 10 percent earned more than $27.23. Median hourly earnings in the telephone communications industry were $22.88 in 2000.

Median hourly earnings of radio mechanics in 2000 were $15.86. The middle 50 percent earned between $12.57 and $20.60. The bottom 10 percent earned less than $9.39, whereas the top 10 percent earned more than $25.62.

Related Occupations

Related occupations that work with electronic equipment include broadcast and sound engineering technicians and radio operators; computer, automated teller, and office machine repairers; electronic home entertainment equipment installers and repairers; and electrical and electronics installers and repairers. Engineering technicians also may repair electronic equipment as part of their duties.

Sources of Additional Information

For information on career opportunities, contact:

- International Brotherhood of Electrical Workers, Telecommunications Department, 1125 15th St. N.W., Room 807, Washington, DC 20005
- Communications Workers of America, 501 3rd St. N.W., Washington, DC 20001. Internet: http://www.cwa-union.org

For information on careers and schools, contact:

- Electronics Technicians Association International, 502 North Jackson, Greencastle, IN 46135
- National Association of Radio and Telecommunications engineers, P.O. Box 678, Medway, MA 02053. Internet: http://www.narte.org

Radiologic Technologists and Technicians

O*NET 29-2034.01, 29-2034.02

Significant Points

- Faster-than-average growth will arise from an increase in the number of middle-aged and older persons who are the primary users of diagnostic procedures.
- Although hospitals will remain the primary employer of radiologic technologists and technicians, a greater number of new jobs will be found in offices and clinics of physicians, including diagnostic imaging centers.
- Radiologic technologists and technicians with cross training in nuclear medicine technology or other modalities will have the best prospects.

Nature of the Work

Radiologic technologists and technicians take X rays and administer nonradioactive materials into patients' blood streams for diagnostic purposes. Some specialize in diagnostic imaging technologies such as computed tomography (CT) and magnetic resonance imaging (MRI).

In addition to radiologic technologists and technicians, others who assist in diagnostic imaging procedures include cardiovascular technologists and technicians, diagnostic medical sonographers, and nuclear medicine technologists.

Radiologic technologists and technicians, also referred to as *radiographers*, produce X-ray films (radiographs) of parts of the human body for use in diagnosing medical problems. They prepare patients for radiologic examinations by explaining the procedure, removing articles such as jewelry, through which X rays cannot pass, and positioning patients so that the parts of the body can be appropriately radiographed. To prevent unnecessary radiation exposure, they surround the exposed area with radiation protection devices, such as lead shields, or limit the size of the X-ray beam. Radiographers position radiographic equipment at the correct angle and height over the appropriate area of a patient's body. Using instruments similar to a measuring tape, they may measure the thickness of the section to be radiographed and set controls on the X-ray machine to produce radiographs of the appropriate density, detail, and contrast. They place the X-ray film under the part of the patient's body to be examined and make the exposure. They then remove the film and develop it.

Experienced radiographers may perform more complex imaging procedures. For fluoroscopies, radiographers prepare a solution of contrast medium for the patient to drink, allowing the radiologist, a physician who interprets radiographs, to see soft tissues in the body. Some radiographers, called *CT technologists*, operate computerized tomography scanners to produce cross sectional images of patients. Others operate machines using strong magnets and radio waves rather than radiation to create an image and are called *magnetic resonance imaging (MRI) technologists*.

Radiologic technologists and technicians must follow physicians' orders precisely and conform to regulations concerning use of radiation to protect themselves, their patients, and coworkers from unnecessary exposure.

In addition to preparing patients and operating equipment, radiologic technologists and technicians keep patient records and adjust and maintain equipment. They also may prepare work schedules, evaluate equipment purchases, or manage a radiology department.

Working Conditions

Most full-time radiologic technologists and technicians work about 40 hours a week; they may have evening, weekend, or on-call hours. Opportunities for part-time and shift work are also available.

Because technologists and technicians are on their feet for long periods and may lift or turn disabled patients, physical stamina is important. Technologists and technicians work at diagnostic machines but may also do some procedures at patients' bedsides. Some travel to patients in large vans equipped with sophisticated diagnostic equipment.

Although potential radiation hazards exist in this occupation, they are minimized by the use of lead aprons, gloves, and other shielding devices, as well as by instruments monitoring radiation exposure. Technologists and technicians wear badges measuring radiation levels in the radiation area, and detailed records are kept on their cumulative lifetime dose.

Employment

Radiologic technologists and technicians held about 167,000 jobs in 2000. About 1 in 5 worked part-time. More than half of all jobs are in hospitals. Most of the rest are in physicians' offices and clinics, including diagnostic imaging centers.

Training, Other Qualifications, and Advancement

Preparation for this profession is offered in hospitals, colleges and universities, vocational-technical institutes, and the U.S. Armed Forces. Hospitals, which employ most radiologic technologists and technicians, prefer to hire those with formal training.

Formal training programs in radiography range in length from 1 to 4 years and lead to a certificate, associate degree, or bachelor's degree. Two-year associate degree programs are most prevalent.

Some 1-year certificate programs are available for experienced radiographers or individuals from other health occupations, such as medical technologists and registered nurses, who want to change fields or specialize in computerized tomography or magnetic resonance imaging. A bachelor's or master's degree in one of the radiologic technologies is desirable for supervisory, administrative, or teaching positions.

The Joint Review Committee on Education in Radiologic Technology accredits most formal training programs for this field. They accredited 584 radiography programs in 2000. Radiography programs require, at a minimum, a high school diploma or the equivalent. High school courses in mathematics, physics, chemistry, and biology are helpful. The programs provide both classroom and clinical instruction in anatomy and physiology, patient care procedures, radiation physics, radiation protection, principles of imaging, medical terminology, positioning of patients, medical ethics, radiobiology, and pathology.

In 1981, Congress passed the Consumer-Patient Radiation Health and Safety Act, which aims to protect the public from the hazards of unnecessary exposure to medical and dental radiation by ensuring operators of radiologic equipment are properly trained. Under the act, the federal government sets voluntary standards that the states, in turn, may use for accrediting training programs and certifying individuals who engage in medical or dental radiography.

In 1999, 35 states and Puerto Rico licensed radiologic technologists and technicians. The American Registry of Radiologic Technologists (ARRT) offers voluntary registration in radiography. To be eligible for registration, technologists generally must graduate from an accredited program and pass an examination. Many employers prefer to hire registered radiographers. To be recertified, radiographers must complete 24 hours of continuing education every other year.

Radiologic technologists and technicians should be sensitive to patients' physical and psychological needs. They must pay attention to detail, follow instructions, and work as part of a team. In addition, operating complicated equipment requires mechanical ability and manual dexterity.

With experience and additional training, staff technologists may become specialists, performing CT scanning, angiography, and magnetic resonance imaging. Experienced technologists may also be promoted to supervisor, chief radiologic technologist, and—ultimately—department administrator or director. Depending on the institution, courses or a master's degree in business or health administration may be necessary for the director's position. Some technologists progress by becoming instructors or directors in radiologic technology programs; others take jobs as sales representatives or instructors with equipment manufacturers.

Job Outlook

Employment of radiologic technologists and technicians is expected to grow faster than the average for all occupations through 2010, as the population grows and ages, increasing the demand for diagnostic imaging. Opportunities are expected to be favorable. Some employers report shortages of radiologic technologists and technicians. Imbalances between the supply of qualified workers and demand should spur efforts to attract and retain qualified radiologic technologists and technicians. For example, employers may provide more flexible training programs, or improve compensation and working conditions.

Although physicians are enthusiastic about the clinical benefits of new technologies, the extent to which they are adopted depends largely on cost and reimbursement considerations. For example, digital imaging technology can improve quality and efficiency, but remains expensive. Some promising new technologies may not come into widespread use because they are too expensive and third-party payers may not be willing to pay for their use.

Radiologic technologists who are educated and credentialed in more than one type of diagnostic imaging technology, such as radiography and sonography or nuclear medicine, will have better employment opportunities as employers look for new ways to control costs. In hospitals, multi-skilled employees will be the most sought after, as hospitals respond to cost pressures by continuing to merge departments.

Hospitals will remain the principal employer of radiologic technologists and technicians. However, a greater number of new jobs will be found in offices and clinics of physicians, including diagnostic imaging centers. Health facilities such as these are expected to grow very rapidly through 2010 due to the strong shift toward outpatient care, encouraged by third-party payers and made possible by technological advances that permit more procedures to be performed outside the hospital. Some job openings will also arise from the need to replace technologists and technicians who leave the occupation.

Earnings

Median annual earnings of radiologic technologists and technicians were $36,000 in 2000. The middle 50 percent earned between $30,220 and $43,380. The lowest 10 percent earned less than $25,310, and the highest 10 percent earned more than $52,050. Median annual earnings in the industries employing the largest numbers of radiologic technologists and technicians in 2000 were:

Medical and dental laboratories	$39,400
Hospitals	36,280
Offices and clinics of medical doctors	34,870

Related Occupations

Radiologic technologists and technicians operate sophisticated equipment to help physicians, dentists, and other health practitioners diagnose and treat patients. Workers in related occupations include cardiovascular technologists and technicians, clinical laboratory technologists and technicians, diagnostic medical sonographers, nuclear medicine technologists, radiation therapists, and respiratory therapists.

Sources of Additional Information

For career information, send a stamped, self-addressed business size envelope with your request to:

- American Society of Radiologic Technologists, 15000 Central Ave. S.E., Albuquerque, NM 87123-3917. Internet: http://www.asrt.org/asrt.htm

For the current list of accredited education programs in radiography, write to:

- Joint Review Committee on Education in Radiologic Technology, 20 N. Wacker Dr., Suite 600, Chicago, IL 60606-2901. Internet: http://www.jrcert.org

For information on certification, contact:

- American Registry of Radiologic Technologists, 1255 Northland Dr., St. Paul, MN 55120-1155. Internet: http://www.arrt.org

Registered Nurses

O*NET 29-1111.00

Significant Points

- The largest healthcare occupation, with more than 2 million jobs.
- One of the 10 occupations projected to have the largest numbers of new jobs.
- Job opportunities are expected to be very good.
- Earnings are above average, particularly for advanced practice nurses, who have additional education or training.

Nature of the Work

Registered nurses (RNs) work to promote health, prevent disease, and help patients cope with illness. They are advocates and health educators for patients, families, and communities. When providing direct patient care, they observe, assess, and record symptoms, reactions, and progress; assist physicians during treatments and examinations; administer medications; and assist in convalescence and rehabilitation. RNs also develop and manage nursing care plans; instruct patients and their families in proper care; and help individuals and groups take steps to improve or maintain their health. While state laws govern the tasks that RNs may perform, it is usually the work setting that determines their daily job duties.

Hospital nurses form the largest group of nurses. Most are staff nurses, who provide bedside nursing care and carry out medical regimens. They also may supervise licensed practical nurses and nursing aides. Hospital nurses usually are assigned to one area, such as surgery, maternity, pediatrics, emergency room, intensive care, or treatment of cancer patients. Some may rotate among departments.

Office nurses care for outpatients in physicians' offices, clinics, surgicenters, and emergency medical centers. They prepare patients for and assist with examinations, administer injections and medications, dress wounds and incisions, assist with minor surgery, and maintain records. Some also perform routine laboratory and office work.

Nursing home nurses manage nursing care for residents with conditions ranging from a fracture to Alzheimer's disease. Although they often spend much of their time on administrative and supervisory tasks, RNs also assess residents' health condition, develop treatment plans, supervise licensed practical nurses and nursing aides, and perform difficult procedures such as starting intravenous fluids. They also work in specialty-care departments, such as long-term rehabilitation units for patients with strokes and head injuries.

Home health nurses provide periodic services to patients at home. After assessing patients' home environments, they care for and instruct patients and their families. Home health nurses care for a broad range of patients, such as those recovering from illnesses and accidents, cancer, and childbirth. They must be able to work independently, and may supervise home health aides.

Public health nurses work in government and private agencies and clinics, schools, retirement communities, and other community settings. They focus on populations, working with individuals, groups, and families to improve the overall health of communities. They also work as partners with communities to plan and implement programs. Public health nurses instruct individuals, families, and other groups regarding health issues, disease prevention, nutrition, and childcare. They arrange for immunizations, blood pressure testing, and other health screening. These nurses also work with community leaders, teachers, parents, and physicians in community health education.

Occupational health or *industrial nurses* provide nursing care at work sites to employees, customers, and others with minor injuries and illnesses. They provide emergency care, prepare accident reports, and arrange for further care if necessary. They also offer health counseling, assist with health examinations and inoculations, and assess work environments to identify potential health or safety problems.

Head nurses or *nurse supervisors* direct nursing activities. They plan work schedules and assign duties to nurses and aides, provide or arrange for training, and visit patients to observe nurses and to ensure the proper delivery of care. They also may see that records are maintained and equipment and supplies are ordered.

At the advanced level, *nurse practitioners* provide basic primary healthcare. They diagnose and treat common acute illnesses and injuries. Nurse practitioners also can prescribe medications, but certification and licensing requirements vary by state. Other advanced practice nurses include *clinical nurse specialists, certified registered nurse anesthetists,* and *certified nurse-midwives*. Advanced practice nurses must meet higher educational and clinical practice requirements beyond the basic nursing education and licensing required of all RNs.

Working Conditions

Most nurses work in well-lighted, comfortable healthcare facilities. Home health and public health nurses travel to patients' homes, schools, community centers, and other sites. Nurses may spend considerable time walking and standing. They need emotional stability to cope with human suffering, emergencies, and other stresses. Patients in hospitals and nursing homes require 24-hour care; consequently, nurses in these institutions may work nights, weekends, and holidays. RNs also may be on-call (available to work on short notice). Office, occupational health, and public health nurses are more likely to work regular business hours. Almost 1 in 10 RNs held more than one job in 2000.

Nursing has its hazards, especially in hospitals, nursing homes, and clinics where nurses may care for individuals with infectious diseases. Nurses must observe rigid guidelines to guard against disease and other dangers, such as those posed by radiation, chemicals used for sterilization of instruments, and anesthetics. In addition, they are vulnerable to back injury when moving patients, shocks from electrical equipment, and hazards posed by compressed gases.

Employment

As the largest healthcare occupation, registered nurses held about 2.2 million jobs in 2000. About 3 out of 5 jobs were in hospitals, in inpatient and outpatient departments. Others were mostly in offices and clinics of physicians and other health practitioners, home healthcare agencies, nursing homes, temporary help agencies, schools, and government agencies. The remainder worked in residential care facilities, social service agencies, religious organizations, research facilities, management and public relations firms, insurance agencies, and private households. About 1 out of 4 RNs worked part-time.

Training, Other Qualifications, and Advancement

In all states and the District of Columbia, students must graduate from an approved nursing program and pass a national licensing examination to obtain a nursing license. Nurses may be licensed in more than one state, either by examination, by endorsement of a license issued by another state, or through a multistate licensing agreement. All states require periodic license renewal, which may involve continuing education.

There are three major educational paths to registered nursing: associate degree in nursing (ADN), bachelor of science degree in nursing (BSN), and diploma. ADN programs, offered by community and junior colleges, take about 2 to 3 years. About half of the 1,700 RN programs in 2000 were at the ADN level. BSN programs, offered by colleges and universities, take 4 or 5 years. More than one-third of all programs in 2000 offered degrees at the bachelor's level. Diploma programs, administered in hospitals, last 2 to 3 years. Only a small number of programs offer diploma-level degrees. Generally, licensed graduates of any of the three program types qualify for entry-level positions as staff nurses.

Many ADN and diploma-educated nurses later enter bachelor's programs to prepare for a broader scope of nursing practice. They can often find a staff nurse position and then take advantage of tuition reimbursement programs to work toward a BSN.

Individuals considering nursing should carefully weigh the pros and cons of enrolling in a BSN program because, if they do so, their advancement opportunities usually are broader. In fact, some career paths are open only to nurses with bachelor's or advanced degrees. A bachelor's degree is often necessary for administrative positions, and it is a prerequisite for admission to graduate nursing programs in research, consulting, teaching, or a clinical specialization.

Nursing education includes classroom instruction and supervised clinical experience in hospitals and other health facilities. Students take courses in anatomy, physiology, microbiology, chemistry, nutrition, psychology and other behavioral sciences, and nursing. Coursework also includes the liberal arts.

Supervised clinical experience is provided in hospital departments such as pediatrics, psychiatry, maternity, and surgery. A growing number of programs include clinical experience in nursing homes, public health departments, home health agencies, and ambulatory clinics.

Nurses should be caring and sympathetic. They must be able to accept responsibility, direct or supervise others, follow orders precisely, and determine when consultation is required.

Experience and good performance can lead to promotion to more responsible positions. Nurses can advance, in management, to assistant head nurse or head nurse. From there, they can advance to assistant director, director, and vice president. Increasingly, management-level nursing positions require a graduate degree in nursing or health services administration. They also require leadership, negotiation skills, and good judgment. Graduate programs preparing executive-level nurses usually last 1 to 2 years.

Within patient care, nurses can advance to clinical nurse specialist, nurse practitioner, certified nurse-midwife, or certified registered nurse anesthetist. These positions require 1 or 2 years of graduate education, leading to a master's degree or, in some instances, to a certificate.

Some nurses move into the business side of healthcare. Their nursing expertise and experience on a healthcare team equip them to manage ambulatory, acute, home health, and chronic care services. Healthcare corporations employ nurses for health planning and development, marketing, and quality assurance. Other nurses work as college and university faculty or do research.

Job Outlook

Job opportunities for RNs are expected to be very good. Employment of registered nurses is expected to grow faster than the average for all occupations through 2010, and because the occupation is very large, many new jobs will result. Thousands of job openings also will result from the need to replace experienced nurses who leave the occupation, especially as the median age of the registered nurse population continues to rise.

Some states report current and projected shortages of RNs, primarily due to an aging RN workforce and recent declines in nursing school enrollments. Imbalances between the supply of and demand for qualified workers should spur efforts to attract and retain qualified RNs. For example, employers may restructure workloads, improve compensation and working conditions, and subsidize training or continuing education.

Faster-than-average growth will be driven by technological advances in patient care, which permit a greater number of medical problems to be treated, and an increasing emphasis on preventive care. In addition, the number of older people, who are much more likely than younger people to need nursing care, is projected to grow rapidly.

Employment in hospitals, the largest sector, is expected to grow more slowly than in other healthcare sectors. While the intensity of nursing care is likely to increase, requiring more nurses per patient, the number of inpatients (those who remain in the hospital for more than 24 hours) is not likely to increase much. Patients are being discharged earlier and more procedures are being done on an outpatient basis, both in and outside hospitals. However, rapid growth is expected in hospital outpatient facilities, such as those providing same-day surgery, rehabilitation, and chemotherapy.

Employment in home healthcare is expected to grow rapidly. This is in response to the growing number of older persons with functional disabilities, consumer preference for care in the home, and technological advances that make it possible to bring increasingly complex treatments into the home. The type of care demanded will require nurses who are able to perform complex procedures.

Employment in nursing homes is expected to grow faster than average due to increases in the number of elderly, many of whom require long-term care. In addition, the financial pressure on hospitals to discharge patients as soon as possible should produce more nursing home admissions. Growth in units that provide specialized long-term rehabilitation for stroke and head injury patients or that treat Alzheimer's victims also will increase employment.

An increasing proportion of sophisticated procedures, which once were performed only in hospitals, are being performed in physicians' offices and clinics, including ambulatory surgicenters and emergency medical centers. Accordingly, employment is expected to grow faster than average in these places as healthcare in general expands.

In evolving integrated healthcare networks, nurses may rotate among employment settings. Because jobs in traditional hospital nursing positions are no longer the only option, RNs will need to be flexible. Opportunities should be excellent, particularly for nurses with advanced education and training.

Earnings

Median annual earnings of registered nurses were $44,840 in 2000. The middle 50 percent earned between $37,870 and $54,000. The lowest 10 percent earned less than $31,890, and the highest 10 percent earned more than $64,360. Median annual earnings in the industries employing the largest numbers of registered nurses in 2000 were as follows:

Personnel supply services	$46,860
Hospitals	45,780
Home healthcare services	43,640
Offices and clinics of medical doctors	43,480
Nursing and personal care facilities	41,330

Many employers offer flexible work schedules, childcare, educational benefits, and bonuses.

Related Occupations

Workers in other healthcare occupations with responsibilities and duties related to those of registered nurses are emergency medical technicians and paramedics, occupational therapists, physical therapists, physician assistants, and respiratory therapists.

Sources of Additional Information

For information on a career as a registered nurse and nursing education, contact:

- National League for Nursing, 61 Broadway, New York, NY 10006. Internet: http://www.nln.org

For a list of BSN and graduate nursing programs, write to:

- American Association of Colleges of Nursing, 1 Dupont Circle NW, Suite 530, Washington, DC 20036. Internet: http://www.aacn.nche.edu

Information on registered nurses also is available from:

- American Nurses Association, 600 Maryland Ave. SW, Washington, DC 20024-2571. Internet: http://www.nursingworld.org

Respiratory Therapists

O*NET 29-1126.00, 29-2054.00

Significant Points

- Hospitals will continue to employ more than 8 out of 10 respiratory therapists, but a growing number of therapists will work in respiratory therapy clinics, nursing homes, home health agencies, and firms that supply respiratory equipment for home use.

- Job opportunities will be best for therapists with cardiopulmonary care skills or experience working with newborns and infants.

Nature of the Work

Respiratory therapists and respiratory therapy technicians (also known as *respiratory care practitioners)* evaluate, treat, and care for patients with breathing disorders. *Respiratory therapists* assume primary responsibility for all respiratory care treatments, including the supervision of respiratory therapy technicians. *Respiratory therapy technicians* provide specific, well-defined respiratory care procedures under the direction of respiratory therapists and physicians. In clinical practice, many of the daily duties of therapists and technicians overlap, although therapists generally have more experience than technicians. In this statement, the term *respiratory therapists* includes both respiratory therapists and respiratory therapy technicians.

To evaluate patients, respiratory therapists test the capacity of the lungs and analyze oxygen and carbon dioxide concentration. They also measure the patient's potential of hydrogen (pH), which indicates the acidity or alkalinity level of the blood. To measure lung capacity, patients breathe into an instrument that measures the volume and flow of oxygen during inhalation and exhalation. By comparing the reading with the norm for the patient's age, height, weight, and sex, respiratory therapists can determine whether lung deficiencies exist. To analyze oxygen, carbon dioxide, and pH levels, therapists draw an arterial blood sample, place it in a blood gas analyzer, and relay the results to a physician.

Respiratory therapists treat all types of patients, ranging from premature infants whose lungs are not fully developed, to elderly people whose lungs are diseased. These workers provide temporary relief to patients with chronic asthma or emphysema, as well as emergency care to patients who are victims of a heart attack, stroke, drowning, or shock.

To treat patients, respiratory therapists use oxygen or oxygen mixtures, chest physiotherapy, and aerosol medications. To increase a patient's concentration of oxygen, therapists place an oxygen mask or nasal cannula on a patient and set the oxygen flow at the level prescribed by a physician. Therapists also connect patients who cannot breathe on their own to ventilators that deliver pressurized oxygen into the lungs. They insert a tube into a patient's trachea, or windpipe; connect the tube to the ventilator; and set the rate, volume, and oxygen concentration of the oxygen mixture entering the patient's lungs.

Therapists regularly check on patients and equipment. If the patient appears to be having difficulty, or if the oxygen, carbon dioxide, or pH level of the blood is abnormal, they change the ventilator setting according to the doctor's order or check equipment for mechanical problems. In homecare, therapists teach patients and their families to use ventilators and other life support systems. Additionally, they visit several times a month to inspect and clean equipment and ensure its proper use and make emergency visits, if equipment problems arise.

Respiratory therapists perform chest physiotherapy on patients to remove mucus from their lungs and make it easier for them to breathe. For example, during surgery, anesthesia depresses respiration, so this treatment may be prescribed to help get the patient's lungs back to normal and to prevent congestion. Chest physiotherapy also helps patients suffering from lung diseases, such as cystic fibrosis, that cause mucus to collect in the lungs. In this procedure, therapists place patients in positions to help drain mucus, thump and vibrate patients' rib cages, and instruct them to cough.

Respiratory therapists also administer aerosols—liquid medications suspended in a gas that forms a mist which is inhaled—and teach patients how to inhale the aerosol properly to assure its effectiveness.

In some hospitals, therapists perform tasks that fall outside their traditional role. Tasks are expanding into cardiopulmonary procedures like electrocardiograms and stress testing, as well as other tasks like drawing blood samples from patients. Therapists also keep records of materials used and charges to patients.

Working Conditions

Respiratory therapists generally work between 35 and 40 hours a week. Because hospitals operate around the clock, therapists may work evenings, nights, or weekends. They spend long periods standing and walking between patients' rooms. In an emergency, therapists work under a great deal of stress.

Because gases used by respiratory therapists are stored under pressure, they are potentially hazardous. However, adherence to safety precautions and regular maintenance and testing of equipment minimize the risk of injury. As in many other health occupations, respiratory therapists run a risk of catching infectious diseases, but carefully following proper procedures minimizes this risk.

Employment

Respiratory therapists held about 110,000 jobs in 2000. More than 4 out of 5 jobs were in hospital departments of respiratory care, anesthesiology, or pulmonary medicine. Respiratory therapy clinics, offices of physicians, nursing homes, and firms that supply respiratory equipment for home use accounted for most of the remaining jobs.

Training, Other Qualifications, and Advancement

Formal training is necessary for entry to this field. Training is offered at the postsecondary level by medical schools, colleges and universities, trade schools, vocational-technical institutes, and the armed forces. Formal training programs vary in length and in the credential or degree awarded. Some programs award associate or bachelor's degrees and prepare graduates for jobs as registered respiratory therapists (RRTs). Other, shorter programs award certificates and lead to jobs as entry-level certified respiratory therapists (CRTs). According to the Committee on Accreditation for Respiratory Care (CoARC), there were 334 accredited RRT programs and 102 accredited CRT programs in the United States in 2000.

Areas of study for respiratory therapy programs include human anatomy and physiology, chemistry, physics, microbiology, and mathematics. Technical courses deal with procedures, equipment, and clinical tests.

More than 40 states license respiratory care personnel. Aspiring respiratory care practitioners should check on licensure requirements with the board of respiratory care examiners for the state in which they plan to work.

The National Board for Respiratory Care (NBRC) offers voluntary certification and registration to graduates of CoARC-accredited programs. Two credentials are awarded to respiratory therapists who satisfy the requirements: Registered Respiratory Therapist (RRT) and Certified Respiratory Therapist (CRT). Graduates from 2- and 4-year programs in respiratory therapy may take the CRT examination. CRTs who meet education and experience requirements can take two separate examinations, leading to the award of the RRT. Either the CRT or RRT examination is the standard in the states requiring licensure.

Most employers require applicants for entry-level or generalist positions to hold the CRT or be eligible to take the certification examination. Supervisory positions and those in intensive care specialties usually require the RRT (or RRT eligibility).

Therapists should be sensitive to patients' physical and psychological needs. Respiratory care practitioners must pay attention to detail, follow instructions, and work as part of a team. In addition, operating complicated equipment requires mechanical ability and manual dexterity.

High school students interested in a career in respiratory care should take courses in health, biology, mathematics, chemistry, and physics. Respiratory care involves basic mathematical problem solving and an understanding of chemical and physical principles. For example, respiratory care workers must be able to compute medication dosages and calculate gas concentrations.

Respiratory therapists advance in clinical practice by moving from care of general to critical patients who have significant problems in other organ systems, such as the heart or kidneys. Respiratory therapists, especially those with 4-year degrees, may also advance to supervisory or managerial positions in a respiratory therapy department. Respiratory therapists in home care and equipment rental firms may become branch managers. Some respiratory therapists advance by moving into teaching positions.

Job Outlook

Job opportunities are expected to remain good. Employment of respiratory therapists is expected to increase faster than the average for all occupations through the year 2010, because of substantial growth of the middle-aged and elderly population—a development that will heighten the incidence of cardiopulmonary disease.

Older Americans suffer most from respiratory ailments and cardiopulmonary diseases such as pneumonia, chronic bronchitis, emphysema, and heart disease. As their numbers increase, the need for respiratory therapists will increase, as well. In addition, advances in treating victims of heart attacks, accident victims, and premature infants (many of whom are dependent on a ventilator during part of their treatment) will increase the demand for the services of respiratory care practitioners.

Opportunities are expected to be favorable for respiratory therapists with cardiopulmonary care skills and experience working with infants.

Although hospitals will continue to employ the vast majority of therapists, a growing number of therapists can expect to work outside of hospitals in respiratory therapy clinics, offices of physicians, nursing homes, or homecare.

Earnings

Median annual earnings of respiratory therapists were $37,680 in 2000. The middle 50 percent earned between $32,140 and $43,430. The lowest 10 percent earned less than $28,620, and the highest 10 percent earned more than $50,660. In hospitals, median annual earnings of respiratory therapists were $38,040 in 2000.

Median annual earnings of respiratory therapy technicians were $32,860 in 2000. The middle 50 percent earned between $27,280 and $39,740. The lowest 10 percent earned less than $22,830, and the highest 10 percent earned more than $46,800. Median annual earnings of respiratory therapy technicians employed in hospitals were $32,830 in 2000.

Related Occupations

Respiratory therapists, under the supervision of a physician, administer respiratory care and life support to patients with heart and lung difficulties. Other workers who care for, treat, or train people to improve their physical condition include registered nurses, occupational therapists, physical therapists, and radiation therapists.

Sources of Additional Information

Information concerning a career in respiratory care is available from:

- American Association for Respiratory Care, 11030 Ables Ln., Dallas, TX 75229-4593. Internet: http://www.aarc.org

For the current list of CoARC-accredited educational programs for respiratory care practitioners, write to:

- Committee on Accreditation for Respiratory Care, 1248 Harwood Rd., Bedford, TX 76021-4244

Information on gaining credentials in respiratory care and a list of state licensing agencies can be obtained from:

- The National Board for Respiratory Care, Inc., 8310 Nieman Rd., Lenexa, KS 66214-1579. Internet: http://www.nbrc.org

Sales Engineers

O*NET 41-9031.00

Significant Points

- A bachelor's degree in engineering is required; many sales engineers have previous work experience in an engineering specialty.
- Projected employment growth stems from the increasing variety and number of goods to be sold.
- Employment opportunities and earnings may fluctuate from year to year.

Nature of the Work

Many products and services, especially those purchased by large companies and institutions, are highly complex. Sales engineers, using their engineering skills, help customers determine which products or services provided by the sales engineer's employer best suit their needs. Sales engineers—who also may be called manufacturers' agents, sales representatives, or technical sales support workers—often work with both the customer and the production, engineering, or research and development departments of their company or of independent firms to determine how products and services could be designed or modified to best suit the customer's needs. They also may advise the customer on how to best utilize the products or services being provided.

Selling, of course, is an important part of the job. Sales engineers use their technical skills to demonstrate to potential customers how and why the products or services they are selling would suit the customer better than competitors' products. Often, there may not be a directly competitive product. In these cases, the job of the sales engineer is to demonstrate to the customer the usefulness of the product or service—for example, how much new production machinery would save the customer.

Most sales engineers have a bachelor's degree in engineering and some have previous work experience in an engineering specialty before becoming a sales engineer. Engineers apply the theories and principles of science and mathematics to technical problems. Their work is the link between scientific discoveries and commercial applications. Many sales engineers specialize in an area related to an engineering specialty. For example, sales engineers selling chemical products may have a background as a chemical engineer while those selling electrical products may have a degree in electrical engineering.

Many of the job duties of sales engineers are similar to those of other salespersons. They must interest the client in purchasing their products, many of which are durable manufactured products such as turbines. Sales engineers are often teamed with other salespersons who concentrate on the marketing and sales, enabling the sales engineer to concentrate on the technical aspects of the job. By working as a sales team, each member is able to utilize his or her strengths and knowledge.

Sales engineers tend to employ selling techniques that are different from those used by most other sales workers. They may use a "consultative" style; that is, they focus on the client's problem and show how it could be solved or mitigated with their product or service. This selling style differs from the "benefits and features" method, whereby the product is described and the customer is left to decide how the product would be useful.

In addition to maintaining current clients and attracting new ones, sales engineers help clients work out any problems that arise when the product is installed, and may continue to serve as a liaison between the client and their company. In addition, due to their familiarity with the client's needs, sales engineers may help identify and develop potential new products.

Sales engineers may work directly for manufacturers or service-providers, or in small independent firms. In an independent firm, they may sell a complimentary line of products from several different suppliers, in which case they are paid on a commission basis.

Working Conditions

Many sales engineers work more than 40 hours per week to meet sales goals and their clients' needs. Selling can be stressful because sales engineers' income and job security often directly depend on their success in sales and customer service.

Some sales engineers have large territories and travel extensively. Because sales regions may cover several states, they may be away from home for several days or even weeks at a time. Others work near their "home base" and travel mostly by automobile. However, international travel is becoming more important to secure contracts with foreign customers.

Although the hours may be long and are often irregular, many sales engineers have the freedom to determine their own schedule. Consequently, they often can arrange their appointments so they can have time off when they want it. However, most independent sales workers do not earn any income while on vacation.

Employment

Sales engineers held about 85,000 jobs in 2000. Almost two-thirds were in durable goods manufacturing industries—for example, industrial machinery and equipment, measuring and controlling devices, or electronic and other electrical equipment—and wholesale trade, including machinery, equipment, and supplies. Services and nondurable goods manufacturing industries employed most of the remaining sales engineers.

Unlike many other sales occupations, very few sales engineers are self-employed.

Training, Other Qualifications, and Advancement

A bachelor's degree in engineering is required to become a sales engineer. However, some workers with previous experience in sales combined with technical experience or training sometimes hold the title of sales engineer. Also, workers who have a degree in a science, such as chemistry, or even a degree in business with little or no previous sales experience, may be termed sales engineers.

Admissions requirements for undergraduate engineering schools include a solid background in mathematics (algebra, geometry, trigonometry, and calculus), physical sciences (biology, chemistry, and physics), and courses in English, social studies, humanities, and computer science. University programs vary in content. For example, some programs emphasize industrial practices, preparing students for a job in industry, whereas others are more theoretical and prepare students for graduate school. Therefore, students should investigate curricula and check accreditations carefully before selecting a college. Once a college has been selected, a student must choose an area of engineering in which to specialize. Some programs offer a general engineering curriculum; students then specialize in graduate school or on the job. Most engineering degrees are granted in electrical, mechanical, or civil engineering. However, engineers trained in one branch may work in related branches.

Many sales engineers first worked as engineers. For some, the engineering experience was necessary to obtain the technical background needed to effectively sell their employers' products or services. Others moved into the occupation because it offered better earnings and advancement potential or because they were looking for a new challenge.

New graduates with engineering degrees may need sales experience and training to obtain employment directly as a sales engineer. This may involve teaming with a sales mentor who is familiar with the business practices, customers, and company procedures and culture. After the training period has been completed, the sales engineer may continue to partner with someone who lacks technical skills, yet excels in the art of sales.

Promotion may include a higher commission rate, larger sales territory, or promotion to supervisor or marketing manager. In other cases, sales engineers may leave their companies and form a small independent firm that may offer higher commissions and more freedom. Independent firms tend to be small, although relatively few sales engineers are self-employed.

It is important for sales engineers to continue their education throughout their careers because much of their value to their employer depends on their knowledge of the latest technology. Sales engineers in high-technology areas, such as information technology or advanced electronics, may find that technical knowledge can become obsolete rapidly.

Job Outlook

Employment of sales engineer is expected to grow about as fast as the average for all occupations through the year 2010. Projected employment growth stems from the increasing variety and number of goods to be sold. Competitive pressures and advancing technology will force companies to improve and update product designs more frequently and to optimize their manufacturing and sales processes.

Employment opportunities and earnings may fluctuate from year to year because sales are affected by changing economic conditions, legislative issues, and consumer preferences. Prospects will be best for those with the appropriate knowledge or technical expertise, as well as the personal traits necessary for successful sales work.

While most job openings will be new positions created as companies expand their sales forces, a relatively small number of openings will arise each year from the need to replace sales workers who transfer to other occupations or leave the labor force.

Earnings

Compensation methods vary significantly by the type of firm and product sold. Most employers use a combination of salary and commission or salary plus bonus. Commissions usually are based on the amount of sales, whereas bonuses may depend on individual performance, on the performance of all sales workers in the group or district, or on the company's performance. Earnings from commissions and bonuses may vary greatly from year to year, depending on sales ability, the demand for the company's products or services, and the overall economy.

Median annual earnings of sales engineers, including commission, were $56,520 in 2000. The middle 50 percent earned between $44,240 and $76,230 a year. The lowest 10 percent earned less than $33,930 and the highest 10 percent earned more than $95,560 a year. Median annual earnings in the industries employing the largest number of sales engineers in 2000 were as follows:

Electrical goods	$67,100
Computer and data processing services	60,810
Professional and commercial equipment	49,860

In addition to their earnings, sales engineers who work for manufacturers are usually reimbursed for expenses such as transportation, meals, hotels, and customer entertainment. In addition to typical benefits, sales engineers often get personal use of a company car and frequent-flyer mileage. Some companies offer incentives such as free vacation trips or gifts for outstanding performance. Sales engineers who work in independent firms may have higher but less stable earnings and, often, relatively few benefits.

Related Occupations

Sales engineers must have sales ability and knowledge of the products they sell, as well as technical and analytical skills. Other occupations that require similar skills include advertising, marketing, promotions, public relations, and sales managers; engineers; insurance sales agents; purchasing managers, buyers, and purchasing agents; real estate brokers and sales agents; sales representatives, wholesale and manufacturing; and securities, commodities, and financial services sales agents.

Sources of Additional Information

For more information about becoming a sales engineer, contact:

- Manufacturers' Agents National Association, P.O. Box 3467, Laguna Hills, CA 92654-3467. Internet: http://www.manaonline.org

Career and certification information is available from:

- Manufacturers' Representatives Educational Research Foundation, P.O. Box 247, Geneva, IL 60134. Internet: http://www.mrerf.org

Science Technicians

O*NET 19-4011.01, 19-4011.02, 19-4021.00, 19-4031.00, 19-4041.01, 19-4041.02, 19-4051.01, 19-4051.02, 19-4091.00, 19-4092.00, 19-4093.00

Significant Points

- Science technicians in production jobs often work in 8-hour shifts around the clock.
- Job opportunities are expected to be best for qualified graduates of science technician training programs or applied science technology programs.

Nature of the Work

Science technicians use the principles and theories of science and mathematics to solve problems in research and development and to help invent and improve products and processes. However, their jobs are more practically oriented than those of scientists. Technicians set up, operate, and maintain laboratory instruments, monitor experiments, make observations, calculate and record results, and often develop conclusions. They must keep detailed logs of all their work-related activities. Those who work in production monitor manufacturing processes and may be involved in ensuring quality by testing products for proper proportions of ingredients, purity, or for strength and durability.

As laboratory instrumentation and procedures have become more complex in recent years, the role of science technicians in research and development has expanded. In addition to performing routine tasks, many technicians also develop and adapt laboratory procedures to achieve the best results, interpret data, and devise solutions to problems, under the direction of scientists. Moreover, technicians must master the laboratory equipment so that they can adjust settings when necessary and recognize when equipment is malfunctioning.

The increasing use of robotics to perform many routine tasks has freed technicians to operate more sophisticated laboratory equipment. Science technicians make extensive use of computers, computer-interfaced equipment, robotics, and high-technology industrial applications, such as biological engineering.

Most science technicians specialize, learning skills and working in the same disciplines as scientists. Occupational titles, therefore, tend to follow the same structure as scientists. *Agricultural technicians* work with agricultural scientists in food, fiber, and animal research, production, and processing. Some conduct tests and experiments to improve the yield and quality of crops or to increase the resistance of plants and animals to disease, insects, or other hazards. Other agricultural technicians do animal breeding and nutrition work. *Food science technicians* assist food scientists and technologists in research and development, production technology, and quality control. For example, food science technicians may conduct tests on food additives and preservatives to ensure FDA compliance on factors such as color, texture, and nutrients. They analyze, record, and compile test results; order supplies to maintain laboratory inventory; and clean and sterilize laboratory equipment.

Biological technicians work with biologists studying living organisms. Many assist scientists who conduct medical research—helping to find a cure for cancer or AIDS, for example. Those who work in pharmaceutical companies help develop and manufacture medicinal and pharmaceutical preparations. Those working in the field of microbiology generally work as lab assistants, studying living organisms and infectious agents. Biological technicians also analyze organic substances, such as blood, food, and drugs, and some examine evidence in criminal investigations. Biological technicians working in biotechnology labs use the knowledge and techniques gained from basic research by scientists, including gene splicing and recombinant DNA, and apply these techniques in product development.

Chemical technicians work with chemists and chemical engineers, developing and using chemicals and related products and equipment. Most do research and development, testing, or other laboratory work. For example, they might test packaging for design, integrity of materials, and environmental acceptability; assemble and operate new equipment to develop new products; monitor product quality; or develop new production techniques. Some chemical technicians collect and analyze samples of air and water to monitor pollution levels. Those who focus on basic research might produce compounds through complex organic synthesis. Chemical technicians within chemical plants are also referred to as *process technicians*. They may operate equipment, monitor plant processes and analyze plant materials.

Environmental science and protection technicians perform laboratory and field tests to monitor environmental resources and determine the contaminants and sources of pollution. They may collect samples for testing or be involved in abating, controlling, or remediating sources of environmental pollutants. Some are responsible for waste management operations, control and management of hazardous materials inventory, or general activities involving regulatory compliance. There is a growing emphasis on pollution prevention activities.

Forensic science technicians investigate crimes by collecting and analyzing physical evidence. Often, they specialize in areas such as DNA analysis or firearm examination, performing tests on weapons or substances, such as fiber, hair, tissue, or body fluids to determine significance to the investigation. They also prepare reports to document their findings and the laboratory techniques used. When criminal cases come to trial, forensic science technicians often provide testimony, as expert witnesses, on specific laboratory findings by identifying and classifying substances, materials, and other evidence collected at the crime scene.

Forest and conservation technicians compile data on the size, content, and condition of forest land tracts. These workers travel through sections of forest to gather basic information, such as species and population of trees, disease and insect damage, tree seedling mortality, and conditions that may cause fire danger. Forest and conservation technicians also train and lead forest and conservation workers in seasonal activities, such as planting tree seedlings, putting out forest fires, and maintaining recreational facilities.

Geological and petroleum technicians measure and record physical and geologic conditions in oil or gas wells, using instruments lowered into wells or by analysis of the mud from wells. In oil and gas exploration, these technicians collect and examine geological data or test geological samples to determine petroleum and mineral content. Some petroleum technicians, called *scouts*, collect information about oil and gas well drilling operations, geological and geophysical prospecting, and land or lease contracts.

Nuclear technicians operate nuclear test and research equipment, monitor radiation, and assist nuclear engineers and physicists in research. Some also operate remote control equipment to manipulate radioactive materials or materials to be exposed to radioactivity.

Other science technicians collect weather information or assist oceanographers.

Working Conditions

Science technicians work under a wide variety of conditions. Most work indoors, usually in laboratories, and have regular hours. Some occasionally work irregular hours to monitor experiments that can not be completed during regular working hours. Production technicians often work in 8-hour shifts around the clock. Others, such as agricultural, forest and conservation, geological and petroleum, and environmental science and protection technicians, perform much of their work outdoors, sometimes in remote locations.

Some science technicians may be exposed to hazards from equipment, chemicals, or toxic materials. Chemical technicians sometimes work with toxic chemicals or radioactive isotopes, nuclear technicians may be exposed to radiation, and biological technicians sometimes work with disease-causing organisms or radioactive agents. Forensic science technicians often are exposed to human body fluids and firearms. However, these working conditions pose little risk, if proper safety procedures are followed. For forensic science technicians, collecting evidence from crime scenes can be distressing and unpleasant.

Employment

Science technicians held a total of about 198,000 jobs in 2000. Employment was distributed as follows:

Chemical technicians	73,000
Biological technicians	41,000
Environmental science and protection technicians, including health	27,000
Forest and conservation technicians	18,000
Agricultural and food science technicians	18,000
Geological and petroleum technicians	10,000
Forensic science technicians	6,400
Nuclear technicians	3,300

Chemical technicians held jobs in a wide range of manufacturing and service industries, but were concentrated in chemical manufacturing, where they held over 30,000 jobs. A significant number, 12,000, worked in research and testing firms. About 45 percent of biological technicians also worked in research and testing firms. Most of the rest of biological technicians worked in drug manufacturing or for federal, state, or local governments. Significant numbers of environmental science and protection technicians also worked for state and local governments and research and testing services. Others worked for engineering and architectural services and management and public relations firms. Nearly 68 percent of forest and conservation technicians held jobs in federal government; another 22 percent worked for state governments. More than 30 percent of agricultural and food science technicians worked for food processing companies; the rest worked for research and testing firms, state governments, and non-veterinary animal services. More than 47 percent of geological and petroleum technicians worked for oil and gas extraction companies, and forensic science technicians worked primarily for state and local governments.

Training, Other Qualifications, and Advancement

There are several ways to qualify for a job as a science technician. Many employers prefer applicants who have at least 2 years of specialized training or an associate degree in applied science or science-related technology. Because employers' preferences vary, however, some science technicians have a bachelor's degree in chemistry, biology, or forensic science, or have taken several science and math courses at 4-year colleges.

Many technical and community colleges offer associate degrees in a specific technology or a more general education in science and mathematics. A number of 2-year associate degree programs are designed to provide easy transfer to a 4-year college or university, if desired. Technical institutes usually offer technician training, but provide less theory and general education than technical or community colleges. The length of programs at technical institutes varies, although 1-year certificate programs and 2-year associate degree programs are common.

About 20 colleges or universities offer bachelor's degree programs in forensic technology, often with an emphasis in a specialty area, such as criminalistics, pathology, jurisprudence, odontology, or toxicology. In contrast to some other science technician positions that require only a 2-year degree, a 4-year degree in forensics science is usually necessary to work in the field. Forestry and conservation technicians can choose from 23 associate degree programs accredited by the Society of American Foresters.

Some schools offer cooperative-education or internship programs, allowing students the opportunity to work at a local company or other workplace, while attending classes in alternate terms. Participation in such programs can significantly enhance a student's employment prospects.

Persons interested in careers as science technicians should take as many high school science and math courses as possible. Science courses taken beyond high school, in an associate or bachelor's program, should be laboratory oriented, with an emphasis on bench skills. Because computers and computer-interfaced equipment often are used in research and development laboratories, technicians should have strong computer skills. Communication skills are also important; technicians often are required to report their findings both through speaking and in writing. Additionally, technicians should be able to work well with others, because teamwork is common. Organizational ability, an eye for detail, and skill in interpreting scientific results are also important.

Prospective science technicians can acquire good career preparation through 2-year formal training programs that combine the teaching of scientific principles and theory with practical hands-on application in a laboratory setting with up-to-date equipment. Graduates of 4-year bachelor's degree programs in science who have considerable experience in laboratory-based courses, have completed internships, or held summer jobs in laboratories, are also well-qualified for science technician positions and are preferred by some employers. However, those with a bachelor's degree who accept technician jobs generally cannot find employment that uses their advanced academic education.

Technicians usually begin work as trainees in routine positions, under the direct supervision of a scientist or a more experienced technician. Job candidates whose training or educational background encompasses extensive hands-on experience with a variety of laboratory equipment, including computers and related equipment, usually require a short period of on-the-job training. As they gain experience, technicians take on more responsibility and carry out assignments under only general supervision, and some eventually become supervisors. However, technicians employed at universities often have their fortunes tied to particular professors; when professors retire or leave, these technicians face uncertain employment prospects.

Job Outlook

Overall employment of science technicians is expected to increase about as fast as the average for all occupations through the year 2010. Continued growth of scientific and medical research, as well as the development and production of technical products, should stimulate demand for science technicians in many industries. In particular, the growing number of agricultural and medicinal products developed from using biotechnology techniques will increase the need for biological technicians. Also, stronger competition among drug companies and an aging population are expected to contribute to the need for innovative and improved drugs, further spurring demand for biological technicians. Fastest employment growth of biological technicians should occur in the drug manufacturing industry and research and testing service firms.

The chemical and drug industry, the major employers of chemical technicians, should face stable demand for new and better pharmaceuticals and personal care products. To meet these demands, chemical and drug manufacturing firms are expected to continue to devote money to research and development, either through in-house teams or outside contractors, spurring employment growth of chemical technicians. An increasing focus on quality assurance will further stimulate demand for these workers. However, growth will be moderated somewhat by an expected slowdown in overall employment in the chemical industry.

Overall employment growth of science technicians will also be fueled by demand for environmental technicians to help regulate waste products; to collect air, water, and soil samples for measuring levels of pollutants; to monitor compliance with environmental regulations; and to clean up contaminated sites.

Demand for forest and conservation technicians at the federal and state government levels will result from a continuing emphasis on sustainability issues, such as environmental protection and responsible land management. However, employment growth may be moderated by downsizing in the federal government and continuing reductions in timber harvesting on federal lands.

Agricultural and food science technicians will be needed to assist agricultural scientists in biotechnology research as it becomes increasingly important to balance greater agricultural output with protection and preservation of soil, water, and the ecosystem. Jobs for forensic science technicians are expected to grow slowly. Crime scene technicians who work for state public safety departments may experience favorable employment prospects if the number of qualified applicants remains low.

Job opportunities are expected to be best for qualified graduates of science technician training programs or applied science technology programs who are well-trained on equipment used in industrial and government laboratories and production facilities. As the instrumentation and techniques used in industrial research, development, and production become increasingly more complex, employers are seeking well-trained individuals with highly developed technical and communication skills.

Along with opportunities created by growth, many job openings should arise from the need to replace technicians who retire or leave the labor force for other reasons. During periods of economic recession, layoffs of science technicians may occur.

Earnings

Median hourly earnings of science technicians in 2000 were as follows:

Occupation	Earnings
Nuclear technicians	$28.44
Forensic science technicians	18.04
Geological and petroleum technicians	17.55
Chemical technicians	17.05
Environmental science and protection technicians, including health	16.26
Biological technicians	15.16
Forest and conservation technicians	14.22
Agricultural and food science technicians	13.02

In the federal government in 2001, science technicians started at $17,483, $19,453, or $22,251, depending on education and experience. Beginning salaries were slightly higher in selected areas of the country where the prevailing local pay level was higher. The average annual salary for biological science technicians in nonsupervisory, supervisory, and managerial positions employed by the federal government in 2001 was $32,753; for physical science technicians, $42,657; for geodetic technicians, $53,143; for hydrologic technicians, $39,518; and for meteorologic technicians, $48,630.

Related Occupations

Other technicians who apply scientific principles at a level usually taught in 2-year associate degree programs include engineering technicians, broadcast technicians and sound engineering technicians and radio operators, drafters, and various health technologists and technicians, including clinical laboratory technologists and technicians, diagnostic medical sonographers, and radiologic technologists and technicians.

Sources of Additional Information

For information about a career as a chemical technician, contact:

- American Chemical Society, Education Division, Career Publications, 1155 16th St. N.W., Washington, DC 20036. Internet: http://www.acs.org

For career information and a list of undergraduate, graduate, and doctoral programs in forensics sciences, contact:

- American Academy of Forensic Sciences, P.O. BOX 669, Colorado Springs, CO, 80901. Internet: http://www.aafs.org

For information on forestry technicians and lists of schools offering education in forestry, send a self-addressed, stamped business envelope to:

- Society of American Foresters, 5400 Grosvenor Ln., Bethesda, MD 20814. Internet: http://www.safnet.org

Semiconductor Processors

O*NET 51-9141.00

Significant Points

- Semiconductor processors is the only production occupation whose employment expected to grow much is expected to grow faster than the average for all occupations.
- An associate degree in a relevant curriculum is increasingly required.

Nature of the Work

Electronic semiconductors—also known as computer chips, microchips, or integrated chips—are the miniature but powerful brains of high technology equipment. They are comprised of a myriad of tiny aluminum wires and electric switches, which manipulate the flow of electrical current. Semiconductor processors are responsible for many of the steps necessary to manufacture each semiconductor that goes into a personal computer, missile guidance system, and a host of other electronic equipment.

Semiconductor processors manufacture semiconductors in disks about the size of dinner plates. These disks, called wafers, are thin slices of silicon on which the circuitry of the microchips is layered. Each wafer is eventually cut into dozens of individual chips.

Semiconductor processors make wafers using photolithography, a printing process for creating plates from photographic images. Operating automated equipment, workers imprint precise microscopic patterns of the circuitry on the wafers, etch out the patterns with acids, and replace the patterns with metals that conduct electricity. Then the wafers receive a chemical bath to make them smooth, and the imprint process begins again on a new layer with the next pattern. Wafers usually have from 8 to 20 such layers of microscopic, three-dimensional circuitry.

Semiconductors are produced in semiconductor fabricating plants, or "fabs." Within fabs, the manufacture and cutting of wafers to create semiconductors takes place in "clean rooms." Clean rooms are production areas that must be kept free of any airborne matter, because the least bit of dust can damage a semiconductor. All semiconductor processors working in clean rooms—both operators and technicians—must wear special lightweight outer garments known as "bunny suits." Bunny suits fit over clothing to prevent lint and other particles from contaminating semiconductor processing work sites.

Operators, who make up the majority of the workers in clean rooms, start and monitor the sophisticated equipment that performs the various tasks during the many steps of the semiconductor production sequence. They spend a great deal of time at computer terminals, monitoring the operation of the equipment to ensure that each of the tasks in the production of the wafer is performed correctly. They also may transfer wafer carriers from one development station to the next; in newer fabs, the lifting of heavy wafer carriers and the constant monitoring for quality control are increasingly being automated, however.

Once begun, production of semiconductor wafers is continuous. Operators work to the pace of the machinery that has largely automated the production process. Operators are responsible for keeping the automated machinery within proper operating parameters.

Technicians account for a smaller percentage of the workers in clean rooms, but they troubleshoot production problems and make equipment adjustments and repairs. They also take the lead in assuring quality control and in maintaining equipment. In order to prevent the need for repairs, technicians perform diagnostic analyses and run computations. For example, technicians may determine if a flaw in a chip is due to contamination and peculiar to that wafer, or if the flaw is inherent in the manufacturing process.

Working Conditions

The work pace in clean rooms is deliberately slow. Limited movement keeps the air in cleanrooms as free as possible of dust and other particles, which can destroy semiconductors during production. Because the machinery sets operators' rate of work in the largely automated production process, workers keep an easy-going pace. Although workers spend some time alone monitoring equipment, operators and technicians spend much of their time working in teams.

Technicians are on their feet most of the day, walking through the clean room to oversee production activities. Operators spend a great deal of time sitting or standing at workstations, monitoring computer readouts and gauges. Sometimes, they must retrieve wafers from one station and take them to another.

The temperature in the clean rooms must be kept within narrow ranges, usually a comfortable 72 degrees Fahrenheit. Although bunny suits cover virtually the entire body, except perhaps the eyes, their light-weight fabric keeps the temperature inside fairly comfortable as well. However, entry and exit of workers in bunny suits from the clean room is controlled to minimize contamination, and workers must be reclothed in a clean suit and decontaminated each time they return to the clean room.

The work environment of semiconductor fabricating plants is one of the safest in any industry. Measures taken to avoid contamination of the wafers lead to more than just antiseptically clean rooms—they result in a work environment nearly free of conditions that cause occupational illnesses and accidents.

Semiconductor fabricating plants operate around the clock. For this reason, night and weekend work is common. In some plants, workers maintain standard 8-hour shifts, 5 days a week. In other plants, employees are on duty for 12-hour shifts to minimize the disruption of clean room operations brought about by shift changes. Managers in some plants allow workers to alternate schedules for equitable distribution of the "graveyard" shift.

Employment

Electronic semiconductor processors held 52,000 jobs in 2000. Nearly all of them were employed in facilities that manufacture electronic components and accessories, although a small percentage worked in plants that primarily manufacture computers and office equipment.

Training, Other Qualifications, and Advancement

People interested in becoming semiconductor processors—either operators or technicians—need a solid background in mathematics and physical sciences. In addition to their application to the complex manufacturing processes performed in fabs, math and science knowledge are essentials for pursuing higher education in semiconductor technology; and knowledge of both subjects is one of the best ways to advance in the semiconductor fabricating field.

Semiconductor processor workers must also be able to think analytically and critically to anticipate problems and avoid costly mistakes. Communication skills also are vital, as workers must be able to convey their thoughts and ideas both orally and in writing.

Employers prefer to hire persons who have completed associate degree programs for semiconductor processor jobs. A high school diploma or equivalent is the minimum requirement for entry-level operator jobs in semiconductor fabrication plants. Although completion of a 1-year certificate program in semiconductor technology offered by some community colleges is an asset, technicians must have at least an associate degree in electronics technology or a related field.

Degree or certificate candidates who get hands-on training while attending school look even more attractive to prospective employers. Semiconductor technology programs in a growing number of community colleges include an internship at a semiconductor fabricating plant; many students in these programs already hold full- or part-time jobs in the industry and work toward degrees in semiconductor technology in their spare time to update their skills or qualify for promotion to technician jobs. In addition, to ensure that operators and technicians keep their skills current, most employers provide 40 hours of formal training annually. Some employers also provide financial assistance to employees who want to earn associate and bachelor's degrees.

Summer and part-time employment provide another option for getting started in the field for those who live near a semiconductor processing plant. Students often are hired to work during the summer, and some students are allowed to continue working part-time during the school year. Students in summer and part-time semiconductor processor jobs learn what education they need to prosper in the field. They also gain valuable experience that may lead to full-time employment after graduation.

Some semiconductor processing technicians transfer to sales engineer jobs with suppliers of the machines that manufacture the semiconductors or become field support personnel.

Job Outlook

Between 2000 and 2010, employment of semiconductor processors is projected to increase faster than the average for all occupations. Besides the creation of new jobs, additional openings will result from the need to replace workers who leave the occupation. Growing demand for semiconductors and semiconductor processors will stem from the many existing and future applications for semiconductors in computers, appliances, machinery, vehicles, cell phones and other telecommunications devices, and other equipment. Job prospects should be best for people with postsecondary education in electronics or semiconductor technology.

The electronic components and accessories industry is projected to be one of the most rapidly growing manufacturing industries. Moreover, industry development of semiconductors made from better materials means that semiconductors will become even smaller, more powerful, and more durable. For example, the industry has begun producing a new generation of microchips, made with copper rather than aluminum wires, which will better conduct electricity. Also, technology to develop chips based on plastic, rather than on silicon, will make computers durable enough to be used in a variety of applications in which they could not easily have been used previously. These technological developments and new applications will lead to employment growth in the industry and more semiconductor processor jobs.

Earnings

Median hourly earnings of electronic semiconductor processors were $12.23 in 2000. The middle 50 percent earned between $10.02 and $15.36 an hour. The lowest 10 percent earned less than $8.85, and the top 10 percent earned more than $19.10 an hour.

Technicians with an associate degree in electronics or semiconductor technology generally started at higher salaries than those with less education. Almost one fourth of all electronic semiconductor processors belong to a union, considerably higher than the rate for all occupations.

Related Occupations

Electronic semiconductor processors do production work that resembles the work of precision assemblers and fabricators of electrical and electronic equipment. Also, many electronic semiconductor processors have academic training in semiconductor technology, which emphasizes scientific and engineering principles. Other occupations that require some college or postsecondary vocational training emphasizing such principles are engineering technicians, electrical and electronics engineers, and science technicians.

Sources of Additional Information

For more information on semiconductor processor careers, contact:

- Semiconductor Industry Association, 181 Metro Dr., Suite 450, San Jose, CA 95110. Internet: http://www.semichips.org
- SEMATECH, 2706 Montopolis Dr., Austin, TX 78741. Maricopa Advanced Technology Education Center (MATEC), 2323 West 14th St., Suite 540, Tempe, AZ 85281. Internet: http://matec.org

Social Scientists, Other

O*NET 19-3041.00, 19-3091.01, 19-3091.02, 19-3092.00, 19-3093.00, 19-3094.00

Significant Points

- Educational attainment of social scientists is among the highest of all occupations.
- Job opportunities are expected to be best in social service agencies, research and testing services, and management consulting firms.

Nature of the Work

The major social science occupations covered in this statement include anthropologists, archaeologists, geographers, historians, political scientists, and sociologists.

Social scientists study all aspects of society—from past events and achievements to human behavior and relationships between groups. Their research provides insights that help us understand different ways in which individuals and groups make decisions, exercise power, and respond to change. Through their studies and analyses, social scientists suggest solutions to social, business, personal, governmental, and environmental problems.

Research is a major activity for many social scientists. They use various methods to assemble facts and construct theories. Applied research usually is designed to produce information that will enable people to make better decisions or manage their affairs more effectively. Interviews and surveys are widely used to collect facts, opinions, or other information. Information collection takes many forms including living and working among the population being studied; field investigations, the analysis of historical records and documents; experiments with human or animal subjects in a laboratory; administration of standardized tests and questionnaires; and preparation and interpretation of maps and computer graphics. The work of the major specialties in social science—other than psychologists, economists, and urban and regional planners—varies greatly, although, specialists in one field may find that their research overlaps work being conducted in another discipline.

Anthropologists study the origin and the physical, social, and cultural development and behavior of humans. They may study the way of life, archaeological remains, language, or physical characteristics of people in various parts of the world. Some compare the customs, values, and social patterns of different cultures. Anthropologists usually concentrate in sociocultural anthropology, archaeology, linguistics, or biological-physical anthropology. *Sociocultural anthropologists* study customs, cultures, and social lives of groups in settings that vary from unindustrialized societies to modern urban centers.

Archaeologists recover and examine material evidence, such as ruins, tools, and pottery remaining from past human cultures in order to determine the history, customs, and living habits of earlier civilizations. *Linguistic anthropologists* study the role and changes over time of language in various cultures. *Biological-physical anthropologists* study the evolution of the human body, look for the earliest evidences of human life, and analyze how culture and biology influence one another. Most anthropologists specialize in one particular region of the world.

Geographers analyze distributions of physical and cultural phenomena on local, regional, continental, and global scales. *Economic geographers* study the distribution of resources and economic activities. *Political geographers* are concerned with the relationship of geography to political phenomena, whereas *cultural geographers* study the geography of cultural phenomena. *Physical geographers* study variations in climate, vegetation, soil, and landforms, and their implications for human activity. *Urban and transportation geographers* study cities and metropolitan areas, while *regional geographers* study the physical, economic, political, and cultural characteristics of regions, ranging in size from a congressional district to entire continents. *Medical geographers* study healthcare delivery systems, epidemiology (the study of the causes and control of epidemics), and the effect of the environment on health. (Some occupational classification systems include geographers under physical scientists rather than social scientists.)

Historians research, analyze, and interpret the past. They use many sources of additional information in their research, including government and institutional records, newspapers and other periodicals, photographs, interviews, films, and unpublished manuscripts such as personal diaries and letters. Historians usually specialize in a country or region; a particular time period; or a particular field, such as social, intellectual, cultural, political, or diplomatic history. *Biographers* collect detailed information on individuals. Other historians help study and preserve archival materials, artifacts, and historic buildings and sites.

Political scientists study the origin, development, and operation of political systems and public policy. They conduct research on a wide range of subjects such as relations between the United States and other countries, the institutions and political life of nations, the politics of small towns or a major metropolis, or the decisions of the U.S. Supreme Court. Studying topics such as public opinion, political decision-making, ideology, and public policy, they analyze the structure and operation of governments as well as various political entities. Depending on the topic, a political scientist might conduct a public opinion survey, analyze election results, analyze public documents, or interview public officials.

Sociologists study society and social behavior by examining the groups and social institutions people form, as well as various social, religious, political, and business organizations. They also study the behavior and interaction of groups, trace their origin and growth, and analyze the influence of group activities on individual members. They are concerned with the characteristics of social groups, organizations, and institutions; the ways individuals are affected by each other and by the groups to which they belong; and the effect of social traits such as sex, age, or race on a person's daily life. The results of sociological research aid educators, lawmakers, administrators, and others interested in resolving social problems and formulating public policy.

Most sociologists work in one or more specialties, such as social organization, stratification, and mobility; racial and ethnic relations; education; family; social psychology; urban, rural, political, and comparative sociology; sex roles and relations; demography; gerontology; criminology; or sociological practice.

Working Conditions

Most social scientists have regular hours. Generally working behind a desk, either alone or in collaboration with other social scientists, they read and write research reports. Many experience the pressures of writing and publishing articles, deadlines and tight schedules, and sometimes they must work overtime, for which they usually are not reimbursed. Social scientists often work as an integral part of a research team, where good communications skills are important. Travel may be necessary to collect information or attend meetings. Social scientists on foreign assignment must adjust to unfamiliar cultures, climates, and languages.

Some social scientists do fieldwork. For example, anthropologists, archaeologists, and geographers may travel to remote areas, live among the people they study, learn their languages, and stay for long periods at the site of their investigations. They may work under rugged conditions, and their work may involve strenuous physical exertion.

Social scientists employed by colleges and universities usually have flexible work schedules, often dividing their time among teaching, research and writing, consulting, or administrative responsibilities.

Employment

Social scientists held about 15,000 jobs in 2000. Many worked as researchers, administrators, and counselors for a wide range of employers, including federal, state, and local governments, educational institutions, social service agencies, research and testing services, and management consulting firms. Other employers include international organizations, associations, museums, and historical societies.

Many additional individuals with training in a social science discipline teach in colleges and universities, and in secondary and elementary schools. The proportion of social scientists that teach varies by specialty—for example, the academic world usually is a more important source of jobs for graduates in history than for graduates in the other fields of study.

Training, Other Qualifications, and Advancement

Educational attainment of social scientists is among the highest of all occupations. The PhD or equivalent degree is a minimum requirement for most positions in colleges and universities and is important for advancement to many top-level nonacademic research and administrative posts. Graduates with master's degrees in applied specialties usually have better professional opportunities outside of colleges and universities, although the situation varies by field. Graduates with a master's degree in a social science may qualify for teaching positions in community colleges. Bachelor's degree holders have limited opportunities and in most social science occupations do not qualify for "professional" positions. The bachelor's degree does, however, provide a suitable background for many different kinds of entry-level jobs, such as research assistant, administrative aide, or management or sales trainee. With the addition of sufficient education courses, social science graduates also can qualify for teaching positions in secondary and elementary schools.

Training in statistics and mathematics is essential for many social scientists. Mathematical and quantitative research methods increasingly are used in geography, political science, and other fields. The ability to use computers for research purposes is mandatory in most disciplines.

Depending on their jobs, social scientists may need a wide range of personal characteristics. Because they constantly seek new information about people, things, and ideas, intellectual curiosity and creativity are fundamental personal traits. The ability to think logically and methodically is important to a political scientist comparing, for example, the merits of various forms of government. Objectivity, open-mindedness, and systematic work habits are important in all kinds of social science research. Perseverance is essential for an anthropologist, who might spend years accumulating artifacts from an ancient civilization. Excellent written and oral communication skills are essential for all these professionals.

Job Outlook

Overall employment of social scientists is expected to grow about as fast as the average for all occupations through 2010. Prospects are best for those with advanced degrees, and usually are better in disciplines such as sociology, anthropology, and archaeology, which offer more opportunities in nonacademic settings.

Government agencies, social service organizations, marketing, research and consulting firms, and a wide range of businesses seek social science graduates, although often in jobs with titles unrelated to their academic discipline. Social scientists will face stiff competition for academic positions. However, the growing importance and popularity of social science subjects in secondary schools is strengthening the demand for social science teachers at that level.

Candidates seeking positions as social scientists can expect to encounter competition in many areas of social science. Some social science graduates, however, will find good employment opportunities in areas outside traditional social science, often in related jobs that require good research, communication, and quantitative skills.

Earnings

Median annual earnings of all other social scientists (excluding economists, psychologists, and urban and regional planners) were $48,330 in 2000. Anthropologists and archeologists had median annual earnings of $36,040; geographers, $46,690; historians, $39,860; political scientists, $81,040; and sociologists, $45,670.

In the federal government, social scientists with a bachelor's degree and no experience could start at $21,900 or $27,200 a year in 2001, depending on their college records. Those with a master's degree could start at $33,300, and those with a PhD degree could begin at $40,200, while some individuals with experience and an advanced degree could start at $48,200. Beginning salaries were slightly higher in selected areas of the country where the prevailing local pay level was higher.

Related Occupations

A number of occupations require training and personal qualities similar to those of social scientists. These include computer programmers; computer software engineers; counselors; lawyers; mathematicians; news analysts, reporters, and correspondents; postsecondary teachers; social workers; statisticians; and systems analysts.

Sources of Additional Information

For information about careers in anthropology, contact:

- The American Anthropological Association, 4350 N. Fairfax Dr., Suite 640, Arlington, VA 22203-1620. Internet: http://www.aaanet.org

For information about careers in archaeology, contact:

- Society for American Archaeology, 900 2nd St. N.E., Suite 12, Washington, DC 20002-3557. Internet: http://www.saa.org
- Archaeological Institute of America, 656 Beacon St., Boston, MA 02215-2006. Internet: http://www.archaeological.org

For information about careers in geography, contact:

- Association of American Geographers, 1710 16th St. N.W., Washington, DC 20009-3198. Internet: http://www.aag.org

Information on careers for historians is available from:

- American Historical Association, 400 A St. S.E., Washington, DC 20003-3889. Internet: http://www.theaha.org
- Organization of American Historians, 112 North Bryan Ave., Bloomington, IN 47408-4199. Internet: http://www.oah.org
- American Association for State and Local History, 1717 Church St., Nashville, TN 37203-2991. Internet: http://www.aaslh.org

For information about careers in political science, contact:

- National Association of Schools of Public Affairs and Administration, 1120 G St. N.W., Suite 730, Washington, DC 20005-3869. Internet: http://www.naspaa.org

Information about careers in sociology is available from:

- American Sociological Association, 1307 New York Ave. N.W., Suite 700, Washington, DC 20005-4712. Internet: http://www.asanet.org

For information about careers in demography, contact:

- Population Association of America, 8630 Fenton St., Suite 722, Silver Spring, MD 20910-3812. Internet: http://www.popassoc.org

Speech-Language Pathologists and Audiologists

O*NET 29-1121.00, 29-1127.00

Significant Points

- Employment of speech-language pathologists and audiologists is expected to grow rapidly because the growing population in older age groups is prone to medical conditions that result in hearing and speech problems.
- About half work in schools, and most others are employed by healthcare facilities.

- A master's degree in speech-language pathology or audiology is the standard credential.

Nature of the Work

Speech-language pathologists assess, diagnose, treat, and help to prevent speech, language, cognitive, communication, voice, swallowing, fluency, and other related disorders; audiologists identify, assess, and manage auditory, balance, and other neural systems.

Speech-language pathologists work with people who cannot make speech sounds, or cannot make them clearly; those with speech rhythm and fluency problems, such as stuttering; people with voice quality problems, such as inappropriate pitch or harsh voice; those with problems understanding and producing language; those who wish to improve their communication skills by modifying an accent; and those with cognitive communication impairments, such as attention, memory, and problem solving disorders. They also work with people who have oral motor problems causing eating and swallowing difficulties.

Speech and language problems can result from a variety of problems including hearing loss, brain injury or deterioration, cerebral palsy, stroke, cleft palate, voice pathology, mental retardation, or emotional problems. Problems can be congenital, developmental, or acquired. Speech-language pathologists use written and oral tests, as well as special instruments, to diagnose the nature and extent of impairment and to record and analyze speech, language, and swallowing irregularities. Speech-language pathologists develop an individualized plan of care, tailored to each patient's needs. For individuals with little or no speech capability, speech-language pathologists may select augmentative or alternative communication methods, including automated devices and sign language, and teach their use. They teach these individuals how to make sounds, improve their voices, or increase their language skills to communicate more effectively. Speech-language pathologists help patients develop, or recover, reliable communication skills so patients can fulfill their educational, vocational, and social roles.

Most speech-language pathologists provide direct clinical services to individuals with communication or swallowing disorders. In speech and language clinics, they may independently develop and carry out treatment programs. In medical facilities, they may work with physicians, Social Workers, Psychologists, and other therapists. Speech-language pathologists in schools develop individual or group programs, counsel parents, and may assist teachers with classroom activities.

Speech-language pathologists keep records on the initial evaluation, progress, and discharge of clients. This helps pinpoint problems, tracks client progress, and justifies the cost of treatment when applying for reimbursement. They counsel individuals and their families concerning communication disorders and how to cope with the stress and misunderstanding that often accompany them. They also work with family members to recognize and change behavior patterns that impede communication and treatment and show them communication-enhancing techniques to use at home.

Some speech-language pathologists conduct research on how people communicate. Others design and develop equipment or techniques for diagnosing and treating speech problems.

Audiologists work with people who have hearing, balance, and related problems. They use audiometers, computers, and other testing devices to measure the loudness at which a person begins to hear sounds, the ability to distinguish between sounds, and the nature and extent of hearing loss. Audiologists interpret these results and may coordinate them with medical, educational, and psychological information to make a diagnosis and determine a course of treatment.

Hearing disorders can result from a variety of causes including trauma at birth, viral infections, genetic disorders, exposure to loud noise, or aging. Treatment may include examining and cleaning the ear canal, fitting and dispensing hearing aids or other assistive devices, and audiologic rehabilitation (including auditory training or instruction in speech or lip reading). Audiologists may recommend, fit, and dispense personal or large area amplification systems, such as hearing aids and alerting devices. Audiologists provide fitting and tuning of cochlear implants and provide the necessary rehabilitation for adjustment to listening with implant amplification systems. They also measure noise levels in workplaces and conduct hearing protection programs in industry, as well as in schools and communities.

Audiologists provide direct clinical services to individuals with hearing or balance disorders. In audiology (hearing) clinics, they may independently develop and carry out treatment programs. Audiologists, in a variety of settings, work as members of interdisciplinary professional teams in planning and implementing service delivery for children and adults, from birth to old age. Similar to speech-language pathologists, audiologists keep records on the initial evaluation, progress, and discharge of clients. These records help pinpoint problems, track client progress, and justify the cost of treatment, when applying for reimbursement.

Audiologists may conduct research on types of, and treatment for, hearing, balance, and related disorders. Others design and develop equipment or techniques for diagnosing and treating these disorders.

Working Conditions

Speech-language pathologists and audiologists usually work at a desk or table in clean comfortable surroundings. The job is not physically demanding but does require attention to detail and intense concentration. The emotional needs of clients and their families may be demanding. Most full-time speech-language pathologists and audiologists work about 40 hours per week; some work part-time. Those who work on a contract basis may spend a substantial amount of time traveling between facilities.

Employment

Speech-language pathologists and audiologists held about 101,000 jobs in 2000. Speech-language pathologists held about 88,000 jobs; and audiologists held about 13,000. About one-half of jobs for speech-language pathologists and audiologists were in preschools, elementary and secondary schools, or colleges and universities. Others were in offices of speech-language pathologists and audiologists; hospitals; offices of physicians; speech, language, and hearing centers; home health agencies; or other facilities. Audiologists are more likely to be employed in independent healthcare offices, while speech-language pathologists are more likely to work in school settings.

A small number of speech-language pathologists and audiologists are self-employed in private practice. They contract to provide services in schools, physician's offices, hospitals, or nursing homes, or work as consultants to industry.

Training, Other Qualifications, and Advancement

Of the states that regulate licensing (45 for speech-language pathologists and 47 for audiologists), almost all require a master's degree or equivalent. Other requirements are 300 to 375 hours of supervised clinical experience, a passing score on a national examination, and 9 months of postgraduate professional clinical experience. Forty-one states have continuing education requirements for licensure renewal. Medicaid, Medicare, and private health insurers generally require a practitioner to be licensed to qualify for reimbursement.

About 242 colleges and universities offer graduate programs in speech-language pathology. Courses cover anatomy and physiology of the areas of the body involved in speech, language, and hearing; the development of normal speech, language, and hearing; the nature of disorders; acoustics; and psychological aspects of communication. Graduate students also learn to evaluate and treat speech, language, and hearing disorders and receive supervised clinical training in communication disorders.

About 112 colleges and universities offer graduate programs in audiology in the United States. Course work includes anatomy; physiology; basic science; math; physics; genetics; normal and abnormal communication development; auditory, balance and neural systems assessment and treatment; audiologic rehabilitation; and ethics.

Speech-language pathologists can acquire the Certificate of Clinical Competence in Speech-Language Pathology (CCC-SLP) offered by the American Speech-Language-Hearing Association, and audiologists can earn the Certificate of Clinical Competence in Audiology (CCC-A). To earn a CCC, a person must have a graduate degree and 375 hours of supervised clinical experience, complete a 36-week postgraduate clinical fellowship, and pass a written examination. According to the American Speech-Language-Hearing Association, as of 2007, audiologists will need to have a bachelor's degree and complete 75 hours of credit toward a doctoral degree in order to seek certification. As of 2012, audiologists will have to earn a doctoral degree in order to be certified.

Speech-language pathologists and audiologists should be able to effectively communicate diagnostic test results, diagnoses, and proposed treatment in a manner easily understood by their clients. They must be able to approach problems objectively and provide support to clients and their families. Because a client's progress may be slow, patience, compassion, and good listening skills are necessary.

Job Outlook

Employment of speech-language pathologists and audiologists is expected to grow much faster than the average for all occupations through the year 2010. Because hearing loss is strongly associated with aging, rapid growth in the population age 55 and over will cause the number of persons with hearing impairment to increase markedly. In addition, baby boomers are now entering middle age, when the possibility of neurological disorders and associated speech, language, and hearing impairments increases. Medical advances are also improving the survival rate of premature infants and trauma and stroke victims, who then need assessment and possible treatment. In health services facilities, federal legislation imposing limits on reimbursement for therapy services may adversely affect the job market for therapy providers over the near term.

Employment in schools will increase along with growth in elementary and secondary school enrollments, including enrollment of special education students. Federal law guarantees special education and related services to all eligible children with disabilities. Greater awareness of the importance of early identification and diagnosis of speech, language, and hearing disorders will also increase employment.

The number of speech-language pathologists and audiologists in private practice will rise due to the increasing use of contract services by hospitals, schools, and nursing homes. In addition to job openings stemming from employment growth, some openings for speech-language pathologists and audiologists will arise from the need to replace those who leave the occupation.

Earnings

Median annual earnings of speech-language pathologists were $46,640 in 2000. The middle 50 percent earned between $37,670 and $56,980. The lowest 10 percent earned less than $30,720, and the highest 10 percent earned more than $69,980. Median annual earnings in the industries employing the largest numbers of speech-language pathologists in 2000 were as follows:

Hospitals	$49,960
Offices of other health practitioners	47,170
Elementary and secondary schools	43,710

Median annual earnings of audiologists were $44,830 in 2000. The middle 50 percent earned between $37,000 and $55,290. The lowest 10 percent earned less than $30,850, and the highest 10 percent earned more than $68,570.

According to a 2000 survey by the American Speech-Language-Hearing Association, the median annual salary for full-time certified speech-language pathologists who worked 11 or 12 months annually was $44,000; for audiologists, $48,000. For those who worked 9 or 10 months annually, the median annual salary for speech-language pathologists was $41,000; for audiologists, $45,000. Speech-language pathologists with doctorate degrees who worked 11 or 12 months annually earned $62,500; and audiologists, $70,000.

Related Occupations

Speech-language pathologists and audiologists specialize in the prevention, diagnosis, and treatment of speech and language and hearing problems. Workers in related occupations include occupational therapists, optometrists, physical therapists, psychologists, recreational therapists, and rehabilitation counselors.

Sources of Additional Information

State licensing boards can provide information on licensure requirements. State departments of education can supply information on certification requirements for those who wish to work in public schools.

General information on careers in speech-language pathology and audiology is available from:

- American Speech-Language-Hearing Association, 10801 Rockville Pike, Rockville, MD 20852. Internet: http://professional.asha.org

Information on a career in audiology is also available from:

- American Academy of Audiology, 8201 Greensboro Dr., Suite 300, McLean, VA 22102

Stationary Engineers and Boiler Operators

O*NET 51-8021.01, 51-8021.02

Significant Points

- Job opportunities will be best for workers with computer skills.
- Stationary engineers and boiler operators usually acquire their skills through a formal apprenticeship program or informal on-the-job training supplemented by courses at a trade or technical school.
- A license to operate boilers, ventilation, air conditioning, and other equipment is required in most states and cities.

Nature of the Work

Heating, air-conditioning, and ventilation systems keep large buildings comfortable all year long. Industrial plants often have facilities to provide electrical power, steam, or other services. Stationary engineers and boiler operators control and maintain these systems, which include boilers, air-conditioning and refrigeration equipment, diesel engines, turbines, generators, pumps, condensers, and compressors. The equipment that stationary engineers and boiler operators control is similar to equipment operated by locomotive or marine engineers, except that it is not on a moving vehicle.

Stationary engineers and boiler operators start up, regulate, and shut down equipment. They ensure that it operates safely, economically, and within established limits by monitoring meters, gauges, and computerized controls. They manually control equipment and, if necessary, make adjustments. They use hand and power tools to perform repairs and maintenance ranging from a complete overhaul to replacing defective valves, gaskets, or bearings. They also record relevant events and facts concerning operation and maintenance in an equipment log. On steam boilers, for example, they observe, control, and record steam pressure, temperature, water level and chemistry, power output, fuel consumption, and emissions. They watch and listen to machinery and routinely check safety devices, identifying and correcting any trouble that develops.

Stationary engineers and boiler operators may use computers to operate the mechanical systems of new buildings and plants. Engineers and operators monitor, adjust, and diagnose these systems from a central location using a computer linked into the buildings' communications network.

Routine maintenance, such as lubricating moving parts, replacing filters, and removing soot and corrosion that can reduce operating efficiency, is a regular part of the work of stationary engineers and boiler operators. They test boiler water and add chemicals to prevent corrosion and harmful deposits. They also may check the air quality of the ventilation system and make adjustments to keep within mandated guidelines.

In a large building or industrial plant, a stationary engineer may be in charge of all mechanical systems in the building. Engineers may supervise the work of assistant stationary engineers, turbine operators, boiler tenders, and air-conditioning and refrigeration operators and mechanics. Some perform other maintenance duties, such as carpentry, plumbing, and electrical repairs. In a small building or industrial plant, there may be only one stationary engineer.

Working Conditions

Stationary engineers and boiler operators generally have steady, year-round employment. The average workweek is 40 hours. In facilities that operate around the clock, engineers and operators usually work one of three daily 8-hour shifts on a rotating basis. Weekend and holiday work often is required.

Engine rooms, power plants, and boiler rooms usually are clean and well lighted. Even under the most favorable conditions, however, some stationary engineers and boiler operators are exposed to high temperatures, dust, dirt, and high noise levels from the equipment. General maintenance duties also may require contact with oil, grease, or smoke. Workers spend much of the time on their feet. They may also have to crawl inside boilers and work in crouching or kneeling positions to inspect, clean, or repair equipment.

Stationary engineers and boiler operators work around potentially hazardous machinery such as boilers and electrical equipment. They must follow procedures to guard against burns, electric shock, and exposure to hazardous materials such as asbestos or certain chemicals.

Employment

Stationary engineers and boiler operators held about 57,000 jobs in 2000. They worked in a variety of places, including factories, hospitals, hotels, office and apartment buildings, schools, and shopping malls. Some are employed as contractors to a building or plant.

Stationary engineers and boiler operators work throughout the country, generally in the more heavily populated areas in which large industrial and commercial establishments are located.

Training, Other Qualifications, and Advancement

Stationary engineers and boiler operators usually acquire their skills through a formal apprenticeship program or informal on-the-job training supplemented by courses at a trade or technical school. In addition, valuable experience can be obtained in the navy or the merchant marine because marine-engineering plants are similar to many stationary power and heating plants. Most employers prefer to hire persons with at least a high school diploma, or equivalent, due to the increasing complexity of the equipment with which engineers and operators work. Many stationary engineers and boiler operators have some college education. Mechanical aptitude, manual dexterity, and good physical condition also are important.

The International Union of Operating Engineers sponsors apprenticeship programs and is the principal union for stationary engineers and boiler operators. In selecting apprentices, most local labor-management apprenticeship committees prefer applicants with education or training in mathematics, computers, mechanical drawing, machine-shop practice, physics, and chemistry. An apprenticeship usually lasts 4 years and includes 8,000 hours of on-the-job training. In addition, apprentices receive 600 hours of classroom instruction in subjects such as boiler design and operation, elementary physics, pneumatics, refrigeration, air conditioning, electricity, and electronics.

Those who acquire their skills on the job usually start as boiler tenders or helpers to experienced stationary engineers and boiler operators. This practical experience may be supplemented by postsecondary vocational training in computerized controls and instrumentation. However, becoming an engineer or operator without completing a formal apprenticeship program usually requires many years of work experience.

Most large and some small employers encourage and pay for skill-improvement training for their employees. Training almost always is provided when new equipment is introduced or when regulations concerning some aspect of the workers' duties change.

Most states and cities have licensing requirements for stationary engineers and boiler operators. Applicants usually must be at least 18 years of age, reside for a specified period in the state or locality, meet experience requirements, and pass a written examination. A stationary engineer or boiler operator who moves from one state or city to another may have to pass an examination for a new license due to regional differences in licensing requirements.

There are several classes of stationary engineer licenses. Each class specifies the type and size of equipment the engineer can operate without supervision. A licensed first-class stationary engineer is qualified to run a large facility, supervise others, and operate equipment of all types and capacities. An applicant for this license may be required to have a high school education, apprenticeship or on-the-job training, and several years of experience. Licenses below first class limit the types or capacities of equipment the engineer may operate without supervision.

Stationary engineers and boiler operators advance by being placed in charge of larger, more powerful, or more varied equipment. Generally, engineers advance to these jobs as they obtain higher class licenses. Some stationary engineers and boiler operators advance to boiler inspectors, chief plant engineers, building and plant superintendents, or building managers. A few obtain jobs as examining engineers or technical instructors.

Job Outlook

Persons wishing to become stationary engineers and boiler operators may face competition for job openings. Employment opportunities will be best for those with apprenticeship training or vocational school courses covering systems operation using computerized controls and instrumentation.

Employment of stationary engineers and boiler operators is expected to decline through the year 2010. Continuing commercial and industrial development will increase the amount of equipment to be operated and maintained. However, automated systems and computerized controls are making newly installed equipment more efficient, thus reducing the number of jobs needed for its operation. Some job openings will arise from the need to replace experienced workers who transfer to other occupations or leave the labor force. However, turnover in this occupation is low, partly due to its high wages. Consequently, relatively few replacement openings are expected.

Earnings

Median annual earnings of stationary engineers and boiler operators were $40,420 in 2000. The middle 50 percent earned between $31,490 and $51,090 a year. The lowest 10 percent earned less than $24,470, and the highest 10 percent earned more than $61,530 a year. Median annual earnings of stationary engineers and boiler operators in 2000 were $46,600 in local government and $37,680 in hospitals.

Related Occupations

Other workers who monitor and operate stationary machinery include chemical plant and system operators; gas plant operators; petroleum pump system operators, refinery operators, and gaugers; power plant operators, distributors, and dispatchers; and water and wastewater treatment plant and system operators. Other workers who maintain the equipment and machinery in a building or plant are industrial machinery repairers and millwrights.

Sources of Additional Information

Information about apprenticeships, vocational training, and work opportunities is available from state employment service offices, locals of the International Union of Operating Engineers, vocational schools, and state and local licensing agencies.

Specific questions about this occupation should be addressed to:

- International Union of Operating Engineers, 1125 17th St. N.W., Washington, DC 20036. Internet: http://www.iuoe.org
- National Association of Power Engineers, Inc., 1 Springfield St., Chicopee, MA 01013
- Building Owners and Managers Institute International, 1521 Ritchie Hwy., Arnold, MD 21012. Internet: http://www.bomi-edu.org

Statisticians

O*NET 15-2041.00

Significant Points

- Many individuals with degrees in statistics enter jobs that do not have the title "statistician."
- A master's degree in statistics or mathematics is the minimum educational requirement for most jobs with this title.
- Although little or no change is expected in employment of statisticians over the 2000–2010 period, job opportunities should remain favorable for individuals with statistical degrees.

Nature of the Work

Statistics is the scientific application of mathematical principles to the collection, analysis, and presentation of numerical data. Statisticians contribute to scientific inquiry by applying their mathematical knowledge to the design of surveys and experiments; collection, processing, and analysis of data; and interpretation of the results. Statisticians often apply their knowledge of statistical methods to a variety of subject areas, such as biology, economics, engineering, medicine, public health, psychology, marketing, education, and sports. Many applications cannot occur without the use of statistical techniques, such as designing experiments to gain federal approval of a newly manufactured drug.

One technique that is especially useful to statisticians is sampling—obtaining information about a population of people or group of things by surveying a small portion of the total. For example, to determine the size of the audience for particular programs, television-rating services survey only a few thousand families, rather than all viewers. Statisticians decide where and how to gather the data, determine the type and size of the sample group, and develop the survey questionnaire or reporting form. They also prepare instructions for workers who will collect and tabulate the data. Finally, statisticians analyze, interpret, and summarize the data using computer software.

In business and industry, statisticians play an important role in quality control and product development and improvement. In an automobile company, for example, statisticians might design experiments to determine the failure time of engines exposed to extreme weather conditions by running individual engines until failure and breakdown. Working for a pharmaceutical company, statisticians might develop and evaluate the results of clinical trials to determine the safety and effectiveness of new medications. And at a computer software firm, statisticians might help construct new statistical software packages to analyze data more accurately and efficiently. In addition to product development and testing, some statisticians also are involved in deciding what products to manufacture, how much to charge for them, and to whom the products should be marketed. Statisticians also may manage assets and liabilities, determining the risks and returns of certain investments.

Numerous statisticians also are employed by nearly every government agency. Some government statisticians develop surveys that measure population growth, consumer prices, or unemployment. Other statisticians work for scientific, environmental, and agricultural agencies and may help to determine the amount of pesticides in drinking water, the number of endangered species living in a particular area, or the number of people afflicted with a particular disease. Other statisticians are employed in national defense agencies, determining the accuracy of new weapons and defense strategies.

Because statistical specialists are used in so many work areas, specialists who use statistics often have different professional designations. For example, a person using statistical methods on economic data may have the title "econometrician," while statisticians in public health and medicine may hold titles such as "biostatistician," "biometrician," or "epidemiologist."

Working Conditions

Statisticians usually work regular hours in comfortable offices. Some statisticians travel to provide advice on research projects, supervise and set up surveys, or gather statistical data. Some may have duties that vary widely, such as designing experiments or performing fieldwork in various communities. Statisticians who work in academia generally have a mix of teaching and research responsibilities.

Employment

Statisticians held about 19,000 jobs in 2000. One-fifth of these jobs were in the federal government, where statisticians were concentrated in the Departments of Commerce, Agriculture, and Health and Human Services. Most of the remaining jobs were in private industry, especially in the research and testing services and management and public relations industries. In addition, many professionals with a background in statistics were among the 20,000 full-time mathematics faculty in colleges and universities in 2000, according to the American Mathematical Society.

Training, Other Qualifications, and Advancement

Although more employment opportunities are becoming available to well-qualified statisticians with bachelor's degrees, a master's degree in statistics or mathematics is usually the minimum educational requirement for most statistician jobs. Research and academic positions in institutions of higher education, for example, require a graduate degree, usually a doctorate, in statistics. Beginning positions in industrial research often require a master's degree combined with several years of experience.

The training required for employment as an entry-level statistician in the federal government, however, is a bachelor's degree, including at least 15 semester hours of statistics or a combination of 15 hours of mathematics and statistics, if at least 6 semester hours are in statistics. Qualifying as a mathematical statistician in the federal government requires 24 semester hours of mathematics and statistics with a minimum of 6 semester hours in statistics and 12 semester hours in an area of advanced mathematics, such as calculus, differential equations, or vector analysis.

About 80 colleges and universities offered bachelor's degrees in statistics in 2000. Many other schools also offered degrees in mathematics, operations research, and other fields that included a sufficient number of courses in statistics to qualify graduates for some beginning positions in the federal government. Required subjects for statistics majors include differential and integral calculus, statistical methods, mathematical modeling, and probability theory. Additional courses that undergraduates should take include linear algebra, design and analysis of experiments, applied multivariate analysis, and mathematical statistics.

In 2000, approximately 110 universities offered a master's degree program in statistics, and about 60 offered a doctoral degree program. Many other schools also offered graduate-level courses in applied statistics for students majoring in biology, business, economics, education, engineering, psychology, and other fields. Acceptance into graduate statistics programs does not require an undergraduate degree in statistics, although good training in mathematics is essential.

Because computers are used extensively for statistical applications, a strong background in computer science is highly recommended. For positions involving quality and productivity improvement, training in engineering or physical science is useful. A background in biological, chemical, or health science is important for positions involving the preparation and testing of pharmaceutical or agricultural products. Courses in economics and business administration are helpful for many jobs in market research, business analysis, and forecasting.

Good communications skills are important for prospective statisticians in industry, where they often need to explain technical matters to persons without statistical expertise. An understanding of business and the economy also is valuable for those who plan to work in private industry.

Beginning statisticians generally are supervised by an experienced statistician. With experience, they may advance to positions with more technical responsibility and, in some cases, supervisory duties. However, opportunities for promotion increase with advanced degrees. Master's and PhD degree holders usually enjoy independence in their work and become qualified to engage in research, develop statistical methods, or, after a number of years of experience in a particular area, become statistical consultants.

Job Outlook

Little or no change is expected in employment of statisticians over the 2000–2010 period. However, job opportunities should remain favorable for individuals with statistical degrees, although many of these positions will not carry the explicit job title "statistician." This is especially true of jobs that involve the analysis and interpretation of data from other disciplines such as economics, biological science, psychology, or engineering. In addition to the limited number of jobs resulting from growth, a number of openings will become available as statisticians retire, transfer to other occupations, or leave the workforce for other reasons.

Among graduates with a bachelor's or master's degree in statistics, those with a strong background in an allied field, such as finance, engineering, or computer science, should have the best prospects of finding jobs related to their field of study. Federal agencies will hire statisticians in many fields, including demography, agriculture, consumer and producer surveys, Social Security, healthcare, and environmental quality. Competition for entry-level positions in the federal government is expected to be strong for those just meeting the minimum qualification standards for statisticians, because the federal government is one of the few employers that considers a bachelor's degree to be an adequate entry-level qualification. Those who meet state certification requirements may become high school statistics teachers.

Manufacturing firms will hire statisticians with master's and doctoral degrees for quality control of various products, including pharmaceuticals, motor vehicles, chemicals, and food. For example, pharmaceutical firms employ statisticians to assess the safety and effectiveness of new drugs. To address global product competition, motor vehicle manufacturers will need statisticians to improve the quality of automobiles, trucks, and their components by developing and testing new designs. Statisticians with knowledge of engineering and the physical sciences will find jobs in research and development, working with teams of scientists and engineers to help improve design and production processes to ensure consistent quality of newly developed products. Many statisticians also will find opportunities developing statistical software for computer software manufacturing firms.

Business firms will rely heavily on workers with a background in statistics to forecast sales, analyze business conditions, and help solve management problems in order to maximize profits. In addition, consulting firms increasingly will offer sophisticated statistical services to other businesses. Because of the widespread use of computers in this field, statisticians in all industries should have good computer programming skills and knowledge of statistical software.

Earnings

Median annual earnings of statisticians were $51,990 in 2000. The middle 50 percent earned between $37,160 and $69,220. The lowest 10 percent had earnings of less than $28,430, while the highest 10 percent earned more than $86,660.

The average annual salary for statisticians in the federal government in nonsupervisory, supervisory, and managerial positions was $68,900 in 2001, while mathematical statisticians averaged $76,530. According to a 2001 survey by the National Association of Colleges and Employers, starting salary offers for mathematics/statistics graduates with a bachelor's degree averaged $46,466 a year.

Related Occupations

People in numerous occupations work with statistics. Among these are actuaries; mathematicians; operations research analysts; systems analysts, computer scientists, and database administrators; computer programmers; computer software engineers; engineers; economists and market and survey researchers; financial analysts and personal financial advisors; and life, physical, and social science occupations.

Sources of Additional Information

For information about career opportunities in statistics, contact:

- American Statistical Association, 1429 Duke St., Alexandria, VA 22314. Internet: http://www.amstat.org

For more information on doctoral-level careers and training in mathematics, a field closely related to statistics, contact:

- American Mathematical Society, 201 Charles St., Providence, RI 02940. Internet: http://www.ams.org

Information on obtaining a statistician position with the federal government is available from the Office of Personnel Management (OPM) through a telephone-based system. Consult your telephone directory under U.S. Government for a local number or call (912) 757-3000; Federal Relay Service: (800) 877-8339. The first number is not toll free, and charges may result. Information also is available from the OPM Internet site: http://www.usajobs.opm.gov

Surgical Technologists

O*NET 29-2055.00

Significant Points

- Most educational programs for surgical technologists last approximately 1 year and result in a certificate.
- Employment of surgical technologists is expected to grow faster than average as the number of surgical procedures grows.

Nature of the Work

Surgical technologists, also called *scrubs* and *surgical or operating room technicians*, assist in surgical operations under the supervision of surgeons, registered nurses, or other surgical personnel. Surgical technologists are members of operating room teams, which most commonly include surgeons, anesthesiologists, and circulating nurses. Before an operation, surgical technologists help prepare the operating room by setting up surgical instruments and equipment, sterile drapes, and sterile solutions. They assemble both sterile and nonsterile equipment, as well as adjust and check it to ensure it is working properly. Technologists also get patients ready for surgery by washing, shaving, and disinfecting incision sites. They transport patients to the operating room, help position them on the operating table, and cover them with sterile surgical "drapes." Technologists also observe patients' vital signs, check charts, and assist the surgical team with putting on sterile gowns and gloves.

During surgery, technologists pass instruments and other sterile supplies to surgeons and surgeon assistants. They may hold retractors, cut sutures, and help count sponges, needles, supplies, and instruments. Surgical technologists help prepare, care for, and dispose of specimens taken for laboratory analysis and help apply dressings. Some operate sterilizers, lights, or suction machines, and help operate diagnostic equipment.

After an operation, surgical technologists may help transfer patients to the recovery room and clean and restock the operating room.

Working Conditions

Surgical technologists work in clean, well-lighted, cool environments. They must stand for long periods and remain alert during operations. At times they may be exposed to communicable diseases and unpleasant sights, odors, and materials.

Most surgical technologists work a regular 40-hour week, although they may be on call or work nights, weekends and holidays on a rotating basis.

Employment

Surgical technologists held about 71,000 jobs in 2000. Almost three-quarters are employed by hospitals, mainly in operating and delivery rooms. Others are employed in clinics and surgical centers, and in the offices of physicians and dentists who perform outpatient surgery. A few, known as private scrubs, are employed directly by surgeons who have special surgical teams, like those for liver transplants.

Training, Other Qualifications, and Advancement

Surgical technologists receive their training in formal programs offered by community and junior colleges, vocational schools, universities, hospitals, and the military. In 2001, the Commission on Accreditation of Allied Health Education Programs (CAAHEP) recognized 350 accredited programs. High school graduation normally is required for admission. Programs last 9 to 24 months and lead to a certificate, diploma, or associate degree.

Programs provide classroom education and supervised clinical experience. Students take courses in anatomy, physiology, microbiology, pharmacology, professional ethics, and medical terminology. Other studies cover the care and safety of patients during surgery, aseptic techniques, and surgical procedures. Students also learn to sterilize instruments; prevent and control infection; and handle special drugs, solutions, supplies, and equipment.

Technologists may obtain voluntary professional certification from the Liaison Council on Certification for the Surgical Technologist by graduating from a CAAHEP-accredited program and passing a national certification examination. They may then use the designation Certified Surgical Technologist, or CST. Continuing education or reexamination is required to maintain certification, which must be renewed every 6 years. Certification may also be obtained from the National Center for Competency Testing. To qualify to take the exam, candidates follow one of three paths: complete an accredited training program, undergo a 2-year hospital on-the-job training program, or acquire seven years experience working in the field. After passing the exam, individuals may use the designation National Certified Technician O.R. This certification may be renewed every 5 years through either continuing education or reexamination. Most employers prefer to hire certified technologists.

Surgical technologists need manual dexterity to handle instruments quickly. They also must be conscientious, orderly, and emotionally stable to handle the demands of the operating room environment. Technologists must respond quickly and know procedures well to have instruments ready for surgeons without having to be told. They are expected to keep abreast of new developments in the field. Recommended high school courses include health, biology, chemistry, and mathematics.

Technologists advance by specializing in a particular area of surgery, such as neurosurgery or open heart surgery. They also may work as circulating technologists. A circulating technologist is the "unsterile" member of the surgical team who prepares patients; helps with anesthesia; obtains and opens packages for the "sterile" persons to remove the sterile contents during the procedure; interviews the patient before surgery; keeps a written account of the surgical procedure; and answers the surgeon's questions about the patient during the surgery. With additional training, some technologists advance to first assistants, who help with retracting, sponging, suturing, cauterizing bleeders, and closing and treating wounds. Some surgical technologists manage central supply departments in hospitals, or take positions with insurance companies, sterile supply services, and operating equipment firms.

Job Outlook

Employment of surgical technologists is expected to grow faster than the average for all occupations through the year 2010 as the volume of surgery increases. The number of surgical procedures is expected to rise as the population grows and ages. As the "baby boom" generation enters retirement age, the over 50 population will account for a larger portion of the general population. Older people require more surgical procedures. Technological advances, such as fiber optics and laser technology, will also permit new surgical procedures to be performed.

Hospitals will continue to be the primary employer of surgical technologists, although much faster employment growth is expected in offices and clinics of physicians, including ambulatory surgical centers.

Earnings

Median annual earnings of surgical technologists were $29,020 in 2000. The middle 50 percent earned between $24,490 and $34,160. The lowest 10 percent earned less than $20,490, and the highest 10 percent earned more than $40,310. Median annual earnings of surgical technologists in 2000 were $31,190 in offices and clinics of medical doctors and $28,340 in hospitals.

Related Occupations

Other health occupations requiring approximately one year of training after high school include dental assistants, licensed practical and licensed vocational nurses, medical and clinical laboratory technicians, medical assistants, and respiratory therapy technicians.

Sources of Additional Information

For additional information on a career as a surgical technologist and a list of CAAHEP-accredited programs, contact:

- Association of Surgical Technologists, 7108-C South Alton Way, Englewood, CO 80112. Internet: http://www.ast.org

For information on becoming a Certified Surgical Technologist, contact:

- Liaison Council on Certification for the Surgical Technologist, 7790 East Arapahoe Rd., Suite 240, Englewood, CO 80112-1274. Internet: http://www.lcc-st.org/index_ie.htm

For information on becoming a National Certified Technician O.R., contact:

- National Center for Competency Testing, 7007 College Blvd., Suite 250, Overland Park, KS 66211

Surveyors, Cartographers, Photogrammetrists, and Surveying Technicians

O*NET 17-1021.00, 17-1022.00, 17-3031.01, 17-3031.02

Significant Points

- Four out of five are employed in engineering services and in government.
- Computer skills enhance employment opportunities.

Nature of the Work

Measuring and mapping the earth's surface are the responsibilities of several different types of workers. Traditional *land surveyors* establish official land, air space, and water boundaries. They write descriptions of land for deeds, leases, and other legal documents; define air space for airports; and measure construction and mineral sites. Other surveyors provide data relevant to the shape, contour, location, elevation, or dimension of land or land features. *Cartographers* compile geographic, political, and cultural information and prepare maps of large areas. *Photogrammetrists* measure and analyze aerial photographs to prepare detailed maps and drawings. *Surveying technicians* assist land surveyors by operating survey instruments and collecting information in the field, and by performing computations and computer-aided drafting in offices. *Mapping technicians* calculate mapmaking information from field notes. They also draw topographical maps and verify their accuracy.

Land surveyors manage survey parties who measure distances, directions, and angles between points and elevations of points, lines, and contours on, above, and below the earth's surface. They plan the fieldwork, select known survey reference points, and determine the precise location of important features in the survey area. Surveyors research legal records, look for evidence of previous boundaries, and analyze the data to determine the location of boundary lines. They also record the results of the survey, verify the accuracy of data, and prepare plots, maps, and reports. Surveyors who establish boundaries must be licensed by the state in which they work; they are known as Professional Land Surveyors. Professional Land Surveyors are sometimes called to provide expert testimony in court cases concerning surveying matters.

A survey party gathers the information needed by the land surveyor. A typical survey party consists of a party chief and one or more surveying technicians and helpers. The party chief, who may be either a land surveyor or a senior surveying technician, leads day-to-day work activities. Surveying technicians assist the party chief by adjusting and operating surveying instruments, such as the theodolite (used to measure horizontal and vertical angles) and electronic distance-measuring equipment. Surveying technicians or assistants position and hold the vertical rods, or targets, that the theodolite operator sights on to measure angles, distances, or elevations. They also may hold measuring tapes, if electronic distance-measuring equipment is not used. Surveying technicians compile notes, make sketches, and enter the data obtained from surveying instruments into computers. Survey parties may include laborers or helpers who perform less-skilled duties, such as clearing brush from sight lines, driving stakes, or carrying equipment.

New technology is changing the nature of the work of surveyors and surveying technicians. For larger projects, surveyors are increasingly using the Global Positioning System (GPS), a satellite system that precisely locates points on the earth by using radio signals transmitted via satellites. To use this system, a surveyor places a satellite signal receiver—a small instrument mounted on a tripod—on a desired point. The receiver simultaneously collects information from several satellites to establish a precise position. The receiver also can be placed in a vehicle for tracing out road systems. Because receivers now come in different sizes and shapes and the cost of the receivers has fallen, much more surveying work is being done using GPS. Surveyors then must interpret and check the results produced by the new technology.

Cartographers measure, map, and chart the earth's surface, which involves everything from geographical research and data compilation to actual map production. They collect, analyze, and interpret both spatial data (such as latitude, longitude, elevation, and distance) and nonspatial data (such as population density, land use patterns, annual precipitation levels, and demographic characteristics). Cartographers prepare maps in either digital or graphic form, using information provided by geodetic surveys, aerial photographs, and satellite data. *Photogrammetrists* prepare detailed maps and drawings from aerial photographs, usually of areas that are inaccessible, difficult, or less cost-efficient to survey by other methods. *Map editors* develop and verify map contents from aerial photographs and other reference sources. Some states require photogrammetrists to be licensed as Professional Land Surveyors.

Some surveyors perform specialized functions that are closer to those of a cartographer than to those of a traditional surveyor. For example, *geodetic surveyors* use high-accuracy techniques, including satellite observations (remote sensing), to measure large areas of the earth's surface. *Geophysical prospecting surveyors* mark sites for subsurface exploration, usually petroleum related. *Marine or hydrographic surveyors* survey harbors, rivers, and other bodies of water to determine shorelines, topography of the bottom, water depth, and other features.

The work of surveyors and cartographers is changing because of advancements in technology. These advancements include not only the GPS, but also new earth resources data satellites, improved aerial photography, and geographic information systems (GIS)—which are computerized data banks of spatial data. From the older specialties of photogrammetrist and cartographer, a new type of mapping scientist is emerging. The *geographic information specialist* combines the functions of mapping science and surveying into a broader field concerned with the collection and analysis of geographic information.

Working Conditions

Surveyors usually work an 8-hour day, 5 days a week, and may spend a lot of time outdoors. Sometimes they work longer hours during the summer, when weather and light conditions are most suitable for fieldwork. Seasonal demands for longer hours are related to demand for specific surveying services. Home purchases are traditionally related to the start and end of the school year. Construction is related to the materials to be used (concrete and asphalt are restricted by outside temperatures, unlike wood framing). Aerial photography is most effective when the leaves are off the trees.

Land surveyors and technicians engage in active, and sometimes strenuous, work. They often stand for long periods, walk considerable distances, and climb hills with heavy packs of instruments and other equipment. They can also be exposed to all types of weather. Traveling often is part of the job; they may commute long distances, stay overnight, or temporarily relocate near a survey site.

While surveyors can spend considerable time indoors planning surveys, analyzing data, and preparing reports and maps, cartographers spend virtually all of their time in offices and seldom visit the sites they are mapping.

Employment

Surveyors, cartographers, photogrammetrists, and surveying technicians held about 121,000 jobs in 2000. Engineering and architectural services firms employed about 63 percent of these workers. Federal, state, and local governmental agencies employed an additional 16 percent. Major federal government employers are the U.S. Geological Survey (USGS), the Bureau of Land Management (BLM), the Army Corps of Engineers, the Forest Service (USFS), the National Oceanic and Atmospheric Administration (NOAA), the National Imagery and Mapping Agency (NIMA), and the Federal Emergency Management Agency (FEMA). Most surveyors in state and local government work for highway departments and urban planning and redevelopment agencies. Construction firms, mining and oil and gas extraction companies, and public utilities also employ surveyors, cartographers, photogrammetrists, and surveying technicians. About 5,000 were self-employed in 2000.

Training, Other Qualifications, and Advancement

Most people prepare for a career as a licensed surveyor by combining postsecondary school courses in surveying with extensive on-the-job training. However, as technology advances, a 4-year college degree is becoming more of a prerequisite. About 25 universities now offer 4-year programs leading to a BS degree in surveying. Junior and community colleges, technical institutes, and vocational schools offer 1-, 2-, and 3-year programs in both surveying and surveying technology.

All 50 states and all U.S. territories (Puerto Rico, Guam, Marianna Islands, and Virgin Islands) license land surveyors. For licensure, most state licensing boards require that individuals pass a written examination given by the National Council of Examiners for Engineering and Surveying. Most states also require that surveyors pass a written examination prepared by the state licensing board. In addition, they must meet varying standards of formal education and work experience in the field. In the past, many individuals started as members of survey crews and worked their way up to become licensed surveyors with little formal training in surveying. However, because of advancing technology and rising licensing standards, formal education requirements are increasing. At present, most states require some formal post-high school coursework and 10 to 12 years of surveying experience to gain licensure. However, requirements vary among states. Generally, the quickest route to licensure is a combination of 4 years of college, 2 to 4 years of experience (a few states do not require any), and passing the licensing examinations. An increasing number of states require a bachelor's degree in surveying or in a closely related field, such as civil engineering or forestry (with courses in surveying), regardless of the number of years of experience.

High school students interested in surveying should take courses in algebra, geometry, trigonometry, drafting, mechanical drawing, and computer science. High school graduates with no formal training in surveying usually start as apprentices. Beginners with postsecondary school training in surveying usually can start as technicians or assistants. With on-the-job experience and formal training in surveying—either in an institutional program or from a correspondence school—workers may advance to senior survey technician, then to party chief, and in some cases, to licensed surveyor (depending on state licensing requirements).

The National Society of Professional Surveyors, a member organization of the American Congress on Surveying and Mapping, has a voluntary certification program for surveying technicians. Technicians are certified at four levels requiring progressive amounts of experience, in addition to the passing of written examinations. Although not required for state licensure, many employers require certification for promotion to positions with greater responsibilities.

Surveyors should have the ability to visualize objects, distances, sizes, and abstract forms. They must work with precision and accuracy because mistakes can be costly. Members of a survey party must be in good physical condition because they work outdoors and often carry equipment over difficult terrain. They need good eyesight, coordination, and hearing to communicate verbally and manually (using hand signals). Surveying is a cooperative process, so good interpersonal skills and the ability to work as part of a team are important. Good office skills are also essential. Surveyors must be able to research old deeds and other legal documents and prepare reports that document their work.

Cartographers and photogrammetrists usually have a bachelor's degree in a field such as engineering, forestry, geography, or a physical science. Although it is possible to enter these positions through previous experience as a photogrammetric or cartographic technician, most cartographic and photogrammetric technicians now have had some specialized postsecondary school training. With the development of GIS, cartographers and photogrammetrists need additional education and stronger technical skills—including more experience with computers—than in the past.

The American Society for Photogrammetry and Remote Sensing has a voluntary certification program for photogrammetrists. To qualify for this professional distinction, individuals must meet work experience standards and pass an oral or written examination.

Job Outlook

Overall employment of surveyors, cartographers, photogrammetrists, and surveying technicians is expected to grow about as fast as the average for all occupations through the year 2010. The widespread availability and use of advanced technologies, such as GPS, GIS, and remote sensing, are increasing both the accuracy and productivity of survey, photogrammetric, and mapping work. However, job openings will continue to result from the need to replace workers who transfer to other occupations or leave the labor force altogether.

Prospects will be best for surveying and mapping technicians, whose numbers are expected to grow slightly faster than the average for all occupations through 2010. The short training period needed to learn to operate the equipment, the current lack of any formal testing or licensing, and the relatively lower wages all make for a healthy demand for these technicians, as well as for a readily available supply.

As technologies become more complex, opportunities will be best for surveyors, cartographers, and photogrammetrists who have at least a bachelor's degree and strong technical skills. Increasing demand for geographic data, as opposed to traditional surveying services, will mean better opportunities for cartographers and photogrammetrists involved in the development and use of geographic and land information systems. New technologies, such as GPS and GIS, also may enhance employment opportunities for surveyors and surveying technicians who have the educational background enabling them to use these systems, but upgraded licensing requirements will continue to limit opportunities for professional advancement for those with less education.

Opportunities for surveyors, cartographers, and photogrammetrists should remain concentrated in engineering, architectural, and surveying services firms. However, nontraditional areas such as urban planning and natural resource exploration and mapping also should provide areas of employment growth, particularly with regard to producing maps for management of natural emergencies and updating maps with the newly available technology. Continued growth in construction through 2010 should require surveyors to lay out streets, shopping centers, housing developments, factories, office buildings, and recreation areas, while setting aside flood plains, wetlands, wildlife habitats and environmentally sensitive areas for protection. However, employment may fluctuate from year to year along with construction activity, or mapping needs for land and resource management.

Earnings

Median annual earnings of surveyors were $36,700 in 2000. The middle 50 percent earned between $26,480 and $49,030. The lowest 10 percent earned less than $19,570, and the highest 10 percent earned more than $62,980.

Median annual earnings of cartographers and photogrammetrists were $39,410 in 2000. The middle 50 percent earned between $29,200 and $51,930. The lowest 10 percent earned less than $23,560 and the highest 10 percent earned more than $64,780.

Median hourly earnings of surveying and mapping technicians were $13.48 in 2000. The middle 50 percent of all surveying technicians earned between $10.46 and $17.81 in 2000. The lowest 10 percent earned less than $8.45, and the highest 10 percent earned more than $22.40. Median hourly earnings of surveying and mapping technicians employed in engineering and architectural services were $12.39 in 2000, while those employed by local governments had median hourly earnings of $15.77.

In 2001, land surveyors in nonsupervisory, supervisory, and managerial positions in the federal government earned an average salary of $57,416; cartographers, $62,369; geodetic technicians, $53,143; surveying technicians, $34,623; and cartographic technicians, $40,775.

Related Occupations

Surveying is related to the work of civil engineers, architects, and landscape architects, because an accurate survey is the first step in land development and construction projects. Cartography and geodetic surveying are related to the work of environmental scientists and geoscientists, who study the earth's internal composition, surface, and atmosphere. Cartography also is related to the work of geographers and urban and regional planners, who study and decide how the earth's surface is to be used.

Sources of Additional Information

Information about career opportunities, licensure requirements, and the surveying technician certification program is available from:

- National Society of Professional Surveyors, Suite #403, 6 Montgomery Village Ave., Gaithersburg, MD 20879. Internet: http://www.acsm.net/nsps/index.html

Information on a career as a geodetic surveyor is available from:

- American Association of Geodetic Surveying (AAGS), Suite #403, 6 Montgomery Village Ave., Gaithersburg, MD 20879. Internet: http://www.acsm.net

General information on careers in photogrammetry and remote sensing is available from:

- ASPRS: The Imaging and Geospatial Information Society, 5410 Grosvenor Lane, Suite 210, Bethesda, MD 20814-2160. Internet: http://www.asprs.org

Systems Analysts, Computer Scientists, and Database Administrators

O*NET 15-1011.00, 15-1051.00, 15-1061.00, 15-1081.00

Significant Points

- As computer applications expand, systems analysts, computer scientists, and database administrators are projected to be the among the fastest growing occupations.
- Relevant work experience and a bachelor's degree are prerequisites for many jobs; for more complex jobs, a graduate degree is preferred.

Nature of the Work

The rapid spread of computers and information technology has generated a need for highly trained workers to design and develop new hardware and software systems and to incorporate new technologies. These workers—computer systems analysts, computer scientists, and database administrators—include a wide range of computer specialists. Job tasks and occupational titles used to describe these workers evolve rapidly, reflecting new areas of specialization or changes in technology, as well as the preferences and practices of employers.

Systems analysts solve computer problems and enable computer technology to meet individual needs of an organization. They help an organization realize the maximum benefit from its investment in equipment, personnel, and business processes. This process may include planning and developing new computer systems or devising ways to apply existing systems' resources to additional operations. Systems analysts may design new systems, including both hardware and software, or add a new software application to harness more of the computer's power. Most systems analysts work with a specific type of system that varies with the type of organization for which they work—for example, business, accounting, or financial systems, or scientific and engineering systems. Some systems analysts also are referred to as *systems developers* or *systems architects*.

Analysts begin an assignment by discussing the systems problem with managers and users to determine its exact nature. They define the goals of the system and divide the solutions into individual steps and separate procedures. Analysts use techniques such as structured analysis, data modeling, information engineering, mathematical model building, sampling, and cost accounting to plan the system. They specify the inputs to be accessed by the system, design the processing steps, and format the output to meet the users' needs. They also may prepare cost-benefit and return-on-investment analyses to help management decide whether implementing the proposed system will be financially feasible.

When a system is accepted, analysts determine what computer hardware and software will be needed to set it up. They coordinate tests and observe initial use of the system to ensure it performs as planned. They prepare specifications, work diagrams, and structure charts for Computer Programmers to follow and then work with them to "debug," or eliminate errors from, the system. Analysts, who do more in-depth testing of products, may be referred to as *software quality assurance analysts*. In addition to running tests, these individuals diagnose problems, recommend solutions, and determine if program requirements have been met.

In some organizations, *programmer-analysts* design and update the software that runs a computer. Because they are responsible for both programming and systems analysis, these workers must be proficient in both areas. As this becomes more commonplace, these analysts increasingly work with object-oriented programming languages, as well as client/server applications development, and multimedia and Internet technology.

One obstacle associated with expanding computer use is the need for different computer systems to communicate with each other. Because of the importance of maintaining up-to-date information—accounting records, sales figures, or budget projections, for example—systems analysts work on making the computer systems within an organization compatible so that information can be shared. Many systems analysts are involved with "networking," connecting all the computers internally—in an individual office, department, or establishment—or externally, because many organizations now rely on e-mail or the Internet. A primary goal of networking is to allow users to retrieve data and information from a mainframe computer or a server and use it on their machine. Analysts must design the hardware and software to allow free exchange of data, custom applications, and the computer power to process it all.

Networks come in many variations and *network systems and data communications analysts* analyze, design, test, and evaluate systems such as local area networks (LAN), wide area networks (WAN), Internet, Intranets, and other data communications systems. These analysts perform network modeling, analysis and planning; they also may research related products and make necessary hardware and software recommendations. *Telecommunications specialists* focus on the interaction between computer and communications equipment.

The growth of the Internet and expansion of the World Wide Web, the graphical portion of the Internet, have generated a variety of occupations related to design, development, and maintenance of Web sites and their servers. For example, *webmasters* are responsible for all technical aspects of a Web site, including performance issues such as speed of access, and for approving site content. *Internet developers* or *web developers*, also called *web designers*, are responsible for day-to-day site design and creation.

Computer scientists work as theorists, researchers, or inventors. Their jobs are distinguished by the higher level of theoretical expertise and innovation they apply to complex problems and the creation or application of new technology. Those employed by academic institutions work in areas ranging from complexity theory, to hardware, to programming language design. Some work on multidisciplinary projects, such as developing and advancing uses of virtual reality, in human-computer interaction, or in robotics. Their counterparts in private industry work in areas such as applying theory, developing specialized languages or information technologies, or designing programming tools, knowledge-based systems, or even computer games.

With the Internet and electronic business creating tremendous volumes of data, there is growing need to be able to store, manage, and extract data effectively. *Database administrators* work with database management systems software and determine ways to organize and store data. They determine user requirements, set up computer databases, and test and coordinate changes. It is the responsibility of an organization's database administrator to ensure performance, understand the platform the database runs on, and add new users. Because they also may design and implement system security, database administrators often plan and coordinate security measures. With the volume of sensitive data generated every second growing rapidly, data integrity, backup, and keeping databases secure have become an increasingly important aspect of the job for database administrators.

Working Conditions

Systems analysts, computer scientists, and database administrators normally work in offices or laboratories in comfortable surroundings. They usually work about 40 hours a week, the same as many other professional or office workers. However, evening or weekend work may be necessary to meet deadlines or solve specific problems. Given the technology available today, telecommuting is common for computer professionals. As networks expand, more work can be done from remote locations using modems, laptops, electronic mail, and the Internet.

Like other workers who spend long periods in front of a computer terminal typing on a keyboard, they are susceptible to eye strain, back discomfort, and hand and wrist problems such as carpal tunnel syndrome or cumulative trauma disorder.

Employment

Systems analysts, computer scientists, and database administrators held about 887,000 jobs in 2000, including about 71,000 who were self-employed. Employment was distributed among the following detailed occupations:

Computer system analysts	431,000
Network systems and data communications analysts	119,000
Database administrators	106,000
Computer and information scientists, research	28,000
All other computer specialists	203,000

Although they are increasingly employed in every sector of the economy, the greatest concentration of these workers is in the computer and data processing services industry. Firms in this industry provide nearly every service related to commercial computer use on a contract basis. Services include systems integration, networking, and reengineering; data processing and preparation; information retrieval, including on-line databases and Internet; onsite computer facilities management; development and management of databases; and a variety of specialized consulting. Many systems analysts, computer scientists, and database administrators work for other employers, such as government, manufacturers of computer and related electronic equipment, insurance companies, financial institutions, and universities.

A growing number of computer specialists, such as systems analysts and network and data communications analysts, are employed on a temporary or contract basis. Many of them are self-employed, working independently as contractors or self-employed consultants. For example, a company installing a new computer system may need the services of several systems analysts just to get the system running. Because not all of them would be needed once the system is functioning, the company might contract with systems analysts or a temporary help agency or consulting firm. Such jobs may last from several months up to 2 years or more. This growing practice enables companies to bring in people with the exact skills they need to complete a particular project, rather than having to spend time or money training or retraining existing workers. Often, experienced consultants then train a company's in-house staff as a project develops.

Training, Other Qualifications, and Advancement

Rapidly changing technology means an increasing level of skill and education demanded by employers. Companies are looking for professionals with a broader background and range of skills, including not only technical knowledge, but also communication and other interpersonal skills. This shift from requiring workers to possess solely sound technical knowledge emphasizes workers who can handle various responsibilities. While there is no universally accepted way to prepare for a job as a systems analyst, computer scientist, or database administrator, most employers place a premium on some formal college education. A bachelor's degree is a prerequisite for many jobs; however, some jobs may require only a 2-year degree. Relevant work experience also is very important. For more technically complex jobs, persons with graduate degrees are preferred.

For systems analyst, programmer analyst, as well as database administrator positions, many employers seek applicants who have a bachelor's degree in computer science, information science, or management information systems (MIS). MIS programs usually are part of the business school or college. These programs differ considerably from computer science programs, emphasizing business and management-oriented coursework and business computing courses. Many employers increasingly seek individuals with a master's degree in business administration (MBA) with a concentration in information systems, as more firms move their business to the Internet. For some networks systems and data communication analysts, such as webmasters, an associate degree or certificate generally is sufficient, although ore advanced positions might require a computer-related bachelor's degree. For computer and information scientists, a doctoral degree generally is required due to the highly technical nature of their work.

Despite the preference towards technical degrees, persons with degrees in a variety of majors find employment in these computer occupations. The level of education and type of training employers require depend on their needs. One factor affecting these needs is changes in technology. As demonstrated by the current demand for workers with skills related to the Internet, employers often scramble to find workers capable of implementing "hot" new technologies. Another factor driving employers' needs is the time frame in which a project must be completed.

Most community colleges and many independent technical institutes and proprietary schools offer an associate degree in computer science or a related information technology field. Many of these programs may be more geared toward meeting the needs of local businesses and are more occupation-specific than those designed for a 4-year degree. Some jobs may be better suited to the level of training these programs offer. Employers usually look for people who have broad knowledge and experience related to computer systems and technologies, strong problem-solving and analytical skills, and good interpersonal skills. Courses in computer science or systems design offer good preparation for a job in these computer occupations. For jobs in a business environment, employers usually want systems analysts to have business management or closely related skills, while a background in the physical sciences, applied mathematics, or engineering is preferred for work in scientifically oriented organizations. Art or graphic design skills may be desirable for webmasters or Web developers.

Job seekers can enhance their employment opportunities by participating in internship or co-op programs offered through their schools. Because many people develop advanced computer skills in one occupation and then transfer those skills into a computer occupation, a related background in the industry in which the job is located, such as financial services, banking, or accounting, can be important. Others have taken computer science courses to supplement their study in fields such as accounting, inventory control, or other business areas. For example, a financial analyst proficient in computers might become a systems analyst or computer support specialist in financial systems development, while a computer programmer might move into a systems analyst job.

Systems analysts, computer scientists, and database administrators must be able to think logically and have good communication skills. They often deal with a number of tasks simultaneously; the ability to concentrate and pay close attention to detail is important. Although these computer specialists sometimes work independently, they often work in teams on large projects. They must be able to communicate effectively with computer personnel, such as programmers and managers, as well as with users or other staff who may have no technical computer background.

Computer scientists employed in private industry may advance into managerial or project leadership positions. Those employed in academic institutions can become heads of research departments or published authorities in their field. Systems analysts may be promoted to senior or lead systems analyst. Those who show leadership ability also can become project managers or advance into management positions such as manager of information systems or chief information officer. Database administrators also may advance into managerial positions such as chief technology officer, based on their experience managing data and enforcing security. Computer specialists with work experience and considerable expertise in a particular subject area or application may find lucrative opportunities as independent consultants or choose to start their own computer consulting firms.

Technological advances come so rapidly in the computer field that continuous study is necessary to keep skills up to date. Employers, hardware and software vendors, colleges and universities, and private training institutions offer continuing education. Additional training may come from professional development seminars offered by professional computing societies.

Technical or professional certification is a way to demonstrate a level of competency or quality in a particular field. Product vendors or software firms also offer certification and may require professionals who work with their products to be certified. Many employers regard these certifications as the industry standard. For example, one method of acquiring enough knowledge to get a job as a database administrator is to become certified in a specific type of database management. Voluntary certification also is available through other organizations. Professional certification may provide a job seeker a competitive advantage.

Job Outlook

Systems analysts, computers scientists, and database administrators are expected to be the among the fastest growing occupations through 2010. Employment of these computer specialists is expected to increase much faster than the average for all occupations as organizations continue to adopt and integrate increasingly sophisticated technologies. Growth will be driven by very rapid growth in computer and data processing services, which is projected to be the fastest growing industry in the U.S. economy. In addition, many job openings will arise annually from the need to replace workers who move into managerial positions or other occupations or who leave the labor force.

The demand for networking to facilitate the sharing of information, the expansion of client/server environments, and the need for computer specialists to use their knowledge and skills in a problem-solving capacity will be major factors in the rising demand for systems analysts, computer scientists, and database administrators. Moreover, falling prices of computer hardware and software should continue to induce more businesses to expand computerized operations and integrate new technologies. In order to maintain a competitive edge and operate more efficiently, firms will continue to demand computer specialists who are knowledgeable about the latest technologies and are able to apply them to meet the needs of businesses.

Increasingly, more sophisticated and complex technology is being implemented across all organizations, which should fuel the demand for these computer occupations. There is a growing demand for system analysts to allow firms to be maximize their efficiency by using available technology. The explosive growth in electronic commerce—doing business on the Internet—and the continuing need to build and maintain databases that store critical information on customers, inventory, and projects is fueling demand for database administrators familiar with the latest technology.

The development of new technologies usually leads to demand for various workers. The expanding integration of Internet technologies by businesses, for example, has resulted in a growing need for specialists who can develop and support Internet and intranet applications. The growth of electronic commerce means more establishments use the Internet to conduct their business online. This translates into a need for information technology professionals who can help organizations use technology to communicate with employees, clients, and consumers. Explosive growth in these areas also is expected to fuel demand for specialists knowledgeable about network, data, and communications security.

As technology becomes more sophisticated and complex, employers demand a higher level of skill and expertise. Individuals with an advanced degree in computer science, computer engineering, or an MBA with a concentration in information systems should enjoy very favorable employment prospects. College graduates with a bachelor's degree in computer science, computer engineering, information science, or management information systems also should enjoy favorable prospects for employment, particularly if they have supplemented their formal education with practical experience. Because employers continue to seek computer specialists who can combine strong technical skills with good interpersonal and business skills, graduates with non-computer science degrees but who have had courses in computer programming, systems analysis, and other information technology areas, also should continue to find jobs in these computer fields. In fact, individuals with the right experience and training can work in these computer occupations regardless of their college major or level of formal education.

Earnings

Median annual earnings of computer systems analysts were $59,330 in 2000. The middle 50 percent earned between $46,980 and $73,210 a year. The lowest 10 percent earned less than $37,460, and the highest 10 percent earned more than $89,040. Median annual earnings in the industries employing the largest numbers of computer systems analysts in 2000 were:

Computer and data processing services	$64,110
Professional and commercial equipment	63,530
Federal government	59,470
Local government	52,490
State government	51,230

Median annual earnings of database administrators were $51,990 in 2000. The middle 50 percent earned between $38,210 and $71,440. The lowest 10 percent earned less than $29,400, and the highest 10 percent earned more than $89,320. In 2000, median annual earnings of database administrators employed in computer and data processing services were $63,710, and in telephone communication, $52,230.

Median annual earnings of network systems and data communication analysts were $54,510 in 2000. The middle 50 percent earned between $42,310 and $69,970. The lowest 10 percent earned less than $33,360, and the highest 10 percent earned more than $88,620. Median annual earnings in the industries employing the largest numbers of network systems and data communications analysts in 2000 were:

Management and public relations	$60,260
Commercial banks	59,910
Computer and data processing services	59,160
Telephone communications	51,780
State government	42,000

Median annual earnings of computer and information scientists, research, were $70,590 in 2000. The middle 50 percent earned between $54,700 and $89,990. The lowest 10 percent earned less than $41,390, and the highest 10 percent earned more than $113,510. Median annual earnings of computer and information scientists employed in computer and data processing services in 2000 were $71,940.

Median annual earnings of all other computer specialists were $50,590 in 2000. Median annual earnings of all other computer specialists employed in computer and data processing services were $51,970, and in professional and commercial equipment, $80,270 in 2000.

According to the National Association of Colleges and Employers, starting offers for graduates with a master's degree in computer science averaged $61,453 in 2001. Starting offers for graduates with a bachelor's degree in computer science averaged $52,723; in computer programming, $48,602; in computer systems analysis, $45,643; in information sciences and systems, $45,182; and in management information systems, $45,585.

According to Robert Half International, starting salaries in 2001 ranged from $72,500 to $105,750 for database administrators. Salaries for Internet-related occupations ranged from $60,500 to $84,250 for security administrators; $58,000 to $82,500 for webmasters; and $56,250 to $76,750 for Internet/Intranet developers.

Related Occupations

Other workers who use logic, and creativity to solve business and technical problems are computer programmers, computer software engineers, computer and information systems managers, financial analysts and personal financial advisors, urban and regional planners, engineers, mathematicians, statisticians, operations research analysts, management analysts, and actuaries.

Sources of Additional Information

Further information about computer careers is available from:

- Association for Computing Machinery (ACM), 1515 Broadway, New York, NY 10036. Internet: http://www.acm.org
- IEEE Computer Society, Headquarters Office, 1730 Massachusetts Ave. N.W., Washington, DC 20036-1992. Internet: http://www.computer.org
- National Workforce Center for Emerging Technologies, 3000 Landerholm Circle S.E., Bellevue, WA 98007. Internet: http://www.nwcet.org

Information about becoming a Certified Computing Professional is available from:

- Institute for Certification of Computing Professionals (ICCP), 2350 East Devon Ave., Suite 115, Des Plaines, IL 60018. Internet: http://www.iccp.org

Television, Video, and Motion Picture Camera Operators and Editors

O*NET 27-4031.00, 27-4032.00

Significant Points

- Technical expertise, a "good eye," imagination, and creativity are essential.
- Keen competition for job openings is expected, because many talented peopled are attracted to the field.
- About one-fourth of camera operators are self-employed.

Nature of the Work

Television, video, and motion picture camera operators produce images that tell a story, inform or entertain an audience, or record an event. *Film and video editors* edit soundtracks, film, and video for the motion picture, cable, and broadcast television industries. Some camera operators do their own editing.

Making commercial quality movies and video programs requires technical expertise and creativity. Producing successful images requires choosing and presenting interesting material, selecting appropriate equipment, and applying a good eye and steady hand to assure smooth natural movement of the camera.

Camera operators use television, video, or motion picture cameras to shoot a wide range of subjects, including television series, studio programs, news and sporting events, music videos, motion pictures, documentaries, and training sessions. Some film or videotape private ceremonies and special events. Those who record images on videotape are often called *videographers*. Many are employed by independent television stations, local affiliates, large cable and television networks, or smaller, independent production companies. *Studio camera operators* work in a broadcast studio and usually videotape their subjects from a fixed position. *News camera operators*, also called *electronic news gathering (ENG) operators*, work as part of a reporting team, following newsworthy events as they unfold. To capture live events, they must anticipate the action and act quickly. ENG operators may need to edit raw footage on the spot for relay to a television affiliate for broadcast.

Camera operators employed in the entertainment field use motion picture cameras to film movies, television programs, and commercials. Those who film motion pictures are also known as *cinematographers*. Some specialize in filming cartoons or special effects. They may be an integral part of the action, using cameras in any of several different camera mounts. For example, the camera operator can be stationary and shoot whatever passes in front of the lens, or the camera can be mounted on a track, with the camera operator responsible for shooting the scene from different angles or directions. Other camera operators sit on cranes and follow the action, while crane operators move them into position. *Steadicam operators* mount a harness and carry the camera on their shoulders to provide a more solid picture while they move about the action. Camera operators who work in the entertainment field often meet with directors, actors, editors, and camera assistants to discuss ways of filming, editing, and improving scenes.

Working Conditions

Working conditions for camera operators and editors vary considerably. Those employed in government, television and cable networks, and advertising agencies usually work a 5-day, 40-hour week. On the other hand, ENG operators often work long, irregular hours and must be available to work on short notice. Camera operators and editors working in motion picture production also may work long, irregular hours.

ENG operators and those who cover major events, such as conventions or sporting events, frequently travel locally, stay overnight on assignments, or travel to distant places for longer periods. Camera operators filming television programs or motion pictures may travel to film on location.

Some camera operators work in uncomfortable or even dangerous surroundings, especially ENG operators covering accidents, natural disasters, civil unrest, or military conflicts. Many camera operators must wait long hours in all kinds of weather for an event to take place and stand or walk for long periods while carrying heavy equipment. ENG operators often work under strict deadlines.

Employment

Television, video, and motion picture camera operators held about 27,000 jobs in 2000; and film and video editors held about 16,000. One-fourth of camera operators were self-employed. Some self-employed camera operators contracted with television networks, documentary or independent filmmakers, advertising agencies, or trade show or convention sponsors to do individual projects for a predetermined fee, often at a daily rate.

Most salaried camera operators were employed by television broadcasting stations or motion picture studios. Half of the salaried film and video editors worked for motion picture studios. Most camera operators and editors worked in metropolitan areas.

Training, Other Qualifications, and Advancement

Employers usually seek applicants with a "good eye," imagination, and creativity, as well as a good technical understanding of camera operation. Camera operators and editors usually acquire their skills through on-the-job training or formal postsecondary training at vocational schools, colleges, universities, or photographic institutes. Formal education may be required for some positions.

Many universities, community and junior colleges, vocational-technical institutes, and private trade and technical schools offer courses in camera operation and videography. Basic courses cover equipment, processes, and techniques. Bachelor's degree programs, especially those including business courses, provide a well-rounded education.

Individuals interested in camera operations should subscribe to videographic newsletters and magazines, join clubs, and seek summer or part-time employment in cable and television networks, motion picture studios, or camera and video stores.

Camera operators in entry-level jobs learn to set up lights, cameras, and other equipment. They may receive routine assignments requiring camera adjustments or decisions on what subject matter to capture. Camera operators in the film and television industries usually are hired for a project based on recommendations from individuals such as producers, directors of photography, and camera assistants from previous projects, or through interviews with the producer. ENG and studio camera operators who work for television affiliates usually start in small markets to gain experience.

Camera operators need good eyesight, artistic ability, and hand-eye coordination. They should be patient, accurate, and detail oriented. Camera operators also should have good communication skills, and, if needed, the ability to hold a camera by hand for extended periods.

Camera operators who operate their own businesses, or freelance, need business skills as well as talent. These individuals must know how to submit bids; write contracts; get permission to shoot on locations that normally are not open to the public; obtain releases to use film or tape of people; price their services; secure copyright protection for their work; and keep financial records.

With increased experience, operators may advance to more demanding assignments or positions with larger or network television stations. Advancement for ENG operators may mean moving to larger media markets. Other camera operators and editors may become directors of photography for movie studios, advertising agencies, or television programs. Some teach at technical schools, film schools, or universities.

Job Outlook

Camera operators and editors can expect keen competition for job openings because the work is attractive to many people. The number of individuals interested in positions as videographers and movie camera operators usually is much greater than the number of openings. Those who succeed in landing a salaried job or attracting enough work to earn a living by freelancing are likely to be the most creative, highly motivated, able to adapt to rapidly changing technologies, and adept at operating a business. Related work experience or job-related training also is beneficial to prospective camera operators.

Employment of camera operators and editors is expected to grow faster than the average for all occupations through 2010. Rapid expansion of the entertainment market, especially motion picture production and distribution, will spur growth of camera operators. In addition, computer and Internet services provide new outlets for interactive productions. Camera operators will be needed to film made-for-the-Internet broadcasts such as live music videos, digital movies, sports, and general information or entertainment programming. These images can be delivered directly into the home either on compact discs or over the Internet. Modest growth also is expected in radio and television broadcasting.

Earnings

Median annual earnings for television, video, and motion picture camera operators were $27,870 in 2000. The middle 50 percent earned between $19,230 and $44,150. The lowest 10 percent earned less than $14,130, and the highest 10 percent earned more than $63,690. Median annual earnings were $31,560 in motion picture production and services and $23,470 in radio and television broadcasting.

Median annual earnings for film and video editors were $34,160 in 2000. The middle 50 percent earned between $24,800 and $52,000. The lowest 10 percent earned less than $18,970, and the highest 10 percent earned more than $71,280. Median annual earnings were $36,770 in motion picture production and services, the industry employing the largest numbers of film and video editors.

Many camera operators who work in film or video are freelancers; their earnings tend to fluctuate each year. Because most freelance camera operators purchase their own equipment, they incur considerable expense acquiring and maintaining cameras and accessories.

Related Occupations

Related arts and media occupations include artists and related workers, broadcast and sound engineering technicians and radio operators, designers, and photographers.

Sources of Additional Information

Information about career and employment opportunities for camera operators and film and video editors is available from local offices of state employment service agencies, local offices of the relevant trade unions, and local television and film production companies who employ these workers.

Tool and Die Makers

O*NET 51-4111.00

Significant Points

- Most tool and die makers train for 4 or 5 years in apprenticeships or postsecondary programs; employers typically recommend apprenticeship training.
- Job seekers with the appropriate skills and background should enjoy excellent opportunities.

Nature of the Work

Tool and die makers are among the most highly skilled production workers in the economy. These workers produce tools, dies, and special guiding and holding devices that enable machines to manufacture a variety of products we use daily—from clothing and furniture to heavy equipment and parts for aircraft.

Toolmakers craft precision tools that are used to cut, shape, and form metal and other materials. They also produce jigs and fixtures (devices that hold metal while it is bored, stamped, or drilled) and gauges and other measuring devices. Die makers construct metal forms (dies) that are used to shape metal in stamping and forging operations. They also make metal molds for diecasting and for molding plastics, ceramics, and composite materials. In addition to developing, designing and producing new tools and dies, these workers also may repair worn or damaged tools, dies, gauges, jigs, and fixtures.

To perform these functions, tool and die makers employ many types of machine tools and precision measuring instruments. They also must be familiar with the machining properties, such as hardness and heat tolerance, of a wide variety of common metals and alloys. As a result, tool and die makers are knowledgeable in machining operations, mathematics, and blueprint reading. In fact, tool and die makers often are considered highly specialized machinists.

Working from blueprints, tool and die makers first must plan the sequence of operations necessary to manufacture the tool or die. Next, they measure and mark the pieces of metal that will be cut to form parts of the final product. At this point, tool and die makers cut, drill, or bore the part as required, checking to ensure that the final product meets specifications. Finally, these workers assemble the parts and perform finishing jobs such as filing, grinding, and polishing surfaces.

Modern technology is changing the ways in which tool and die makers perform their jobs. Today, for example, these workers often use computer-aided design (CAD) to develop products and parts. Specifications entered into computer programs can be used to electronically develop drawings for the required tools and dies. Numerical tool and process control programmers use computer-aided manufacturing (CAM) programs to convert electronic drawings into computer programs that contain a sequence of cutting tool operations. Once these programs are developed, computer numerically controlled (CNC) machines follow the set of instructions contained in the program to produce the part. (Computer-controlled machine tool operators or machinists normally operate CNC machines; however, tool and die makers are trained in both operating CNC machines and writing CNC programs, and they may perform either task. CNC programs are stored electronically for future use, saving time and increasing worker productivity.) Next, tool and die makers assemble the different parts into a functioning machine. They file, grind, shim, and adjust the different parts to properly fit them together. Finally, the tool and die makers set up a test run using the tools or dies they have made to make sure that the manufactured parts meet specifications. If problems occur, tool and die makers compensate by adjusting the tools or dies.

Working Conditions

Tool and die makers usually work in toolrooms. These areas are quieter than the production floor because there are fewer machines in use at one time. They also are generally clean and cool to minimize heat-related expansion of metal workpieces and to accommodate the growing number of computer-operated machines. To minimize the exposure of workers to moving parts, machines have guards and shields. Many computer-controlled machines are totally enclosed, minimizing the exposure of workers to noise, dust, and the lubricants used to cool workpieces during machining. Tool and die makers also must follow safety rules and wear protective equipment, such as safety glasses to shield against bits of flying metal, earplugs to protect against noise, and gloves and masks to reduce exposure to hazardous lubricants and cleaners. These workers also need stamina because they often spend much of the day on their feet and may do moderately heavy lifting.

Companies employing tool and die makers have traditionally operated only one shift per day. Overtime and weekend work are common, especially during peak production periods.

Employment

Tool and die makers held about 130,000 jobs in 2000. Most worked in industries that manufacture metalworking machinery and equipment, metal forgings and stampings, motor vehicles, miscellaneous plastics products, and aircraft and parts. Although they are found throughout the country, jobs are most plentiful in the midwest, northeast, and west, where many of the metalworking industries are located.

Training, Other Qualifications, and Advancement

Most tool and die makers learn their trade through 4 or 5 years of education and training in formal apprenticeships or postsecondary programs. The best way to learn all aspects of tool and die making, according to most employers, is a formal apprenticeship program that combines classroom instruction and job experience. A growing number of tool and die makers receive most of their formal classroom training from community and technical colleges, sometimes in conjunction with an apprenticeship program.

Tool and die maker trainees learn to operate milling machines, lathes, grinders, spindles, and other machine tools. They also learn to use handtools for fitting and assembling gauges, and other mechanical and metal-forming equipment. In addition, they study metalworking processes, such as heat treating and plating. Classroom training usually consists of mechanical drawing, tool designing, tool programming, blueprint reading, and mathematics courses, including algebra, geometry, trigonometry, and basic statistics. Tool and die makers increasingly must have good computer skills to work with CAD technology and CNC machine tools.

Workers who become tool and die makers without completing formal apprenticeships generally acquire their skills through a combination of informal on-the-job training and classroom instruction at a vocational school or community college. They often begin as machine operators and gradually take on more difficult assignments. Many machinists become tool and die makers.

Because tools and dies must meet strict specifications—precision to one ten-thousandth of an inch is common—the work of tool and die makers requires a high degree of patience and attention to detail. Good eyesight is essential. Persons entering this occupation also should be mechanically inclined, able to work and solve problems independently, and capable of doing work that requires concentration and physical effort.

There are several ways for skilled workers to advance. Some move into supervisory and administrative positions in their firms; many obtain their college degree and go into engineering or tool design Some may start their own shops.

Job Outlook

Applicants with the appropriate skills and background should enjoy excellent opportunities for tool and die maker jobs. The number of workers receiving training in this occupation is expected to continue to be fewer than the number of openings created each year by tool and die makers who retire or transfer to other occupations. As more of these highly skilled workers retire, employers in certain parts of the country report difficulty attracting well-trained applicants. A major factor limiting the number of people entering the occupation is that many young people who have the educational and personal qualifications necessary to learn tool and die making may prefer to attend college or may not wish to enter production-related occupations.

Despite expected excellent employment opportunities, little or no change in employment of tool and die makers is projected over the 2000–2010 period because advancements in automation, including CNC machine tools and computer-aided design, should improve worker productivity, thus limiting employment. On the other hand, tool and die makers play a key role in the operation of many firms. As firms invest in new equipment, modify production techniques, and implement product design changes more rapidly, they will continue to rely heavily on skilled tool and die makers for retooling.

Earnings

Median hourly earnings of tool and die makers were $19.76 in 2000. The middle 50 percent earned between $15.67 and $24.45. The lowest 10 percent had earnings of less than $12.44, while the top 10 percent earned more than $28.88. Median hourly earnings in the manufacturing industries employing the largest number of tool and die makers in 2000 are shown in the following table:

Motor vehicles and equipment	$25.76
Aircraft and parts	22.17
Metal forgings and stampings	21.52
Metalworking machinery	18.99
Miscellaneous plastics products, not elsewhere classified	18.92

Related Occupations

The occupations most closely related to the work of tool and die makers are other machining occupations. These include machinists; computer-control programmers and operators; and machine setters, operators, and tenders—metal and plastic.

Another occupation that requires precision and skill in working with metal is welding, soldering, and brazing workers.

Sources of Additional Information

For information about careers in tool and die making, contact:

- Precision Machine Products Association, 6700 West Snowville Rd., Brecksville, OH 44141-3292. Internet: http://www.pmpa.org
- National Tooling and Metalworking Association, 9300 Livingston Rd., Ft. Washington, MD 20744. Internet: http://www.ntma.org
- PMA Educational Foundation, 6363 Oak Tree Blvd., Independence, OH 44131-2500. Internet: http://www.pmaef.org

Water and Liquid Waste Treatment Plant and System Operators

O*NET 51-8031.00

Significant Points

- Employment is concentrated in local government and private water supply and sanitary services companies.
- Postsecondary training is increasingly an asset as the number of regulated contaminants grows and treatment plants become more complex.
- Operators must pass exams certifying that they are capable of overseeing various treatment processes.

Nature of the Work

Clean water is essential for everyday life. *Water treatment plant and system operators* treat water so that it is safe to drink. *Liquid waste treatment plant and system operators*, also known as wastewater treatment plant and system operators, remove harmful pollutants from domestic and industrial liquid waste so that it is safe to return to the environment.

Water is pumped from wells, rivers, and streams to water treatment plants, where it is treated and distributed to customers. Liquid waste travels through customers' sewer pipes to liquid waste treatment plants, where it is treated and returned to streams, rivers, and oceans, or reused for irrigation and landscaping. Operators in both types of plants control processes and equipment to remove or destroy harmful materials, chemical compounds, and microorganisms from the water. They also control pumps, valves, and other processing equipment to move the water or liquid waste through the various treatment processes, and dispose of the removed waste materials.

Operators read, interpret, and adjust meters and gauges to make sure plant equipment and processes are working properly. They operate chemical-feeding devices, take samples of the water or liquid waste, perform chemical and biological laboratory analyses, and adjust the amount of chemicals, such as chlorine, in the water. They use a variety of instruments to sample and measure water quality, and common hand and power tools to make repairs. Operators also make minor repairs to valves, pumps, and other equipment.

Water and liquid waste treatment plant and system operators increasingly rely on computers to help monitor equipment, store sampling results, make process-control decisions, schedule and record maintenance activities, and produce reports. When problems occur, operators may use their computers to determine the cause of the malfunction and its solution.

Occasionally operators must work under emergency conditions. A heavy rainstorm, for example, may cause large amounts of liquid waste to flow into sewers, exceeding a plant's treatment capacity. Emergencies also can be caused by conditions inside a plant, such as chlorine gas leaks or oxygen deficiencies. To handle these conditions, operators are trained to make an emergency management response and use special safety equipment and procedures to protect public health and the facility. During these periods, operators may work under extreme pressure to correct problems as quickly as possible. These periods may create dangerous working conditions, and operators must be extremely cautious.

The specific duties of plant operators depend on the type and size of plant. In smaller plants, one operator may control all machinery, perform tests, keep records, handle complaints, and do repairs and maintenance. A few operators may handle both a water treatment and a liquid waste treatment plant. In larger plants with many employees, operators may be more specialized and only monitor one process. The staff also may include chemists, engineers, laboratory technicians, mechanics, helpers, supervisors, and a superintendent.

Water pollution standards have become increasingly stringent since adoption of two major federal environmental statutes: the Clean Water Act of 1972, which implemented a national system of regulation on the discharge of pollutants; and the Safe Drinking Water Act of 1974, which established standards for drinking water. Industrial facilities sending their wastes to municipal treatment plants must meet certain minimum standards to ensure that the wastes have been adequately pretreated and will not damage municipal treatment facilities. Municipal water treatment plants also must meet stringent drinking water standards. The list of contaminants regulated by these statutes has grown over time. For example, the 1996 Safe Drinking Water Act Amendments include standards for the monitoring of cryptosporidium and giardia, two biological organisms that cause health problems. Operators must be familiar with the guidelines established by federal regulations and how they affect their plant. In addition to federal regulations, operators also must be aware of any guidelines imposed by the state or locality in which the plant operates.

Working Conditions

Water and liquid waste treatment plant and system operators work both indoors and outdoors, and may be exposed to noise from machinery and unpleasant odors. Operators' work is physically demanding and often is performed in unclean locations. They must pay close attention to safety procedures for they may be confronted with hazardous conditions, such as slippery walkways, dangerous gases, and malfunctioning equipment. Plants operate 24 hours a day, 7 days a week; therefore, operators work one of three 8-hour shifts, including weekends and holidays, on a rotational basis. Operators may be required to work overtime.

Employment

Water and liquid waste treatment plant and system operators held about 88,000 jobs in 2000. Most worked for local governments. Some worked for private water supply and sanitary services companies, which increasingly provide operation and management services to local governments on a contract basis.

Water and liquid waste treatment plant and system operators are employed throughout the country, but most jobs are in larger towns and cities. Although nearly all work full time, those who work in small towns may only work part-time at the treatment plant—the remainder of their time may be spent handling other municipal duties.

Training, Other Qualifications, and Advancement

A high school diploma usually is required to become a water or liquid waste treatment plant operator. Operators need mechanical aptitude and should be competent in basic mathematics, chemistry, and biology. They must have the ability to apply data to formulas of treatment requirements, flow levels, and concentration levels. Some basic familiarity with computers also is necessary because of the trend toward computer-controlled equipment and more sophisticated instrumentation. Certain positions—particularly in larger cities and towns—are covered by civil service regulations. Applicants for these positions may be required to pass a written examination testing mathematics skills, mechanical aptitude, and general intelligence.

Completion of an associate degree or 1-year certificate program in water quality and liquid waste treatment technology increases an applicant's chances for employment and promotion because plants are becoming more complex. Offered throughout the country, these programs provide a good general knowledge of water and liquid waste treatment processes, as well as basic preparation for becoming an operator.

Trainees usually start as attendants or operators-in-training and learn their skills on the job under the direction of an experienced operator. They learn by observing and doing routine tasks such as recording meter readings; taking samples of liquid waste and sludge; and performing simple maintenance and repair work on pumps, electric motors, valves, and other plant equipment. Larger treatment plants generally combine this on-the-job training with formal classroom or self-paced study programs.

The Safe Drinking Water Act Amendments of 1996, enforced by the U.S. Environmental Protection Agency, specify national minimum standards for certification and recertification of operators of community and nontransient, noncommunity water systems. As a result, operators must pass an examination to certify that they are capable of overseeing liquid waste treatment plant operations. There are different levels of certification depending on the operator's experience and training. Higher certification levels qualify the operator for a wider variety of treatment processes. Certification requirements vary by state and by size of treatment plants. Although relocation may mean having to become certified in a new location, many states accept other states' certifications.

Most state drinking water and water pollution control agencies offer training courses to improve operators' skills and knowledge. These courses cover principles of treatment processes and process control, laboratory procedures, maintenance, management skills, collection systems, safety, chlorination, sedimentation, biological treatment, sludge treatment and disposal, and flow measurements. Some operators take correspondence courses on subjects related to water and liquid waste treatment, and some employers pay part of the tuition for related college courses in science or engineering.

As operators are promoted, they become responsible for more complex treatment processes. Some operators are promoted to plant supervisor or superintendent; others advance by transferring to a larger facility. Postsecondary training in water and liquid waste treatment, coupled with increasingly responsible experience as an operator, may be sufficient to qualify for superintendent of a small plant, where a superintendent also serves as an operator. However, educational requirements are rising as larger, more complex treatment plants are built to meet new drinking water and water pollution control standards. With each promotion, the operator must have greater knowledge of federal, state, and local regulations. Superintendents of large plants generally need an engineering or science degree.

A few operators get jobs with state drinking water or water pollution control agencies as technicians, who monitor and provide technical assistance to plants throughout the state. Vocational-technical school or community college training generally is preferred for technician jobs. Experienced operators may transfer to related jobs with industrial liquid waste treatment plants, water or liquid waste treatment equipment and chemical companies, engineering consulting firms, or vocational-technical schools.

Job Outlook

Employment of water and liquid waste treatment plant and system operators is expected to grow as fast as the average for all occupations through the year 2010. Because the number of applicants in this field is normally low, job prospects will be good for qualified applicants.

The increasing population and growth of the economy are expected to boost demand for essential water and liquid waste treatment services. As new plants are constructed to meet this demand, employment of water and liquid waste treatment plant and system operators will increase. In addition, many job openings will occur as experienced operators transfer to other occupations or leave the labor force.

Local governments are the largest employers of water and liquid waste treatment plant and system operators. However, federal certification requirements have increased reliance on private firms specializing in the operation and management of water and liquid waste treatment facilities. As a result, employment in privately owned facilities will grow faster than the average. Increased pretreatment activity by manufacturing firms also will create new job opportunities.

Earnings

Median annual earnings of water and liquid waste treatment plant and system operators were $31,380 in 2000. The middle 50 percent earned between $24,390 and $39,530. The lowest 10 percent earned less than $19,120, and the highest 10 percent earned more than $47,370. Median annual earnings of water and liquid waste treatment plant and systems operators in 2000 were $31,120 in local government and $29,810 in water supply.

In addition to their annual salaries, water and liquid waste treatment plant and system operators usually receive benefits that may include health and life insurance, a retirement plan, and educational reimbursement for job-related courses.

Related Occupations

Other workers whose main activity consists of operating a system of machinery to process or produce materials include chemical plant and system operators; gas plant operators; petroleum pump system operators, refinery operators, and gaugers; power plant operators, distributors, and dispatchers; and stationary engineers and boiler operators.

Sources of Additional Information

For information on employment opportunities, contact state or local water pollution control agencies, state water and liquid waste operator associations, state environmental training centers, or local offices of the state employment service.

For information on certification, contact:

- Association of Boards of Certification, 208 Fifth St., Ames, IA 50010-6259. Internet: http://www.abccert.org

For educational information related to a career as a water or liquid waste treatment plant and system operator, contact:

- American Water Works Association, 6666 West Quincy Ave., Denver, CO 80235
- Water Environment Federation, 601 Wythe St., Alexandria, VA 22314-1994. Internet: http://www.wef.org

Writers and Editors

O*NET 27-3041.00, 27-3042.00, 27-3043.01, 27-3043.02, 27-3043.04

Significant Points

- Most jobs require a college degree either in the liberal arts—communications, journalism, and English are preferred—or a technical subject for technical writing positions.
- Competition is expected to be less for lower paying, entry-level jobs at small daily and weekly newspapers, trade publications, and radio and television broadcasting stations in small markets.
- Persons who fail to gain better paying jobs or earn enough as independent writers usually are able to transfer readily to communications-related jobs in other occupations.

Nature of the Work

Writers and editors communicate through the written word. Writers and editors generally fall into one of three categories. *Writers and authors* develop original fiction and nonfiction for books, magazines and trade journals, newspapers, online publications, company newsletters, radio and television broadcasts, motion pictures, and advertisements. *Technical writers* develop scientific or technical materials, such as scientific and medical reports, equipment manuals, appendices, or operating and maintenance instructions. They also may assist in layout work. *Editors* select and prepare material for publication or broadcast and review and prepare a writer's work for publication or dissemination.

Nonfiction writers either select a topic or are assigned one, often by an editor or publisher. Then, they gather information through personal observation, library and Internet research, and interviews. Writers select the material they want to use, organize it, and use the written word to express ideas and convey information. Writers also revise or rewrite sections, searching for the best organization or the right phrasing.

Creative writers, poets, and lyricists, including novelists, playwrights, and screenwriters, create original works (such as prose, poems, plays, and song lyrics) for publication or performance. Some works may be commissioned (at the request of a sponsor); others may be written for hire (based on completion of a draft or an outline). *Copy writers* prepare advertising copy for use by publication or broadcast media, or to promote the sale of goods and services. *Newsletter writers* produce information for distribution to association members, corporate employees, organizational clients, or the public. Writers and authors also construct crossword puzzles and prepare speeches.

Technical writers put scientific and technical information into easily understandable language. They prepare scientific and technical reports, operating and maintenance manuals, catalogs, parts lists, assembly instructions, sales promotion materials, and project proposals. They also plan and edit technical reports and oversee preparation of illustrations, photographs, diagrams, and charts. *Science and medical writers* prepare a range of formal documents presenting detailed information on the physical or medical sciences. They impart research findings for scientific or medical professions, organize information for advertising or public relations needs, and interpret data and other information for a general readership.

Many writers prepare material directly for the Internet. For example, they may write for electronic newspapers or magazines, create short fiction, or produce technical documentation only available online. Also, they may write the text of Web sites. These writers should be knowledgeable about graphic design, page layout and desktop publishing software. Additionally, they should be familiar with interactive technologies of the Web so they can blend text, graphics, and sound together.

Freelance writers sell their work to publishers, publication enterprises, manufacturing firms, public relations departments, or advertising agencies. Sometimes, they contract with publishers to write a book or article. Others may be hired on a job-basis to complete specific assignments such as writing about a new product or technique.

Editors review, rewrite, and edit the work of writers. They may also do original writing. An editor's responsibilities vary depending on the employer and type and level of editorial position held. In the publishing industry, an editor's primary duties are to plan the contents of books, technical journals, trade magazines, and other general interest publications. Editors decide what material will appeal to readers, review and edit drafts of books and articles, offer comments to improve the work, and suggest possible titles. Additionally, they oversee the production of the publications.

Major newspapers and newsmagazines usually employ several types of editors. The *executive editor* oversees *assistant editors* who have responsibility for particular subjects, such as local news, international news, feature stories, or sports. Executive editors generally have the final say about what stories are published and how they are covered. The *managing editor* usually is responsible for the daily operation of the news department. *Assignment editors* determine which reporters will cover a given story. *Copy editors* mostly review and edit a reporter's copy for accuracy, content, grammar, and style.

In smaller organizations, like small daily or weekly newspapers or membership newsletter departments, a single editor may do everything or share responsibility with only a few other people. Executive and managing editors typically hire writers, reporters, or other employees. They also plan budgets and negotiate contracts with freelance writers, sometimes called "stringers" in the news industry. In broadcasting companies, *program directors* have similar responsibilities.

Editors and program directors often have assistants. Many assistants, such as copy editors or *production assistants*, hold entry-level jobs. They review copy for errors in grammar, punctuation, and spelling, and check copy for readability, style, and agreement with editorial policy. They suggest revisions, such as changing words or rearranging sentences to improve clarity or accuracy. They also do research for writers and verify facts, dates, and statistics. Production assistants arrange page layouts of articles, photographs, and advertising; compose headlines; and prepare copy for printing. *Publication assistants* who work for publishing houses may read and evaluate manuscripts submitted by freelance writers, proofread printers' galleys, or answer letters about published material. Production assistants on small papers or in radio stations compile articles available from wire services or the Internet, answer phones, and make photocopies.

Most writers and editors use personal computers or word processors. Many use desktop or electronic publishing systems, scanners, and other electronic communications equipment.

Working Conditions

Some writers and editors work in comfortable, private offices; others work in noisy rooms filled with the sound of keyboards and computer printers as well as the voices of other writers tracking down information over the telephone. The search for information sometimes requires travel to diverse workplaces, such as factories, offices, or laboratories, but many have to be content with telephone interviews, the library, and the Internet.

For some writers, the typical workweek runs 35 to 40 hours. However, writers occasionally may work overtime to meet production deadlines. Those who prepare morning or weekend publications and broadcasts work some nights and weekends. Freelance writers generally work more flexible hours, but their schedules must conform to the needs of the client. Deadlines and erratic work hours, often part of the daily routine for these jobs, may cause stress, fatigue, or burnout.

Changes in technology and electronic communications also affect a writer's work environment. For example, laptops allow writers to work from home or while on the road. Writers and editors who use computers for extended periods may experience back pain, eyestrain, or fatigue.

Employment

Writers and editors held about 305,000 jobs in 2000. About 126,000 jobs were for writers and authors; 57,000 were for technical writers; and 122,000 were for editors. Nearly one-fourth of jobs for writers and editors were salaried positions with newspapers, magazines, and book publishers. Substantial numbers, mostly technical writers, work for computer software firms. Other salaried writers and editors work in educational facilities, advertising agencies, radio and television broadcasting studios, public relations firms, and business and nonprofit organizations, such as professional associations, labor unions, and religious organizations. Some develop publications and technical materials for government agencies or write for motion picture companies.

Jobs with major book publishers, magazines, broadcasting companies, advertising agencies, and public relations firms are concentrated in New York, Chicago, Los Angeles, Boston, Philadelphia, and San Francisco. Jobs with newspapers, business and professional journals, and technical and trade magazines are more widely dispersed throughout the country.

Thousands of other individuals work as freelance writers, earning some income from their articles, books, and less commonly, television and movie scripts. Most support themselves with income derived from other sources.

Training, Other Qualifications, and Advancement

A college degree generally is required for a position as a writer or editor. Although some employers look for a broad liberal arts background, most prefer to hire people with degrees in communications, journalism, or English. For those who specialize in a particular area, such as fashion, business, or legal issues, additional background in the chosen field is expected. Knowledge of a second language is helpful for some positions.

Technical writing requires a degree in, or some knowledge about, a specialized field—engineering, business, or one of the sciences, for example. In many cases, people with good writing skills can learn specialized knowledge on the job. Some transfer from jobs as technicians, scientists, or engineers. Others begin as research assistants, or trainees in a technical information department, develop technical communication skills, and then assume writing duties.

Writers and editors must be able to express ideas clearly and logically and should love to write. Creativity, curiosity, a broad range of knowledge, self-motivation, and perseverance also are valuable. Writers and editors must demonstrate good judgment and a strong sense of ethics in deciding what material to publish. Editors also need tact and the ability to guide and encourage others in their work.

For some jobs, the ability to concentrate amid confusion and to work under pressure is essential. Familiarity with electronic publishing, graphics, and video production equipment increasingly is needed. Online newspapers and magazines require knowledge of computer software used to combine online text with graphics, audio, video, and 3-D animation.

High school and college newspapers, literary magazines, community newspapers, and radio and television stations all provide valuable, but sometimes unpaid, practical writing experience. Many magazines, newspapers, and broadcast stations have internships for students. Interns write short pieces, conduct research and interviews, and learn about the publishing or broadcasting business.

In small firms, beginning writers and editors hired as assistants may actually begin writing or editing material right away. Opportunities for advancement can be limited, however. In larger businesses, jobs usually are more formally structured. Beginners generally do research, fact-checking, or copy editing. They take on full-scale writing or editing duties less rapidly than do the employees of small companies. Advancement often is more predictable, although it comes with the assignment of more important articles.

Job Outlook

Employment of writers and editors is expected to increase faster than the average for all occupations through the year 2010. Employment of salaried writers and editors for newspapers, periodicals, book publishers, and nonprofit organizations is expected to increase as demand grows for their publications. Magazines and other periodicals increasingly are developing market niches, appealing to readers with special interests. Also, online publications and services are growing in number and sophistication, spurring the demand for writers and editors. Businesses and organizations are developing newsletters and Internet Web sites and more companies are experimenting with publishing materials directly for the Internet. Advertising and public relations agencies, which also are growing, should be another source of new jobs. Demand for technical writers and writers with expertise in specialty areas, such as law, medicine, or economics, is expected to increase because of the continuing expansion of scientific and technical information and the need to communicate it to others.

In addition to job openings created by employment growth, many openings will occur as experienced workers retire, transfer to other occupations, or leave the labor force. Replacement needs are relatively high in this occupation; many freelancers leave because they cannot earn enough money.

Despite projections of fast employment growth and numerous replacement needs, the outlook for most writing and editing jobs is expected to be competitive. Many people with writing or journalism training are attracted to the occupation. Opportunities should be best for technical writers and those with training in a specialized field. Rapid growth and change in the high technology and electronics industries result in a greater need for people to write users' guides, instruction manuals, and training materials. Developments and discoveries in the law, science, and technology generate demand for people to interpret technical information for a more general audience. This work requires people who are not only technically skilled as writers, but also familiar with the subject area. Also, individuals with the technical skills for working on the Internet may have an advantage finding a job as a writer or editor.

Opportunities for editing positions on small daily and weekly newspapers and in small radio and television stations, where the pay is low, should be better than those in larger media markets. Some small publications hire freelance copy editors as backup for staff editors or as additional help with special projects. Aspiring writers and editors benefit from academic preparation in another discipline as well, either to qualify them as writers specializing in that discipline or as a career alternative if they are unable to get a job in writing.

Earnings

Median annual earnings for salaried writers and authors were $42,270 in 2000. The middle 50 percent earned between $29,090 and $57,330. The lowest 10 percent earned less than $20,290, and the highest 10 percent earned more than $81,370. Median annual earnings were $26,470 in the newspaper industry.

Median annual earnings for salaried technical writers were $47,790 in 2000. The middle 50 percent earned between $37,280 and $60,000. The lowest 10 percent earned less than $28,890, and the highest 10 percent earned more than $74,360. Median annual earnings in computer and data processing services were $51,220.

Median annual earnings for salaried editors were $39,370 in 2000. The middle 50 percent earned between $28,880 and $54,320. The lowest 10 percent earned less than $22,460, and the highest 10 percent earned more than $73,330. Median annual earnings in the industries employing the largest numbers of editors were as follows:

Computer and data processing services	$45,800
Periodicals	42,560
Newspapers	37,560
Books	37,550

Related Occupations

Writers and editors communicate ideas and information. Other communications occupations include announcers; interpreters and translators; news analysts, reporters, and correspondents; and public relations specialists.

Sources of Additional Information

For information on careers in technical writing, contact:

- Society for Technical Communication, Inc., 901 N. Stuart St., Suite 904, Arlington, VA 22203. Internet: http://www.stc.org

For information on union wage rates for newspaper and magazine editors, contact:

- The Newspaper Guild-CWA, Research and Information Department, 501 Third St. NW, Suite 250, Washington, DC 20001

Section Two

THE QUICK JOB SEARCH

Seven Steps to Getting a Good Job in Less Time

The Complete Text of a Results-Oriented Booklet by Michael Farr

Millions of job seekers have found better jobs faster using the techniques in The Quick Job Search. *So can you!* The Quick Job Search *covers the essential steps proven to cut job search time in half and is used widely by job search programs throughout North America. Topics include how to identify your key skills, define and research your ideal job, write a great resume, use the most effective job search methods, get more interviews, answer tough interview questions, and much more. While it is a section in this book,* The Quick Job Search *is available as a separate booklet and in an expanded form as* Seven Steps to Getting a Job Fast. *At the end of this section, I've included several professionally written resumes for you to use as examples when writing your resume. These resumes would be suitable for use when applying for some of America's top computer and technical jobs.*

The Quick Job Search Is Short, But It May Be All You Need

While *The Quick Job Search* is short, it covers the basics on how to explore career options and conduct an effective job search. While these topics can seem complex, I have found some simple truths about job searches:

- ★ If you are going to work, you might as well define what you really want to do and are good at.
- ★ If you are looking for a job, you might as well use techniques that will reduce the time it takes to find one—and that help you get a better job than you might otherwise.

That's what I emphasize in this little book.

Trust Me—Do the Worksheets. I know you will resist completing the worksheets. But trust me, they are worth your time. Doing them will give you a better sense of what you are good at, what you want to do, and how to go about getting it. You will also most likely get more interviews and present yourself better. Is this worth giving up a night of TV? Yes, I think so. Once you finish this minibook and its activities, you will have spent more time planning your career and will know more about finding a job than most people do.

Why Such a Short Book? I've taught job seeking skills for many years, and I've written longer and more detailed books than this one. Yet I have often been asked to tell someone, in a few minutes or hours, the most important things they should do in their career planning or job search. Instructors and counselors also ask the same question because they have only a short time to spend with folks they're trying to help. I've given this a lot of thought, and the seven topics in this minibook are the ones I think are most important to know.

This minibook is short enough to scan in a morning and conduct a more effective job search that afternoon. Doing all the activities would take more time but would prepare you far better. Of course, you can learn more about all of the topics it covers, but this little book may be all you need.

You can't just read about getting a job. The best way to get a job is to go out and get interviews! And the best way to get interviews is to make a job out of getting a job.

After many years of experience, I have identified just seven basic things you need to do that make a big difference in your job search. Each will be covered and expanded on in this minibook.

1. Identify your key skills.
2. Define your ideal job.
3. Learn the two most effective job search methods.
4. Write a superior resume.
5. Organize your time to get two interviews a day.
6. Dramatically improve your interviewing skills.
7. Follow up on all leads.

We all have thousands of skills. *Consider the many skills required to do even a simple thing like ride a bike or bake a cake. But, of all the skills you have, employers want to know those key skills you have for the job they need done. You must clearly identify these key skills, and then emphasize them in interviews.*

Step 1

Identify Your Key Skills and Develop a "Skills Language" to Describe Yourself

One employer survey found that about 90 percent of the people interviewed by the employers did not present the skills they had to do the job they sought. They could not answer the basic question "Why should I hire you?"

Knowing and describing your skills are essential to doing well in interviews. This same knowledge is important in deciding what type of job you will enjoy and do well. For these reasons, I consider identifying your skills a necessary part of a successful career plan or job search.

The Three Types of Skills

Most people think of their "skills" as job-related skills, such as using a computer. But we all have other types of skills that are important for success on a job—and that are important to employers. The triangle at right presents skills in three groups, and I think that this is a very useful way to consider skills for our purposes.

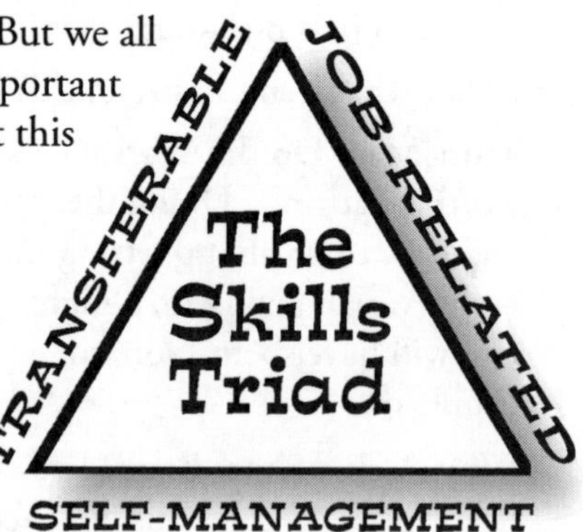

The Skills Triad

Let's review these three types of skills—self-management, transferable, and job-related—and identify those that are most important to you.

Self-Management Skills

Write down three things about yourself that you think make you a good worker.

Your "Good Worker" Traits

1. ____________________

2. ____________________

3. ____________________

You just wrote down the most important things for an employer to know about you! They describe your basic personality and your ability to adapt to new environments. They are some of the most important skills to emphasize in interviews, yet most job seekers don't realize their importance—and don't mention them.

Review the Self-Management Skills Checklist that follows and put a check mark beside any skills you have. The key self-management skills listed first cover abilities that employers find particularly important. If one or more of the key self-management skills apply to you, mentioning them in interviews can help you greatly.

Self-Management Skills Checklist

Check Your Key Self-Management Skills—Employers Value These Highly

______ accept supervision

______ get along with coworkers

______ get things done on time

______ good attendance

______ hard worker

______ honest

______ productive

______ punctual

Check Your Other Self-Management Skills

______ able to coordinate

______ ambitious

______ assertive

______ capable

______ cheerful

______ competent

______ complete assignments

______ conscientious

______ creative

______ dependable

______ discreet

______ eager

______ efficient

______ energetic

______ enthusiastic

______ expressive

______ flexible

______ formal

______ friendly

______ good-natured

______ helpful

______ humble

______ imaginative

______ independent

______ industrious

______ informal

______ intelligent

______ intuitive

______ learn quickly

______ loyal

______ mature

______ methodical

______ modest

______ motivated

______ natural

______ open-minded

______ optimistic

______ original

______ patient

______ persistent

______ physically strong

______ practice new skills

(continues)

(continued)

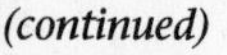

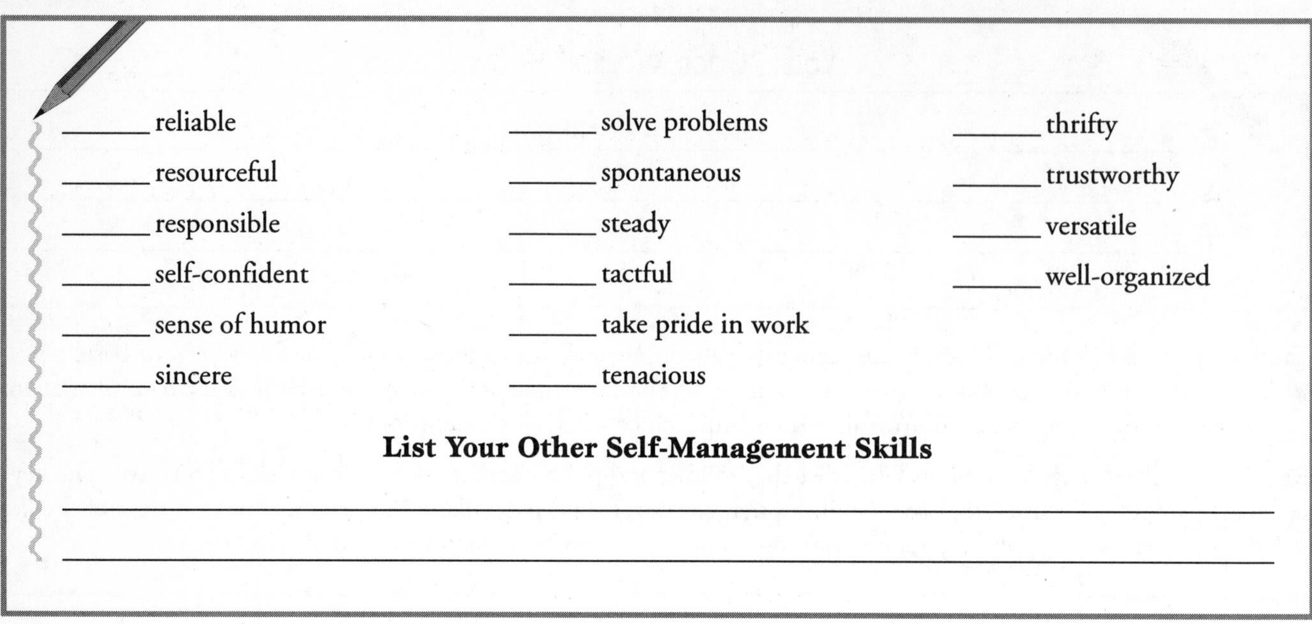

______ reliable	______ solve problems	______ thrifty
______ resourceful	______ spontaneous	______ trustworthy
______ responsible	______ steady	______ versatile
______ self-confident	______ tactful	______ well-organized
______ sense of humor	______ take pride in work	
______ sincere	______ tenacious	

List Your Other Self-Management Skills

____

____

After you are done with the list, circle the five skills you feel are most important and list them in the box that follows.

YOUR TOP FIVE SELF-MANAGEMENT SKILLS

1. ____
2. ____
3. ____
4. ____
5. ____

Note

When thinking about their skills, some people find it helpful to complete the Essential Job Search Data Worksheet that starts on page 326. It organizes skills and accomplishments from previous jobs and other life experiences. Take a look at it and decide whether to complete it now or later.

Transferable Skills

We all have skills that can transfer from one job or career to another. For example, the ability to organize events could be used in a variety of jobs and may be essential for success in certain occupations. Your mission is to find a job that requires the skills you have and enjoy using.

In the following list, put a check mark beside the skills you have. You may have used them in a previous job or in some non-work setting.

Quip

It's not bragging if it's true. Using your new skills language may be uncomfortable at first, but employers need to learn about your skills. So practice saying positive things about the skills you have for the job. If you don't, who will?

Transferable Skills Checklist

Check Your Key Transferable Skills—Employers Value These Highly

______ instruct others
______ negotiate
______ manage money, budgets
______ manage people
______ meet deadlines
______ meet the public
______ organize/manage projects
______ public speaking
______ written communication

Check Your Skills for Working with Things

______ assemble things
______ build things
______ construct/repair things
______ drive, operate vehicles
______ good with hands
______ observe/inspect things
______ operate tools, machines
______ repair things
______ use complex equipment
______ use computers

Check Your Skills for Working with Data

______ analyze data
______ audit records
______ budget
______ calculate/compute
______ check for accuracy
______ classify things
______ compare
______ compile
______ count
______ detail-oriented
______ evaluate
______ investigate
______ keep financial records
______ locate information
______ manage money
______ observe/inspect
______ record facts
______ research
______ synthesize
______ take inventory

Check Your Skills for Working with People

______ administer
______ advise
______ care for
______ coach
______ confront others
______ counsel people
______ demonstrate
______ diplomatic
______ help others
______ instruct
______ interview people
______ kind
______ listen
______ negotiate
______ outgoing
______ patient
______ perceptive
______ persuade
______ pleasant
______ sensitive
______ sociable
______ supervise
______ tactful
______ tolerant
______ tough
______ trusting
______ understanding

(continues)

(continued)

Check Your Skills for Working with Words/Ideas

______ articulate
______ communicate verbally
______ correspond with others
______ create new ideas
______ design
______ edit
______ ingenious
______ inventive
______ library research
______ logical
______ public speaking
______ remember information
______ write clearly

Check Your Leadership Skills

______ arrange social functions
______ competitive
______ decisive
______ delegate
______ direct others
______ explain things to others
______ influence others
______ initiate new tasks
______ make decisions
______ manage or direct others
______ mediate problems
______ motivate people
______ negotiate agreements
______ plan events
______ results-oriented
______ risk-taker
______ run meetings
______ self-confident
______ self-motivate
______ solve problems

Check Your Creative/Artistic Skills

______ artistic
______ dance, body movement
______ drawing, art
______ expressive
______ perform, act
______ present artistic ideas

List Your Other Transferable Skills

____

____

When you are finished, circle the five transferable skills you feel are most important for you to use in your next job and list them in the box below.

YOUR TOP FIVE TRANSFERABLE SKILLS

1. ____
2. ____
3. ____
4. ____
5. ____

Job-Related Skills

Job content or job-related skills are those you need to do a particular occupation. A carpenter, for example, needs to know how to use various tools. Before you select job-related skills to emphasize, you must first have a clear idea of the jobs you want. So let's put off developing your job-related skills list until you have defined the job you want. That topic is covered next.

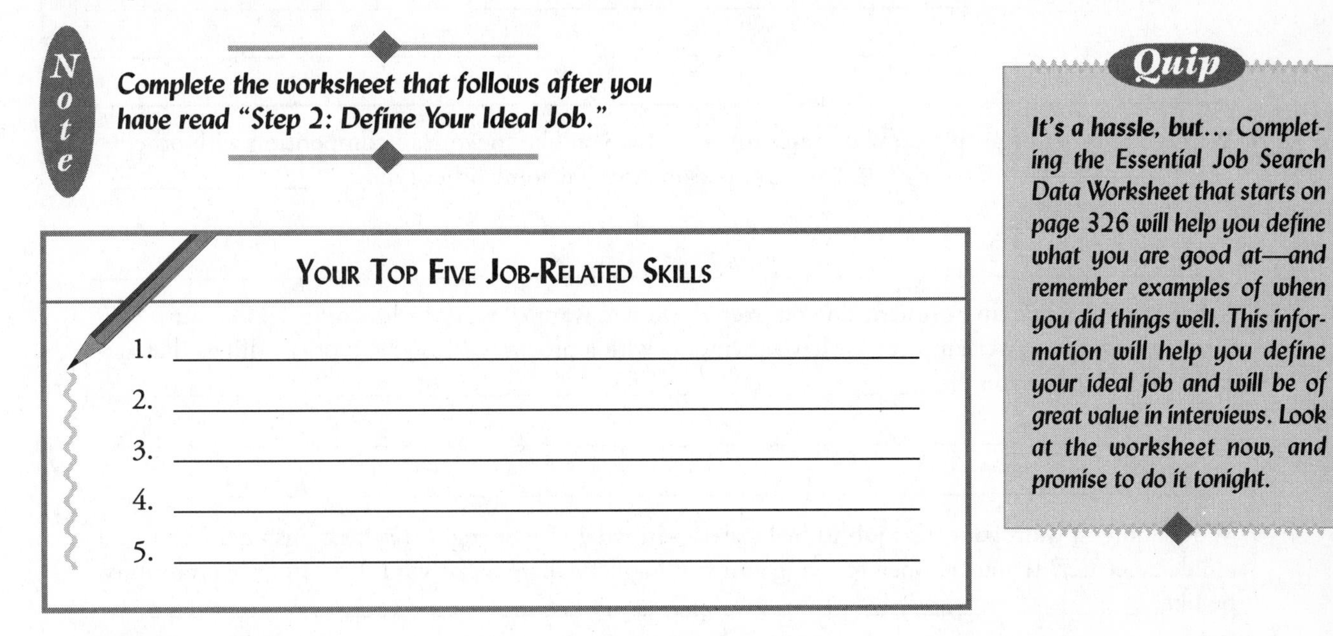

Note

Complete the worksheet that follows after you have read "Step 2: Define Your Ideal Job."

Quip

It's a hassle, but... *Completing the Essential Job Search Data Worksheet that starts on page 326 will help you define what you are good at—and remember examples of when you did things well. This information will help you define your ideal job and will be of great value in interviews. Look at the worksheet now, and promise to do it tonight.*

Your Top Five Job-Related Skills

1. ____________________
2. ____________________
3. ____________________
4. ____________________
5. ____________________

Step 2

Define Your Ideal Job (You Can Compromise on It Later If Needed)

Too many people look for a job without clearly knowing what they are looking for. Before you go out seeking *a* job, I suggest that you first define exactly what you want—*the* job.

Most people think that a job objective is the same as a job title, but it isn't. You need to consider other elements of what makes a job satisfying for you. Then, later, you can decide what that job is called and what industry it might be in.

Eight Factors to Consider in Defining Your Ideal Job

As you try to define your ideal job, consider the following eight important questions. Once you know what you want, your task then becomes finding a position that is as close to your ideal job as possible.

1. **What skills do you want to use?** From the skills lists in Step 1, select the top five skills that you enjoy using and most want to use in your next job. ____________________

____

____

(continues)

(continued)

2. **What type of special knowledge do you have?** Perhaps you know how to fix radios, keep accounting records, or cook food. Write down the things you know from schooling, training, hobbies, family experiences, and other sources. One or more of these knowledges could make you a very special applicant in the right setting. ______________________________

3. **With what types of people do you prefer to work?** Do you like to work in competition with others, or do you prefer hardworking folks, creative personalities, or some other types? ______________________________

4. **What type of work environment do you prefer?** Do you want to work inside, outside, in a quiet place, a busy place, a clean place, or have a window with a nice view? List the types of things that are most important to you. ______________________________

5. **Where do you want your next job to be located—in what city or region?** Near a bus line? Close to a childcare center? If you are open to living and working anywhere, what would your ideal community be like? ______________________________

6. **How much money do you hope to make in your next job?** Many people will take less money if they like a job in other ways—or if they quickly need a job to survive. Think about the minimum you would take as well as what you would eventually like to earn. Your next job will probably pay somewhere in between. ______________________________

7. **How much responsibility are you willing to accept?** Usually, the more money you want to make, the more responsibility you must accept. Do you want to work by yourself, be part of a group, or be in charge? If so, at what level? ______________________________

8. **What things are important or have meaning to you?** Do you have important values you would prefer to include as a basis of the work you do? For example, some people want to work to help others, clean up our environment, build things, make machines work, gain power or prestige, or care for animals or plants. Think about what is important to you and how you might include this in your next job.

____

Is It Possible to Find Your Ideal Job?

Can you find a job that meets all the criteria you just defined? Perhaps. Some people do. The harder you look, the more likely you are to find it. But you will likely need to compromise, so it is useful to know what is *most* important to include in your next job. Go back over your responses to the eight factors and mark those few things that you would most like to have or include in your ideal job. Then write a brief outline of this ideal job below. Don't worry about a job title, or whether you have the experience, or other practical matters yet.

MY IDEAL JOB WOULD BE

____

____

____

____

____

Explore Specific Job Titles and Industries

You might find your ideal job in an occupation you haven't considered yet. And, even if you are sure of the occupation you want, it may be in an industry that's not familiar to you. This combination of occupation and industry forms the basis for your job search, and you should consider a variety of options.

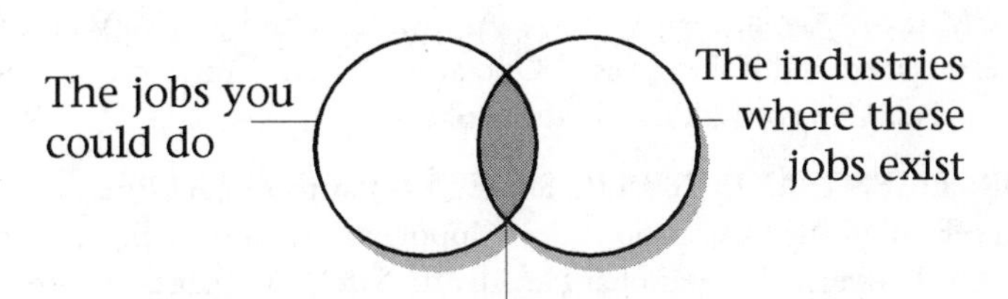

Your ideal job exists in the overlap of those jobs that interest you most *and* of those industries that best meet your needs and interests!

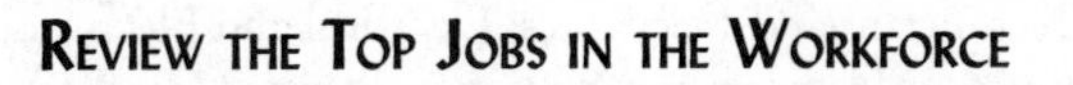

REVIEW THE TOP JOBS IN THE WORKFORCE

There are thousands of job titles, and many jobs are highly specialized, employing just a few people. While one of these more specialized jobs may be just what you want, most work falls within more general job titles that employ large numbers of people.

The list of job titles that follows was developed by the U.S. Department of Labor. It contains approximately 260 major jobs that employ about 85 percent of the U.S. workforce.

The job titles are organized within 14 major groupings called interest areas, presented in all capital letters and bold type. These groupings will help you quickly identify fields most likely to interest you. Job titles are presented in regular type within these groupings.

(continues)

(continued)

Begin with the interest areas that appeal to you most, and underline any job title that interests you. (Don't worry for now about whether you have the experience or credentials to do these jobs.) Then quickly review the remaining interest areas, underlining any job titles there that interest you. Note that some job titles are listed more than once because they fit into more than one interest area. When you have gone through all 14 interest areas, go back and circle the 5 to 10 job titles that interest you most. These are the ones you will want to research in more detail.

1. **ARTS, ENTERTAINMENT, AND MEDIA**—Actors, Producers, and Directors; Announcers; Artists and Related Workers; Athletes, Coaches, Umpires, and Related Workers; Broadcast and Sound Engineering Technicians and Radio Operators; Dancers and Choreographers; Designers; Desktop Publishers; Musicians, Singers, and Related Workers; News Analysts, Reporters, and Correspondents; Photographers; Public Relations Specialists; Television, Video, and Motion Picture Camera Operators and Editors; Writers and Editors
2. **SCIENCE, MATH, AND ENGINEERING**—Actuaries; Aerospace Engineers; Agricultural and Food Scientists; Agricultural Engineers; Architects, Except Landscape and Naval; Atmospheric Scientists; Biological and Medical Scientists; Biomedical Engineers; Chemical Engineers; Chemists and Materials Scientists; Civil Engineers; Computer and Information Systems Managers; Computer-Control Programmers and Operators; Computer Hardware Engineers; Computer Programmers; Computer Software Engineers; Computer Support Specialists and Systems Administrators; Conservation Scientists and Foresters; Construction and Building Inspectors; Drafters; Economists and Market and Survey Researchers; Electrical and Electronics Engineers; Engineering and Natural Sciences Managers; Engineering Technicians; Engineers; Environmental Engineers; Environmental Scientists and Geoscientists; Industrial Engineers, Including Health and Safety; Landscape Architects; Materials Engineers; Mathematicians; Mechanical Engineers; Mining and Geological Engineers, Including Mining Safety Engineers; Nuclear Engineers; Operations Research Analysts; Petroleum Engineers; Physicists and Astronomers; Psychologists; Sales Engineers; Science Technicians; Social Scientists, Other; Statisticians; Surveyors, Cartographers, Photogrammetrists, and Surveying Technicians; Systems Analysts, Computer Scientists, and Database Administrators; Urban and Regional Planners
3. **PLANTS AND ANIMALS**—Agricultural Workers; Animal Care and Service Workers; Farmers, Ranchers, and Agricultural Managers; Fishers and Fishing Vessel Operators; Forest, Conservation, and Logging Workers; Grounds Maintenance Workers; Pest Control Workers; Veterinarians
4. **LAW, LAW ENFORCEMENT, AND PUBLIC SAFETY**—Correctional Officers; Emergency Medical Technicians and Paramedics; Firefighting Occupations; Job Opportunities in the Armed Forces; Judges, Magistrates, and Other Judicial Workers; Lawyers; Occupational Health and Safety Specialists and Technicians; Paralegals and Legal Assistants; Police and Detectives; Private Detectives and Investigators; Security Guards and Gaming Surveillance Officers
5. **MECHANICS, INSTALLERS, AND REPAIRERS**—Aircraft and Avionics Equipment Mechanics and Service Technicians; Automotive Body and Related Repairers; Automotive Service Technicians and Mechanics; Coin, Vending, and Amusement Machine Servicers and Repairers; Computer, Automated Teller, and Office Machine Repairers; Diesel Service Technicians and Mechanics; Electrical and Electronics Installers and Repairers; Electronic Home Entertainment Equipment Installers and Repairers; Elevator Installers and Repairers; Heating, Air-Conditioning, and Refrigeration Mechanics and Installers; Heavy Vehicle and Mobile Equipment Service Technicians and Mechanics; Home Appliance Repairers; Industrial Machinery Installation, Repair, and Maintenance Workers; Line Installers and Repairers; Precision Instrument and Equipment Repairers; Radio and Telecommunications Equipment Installers and Repairers; Small Engine Mechanics
6. **CONSTRUCTION, MINING, AND DRILLING**—Boilermakers; Brickmasons, Blockmasons, and Stonemasons; Carpenters; Carpet, Floor, and Tile Installers and Finishers; Cement Masons, Concrete Finishers, Segmental Pavers, and Terrazzo Workers; Construction Equipment Operators; Construction Laborers; Construction Managers; Drywall Installers, Ceiling Tile Installers, and Tapers; Electricians; Glaziers; Hazardous Materials Removal

Workers; Insulation Workers; Painters and Paperhangers; Pipelayers, Plumbers, Pipefitters, and Steamfitters; Plasterers and Stucco Masons; Roofers; Sheet Metal Workers; Structural and Reinforcing Iron and Metal Workers

7. **TRANSPORTATION**—Air Traffic Controllers; Aircraft Pilots and Flight Engineers; Busdrivers; Rail Transportation Occupations; Taxi Drivers and Chauffeurs; Truckdrivers and Driver/Sales Workers; Water Transportation Occupations

8. **INDUSTRIAL PRODUCTION**—Assemblers and Fabricators; Bookbinders and Bindery Workers; Dental Laboratory Technicians; Food Processing Occupations; Industrial Production Managers; Inspectors, Testers, Sorters, Samplers, and Weighers; Jewelers and Precious Stone and Metal Workers; Machine Setters, Operators, and Tenders—Metal and Plastic; Machinists; Material Moving Occupations; Ophthalmic Laboratory Technicians; Painting and Coating Workers, Except Construction and Maintenance; Photographic Process Workers and Processing Machine Operators; Power Plant Operators, Distributors, and Dispatchers; Prepress Technicians and Workers; Printing Machine Operators; Semiconductor Processors; Stationary Engineers and Boiler Operators; Tool and Die Makers; Water and Liquid Waste Treatment Plant and System Operators; Welding, Soldering, and Brazing Workers; Woodworkers

9. **BUSINESS DETAIL**—Administrative Services Managers; Bill and Account Collectors; Billing and Posting Clerks and Machine Operators; Bookkeeping, Accounting, and Auditing Clerks; Brokerage Clerks; Cargo and Freight Agents; Cashiers; Communications Equipment Operators; Computer Operators; Counter and Rental Clerks; Couriers and Messengers; Court Reporters; Credit Authorizers, Checkers, and Clerks; Customer Service Representatives; Data Entry and Information Processing Workers; Dispatchers; File Clerks; Financial Clerks; Gaming Cage Workers; Human Resources Assistants, Except Payroll and Timekeeping; Information and Record Clerks; Interviewers; Material Recording, Scheduling, Dispatching, and Distributing Occupations, Except Postal Workers; Medical Records and Health Information Technicians; Medical Transcriptionists; Meter Readers, Utilities; Office and Administrative Support Worker Supervisors and Managers; Office Clerks, General; Order Clerks; Payroll and Timekeeping Clerks; Postal Service Workers; Procurement Clerks; Production, Planning, and Expediting Clerks; Receptionists and Information Clerks; Secretaries and Administrative Assistants; Shipping, Receiving, and Traffic Clerks; Stock Clerks and Order Fillers; Tellers; Weighers, Measurers, Checkers, and Samplers, Recordkeeping

10. **SALES AND MARKETING**—Advertising, Marketing, Promotions, Public Relations, and Sales Managers; Demonstrators, Product Promoters, and Models; Insurance Sales Agents; Real Estate Brokers and Sales Agents; Retail Salespersons; Sales Representatives, Wholesale and Manufacturing; Sales Worker Supervisors; Securities, Commodities, and Financial Services Sales Agents; Travel Agents

11. **RECREATION, TRAVEL, AND OTHER PERSONAL SERVICE**—Barbers, Cosmetologists, and Other Personal Appearance Workers; Building Cleaning Workers; Chefs, Cooks, and Food Preparation Workers; Flight Attendants; Food and Beverage Serving and Related Workers; Food Service Managers; Gaming Services Occupations; Hotel, Motel, and Resort Desk Clerks; Lodging Managers; Personal and Home Care Aides; Recreation and Fitness Workers; Reservation and Transportation Ticket Agents and Travel Clerks; Textile, Apparel, and Furnishings Occupations

12. **EDUCATION AND SOCIAL SERVICE**—Archivists, Curators, and Museum Technicians; Childcare Workers; Counselors; Education Administrators; Instructional Coordinators; Library Assistants, Clerical; Library Technicians; Probational Officers and Correctional Treatment Specialists; Protestant Ministers; Rabbis; Roman Catholic Priests; Social and Human Service Assistants; Social Workers; Teacher Assistants; Teachers—Adult Literacy and Remedial and Self-Enrichment Education; Teachers—Postsecondary; Teachers—Preschool, Kindergarten, Elementary, Middle, and Secondary; Teachers—Special Education

13. **GENERAL MANAGEMENT AND SUPPORT**—Accountants and Auditors; Budget Analysts; Claims Adjusters, Appraisers, Examiners, and Investigators; Cost Estimators; Financial Analysts and Personal Financial Advisors; Financial Managers; Funeral Directors; Human Resources, Training, and Labor Relations Managers and Specialists;

(continues)

(continued)

Insurance Underwriters; Loan Officers and Counselors; Management Analysts; Property, Real Estate, and Community Association Managers; Purchasing Managers, Buyers, and Purchasing Agents; Tax Examiners, Collectors, and Revenue Agents; Top Executives

14. **MEDICAL AND HEALTH SERVICES**—Cardiovascular Technologists and Technicians; Chiropractors; Clinical Laboratory Technologists and Technicians; Dental Assistants; Dental Hygienists; Dentists; Diagnostic Medical Sonographers; Dietitians and Nutritionists; Licensed Practical and Licensed Vocational Nurses; Medical and Health Services Managers; Medical Assistants; Nuclear Medicine Technologists; Nursing, Psychiatric, and Home Health Aides; Occupational Therapist Assistants and Aides; Occupational Therapists; Opticians, Dispensing; Optometrists; Pharmacists; Pharmacy Aides; Pharmacy Technicians; Physical Therapist Assistants and Aides; Physical Therapists; Physician Assistants; Physicians and Surgeons; Podiatrists; Radiologic Technologists and Technicians; Recreational Therapists; Registered Nurses; Respiratory Therapists; Speech-Language Pathologists and Audiologists; Surgical Technologists

Note

Thorough descriptions for many of the job titles in the preceding list can be found in this book. You can also find job descriptions on the Internet at http://www.bls.gov/oco/.

The Guide for Occupational Exploration, *Third Edition, provides more information on the interest areas used in the list. This book is published by JIST Works and also describes about 1,000 major jobs, arranged within groupings of related jobs.*

Finally, "A Short List of Additional Resources" at the end of this minibook will give you resources for more job information.

Consider Major Industries

The *Career Guide to Industries,* another book by the U.S. Department of Labor, gives very helpful reviews for the major industries mentioned in the list below. Organized in groups of related industries, this list covers about 75 percent of the nation's workforce. Many libraries and bookstores carry the *Career Guide to Industries,* or you can find the information on the Internet at http://www.bls.gov/oco/cg/.

Underline industries that interest you, and then learn more about the opportunities they present. Note that some industries pay more than others, often for the same skills or jobs, so you should explore a variety of industry options.

GOODS-PRODUCING INDUSTRIES—**Agriculture, Mining, and Construction:** agricultural production; agricultural services; construction; mining and quarrying; oil and gas extraction. **Manufacturing:** aerospace manufacturing; apparel and other textile products; chemical manufacturing, except drugs; drug manufacturing; electronic equipment manufacturing; food processing; motor vehicle and equipment manufacturing; printing and publishing; steel manufacturing; textile mill products.

SERVICE-PRODUCING INDUSTRIES—**Transportation, Communications, and Public Utilities:** air transportation; cable and other pay television services; public utilities; radio and television broadcasting; telecommunications; trucking and warehousing. **Wholesale and Retail Trade:** department, clothing, and accessory stores; eating and drinking places; grocery stores; motor vehicle dealers; wholesale trade. **Finance and Insurance:** banking; insurance; securities and commodities. **Services:** advertising; amusement and recreation services; childcare services; computer and data processing services; educational services; health services; hotels and other lodging places; management and public relations services; motion picture production and distribution; personnel supply services; social services, except child care. **Government:** federal government; state and local government, except education and health.

YOUR TOP JOBS AND INDUSTRIES WORKSHEET

Go back over the lists of job titles and industries. For items 1 and 2 below, list the jobs that interest you most. Then select the industries that interest you most, and list them below in item 3. These are the jobs and industries you should research most carefully. Your ideal job is likely to be found in some combination of these jobs and industries, or in more specialized but related jobs and industries.

1. The 5 job titles that interest you most ______________________________

____

____

____

____

2. The 5 next most interesting job titles ______________________________

____

____

____

____

3. The industries that interest you most______________________________

____

____

____

____

And, Now, We Return to Job-Related Skills

Back on page 295, I suggested that you should first define the job you want, and then identify key job-related skills you have that support your ability to do it. These are the job-related skills to emphasize in interviews.

So, now that you have determined your ideal job, you can pinpoint the job-related skills it requires. If you haven't done so, complete the Essential Job Search Data Worksheet on pages 326–330. It will give you specific skills and accomplishments to highlight.

Yes, completing that worksheet requires time, but doing so will help you clearly define key skills to emphasize in interviews, when what you say matters so much. People who complete that worksheet will do better in their interviews than those who don't.

After you complete the Essential Job Search Data Worksheet, go back to page 295 and write in Your Top Five Job-Related Skills. Include there the job-related skills you have that you would most like to use in your next job.

Use the Most Effective Methods to Reduce Your Job Search Time

Employer surveys found that most employers don't advertise their job openings. They hire people they know, people who find out about the jobs through word of mouth, and people who happen to be in the right place at the right time. While luck plays a part, you can increase your "luck" in finding job openings.

Let's look at the job search methods that people use. The U.S. Department of Labor conducts a regular survey of unemployed people actively looking for work. The survey results are presented below.

Percentage of Unemployed Using Various Job Search Methods

- Contacted employer directly: 64.5%
- Sent out resumes/filled out applications: 48.3%
- Contacted public employment agency: 20.4%
- Placed or answered help wanted ads: 14.5%
- Contacted friends or relatives: 13.5%
- Contacted private employment agency: 6.6%
- Used other active search methods: 4.4%
- Contacted school employment center: 2.3%
- Checked union or professional registers: 1.5%

Source: U.S. Department of Labor, Current Population Survey

Note

This section covers a number of job search methods. Most of the material is presented as information, with a few interactive activities. While each topic is short and reasonably interesting, taking a break now and then will help you absorb it all.

What Job Search Methods Work Best?

The survey shows that most people use more than one job search technique. For example, one person might read want ads, fill out applications, and ask friends for job leads. Others might send out resumes, contact everyone they know from professional contacts, and sign up at employment agencies.

But the survey covered only nine job search methods and, on those, asked only whether the job seeker did or did not use each method. The survey did not cover the Internet, nor did it ask whether a method worked in getting job offers.

Unfortunately, there hasn't been much recent research on the effectiveness of various job search methods. Most of what we know is based on older research and the observations of people who work with job seekers. I'll share what we do know about the effectiveness of job search methods in the content that follows.

Your job search objective. Almost everyone finds a job eventually, so your objective should be to find a good job in less time. The job search methods I emphasize in this minibook will help you do just that.

Getting the Most Out of Less-Effective Job Search Methods

The truth is that every job search method works for some people. But experience and research show that some methods are more effective than others are. Your task in the job search is to spend more of your time using more effective methods—and increase the effectiveness of all the methods you use.

So let's start by looking at some traditional job search methods and how you can increase their effectiveness. Only about one-third of all job seekers get their jobs using one of these methods, but you should consider using them.

Help Wanted Ads

Most jobs are never advertised, and only about 15 percent of all people get their jobs through the want ads. Everyone who reads the paper knows about these openings, so competition is fierce. The Internet also lists many job openings. But, as with want ads, enormous numbers of people view these postings. Some people do get jobs this way, so go ahead and apply. Just be sure to spend most of your time using more effective methods.

Filling Out Applications

Most employers require job seekers to complete an application form. But applications are designed to collect negative information—and employers use applications to screen people out. If, for example, your training or work history is not the best, you will often never get an interview, even if you can do the job.

Completing applications is a more effective approach for young job seekers than for adults. The reason is that there is a shortage of workers for the relatively low-paying and entry-level jobs youth typically seek. As a result, employers are more willing to accept a lack of experience or lack of job skills in youth. Even so, you will get better results by filling out the application if asked to do so and then requesting an interview with the person in charge.

When you complete an application, make it neat and error-free, and do not include anything that could get you screened out. If necessary, leave a problem section blank. It can always be explained after you get an interview.

Employment Agencies

There are three types of employment agencies. One is operated by the government and is free. The others are run as for-profit businesses and charge a fee to either you or an employer. There are advantages and disadvantages to using each.

The government employment service and one-stop centers. Each state has a network of local offices to pay unemployment compensation, provide job leads, and offer other services—at no charge to you or to employers. The service's name varies by state, and it may be called "Job Service," "Department of Labor," "Unemployment Office," "Workforce Development," or another name. Many states have "One-Stop Centers" that provide employment counseling, reference books, computerized career information, job listings, and other resources.

An Internet site at www.doleta.gov/uses will give you information on the programs provided by the government employment service, plus links to other useful sites. Another Internet site, called America's Job Bank at www.ajb.dni.us, allows visitors to see all jobs listed with the government employment service and to search for jobs by region and other criteria.

The government employment service lists only 5 to 10 percent of the available openings nationally, and only about 6 percent of all job seekers get their jobs there. Even so, visit your local office early in your job search. Find out if you qualify for unemployment compensation and learn more about its services. Look into it—the price is right.

Private employment agencies. Private employment agencies are businesses that charge a fee either to you or to the employer who hires you. You will often see their ads in the help wanted section of the newspaper, and many have Web sites. Fees can be from less than one month's pay to 15 percent or more of your annual salary.

Be careful about using fee-based employment agencies. Recent research indicates that more people use and benefit from fee-based agencies than in the past. But realize that relatively few people who register with private agencies get a job through them.

If you use them, ask for interviews where the employer pays the fee. Do not sign an exclusive agreement or be pressured into accepting a job, and continue to actively look for your own leads. You can find these agencies in the phone book's yellow pages, and a government-run Web site at www.ajb.dni.us lists many of them as well.

Temporary agencies. Temporary agencies offer jobs lasting from several days to many months. They charge the employer a bit more than your salary, and then pay you and keep the difference. You pay no direct fee to the agency. Many private employment agencies now provide temporary jobs as well.

Temp agencies have grown rapidly for good reason. They provide employers with short-term help, and employers often use them to find people they may want to hire later. Temp agencies can help you survive between jobs and get experience in different work settings; working for a temp agency may lead to a regular job. They provide a very good option while you look for long-term work, and you may get a job offer while working in a temp job.

School and Employment Services

Contacting a school employment center is one of the job search methods included in the survey presented earlier. Only a small percentage of respondents said they used this option. This is probably because few had the service available.

If you are a student or graduate, find out about these services. Some schools provide free career counseling, resume-writing help, referrals to job openings, career interest tests, reference materials, and other services. Special career programs work with veterans, people with disabilities, welfare recipients, union members, professional groups, and many others. So check out these services and consider using them.

Mailing Resumes and Posting Them on the Internet

Many job search "experts" used to suggest that sending out lots of resumes was a great technique. That advice probably helped sell their resume books, but mailing resumes to people you do not know was never an effective approach. Every so often this would work, but a 95-percent failure rate and few interviews were the more common outcomes, and still are.

While mailing your resume to strangers doesn't make much sense, posting it on the Internet might because

1. It doesn't take much time.
2. Many employers have the potential of finding your resume.

But job searching on the Internet also has its limitations, just like other methods. I'll cover resumes in more detail later and provide tips on using the Internet throughout this minibook.

The Two Job Search Methods That Work Best

The fact is that most jobs are not advertised, so how do *you* find them? The same way that about two-thirds of all job seekers do: networking with people you know (which I call warm contacts), and making direct contacts with an employer (which I call cold contacts). Both of these methods are based on the most important job search rule of all:

THE MOST IMPORTANT JOB SEARCH RULE

DON'T WAIT UNTIL THE JOB OPENS BEFORE CONTACTING THE EMPLOYER!

Employers fill most jobs with people they have met before jobs formally "open." So the trick is to meet people who can hire you before a job is available! Instead of saying, "Do you have any jobs open?" say, "I realize you may not have any openings now, but I would still like to talk to you about the possibility of future openings."

Method 1: Develop a Network of Contacts in Five Easy Stages

One study found that 40 percent of all people located their jobs through a lead provided by a friend, a relative, or an acquaintance. That makes people you know the number one source of job leads—more effective than all the traditional methods combined!

Developing and using your contacts is called "networking," and here's how it works:

1. **Make lists of people you know.** Make a thorough list of anyone you are friendly with. Then make a separate list of all your relatives. These two lists alone often add up to 25 to 100 people or more. Next, think of other groups of people that you have something in common with, such as former coworkers or classmates, members of your social or sports groups, members of your professional association, former employers, and members of your religious group. You may not know many of these people personally or well, but most will help you if you ask them.
2. **Contact them in a systematic way.** Next, contact each of these people. Obviously, some people will be more helpful than others, but any one of them might help you find a job lead.
3. **Present yourself well.** Begin with your friends and relatives. Call and tell them you are looking for a job and need their help. Be as clear as possible about the type of employment you want and the skills and qualifications you have. Look at the sample JIST Cards and phone script on pages 306–307 for good presentation ideas.
4. **Ask them for leads.** It is possible that your contacts will know of a job opening that might interest you. If so, get the details and get right on it! More likely, however, they will not, so you should ask each person these three questions:

The Three Magic Networking Questions

- *Do you know of any openings for a person with my skills?* If the answer is no (which it usually is), then ask...
- *Do you know of someone else who might know of such an opening?* If your contact does, get that name and ask for another one. If he or she doesn't, ask...
- *Do you know of anyone who might know of someone else who might know?* Another good way to ask this is "Do you know someone who knows lots of people?" If all else fails, this will usually get you a name.

5. **Contact these referrals and ask them the same questions.** With each original contact, you can extend your network of acquaintances by hundreds of people. Eventually, one of these people will hire you or refer you to someone who will!

If you are persistent in following these five steps, networking may be the only job search method you need. It works.

Method 2: Contact Employers Directly

It takes more courage, but making direct contact with employers is a very effective job search technique. I call these "cold contacts" because people you don't know in advance will need to "warm up" to your inquiries. Two basic techniques for making cold contacts follow.

Use the yellow pages to find potential employers. Begin by looking at the index in your phone book's yellow pages. For each entry, ask yourself, "Would an organization of this kind need a person with my skills?" If you answer yes, then that organization or business type is a possible target. You can also rate "yes" entries based on your interest, writing an A next to those that seem very interesting, a B next to those that you are not sure of, and a C next to those that aren't interesting at all.

Next, select a type of organization that got a "yes" response, such as "hotels," and turn to that section of the yellow pages. Call each hotel listed and ask to speak to the person who is most likely to hire or supervise you. The sample telephone script on the following page gives you ideas about what to say.

You can easily adapt this approach to the Internet by using sites like www.yellowpages.com to get contacts anywhere in the world.

Drop in without an appointment. You can also walk into a business or organization that interests you and ask to speak to the person in charge. Particularly effective in small businesses, dropping in also works surprisingly well in larger ones. Remember to ask for an interview even if there are no openings now. If your timing is inconvenient, ask for a better time to come back for an interview.

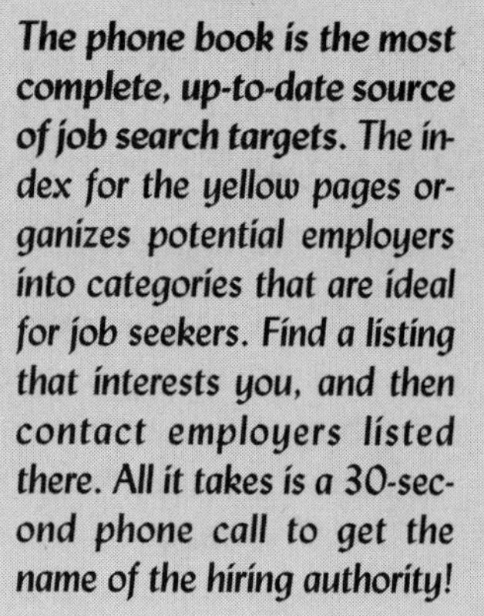

Most Jobs Are with Small Employers

Businesses and organizations with 250 or fewer employees employ about 70 percent of all U.S. workers. They are also the source for as much as 80 percent of the new jobs created each year. They are simply too important to overlook in your job search! Many of them don't have personnel departments, making direct contacts even easier and more effective.

JIST Cards®—An Effective Mini Resume

Look at the sample cards that follow—they are JIST Cards, and they get results. Computer printed or even neatly written on a 3-by-5–inch card, JIST Cards include the essential information employers want to know.

JIST Cards have been used by thousands of job search programs and millions of people. Employers like their direct and timesaving format, and they have been proven as an effective tool to get job leads. Attach one to your resume. Give them to friends, relatives, and other contacts and ask them to pass them along to others who might know of an opening. Enclose them in thank-you notes after an interview. Leave one with employers as a business card. However you get them in circulation, you may be surprised at how well they work.

Sandra Zaremba

Home phone: (219) 232-7608
E-mail: SKZ1128@aol.com

Position: General Office/Clerical

Over two years' work experience, plus one year of training in office practices. Familiar with a variety of computer programs, including word processing, spreadsheets, accounting, and some database and graphic design programs. Am Internet literate and word-process 70 wpm accurately. Can post general ledger and handle payables, receivables, and most accounting tasks. Responsible for daily deposits averaging more than $200,000 monthly. Good interpersonal skills. Can meet strict deadlines and handle pressure well.

Willing to work any hours.

Organized, honest, reliable, and hard working.

```
Jonathan McLaughlin
Answering machine: (509) 674-8736
Cell phone: (509) 541-0981

Objective: Electronics—installation, maintenance, and sales

Skills: Four years of work experience, plus two years advanced
training in electronics. AS degree in Electronics Engineering
Technology. Managed a $500,000/year business while going to
school full time, with grades in the top 25%. Familiar with all
major electronics diagnostic and repair equipment. Hands-on
experience with medical, consumer, communications, business,
and industrial electronics equipment and applications. Good
problem-solving and communication skills. Customer service
oriented.

Willing to do what it takes to get the job done.

Self-motivated, dependable, learn quickly.
```

You can easily create JIST Cards on a computer and print them on card stock you can buy at any office supply store. Or have a few hundred printed cheaply by a local quick print shop. While they are often done as 3-by-5 cards, they can be printed in any size or format.

Use the Phone to Get Job Leads

Once you have created your JIST Card, you can use it as the basis for a phone "script" to make warm or cold calls. Revise your JIST Card content so that it sounds natural when spoken, and then edit it until you can read it out loud in about 30 seconds. Use the sample phone script that follows to help you write your own.

"Hello, my name is Pam Nykanen. I am interested in a position in hotel management. I have four years' experience in sales, catering, and accounting with a 300-room hotel. I also have an associate degree in hotel management, plus one year of experience with the Bradey Culinary Institute. During my employment, I helped double revenues from meetings and conferences and increased bar revenues by 46 percent. I have good problem-solving skills and am good with people. I am also well-organized, hard working, and detail-oriented. When may I come in for an interview?"

Once you have your script, make some practice calls to warm or cold contacts. If making cold calls, contact the person most likely to supervise you. Then present your script just as you practiced it, without stopping.

While it doesn't work every time, most people, with practice, can get one or more interviews in an hour by making such calls.

While the sample script assumes you are calling someone you don't know, it can be changed for warm contacts and referrals. Making cold calls takes courage—but works very well for many who are willing to do it.

Dialing for dollars. The phone can get you more interviews per hour than any other job search tool. But it won't work unless you use it actively.

Overcome phone phobia! Making cold calls takes guts, but job search programs find that most people can get one or more interviews an hour using cold calls. Start by calling people you know and people they refer you to. Then try calls to businesses that don't sound very interesting. As you get better, call more desirable targets.

Using the Internet in Your Job Search

I provide Internet-related tips throughout this minibook, but the Internet deserves a separate section, so here it is.

The Internet has limitations as a job search tool. While many have used it to get job leads, it has not worked well for far more. Too many assume they can simply add their resume to resume databases, and employers will line up to hire them. Just like the older approach of sending out lots of resumes, good things sometimes happen, but not often.

I recommend two points that apply to all job search methods, including using the Internet:

- It is unwise to rely on just one or two methods in conducting your job search.
- It is essential that you use an active rather than a passive approach in your job search.

Tips to Increase Your Internet Effectiveness

I encourage you to use the Internet in your job search but suggest you use it along with other techniques, including direct contacts with employers. The following suggestions can increase the effectiveness of using the Internet in your job search.

If the Internet is new to you. *I recommend a book titled Cyberspace Job Search Kit by Mary Nemnich and Fred Jandt. It covers the basics plus lots of advice on using the Internet for career planning and job seeking.*

Be as specific as possible in the job you seek. This is important in using any job search method and even more so in using the Internet. The Internet is enormous, so it is essential to be as focused as possible in what you are looking for. Narrow your job title or titles to be as specific as possible. Limit your search to specific industries or areas of specialization.

Have reasonable expectations. Success on the Internet is more likely if you understand its limitations. For example, employers trying to find someone with skills in high demand, such as network engineers or nurses, are more likely to use the Internet to recruit job candidates.

Limit your geographic options. If you don't want to move, or would move but only to certain areas, state this preference on your resume and restrict your search to those areas. Many Internet sites allow you to view only those jobs that meet your location criteria.

Create an electronic resume. With few exceptions, resumes submitted on the Internet end up as simple text files with no graphic elements. Employers search databases of many resumes for those that include key words or meet other searchable criteria. So create a simple text resume for Internet use and include on it words likely to be used by employers searching for someone with your abilities. See the resume on page 318 for an example.

Get your resume into the major resume databases. Most Internet employment sites let you add your resume for free, and then charge employers to advertise openings or to search for candidates. These easy-to-use sites often provide all sorts of useful information for job seekers.

Make direct contacts. Visit Web sites of organizations that interest you and learn more about them. Some will post openings, allow you to apply online, or even provide access to staff who can answer your questions. Even if they don't, you can always e-mail a request for the name of the person in charge of the area that interests you and then communicate with that person directly.

Network. You can network online, too, to get names and e-mail addresses of potential employer contacts or other people who might know of someone with a job opening. Look at interest groups, professional association sites, alumni sites, chat rooms, and employer sites—just some of the many creative ways to network with and to interact with people via the Internet.

Useful Internet Sites

Thousands of Internet sites provide information on careers or education. Many have links to other sites that they recommend. Service providers such as America Online (www.aol.com) and the Microsoft Network (www.msn.com) have career information and job listings plus links to other sites. Larger portal sites offer links to recommended career-related sites—AltaVista (www.altavista.com), Lycos (www.lycos.com), and Yahoo (www.yahoo.com) are just a few of these portals. These major career-specific sites can get you started:

- The Riley Guide at www.rileyguide.com
- America's Job Bank at www.ajb.dni.us
- CareerPath.com at www.careerpath.com
- Monster.com at www.monster.com
- The JIST site at www.jist.com

Step 4

Write a Simple Resume Now and a Better One Later

Sending out resumes and waiting for responses is not an effective job-seeking technique. But, many employers *will* ask you for one, and a resume can be a useful tool in your job search. I suggest that you begin with a simple resume you can complete quickly. I've seen too many people spend weeks working on their resume when they could have been out getting interviews instead. If you want a "better" resume, you can work on it on weekends and evenings. So let's begin with the basics.

Tips for Creating a Superior Resume

The following tips make sense for any resume format:

Write it yourself. It's okay to look at other resumes for ideas, but write yours yourself. It will force you to organize your thoughts and background.

Make it error-free. One spelling or grammar error will create a negative impressionist (see what I mean?). Get someone else to review your final draft for any errors. Then review it again because these rascals have a way of slipping in.

Make it look good. Poor copy quality, cheap paper, bad type quality, or anything else that creates a poor appearance will turn off employers to even the best resume content. Get professional help with design and printing if necessary. Many professional resume writers and even print shops offer writing and desktop design services if you need help.

Be brief, be relevant. Many good resumes fit on one page, and few justify more than two. Include only the most important points. Use short sentences and action words. If it doesn't relate to and support the job objective, cut it!

Most jobs are never advertised because employers don't need to advertise or don't want to. Employers trust people referred to them by someone they know far more than they trust strangers. And most jobs are filled by referrals and people that the employer knows, eliminating the need to advertise. So, your job search must involve more than looking at ads.

Be honest. Don't overstate your qualifications. If you end up getting a job you can't handle, who does it help? And a lie can result in your being fired later.

Be positive. Emphasize your accomplishments and results. A resume is no place to be too humble or to display your faults.

Be specific. Instead of saying, "I am good with people," say, "I supervised four people in the warehouse and increased productivity by 30 percent." Use numbers whenever possible, such as the number of people served, percent sales increase, or dollars saved.

You should also know that everyone feels that he or she is a resume expert. Whatever you do, someone will tell you that it's wrong. Remember that a resume is simply a job search tool. You should never delay or slow down your job search because your resume is not "good enough." The best approach is to create a simple and acceptable resume as quickly as possible and then use it. As time permits, create a better one if you feel you must.

Quip

Avoid the resume pile. *Resume experts often suggest that a "dynamite" resume will jump out of the pile. This is old-fashioned advice. It assumes that you are applying to large organizations and for advertised jobs. Today, most jobs are with small employers and are not advertised. My advice is to avoid joining that stack of resumes in the first place, by looking for openings that others overlook.*

Chronological Resumes

Most resumes use a chronological format where the most recent experience is listed first, followed by each previous job. This arrangement works fine for someone with work experience in several similar jobs, but not as well for those with limited experience or for career changers.

Look at the two resumes for Judith Jones that follow. Both use the chronological approach. The first resume would work fine for most job search needs—and it could be completed in about an hour. Notice that the second one includes some improvements. The first resume is good, but most employers would like the additional positive information in the improved resume.

Basic Chronological Resume Example

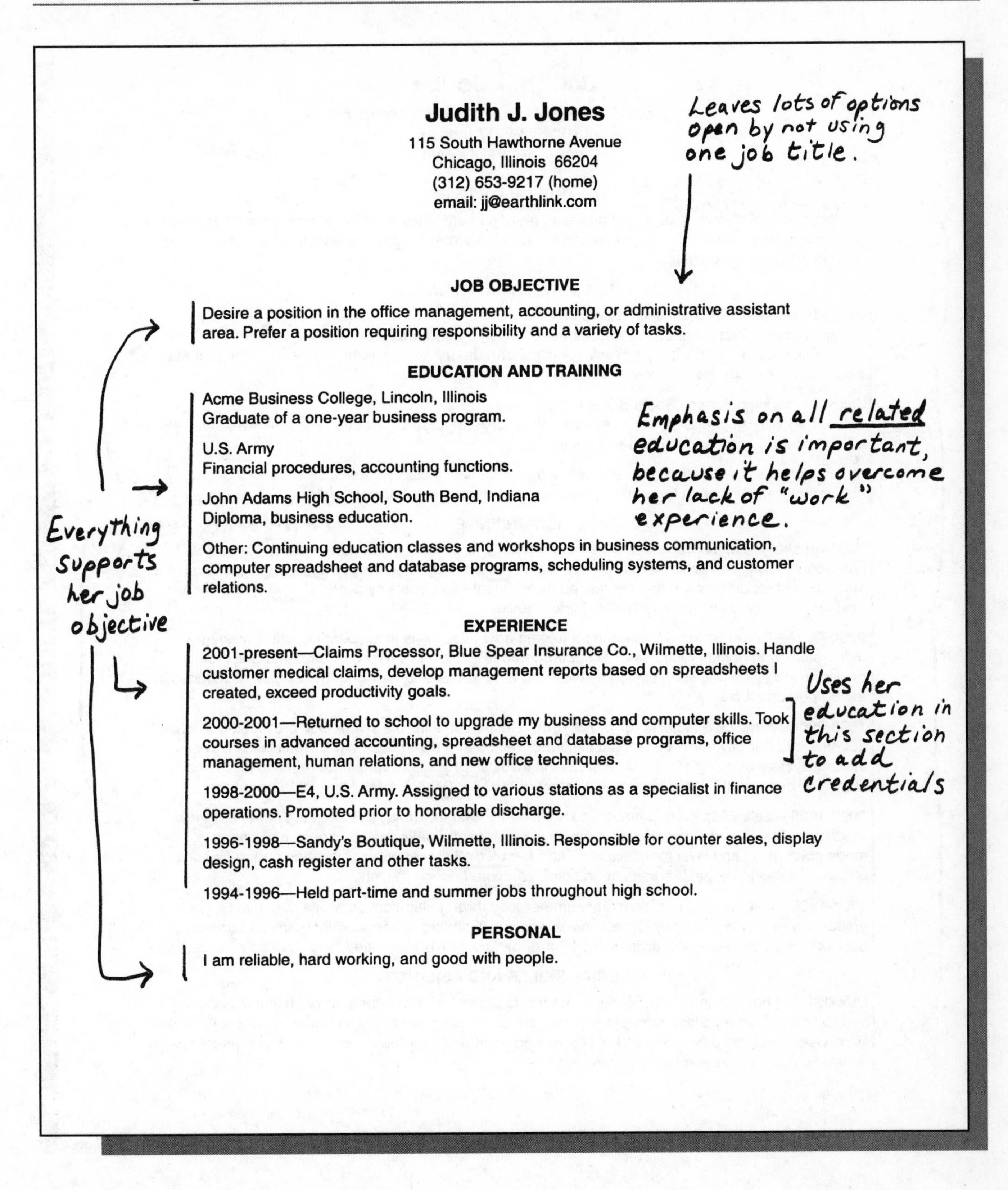

Judith J. Jones

115 South Hawthorne Avenue
Chicago, Illinois 66204
(312) 653-9217 (home)
email: jj@earthlink.com

JOB OBJECTIVE

Desire a position in the office management, accounting, or administrative assistant area. Prefer a position requiring responsibility and a variety of tasks.

EDUCATION AND TRAINING

Acme Business College, Lincoln, Illinois
Graduate of a one-year business program.

U.S. Army
Financial procedures, accounting functions.

John Adams High School, South Bend, Indiana
Diploma, business education.

Other: Continuing education classes and workshops in business communication, computer spreadsheet and database programs, scheduling systems, and customer relations.

EXPERIENCE

2001-present—Claims Processor, Blue Spear Insurance Co., Wilmette, Illinois. Handle customer medical claims, develop management reports based on spreadsheets I created, exceed productivity goals.

2000-2001—Returned to school to upgrade my business and computer skills. Took courses in advanced accounting, spreadsheet and database programs, office management, human relations, and new office techniques.

1998-2000—E4, U.S. Army. Assigned to various stations as a specialist in finance operations. Promoted prior to honorable discharge.

1996-1998—Sandy's Boutique, Wilmette, Illinois. Responsible for counter sales, display design, cash register and other tasks.

1994-1996—Held part-time and summer jobs throughout high school.

PERSONAL

I am reliable, hard working, and good with people.

Improved Chronological Resume Example

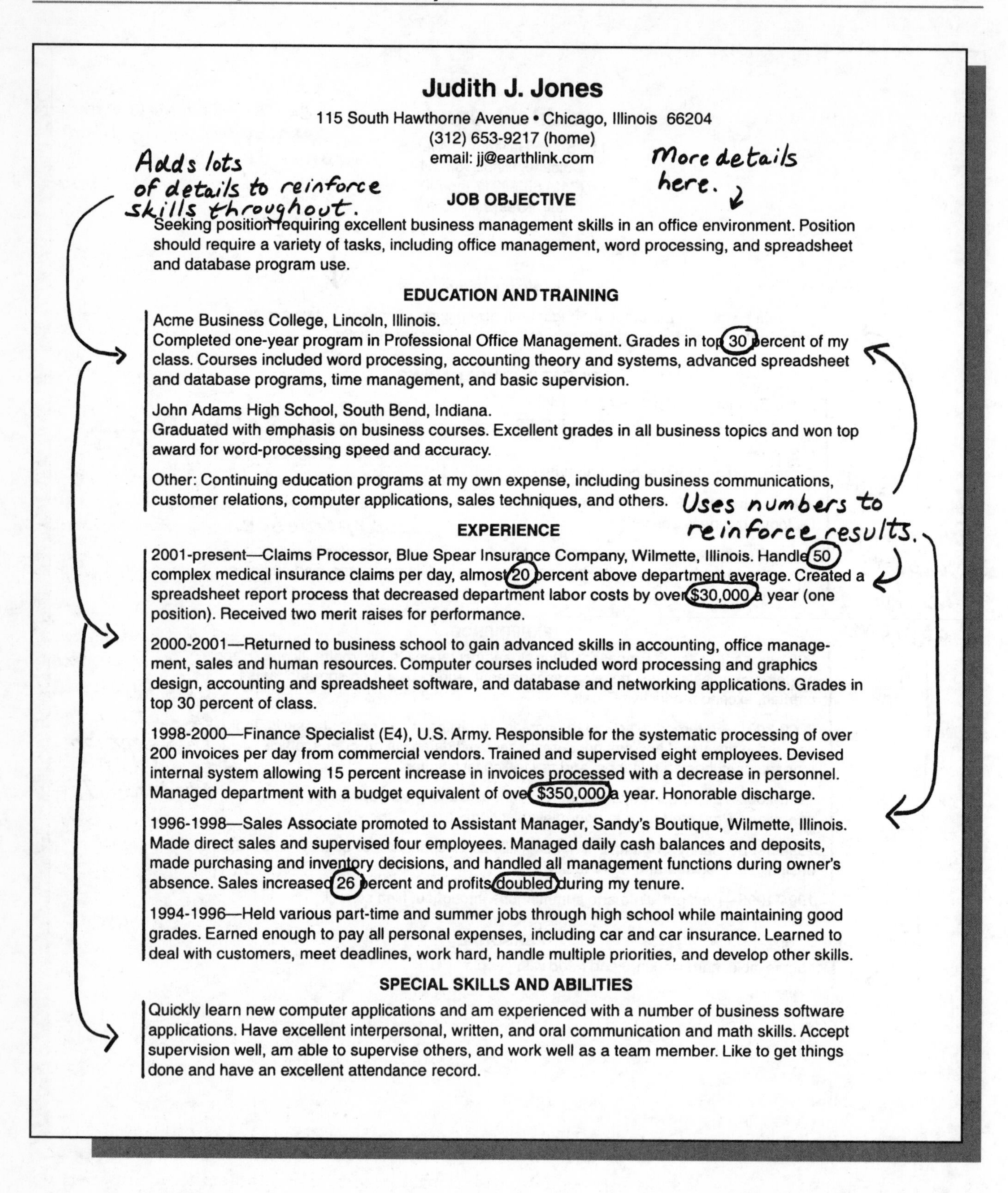

Judith J. Jones

115 South Hawthorne Avenue • Chicago, Illinois 66204
(312) 653-9217 (home)
email: jj@earthlink.com

JOB OBJECTIVE

Seeking position requiring excellent business management skills in an office environment. Position should require a variety of tasks, including office management, word processing, and spreadsheet and database program use.

EDUCATION AND TRAINING

Acme Business College, Lincoln, Illinois.
Completed one-year program in Professional Office Management. Grades in top 30 percent of my class. Courses included word processing, accounting theory and systems, advanced spreadsheet and database programs, time management, and basic supervision.

John Adams High School, South Bend, Indiana.
Graduated with emphasis on business courses. Excellent grades in all business topics and won top award for word-processing speed and accuracy.

Other: Continuing education programs at my own expense, including business communications, customer relations, computer applications, sales techniques, and others.

EXPERIENCE

2001-present—Claims Processor, Blue Spear Insurance Company, Wilmette, Illinois. Handle 50 complex medical insurance claims per day, almost 20 percent above department average. Created a spreadsheet report process that decreased department labor costs by over $30,000 a year (one position). Received two merit raises for performance.

2000-2001—Returned to business school to gain advanced skills in accounting, office management, sales and human resources. Computer courses included word processing and graphics design, accounting and spreadsheet software, and database and networking applications. Grades in top 30 percent of class.

1998-2000—Finance Specialist (E4), U.S. Army. Responsible for the systematic processing of over 200 invoices per day from commercial vendors. Trained and supervised eight employees. Devised internal system allowing 15 percent increase in invoices processed with a decrease in personnel. Managed department with a budget equivalent of over $350,000 a year. Honorable discharge.

1996-1998—Sales Associate promoted to Assistant Manager, Sandy's Boutique, Wilmette, Illinois. Made direct sales and supervised four employees. Managed daily cash balances and deposits, made purchasing and inventory decisions, and handled all management functions during owner's absence. Sales increased 26 percent and profits doubled during my tenure.

1994-1996—Held various part-time and summer jobs through high school while maintaining good grades. Earned enough to pay all personal expenses, including car and car insurance. Learned to deal with customers, meet deadlines, work hard, handle multiple priorities, and develop other skills.

SPECIAL SKILLS AND ABILITIES

Quickly learn new computer applications and am experienced with a number of business software applications. Have excellent interpersonal, written, and oral communication and math skills. Accept supervision well, am able to supervise others, and work well as a team member. Like to get things done and have an excellent attendance record.

Tips for Writing a Simple Chronological Resume

Follow these tips for writing a basic chronological resume.

Name. Use your formal name rather than a nickname if it sounds more professional.

Address and contact information. Avoid abbreviations in your address and include your ZIP code. If you may move, use a friend's address or include a forwarding address. Most employers will not write to you, so you must provide reliable phone numbers and other contact options. Always include your area code in your phone number because you never know where your resume might travel. If you don't have an answering machine, get one—and leave it on whenever you leave home. Make sure your voice message presents you in a professional way. Include alternative ways to reach you, like a cell phone, beeper, and e-mail address.

Job objective. You should almost always have one, even if it is general. In the sample resumes, notice how Judy keeps her options open with her broad job objective. Writing "secretary" or "clerical" might limit her from being considered for jobs she might take.

Education and training. Include any training or education you've had that supports your job objective. If you did not finish a formal degree or program, list what you did complete and emphasize accomplishments. If your experience is not strong, add details here like related courses and extracurricular activities. In the two examples, Judy puts her business schooling in both the education and experience sections. Doing this fills a job gap and allows her to present her training as equal to work experience.

Previous experience. Include the basics like employer name, job title, dates employed, and responsibilities—but emphasize things like specific skills, results, accomplishments, and superior performance.

Personal data. Do not include irrelevant details like height, weight, and marital status—doing so is considered very old-fashioned. But you can include information like hobbies or leisure activities in a special section that directly supports your job objective. The first sample includes a "Personal" section in which Judy lists some of her strengths, which are often not included in a resume.

References. You do not need to list references since employers will ask for them if they want them. If your references are particularly good, you can mention this somewhere—the last section is often a good place. List your references on a separate page and give it to employers who ask. Ask your references what they will say about you and, if positive, if they would write you a letter of recommendation to give to employers.

Tips for an Improved Chronological Resume

Once you have a simple, errorless, and eye-pleasing resume, get on with your job search. There is no reason to delay! If you want to create a better resume in your spare time (evenings or weekends), try these additional tips.

Skip the negatives. *Remember that a resume can get you screened out, but it is up to you to get the interview and the job. So, cut out anything negative in your resume!*

Job objective. A poorly written job objective can limit the jobs an employer might consider you for. Think of the skills you have and the types of jobs you want to do; describe them in general terms. Instead of using a narrow job title like "restaurant manager," you might write "manage a small to mid-sized business."

Education and training. New graduates should emphasize their recent training and education more than those with a few years of related work experience would. A more detailed education and training section might include specific courses you took, and activities or accomplishments that support your job objective or reinforce your key skills. Include other details that reflect how hard you work, such as working your way through school or handling family responsibilities.

Skills and accomplishments. Include things that support your ability to do well in the job you seek now. Even small things count. Maybe your attendance was perfect, you met a tight deadline, or you did the work of others during vacations. Be specific and include numbers—even if you have to estimate them. The improved chronological resume example features a "Special Skills and Abilities" section and more accomplishments and skills. Notice the impact of the numbers to reinforce results.

Job titles. Past job titles may not accurately reflect what you did. For example, your job title may have been "cashier," but you also opened the store, trained new staff, and covered for the boss on vacations. Perhaps "head cashier and assistant manager" would be more accurate. Check with your previous employer if you are not sure.

Promotions. If you were promoted or got good evaluations, say so—"cashier, promoted to assistant manager," for example. A promotion to a more responsible job can be handled as a separate job if doing so results in a stronger resume.

Gaps in employment and other problem areas. Employee turnover is expensive, so few employers want to hire people who won't stay or who won't work out. Gaps in employment, jobs held for short periods, or a lack of direction in the jobs you've held are all things that concern employers. So consider your situation and try to give an explanation of a problem area. Here are a few examples:

2000—Continued my education at...

2002—Traveled extensively throughout the United States.

2000 to present—Self-employed barn painter and widget maker.

2001—Had first child, took year off before returning to work.

Use entire years to avoid displaying an employment gap you can't explain easily. For example "2001 to 2002" can cover a few months of unemployment at the beginning of 2001.

Skills and Combination Resumes

The functional or "skills" resume emphasizes your most important skills, supported by specific examples of how you have used them. This approach allows you to use any part of your life history to support your ability to do the job you want.

While a skills resume can be very effective, it requires more work to create one. And some employers don't like them because they can hide a job seeker's faults (such as job gaps, lack of formal education, or little related work experience) better than a chronological resume. Still, a skills resume may make sense for you.

Look over the sample resumes that follow for ideas. Notice that one resume includes elements of a skills *and* a chronological resume. This so-called "combination" resume makes sense if your previous job history or education and training are positive.

More Resume Examples

A resume is not the most effective tool for getting interviews. A better approach is to make direct contact with those who hire or supervise people with your skills and ask them for an interview, even if no openings exist now. Then send a resume.

Find resume layout and presentation ideas in the four samples that follow.

The chronological resume sample on page 315 focuses on accomplishments through the use of numbers. While Maria's resume does not say so, it is obvious that she works hard and that she gets results.

The skills resume on page 316 is for a recent high school graduate whose only paid work experience was at a fast-food place!

The combination resume on page 317 emphasizes Mary Beth's relevant education and transferable skills because she has no work experience in the field. The resume was submitted by professional resume writer Janet L. Beckstrom of Flint, Michigan (e-mail wordcrafter@voyager.net).

The electronic resume on page 318 is appropriate for scanning or e-mail submission. It has a plain format that is easily read by scanners. It also has lots of key words that increase its chances of being selected when an employer searches a database. The resume is based on one from *Cyberspace Resume Kit* by Mary Nemnich and Fred Jandt and published by JIST Works.

Chronological Resume Emphasizes Results

A simple chronological format with few but carefully chosen words. It has an effective summary at the beginning, and every word supports her job objective.

She emphasizes results!

Maria Marquez

4141 Beachway Road
Redondo Beach, California 90277

Messages: (213) 432-2279
E-mail: mmarq@msn.net

Objective: Management Position in a Major Hotel

Summary of Experience: Four years' experience in sales, catering, banquet services, and guest relations in 300-room hotel. Doubled sales revenues from conferences and meetings. Increased dining room and bar revenues by 44%. Won prestigious national and local awards for increased productivity and services.

Experience:

Park Regency Hotel, Los Angeles, California
Assistant Manager
1999 to Present

- Oversee a staff of 36 including dining room and bar, housekeeping, and public relations operations.
- Introduced new menus and increased dining room revenues by 44%. Gourmet America awarded us their first place Hotel Haute Cuisine award as a result of my efforts.
- Attracted 28% more diners with the first revival of Big Band Cocktail Dances in the Los Angeles area.

Notice her use of numbers to increase impact of statements

Kingsmont Hotel, Redondo Beach, California
Sales and Public Relations
1997 to 1999

- Doubled revenues per month from conferences and meetings.
- Redecorated meeting rooms and updated sound and visual media equipment. Careful scheduling resulted in no lost revenue during this time.
- Instituted staff reward program, which resulted in an upgrade from B- to AAA+ in the *Car and Travel Handbook*.

Bullets here and above increase readability and emphasis.

Education: Associate Degree in Hotel Management from Henfield College of San Francisco. One-year certification program with the Boileau Culinary Institute, where I won the Grand Prize Scholarship.

While Maria had only a few years of related work experience, she used this resume to help her land a very responsible job in a large resort hotel.

Skills Resume for Someone with Limited Work Experience

A skills resume where each skill directly supports the job objective of this recent high school graduate with very limited work experience.

Lisa M. Rhodes
813 Lava Court • Denver, Colorado 81613
Home: (413) 643-2173 (leave message)
Cell phone: (413) 442-1659
Email: lrhodes@netcom.net

Support for the skills comes from all life activities: school, clubs, part-time jobs.

Position Desired

Sales-oriented position in a retail sales or distribution business.

Key skills

Skills and Abilities

Communications Good written and verbal presentation skills. Use proper grammar and have a good speaking voice.

Interpersonal Able to get along well with coworkers and accept supervision. Received positive evaluations from previous supervisors.

Good emphasis on adaptive skills.

Flexible Willing to try new things and am interested in improving efficiency on assigned tasks.

Attention to Detail Concerned with quality. My work is typically orderly and attractive. Like to see things completed correctly and on time.

Hard Working Throughout high school, worked long hours in strenuous activities while attending school full-time. Often handled as many as 65 hours a week in school and other structured activities, while maintaining above-average grades.

Very strong statement.

Good use of numbers

Customer Contacts Routinely handled as many as 500 customer contacts a day (10,000 per month) in a busy retail outlet. Averaged lower than a .001% complaint rate and was given the "Employee of the Month" award in my second month of employment. Received two merit increases. Never absent or late.

Cash Sales Handled over $2,000 a day ($40,000 a month) in cash sales. Balanced register and prepared daily sales summary and deposits.

Reliable Excellent attendance record, trusted to deliver daily cash deposits totaling over $40,000 a month.

Education

Franklin High School. Took advanced English and other classes. Member of award-winning band. Excellent attendance record. Superior communication skills. Graduated in top 30% of class.

Other

Active gymnastics competitor for four years. This taught me discipline, teamwork, how to follow instructions, and hard work. I am ambitious, outgoing, reliable, and willing to work.

Lisa's resume makes it clear that she is talented and hard working.

Combination Resume for a Career Changer

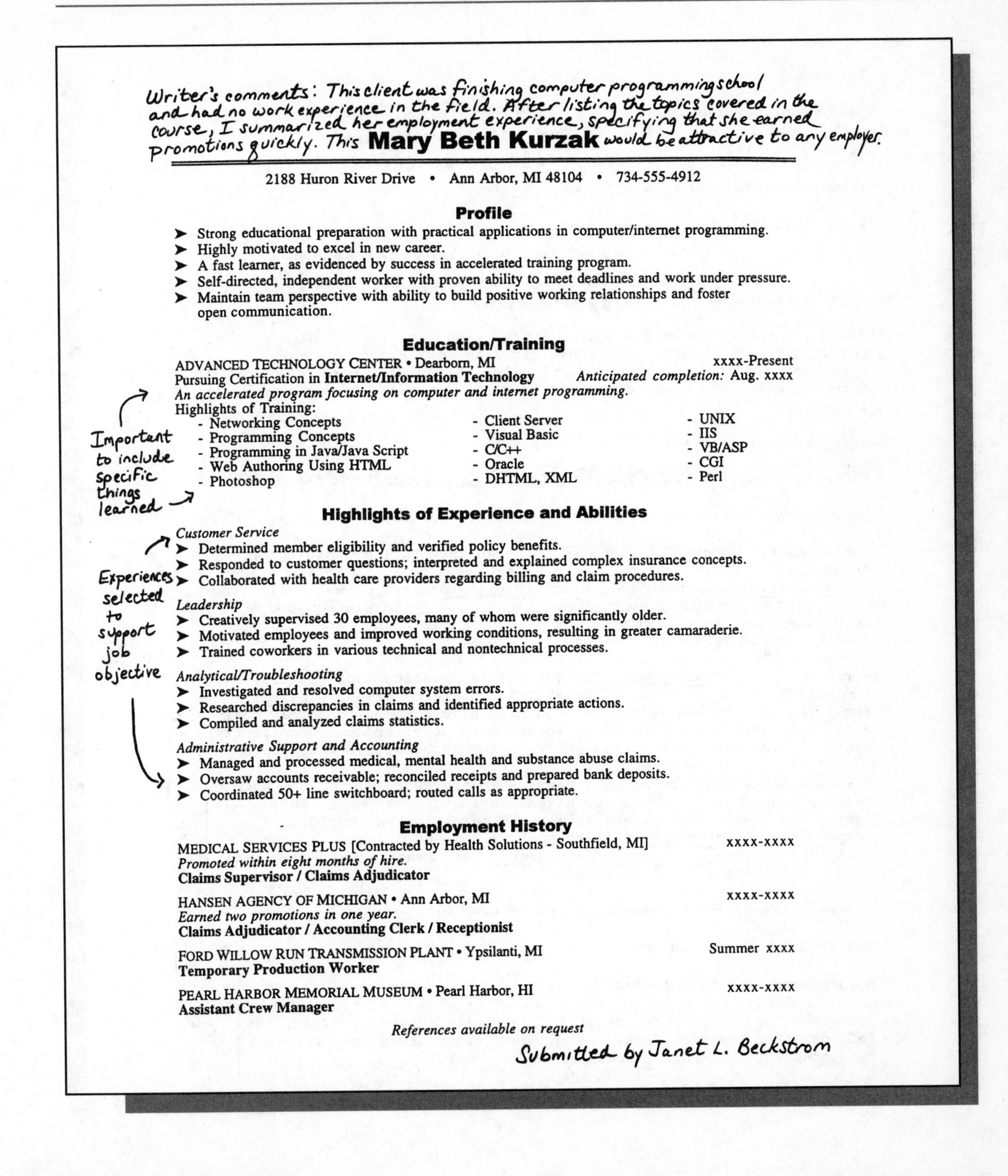

Writer's comments: This client was finishing computer programming school and had no work experience in the field. After listing the topics covered in the course, I summarized her employment experience, specifying that she earned promotions quickly. This would be attractive to any employer.

Mary Beth Kurzak

2188 Huron River Drive • Ann Arbor, MI 48104 • 734-555-4912

Profile

- Strong educational preparation with practical applications in computer/internet programming.
- Highly motivated to excel in new career.
- A fast learner, as evidenced by success in accelerated training program.
- Self-directed, independent worker with proven ability to meet deadlines and work under pressure.
- Maintain team perspective with ability to build positive working relationships and foster open communication.

Education/Training

ADVANCED TECHNOLOGY CENTER • Dearborn, MI — xxxx-Present
Pursuing Certification in **Internet/Information Technology** — *Anticipated completion:* Aug. xxxx
An accelerated program focusing on computer and internet programming.
Highlights of Training:

- Networking Concepts
- Programming Concepts
- Programming in Java/Java Script
- Web Authoring Using HTML
- Photoshop
- Client Server
- Visual Basic
- C/C++
- Oracle
- DHTML, XML
- UNIX
- IIS
- VB/ASP
- CGI
- Perl

Important to include specific things learned

Highlights of Experience and Abilities

Customer Service
- Determined member eligibility and verified policy benefits.
- Responded to customer questions; interpreted and explained complex insurance concepts.
- Collaborated with health care providers regarding billing and claim procedures.

Leadership
- Creatively supervised 30 employees, many of whom were significantly older.
- Motivated employees and improved working conditions, resulting in greater camaraderie.
- Trained coworkers in various technical and nontechnical processes.

Analytical/Troubleshooting
- Investigated and resolved computer system errors.
- Researched discrepancies in claims and identified appropriate actions.
- Compiled and analyzed claims statistics.

Administrative Support and Accounting
- Managed and processed medical, mental health and substance abuse claims.
- Oversaw accounts receivable; reconciled receipts and prepared bank deposits.
- Coordinated 50+ line switchboard; routed calls as appropriate.

Experiences selected to support job objective

Employment History

MEDICAL SERVICES PLUS [Contracted by Health Solutions - Southfield, MI] — xxxx-xxxx
Promoted within eight months of hire.
Claims Supervisor / Claims Adjudicator

HANSEN AGENCY OF MICHIGAN • Ann Arbor, MI — xxxx-xxxx
Earned two promotions in one year.
Claims Adjudicator / Accounting Clerk / Receptionist

FORD WILLOW RUN TRANSMISSION PLANT • Ypsilanti, MI — Summer xxxx
Temporary Production Worker

PEARL HARBOR MEMORIAL MUSEUM • Pearl Harbor, HI — xxxx-xxxx
Assistant Crew Manager

References available on request

Submitted by Janet L. Beckstrom

Electronic Resume

RICHARD JONES

3456 Generic Street

Potomac MD 11721

Phone messages: (301) 927-1189

E-mail: richj@riverview.com

Format to be scanned or e-mailed. No bold, bullets, or italics.

Lots of key words to help get selected by computer searches by employers.

SUMMARY OF SKILLS

Rigger, maintenance mechanic (carpentry, electrical, plumbing, painting), work leader. Read schematic diagrams. Flooring (wood, linoleum, carpet, ceramic and vinyl tile). Plumbing (pipes, fixtures, fire systems). Certified crane and forklift operator, work planner, inspector.

++

EXPERIENCE *Includes results statements and numbers below.*

Total of nine years in the trades—apprentice to work leader.

* MAINTENANCE MECHANIC LEADER, Smithsonian Institution, Washington, DC, April 2001 to present: Promoted to supervise nine staff in all trades, including plumbing, painting, electrical, carpentry, drywall, flooring. Prioritize and schedule work, inspect and approve completed jobs. Responsible for annual budget of $750,000 and assuring that all work done to museum standards and building codes.

* RIGGER / MAINTENANCE MECHANIC, Smithsonian Institution, February 1998 to April 2001: Built and set up exhibits. Operated cranes, rollers, forklifts, and rigged mechanical and hydraulic systems to safely move huge, priceless museum exhibits. Worked with other trades in carpentry, plumbing, painting, electrical and flooring to construct exhibits.

* RIGGER APPRENTICE, Portsmouth Naval Shipyard, Portsmouth, NH, August 1996 to January 1998: Qualified signalman for cranes. Moved and positioned heavy machines and structural parts for shipbuilding. Responsible for safe operation of over $2,000,000 of equipment on a daily basis, with no injury or accidents. Used cranes, skids, rollers, jacks, forklifts and other equipment.

+++

TRAINING AND EDUCATION

HS graduate, top 50% of class.

Additional training in residential electricity, drywall, HV/AC, and refrigeration systems. Certified in heavy crane operation, forklift, regulated waste disposal, industrial blueprints.

Note

Use the information from your completed Essential Job Search Data Worksheet to write your resume.

Step 5

Redefine What Counts As an Interview, Then Get 2 a Day

The average job seeker gets about 5 interviews a month—fewer than 2 a week. Yet many job seekers use the methods in this minibook to get 2 interviews a day. Getting 2 interviews a day equals 10 a week and 40 a month. That's 800 percent more interviews than the average job seeker gets. Who do you think will get a job offer quicker?

But getting 2 interviews a day is nearly impossible unless you redefine what counts as an interview. If you define an interview in a different way, getting 2 a day is quite possible.

The New Definition of an Interview

An interview is any face-to-face contact with someone who has the authority to hire or supervise a person with your skills—even if no opening exists at the time you interview.

If you use this new definition, it becomes *much* easier to get interviews. You can now interview with all sorts of potential employers, not just those who have job openings now. While most other job seekers look for advertised or actual openings, you can get interviews before a job opens up or before it is advertised and widely known. You will be considered for jobs that may soon be created but that others will not know about. And, of course, you can also interview for existing openings like everyone else does.

Spending as much time as possible on your job search and setting a job search schedule are important parts of this step.

Make Your Search a Full-Time Job

Job seekers average fewer than 15 hours a week looking for work. On average, unemployment lasts three or more months, with some people out of work far longer (older workers and higher earners, for example). My many years of experience with job seekers indicate that the more time you spend on your job search each week, the less time you will likely remain unemployed.

Of course, using the more effective job search methods presented in this minibook also helps. Many job search programs that teach job seekers my basic approach of using more effective methods and spending more time looking have proven to cut in half the average time needed to find a job. More importantly, many job seekers also find better jobs using these methods.

So, if you are unemployed and looking for a full-time job, you should plan to look on a full-time basis. It just makes sense to do so, although many do not, or they start out well but quickly get discouraged. Most job seekers simply don't have a structured plan—they have no idea what they are going to do next Thursday. The plan that follows will show you how to structure your job search like a job.

Decide How Much Time You Will Spend Looking for Work Each Week and Day

First and most importantly, decide how many hours you are willing to spend each week on your job search. You should spend a minimum of 25 hours a week on hard-core job search activities with no goofing around. Let me walk you through a simple but effective process to set a job search schedule for each week.

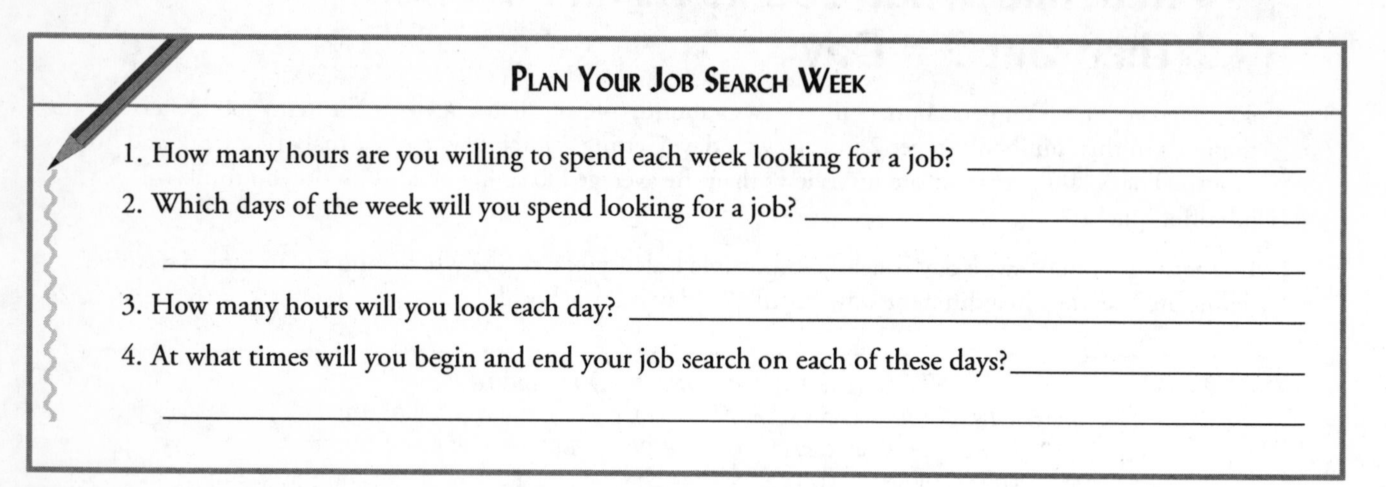

PLAN YOUR JOB SEARCH WEEK

1. How many hours are you willing to spend each week looking for a job? ________________
2. Which days of the week will you spend looking for a job? ________________

____

3. How many hours will you look each day? ________________
4. At what times will you begin and end your job search on each of these days? ________________

____

Create a Specific Daily Job Search Schedule

Having a specific daily schedule is essential because most job seekers find it hard to stay productive each day. The sample daily schedule that follows is the result of years of research into what schedule gets the best results. I tested many schedules in job search programs I ran, and this particular schedule worked best. So use the sample daily schedule for ideas on creating your own, but I urge you to consider using one like this. Why? Because it works.

A Sample Daily Schedule That Works

Time	Activity
7 a.m.	Get up, shower, dress, eat breakfast.
8–8:15 a.m.	Organize work space, review schedule for today's interviews and promised follow-ups, update schedule as needed.
8:15–9 a.m.	Review old leads for follow-up needed today; develop new leads from want ads, yellow pages, the Internet, warm contact lists, and other sources; complete daily contact list.
9–10 a.m.	Make phone calls and set up interviews.
10–10:15 a.m.	Take a break.
10:15–11 a.m.	Make more phone calls, set up more interviews.
11 a.m.–Noon	Send follow-up notes and do other "office" activities as needed.
Noon–1 p.m.	Lunch break, relax.
1–3 p.m.	Go on interviews, make cold contacts in the field.
Evening	Read job search books, make calls to warm contacts not reachable during the day, work on a "better" resume, spend time with friends and family, exercise, relax.

Do It Now: Get a Daily Planner and Write Down Your Job Search Schedule

This is important: If you are not accustomed to using a daily schedule book or electronic planner, promise yourself to get a good one tomorrow. Choose one that allows plenty of space for each day's plan on an hourly basis, plus room for daily to-do listings. Write in your daily schedule in advance; then add interviews as they come. Get used to carrying it with you and use it!

Step 6

Dramatically Improve Your Interviewing Skills

Interviews are where the job search action is. You have to get them; then you have to do well in them. According to surveys of employers, most job seekers do not effectively present the skills they have to do the job. Even worse, most job seekers can't answer one or more problem questions.

This lack of performance in interviews is one reason why employers will often hire a job seeker who does well in the interview over someone with better credentials. The good news is that you can do simple things to dramatically improve your interviewing skills. This section will emphasize interviewing tips and techniques that make the most difference.

Your First Impression May Be the Only One You Make

Some research suggests that if the interviewer forms a negative impression in the first five minutes of an interview, your chances of getting a job offer approach zero. But I know from experience that many job seekers can create a lasting negative impression in seconds. Since a positive first impression is so important, here are some suggestions to help you get off to a good start.

Dress and groom like the interviewer is likely to be dressed—but cleaner! Employer surveys find that almost half of all people's dress or grooming creates an initial negative impression. So this is a big problem. If necessary, get advice on your interviewing outfits from someone who dresses well. Pay close attention to your grooming too—little things do count.

Be early. Leave in plenty of time to be a few minutes early to an interview.

Be friendly and respectful with the receptionist. Doing otherwise will often get back to the interviewer and result in a quick rejection.

Follow the interviewer's lead in the first few minutes. It's often informal small talk but very important for that person to see how you interact. This is a good time to make a positive comment on the organization or even something you see in the office.

Do some homework on the organization before you go. You can often get information on a business and on industry trends from the Internet or a library.

Make a good impression before you arrive. Your resume, e-mails, applications, and other written correspondence create an impression before the interview, so make them professional and error-free.

A Traditional Interview Is Not a Friendly Exchange

In a traditional interview situation, there is a job opening, and you will be one of several applicants for it. In this setting, the employer's task is to eliminate all applicants but one. The interviewer's questions are designed to elicit information that can be used to screen you out. And your objective is to avoid getting screened out. It's hardly an open and honest interaction, is it?

This illustrates yet another advantage of setting up interviews before an opening exists. This eliminates the stress of a traditional interview. Employers are not trying to screen you out, and you are not trying to keep them from finding out stuff about you.

Having said that, knowing how to answer questions that might be asked in a traditional interview is good preparation for any interview you face.

How to Answer Tough Interview Questions

Your answers to a few key problem questions may determine if you get a job offer. There are simply too many possible interview questions to cover one by one. Instead, the 10 basic questions below cover variations of most other interview questions. So, if you can learn to answer these 10 questions well, you will know how to answer most others.

Top 10 Problem Interview Questions

1. Why should I hire you?
2. Why don't you tell me about yourself?
3. What are your major strengths?
4. What are your major weaknesses?
5. What sort of pay do you expect to receive?
6. How does your previous experience relate to the jobs we have here?
7. What are your plans for the future?
8. What will your former employer (or references) say about you?
9. Why are you looking for this type of position, and why here?
10. Why don't you tell me about your personal situation?

The Three-Step Process for Answering Interview Questions

I know this might seem too simple, but the three-step process is easy to remember and can help you create a good answer to most interview questions. The technique has worked for thousands of people, so consider trying it. The three steps are

1. Understand what is really being asked.
2. Answer the question briefly.
3. Answer the real concern.

Step 1. Understand what is really being asked. Most questions are designed to find out about your self-management skills and personality, but interviewers are rarely this blunt. The employer's *real* question is often one or more of the following:

- Can I depend on you?
- Are you easy to get along with?
- Are you a good worker?
- Do you have the experience and training to do the job if we hire you?
- Are you likely to stay on the job for a reasonable period of time and be productive?

Ultimately, if you don't convince the employer that you will stay and be a good worker, it won't matter if you have the best credentials—he or she won't hire you.

Step 2. Answer the question briefly. Present the facts of your particular work experience, but...

- Present them as advantages, not disadvantages.

Many interview questions encourage you to provide negative information. One classic question I included in my list of Top 10 Problem Interview Questions was "What are your major weaknesses?" This is obviously a trick question, and many people are just not prepared for it.

A good response is to mention something that is not very damaging, such as *"I have been told that I am a perfectionist, sometimes not delegating as effectively as I might."* But your answer is not complete until you continue with Step 3.

Step 3. Answer the real concern by presenting your related skills.

- Base your answer on the key skills you have that support the job, and give examples to support these skills.

For example, an employer might say to a recent graduate, "We were looking for someone with more experience in this field. Why should we consider you?" Here is one possible answer:

"I'm sure there are people who have more experience, but I do have more than six years of work experience, including three years of advanced training and hands-on experience using the latest methods and techniques. Because my training is recent, I am open to new ideas and am used to working hard and learning quickly."

In the previous example (about your need to delegate), a good skills statement might be

"I've been working on this problem and have learned to let my staff do more, making sure that they have good training and supervision. I've found that their performance improves, and it frees me up to do other things."

Whatever your situation, learn to answer questions that present you well. It's essential to communicate your skills during an interview, and the three-step process can help you answer problem questions and dramatically improve your responses. It works!

The Most Important Interview Question of All: "Why Should I Hire You?"

This is the most important question of all to answer well. Do you have a convincing argument why someone should hire you over someone else? If you don't, you probably won't get that job you really want. So think carefully about why someone *should* hire you and practice your response. Then, make sure you communicate this in the interview, even if the interviewer never asks the question in a clear way.

Tips on Negotiating Pay—How to Earn a Thousand Dollars a Minute

Remember these few essential ideas when it comes time to negotiate your pay.

THE ONLY TIME TO NEGOTIATE IS AFTER YOU HAVE BEEN OFFERED THE JOB.

Employers want to know how much you want to be paid so that they can eliminate you from consideration. They figure if you want too much, you won't be happy with their job and won't stay. And if you will take too little, they may think you don't have enough experience. So *never* discuss your salary expectations until an employer offers you the job.

IF PRESSED, SPEAK IN TERMS OF WIDE PAY RANGES.

If you are pushed to reveal your pay expectations early in an interview, ask the interviewer what the normal pay range is for this job. Interviewers will often tell you, and you can say that you would consider offers in this range.

If you are forced to be more specific, speak in terms of a wide pay range. For example, if you figure that the company will likely pay from $20,000 to $25,000 a year, say that you would consider "any fair offer in the low to mid-twenties." This statement covers the employer's range *and goes a bit higher*. If all else fails, tell the interviewer that you would consider any reasonable offer.

For this to work, you must know in advance what the job is likely to pay. You can get this information by asking people who do similar work, or from a variety of books and Internet sources of career information.

SALARY NEGOTIATION RULE

THE ONE WHO NAMES A SPECIFIC NUMBER FIRST, LOSES.

Don't Say "No" Too Quickly

Never, ever turn down a job offer during an interview! Instead, thank the interviewer for the offer and ask to consider the offer overnight. You can turn it down tomorrow, saying how much you appreciate the offer and asking to be considered for other jobs that pay better or whatever.

But this is no time to be playing games. If you want the job, you should say so. And it is okay to ask for additional pay or other concessions. But if you simply can't accept the offer, say why and ask the interviewer to keep you in mind for future opportunities. You just never know.

Follow Up on All Contacts

It's a fact: People who follow up with potential employers and with others in their network get jobs faster than those who do not. Here are a few rules to guide you in your job search.

Rules for Effective Follow-Up

- Send a thank-you note or e-mail to every person who helps you in your job search.
- Send the note within 24 hours after speaking with the person.
- Enclose JIST Cards with thank-you notes and all other correspondence.
- Develop a system to keep following up with good contacts.

Thank-You Notes Make a Difference

While thank-you notes can be e-mailed, most people appreciate and are more impressed by a mailed note. Here are some tips about mailed thank-you notes that you can easily adapt to e-mail use.

Thank-you notes can be handwritten or typed on quality paper and matching envelopes. Keep them simple, neat, and error-free. And make sure to include a few copies of your JIST Cards. Here is an example of a simple thank-you note.

2244 Riverwood Avenue
Philadelphia, PA 17963
April 16, XXXX

Ms. Helen A. Colcord
Henderson & Associates, Inc.
1801 Washington Blvd., Suite 1201
Philadelphia, PA 17993

Dear Ms. Colcord:

Thank you for sharing your time with me so generously today. I really appreciated seeing your state-of-the-art computer equipment.

Your advice has already proved helpful. I have an appointment to meet with Mr. Robert Hopper on Friday. As you anticipated, he does intend to add more computer operators in the next few months.

In case you think of someone else who might need a person like me, I'm enclosing another JIST Card. I will let you know how the interview with Mr. Hopper goes.

Sincerely,

William Henderson

William Henderson

Use Job Lead Cards to Help Organize Follow-Ups

If you use contact management software, use it to schedule follow-up activities. But the simple paper system I describe here can work very well or can be adapted for setting up your contact management software.

Use a simple 3-by-5–inch card to record essential information about each person in your network. Buy a 3-by-5–inch card file box and tabs for each day of the month. File the cards under the date you want to contact the person, and the rest is easy.

I've found that staying in touch with a good contact every other week can pay off big. Here's a sample card to give you ideas about creating your own.

Organization: Mutual Health Insurance
Contact person: Anna Tomey
Phone number: (555) 555-2211
Source of lead: Aunt Ruth
Notes: 4/10 called. Anna on vacation. Call back 4/15. 4/15 Interview set 4/20 at 1:30. 4/20 Anna showed me around. They use the same computers we used in school. Sent thank-you note and JIST Card. Call back 5/1. 5/1 Second interview 5/8 at 9 a.m.

In Closing

You won't get a job offer because someone knocks on your door and offers one. Job seeking does involve luck, but you are more likely to get lucky if you are out getting interviews.

I'll close this minibook with a few final tips:

- Approach your job search as if it were a job itself. Create and stick to a daily schedule, and spend at least 25 hours a week looking.
- Follow up on each lead you generate and ask each one for referrals.
- Set out each day to schedule at least two interviews, and use the new definition of an interview, which includes talking to businesses that don't have an opening now.
- Send out lots of thank-you notes and JIST Cards.
- When you want the job, tell the employer that you want it and why the company should hire you over someone else.

Don't get discouraged. There are lots of jobs out there, and someone needs an employee with your skills—your job is to find that someone.

I wish you luck in your job search and your life.

Essential Job Search Data Worksheet

Take some time to complete this worksheet carefully. It will help you write your resume and answer interview questions. You can also photocopy it and take it with you to help complete applications and as a reference throughout your job search. Use an erasable pen or pencil to allow for corrections. Whenever possible, emphasize skills and accomplishments that support your ability to do the job you want. Use extra sheets as needed.

Your name ____________________

Date completed ____________________

Job objective ____________________

Key Accomplishments

List three accomplishments that best prove your ability to do the kind of job you want.

1. ____________________
2. ____________________
3. ____________________

Education and Training

Name of high school(s); years attended ____________________

____

Subjects related to job objective ____________________

____

Related extracurricular activities/hobbies/leisure activities ____________________

____

Accomplishments/things you did well ____________________

____

Specific things you can do as a result ____________________

____

Schools you attended after high school; years attended; degrees/certificates earned ____________________

____

Courses related to job objective ____________________

____

Related extracurricular activities/hobbies/leisure activities ____________________

____

Accomplishments/things you did well ____________________

____

Specific things you can do as a result ____________________

____

Other Training

Include formal or informal learning, workshops, military training, things you learned on-the-job or from hobbies—anything that will help support your job objective. Include specific dates, certificates earned, or other details as needed. ____________________

____

____

____

Work and Volunteer History

List your most recent job first, followed by each previous job. Military experience, unpaid or volunteer work, and work in a family business should be included here, too. If needed, use additional sheets to cover *all* significant paid or unpaid work experiences. Emphasize details that will help support your new job objective! Include numbers to support what you did: the number of people served over one or more years,

(continues)

(continued)

the number of transactions processed, percentage of sales increased, total inventory value you were responsible for, payroll of the staff you supervised, total budget responsible for, and so on. Emphasize results you achieved, using numbers to support them whenever possible. Mentioning these things on your resume and in an interview will help you get the job you want!

Job 1

Dates employed ______________________________

Name of organization ______________________________

Supervisor's name and job title ______________________________

Address ______________________________

____

Phone number/e-mail address/Web site ______________________________

____

What did you accomplish and do well? ______________________________

____

Things you learned, skills you developed or used ______________________________

____

Raises, promotions, positive evaluations, awards ______________________________

____

Computer software, hardware, and other equipment you used ______________________________

____

Other details that might support your job objective ______________________________

____

Job 2

Dates employed ______________________________

Name of organization ______________________________

Supervisor's name and job title ______________________________

Address ______________________________

____

Phone number/e-mail address/Web site ______________________________

____

What did you accomplish and do well? ______________________________

____

Things you learned, skills you developed or used ______________________________

____

Raises, promotions, positive evaluations, awards ______________________________

____

Computer software, hardware, and other equipment you used ____________________

____

Other details that might support your job objective ___________________________

____

Job 3

Dates employed ____

Name of organization ____

Supervisor's name and job title ____

Address ____

____

Phone number/e-mail address/Web site ____________________________________

____

What did you accomplish and do well? _____________________________________

____

Things you learned, skills you developed or used ______________________________

____

Raises, promotions, positive evaluations, awards ______________________________

____

Computer software, hardware, and other equipment you used ____________________

____

Other details that might support your job objective ___________________________

____

References

Think of people who know your work well and will say positive things about your work and character. Past supervisors are best. Contact them and tell them what type of job you want and your qualifications, and ask what they will say about you if contacted by a potential employer. Some employers will not provide references by phone, so ask them for a letter of reference for you in advance. If a past employer may say negative things, negotiate what they will say or get written references from others you worked with there.

(continues)

(continued)

Reference name ______________________________

Position or title ______________________________

Relationship to you ______________________________

Contact information (complete address, phone number, e-mail address) ______________________________

____

Reference name ______________________________

Position or title ______________________________

Relationship to you ______________________________

Contact information (complete address, phone number, e-mail address) ______________________________

____

____

Reference name ______________________________

Position or title ______________________________

Relationship to you ______________________________

Contact information (complete address, phone number, e-mail address) ______________________________

____

____

A Short List of Additional Resources

Thousands of books and uncounted Internet sites provide information on career subjects. Space limitations do not permit me to describe the many good resources available, so I list here some of the most useful ones. Because this is my list, I've included books I've written or that JIST publishes. You should be able to find these and many other resources at libraries, bookstores, and Web book-selling sites, such as Amazon.com.

Resumes and Cover Letter Books

My books. I'm told that *The Quick Resume & Cover Letter Book* is now one of the top-selling resume books at various large bookstore chains. It is very simple to follow, is inexpensive, has good design, and has good sample resumes written by professional resume writers. *America's Top Resumes for America's Top Jobs* includes a wonderfully diverse collection of resumes covering most major jobs and life situations, plus brief but helpful resume-writing advice.

Other books published by JIST. The following titles include many sample resumes written by professional resume writers, as well as good advice: *Cover Letter Magic,* by Enelow and Kursmark; *Cyberspace Resume Kit,* Nemnich and Jandt; *The Edge Resume & Job Search Strategy,* Corbin and Wright; *The Federal Resume Guidebook,* Troutman; *Expert Resumes for Computer and Web Jobs,* Enelow and Kursmark; *Expert Resumes for Teachers and Educators,* Enelow and Kursmark; *Gallery of Best Cover Letters,* Noble; *Gallery of Best Resumes,* Noble; and *Resume Magic,* Whitcomb.

Job Search Books

My books. *The Very Quick Job Search—Get a Better Job in Half the Time* is a thorough book with detailed advice and a "quick" section of key tips you can finish in a few hours. *The Quick Interview & Salary Negotiation Book* also has a section of quick tips likely to make the biggest difference in your job search, as well as sections with more detailed information on problem questions and other topics. *Getting the Job You Really Want* includes many in-the-book activities and good career decision-making and job search advice.

Other books published by JIST. Titles include *Career Success Is Color Blind,* Stevenson; *Cyberspace Job Search Kit,* Nemnich and Jandt; *Inside Secrets to Finding a Teaching Job,* Warner and Bryan; and *Job Search Handbook for People with Disabilities,* Ryan.

Books with Information on Jobs

Primary references. The *Occupational Outlook Handbook* is the source of job titles listed in this minibook. Published by the U.S. Department of Labor and updated every other year, the *OOH* covers about 85 percent of the workforce. A book titled the *O*NET Dictionary of Occupational Titles* has descriptions for over 1,000 jobs based on the O*NET (for Occupational Information Network) database developed by the Department of Labor. The *Enhanced Occupational Outlook Handbook* includes the *OOH* descriptions plus more than 2,000 additional descriptions of related jobs from the O*NET and other sources. The *Guide for Occupational Exploration,* Third Edition, allows you to explore major jobs based on your interests. All of these books are available from JIST.

Other books published by JIST. Here are a few good books that include job descriptions and helpful details on career options: *America's Fastest Growing Jobs; America's Top Jobs for College Graduates; America's Top Jobs for People Without a Four-Year Degree; America's Top Computer and Technical Jobs; America's Top Medical, Education & Human Services Jobs; America's Top White-Collar Jobs; Best Jobs for the 21st Century* (plus versions for people with and without degrees); *Career Guide to America's Top Industries;* and *Guide to America's Federal Jobs.*

Internet Resources

If the Internet is new to you, I recommend a book titled *Cyberspace Job Search Kit* by Mary Nemnich and Fred Jandt. It covers the basics plus offers advice on using the Internet for career planning and job seeking. The *Quick Internet Guide to Career and Education Information* by Anne Wolfinger gives excellent reviews of the most helpful sites and ideas on how to use them. The *Occupational Outlook Handbook*'s job descriptions also include Internet addresses. And www.jist.com lists recommended sites for career, education, and related topics, along with comments on each.

Be aware that some sites provide poor advice, so ask your librarian, instructor, or counselor for suggestions on those best for your needs.

Other Resources

Libraries. Most libraries have the books mentioned here, as well as many other resources. Many also provide Internet access so that you can research online information. Ask the librarian for help finding what you need.

People. People who hold the jobs that interest you are one of the best career information sources. Ask them what they like and don't like about their work, how they got started, and the education or training needed. Most people are helpful and will give advice you can't get any other way.

Career counseling. A good vocational counselor can help you explore career options. Take advantage of this service if it is available to you! Also consider a career-planning course or program, which will encourage you to be more thorough in your thinking.

Sample Resumes for Some of America's Top Computer and Technical Jobs

If you read the previous information in this section, you know that I believe resumes are an over-rated job search tool. Even so, you will probably need one, and you should have a good one.

Unlike some career authors, I do not preach that there is one right way to do a resume. I encourage you to be an individual and to do what you think will work well for you. But, I also know that some resumes are clearly better than others. The following pages contain several resumes that you can use as examples when preparing your resume. All these resumes are from my book *America's Top Resumes for America's Top Jobs,* available through JIST Publishing.

The resumes have these points in common:

- Each is for an occupation described in this book. The resumes are arranged in alphabetic order by job title.
- Each is particularly good in some way. Notes on the resumes point out their features.
- Each was written by a professional resume writer who is a member of one or more professional associations, including the Professional Association of Resume Writers (www.parw.com) or the National Resumes Writers' Association (www.nrwa.com). These writers are highly qualified and hold various credentials. Most are willing to provide help (for a fee) and welcome your contacting them (although this is not a personal endorsement).

Contact Information for Resume Contributors

The following professional resume writers contributed resumes to this section. Their names are listed in alphabetical order. Each entry indicates which resume that person contributed.

Ann Baehr, CPRW
President, Best Resumes
122 Sheridan St.
Brentwood, NY 11717
Phone: (631) 435-1879
Fax: (631) 435-3655
E-mail: resumesbest@earthlink.net
Resumes on pages 336 and 338

Rima Bogardus, CPRW
President, Career Support Services
P.O. Box 3035
Cary, NC 27519
Phone: (919) 380-3770
E-mail: rima@careersupportservices.com
www.careersupportservices.com
Resume on page 334

Arnold G. Boldt, CPRW, JCTC
Arnold-Smith Associates
625 Panorama Trail, Bldg. 2 #200
Rochester, NY 14625
Phone: (585) 383-0350
Fax: (585) 387-0516
E-mail: Arnie@ResumeSOS.com
Resume on page 340

Jewel Bracy DeMaio, CPRW, CEIP
President, A Perfect Resume.com
419 Valley Rd.
Elkins Park, PA 19027
Toll-free phone: (800) 227-5131
Fax: (215) 782-8278
E-mail: mail@aperfectresume.com
www.aperfectresume.com
Resume on page 337

Susan Guarneri, NCC, NCCC, LPC, CPRW, IJCTC, CCM, CEIP
President, Guarneri Associates/Resumagic
1101 Lawrence Rd.
Lawrenceville, NJ 08648
Phone: (609) 771-1669
Fax: (609) 637-0449
E-mail: Resumagic@aol.com
www.resume-magic.com
Resume on page 335

Diana C. LeGere
President, Executive Final Copy
P.O. Box 171311
Salt Lake City, UT 84117
Toll-free phone: (866) 754-5465
Phone: (801) 550-5697
Fax: (626) 602-8715
E-mail: executiveresumes@yahoo.com
www.executivefinalcopy.com
Resume on page 333

Sharon Pierce-Williams, MEd, CPRW
President, TheResume.Doc
609 Lincolnshire Ln.
Findlay, OH 45840
Phone: (419) 422-0228
Fax: (419) 425-1185
E-mail: TheResumeDocSPW@aol.com
Resume on page 339

Automotive Service Technicians and Mechanics

This concise, dynamic resume has been very successful for John. Each time he faxes it to someone, he gets called for an interview within minutes!

JOHN C. TAYLER

Phone: 801-892-6299 • **Pager: 801-892-6380** • 2463 Elm Avenue, Salt Lake City, UT 84109 • JCTayler@hotmail.com

The reader can see immediately how qualified John is within the industry.

ASE CERTIFIED MASTER TECHNICIAN

STATE INSPECTIONS CERTIFIED / COUNTY EMISSIONS CERTIFIED

Master Technician with 20+ years' experience in all phases of Automotive Technology. Proven leader in the industry possessing **expert diagnostic ability.** Proficient in **heavy equipment repair.** Highly motivated team player dedicated to producing outstanding results and increased profitability.

SPECIALIZED TRAINING

- **Advanced Automotive Climate Control,** Auto Parts Plus — 2001
- **Advanced Automotive Tune-up and Electrical Troubleshooting,** Federal Mogul/Champion — 2000
- **Advanced Hydrostatic Drive Diagnostics and Repair,** Bombardier Recreational Product — 2000
- **Advanced Applied Emissions Technical Training,** Salt Lake Community College — 1995
- **Utah Safety Inspection Training,** SLC Emissions Testing — 1995
- **ASE Certifications,** Salt Lake Community College — 1994

AUTOMOTIVE TECHNICAL EXPERTISE

- Full Service Automotive Excellence, providing **Bumper-to-Bumper Expert Diagnostics and Repair.**

- Specializing in major and minor repair for **ALL MAKES and MODELS**: **Japanese, European and Domestic** vehicles, 4x4s, Trucks, RVs, Dump Trucks, Caterpillars, Tractors, Tractor-Trailers, Snow Cats, Diesel, and other heavy equipment.

- Brakes — Clutches — Water Pumps — P/S Pumps — Rack & Pinion Steering — Shocks & Struts
- Front-end Work — 4 x 4 Front-end Work — Precision Engine Rebuilding — Engine Overhaul
- Cooling and Heating Systems — Batteries — Mufflers — Carburetors — CV Joints — Differentials
- Radiators — Alternators — Starters — Fuel Injection Systems — Wiring Repair and Replacement

PROFESSIONAL EXPERIENCE

- Tayler Mobile Automotive, **Owner/Manager,** present
- McNeil's Automotive, **Lead Automotive Technician,** 1996–present
- Snowbird Ski Resort, **Lead Heavy Line Technician,** 1992–1996
- Marathon Automotive, **Lead Automotive Technician,** 1991–1992
- Courtesy Car Company, **Service Manager,** 1985–1991
- Kmart (presently Penske Automotive), **Acting Service Manager,** 1980–1985

Cardiovascular Technologists and Technicians

"Profile" and "Professional Experience" sections emphasize Amanda's ability to handle a heavy patient load while remaining personable and producing accurate test results, which are highly desirable qualities in healthcare technicians.

Amanda C. Jackson, CCT, RDCS

785 Bentwood Drive
Willowick, OH 44095

(440) 906-4431
acjackson@aol.com

PROFILE

Amiable, caring **Echocardiography Technician** with over 10 years' experience • Recognized for reliability and accuracy • Proven ability to develop positive rapport with patients who have a variety of needs, conditions, and backgrounds • Strong mechanical troubleshooting skills • Able to win patient confidence and cooperation • Computer proficient

PROFESSIONAL EXPERIENCE

Echo Technician — 1993–present
Advanced Cardiology Services, Cleveland, OH
Cardiac Diagnostic Center consists of 4 labs and supports 25 physicians.
Echo lab is accredited by ICAEL.

- Perform echocardiograms, stress echoes, Holter monitoring, Holter scanning, thallium stress tests, Cardiolite studies, and pacemaker evaluations. Excellent working knowledge of Agilent 5500 and HP 2000 cardiac ultrasound machines.
- Conduct diagnostic testing on an average of 20 patients per day without compromising accuracy or compassion.
- Developed a reputation as a strong troubleshooter. Often assist co-workers who are having technical problems with equipment.
- Research products, manage inventory, and order medical supplies for 2 labs. Purchase cardiac medications for all 4 labs. Saved facility thousands of dollars per year through wise selection of products.

EKG/Stress Technician — 1990–1993
Mt. Sinai Hospital, Dayton, OH

- Conducted EKGs, stress tests, thallium stress tests, and Holter scanning in various units of the hospital.
- Developed a reputation for quality and compassion. Was encouraged by cardiologists to apply for position at Advanced Cardiology Services.

EKG/Stress Technician — 1988–1990
Cardiovascular Consultants, Philadelphia, PA

- Conducted EKGs, stress tests, and Holter scanning.
- Managed heavy patient load. Became proficient at multitasking with 10 physicians in the office every day.
- Trained new employees.

CREDENTIALS

Registered Diagnostic Cardiac Sonographer (RDCS) — 1999
Certified Cardiographic Technician (CCT) — 1990
BCLS Certified
Currently pursuing ACLS Certification

EDUCATION

Technical Workshop in Cardiac Ultrasound — 1991
Owen-Brown, Dallas, TX

Cardiography certificate GPA 3.9 — 1989
Pennsylvania School of Health Technology, Philadelphia, PA

PROFESSIONAL AFFILIATIONS

Society of Diagnostic Medical Sonographers
American Society of Echocardiography

COMMUNITY INVOLVEMENT

Nursing home volunteer
Nursing home ombudsman (Ohio program)

Computer Programmers

This resume helped Barry receive many interviews for programming positions!

BARRY BARTHOLOMEW

77 Apple Valley Road
Princeton, NJ 08540
609-924-5555
barrybart@penn.rr.com

OBJECTIVE: Senior Systems Programmer

SUMMARY OF QUALIFICATIONS

⇒ Over 6 years' experience in computer operations, promoted twice in that time.
⇒ Computer Programming degree from The Computer Institute.
⇒ Dedicated, resourceful and reliable worker; can be counted on to get the job done.
⇒ Work well independently as well as cooperatively in a team environment.
⇒ Expert troubleshooter; strengths in analysis, logic and accuracy.

Hardware Software	Digital GEMS	IBM PC	AS/400	LAN/WAN Networks
	DEC/VAX VMS	UNIX	OS/VS JCL	COBOL Language
	CMS/XEDIT	Dos 7.0	Windows 95	IBM Assembler Language
	MS Office	OS/2	Sybase	

EXPERIENCE & ACCOMPLISHMENTS

1994–2002 **Computer Operator III/Shift Leader** BLAKE CORPORATION, Newtown, PA

Provided computer systems operations to maintain the central computer system for daily operation of the state lottery.

- Oversaw the shift's triplex systems and activities-resolving problems (such as power shortages, tight deadlines and increasing capabilities) creatively and efficiently.
- Operated, monitored and controlled computers and associated equipment required to do batch processing of business applications in a real-time, on-line environment:
  — accurately performed the daily and weekly batch processing of reports and tapes.
  — ensured the central system maintained the daily time schedule of operations.
- Performed hardware and software testing and monitored systems progress via Digital Alpha Server 2100, digital tape drives, VAX VMS and UNIX software.
- Experienced with software packages, including installing, testing, debugging, maintenance, system enhancements, and new program applications such as installing new lottery games.
- Trained, supervised and evaluated technical staff of 3 computer operators, increasing employee retention by 60% through delegation and collaboration.
- Prepared daily report distributions (liability, system totals) for the state, weekly invoices and monthly cumulative reports as the Shift Supervisor.

1992–1994 **Accounts Payable Data Entry** THE CORNING ORGANIZATION, Trenton, NJ

- Keyed/edited high volume of accounts payable invoices; entered cash receipts and disbursements, general ledger, and job cost adjustments.
- Trained 7 data entry operators to ensure accurate documentation of company files.

EDUCATION

Diploma, Computer Programming, The Computer Institute, Newark, NJ
BS, Public Administration/Business (Cum Laude), University of Tarrytown, Tarrytown, NY

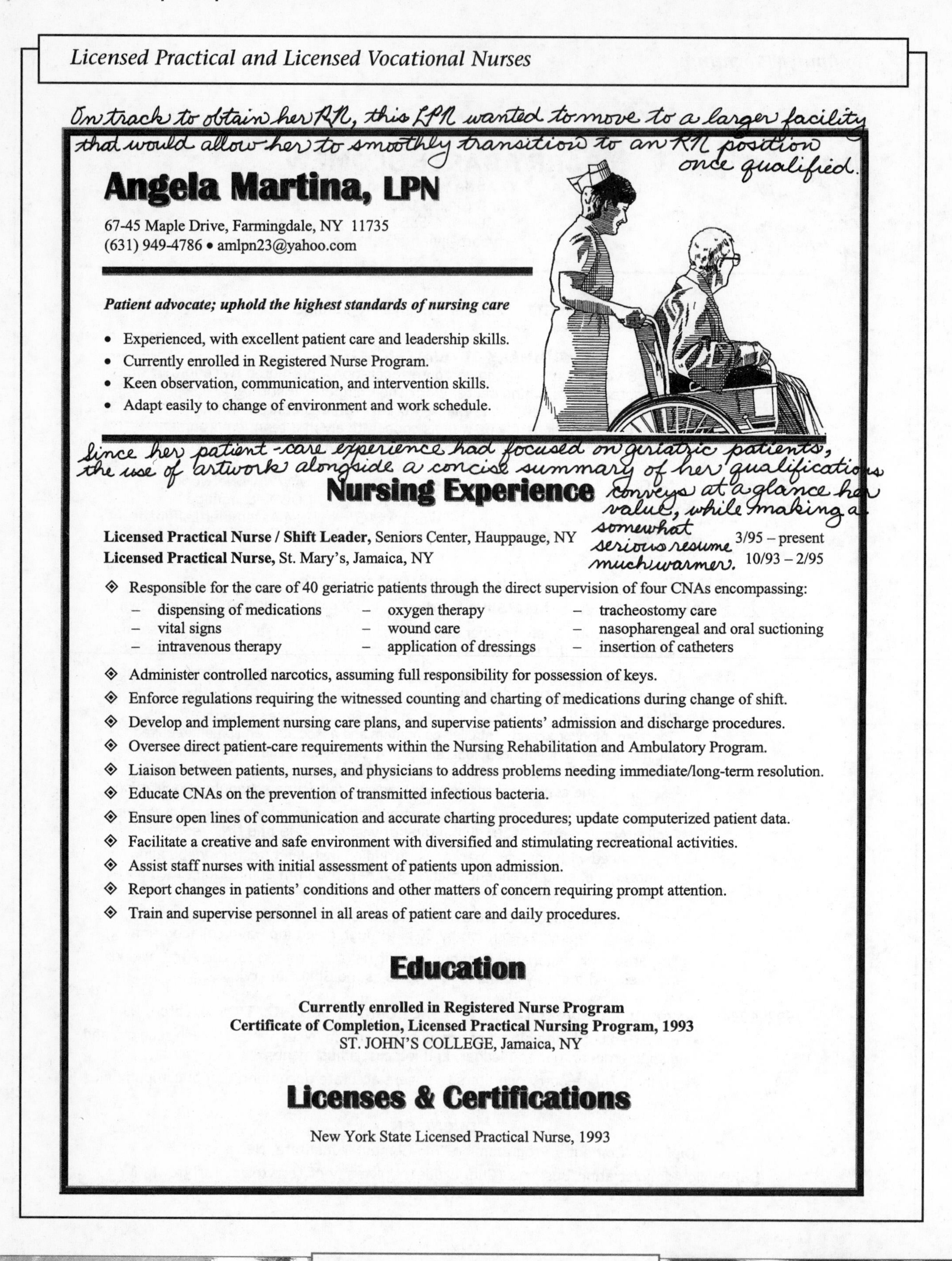

Licensed Practical and Licensed Vocational Nurses

On track to obtain her RN, this LPN wanted to move to a larger facility that would allow her to smoothly transition to an RN position once qualified.

Angela Martina, LPN

67-45 Maple Drive, Farmingdale, NY 11735
(631) 949-4786 • amlpn23@yahoo.com

Patient advocate; uphold the highest standards of nursing care

- Experienced, with excellent patient care and leadership skills.
- Currently enrolled in Registered Nurse Program.
- Keen observation, communication, and intervention skills.
- Adapt easily to change of environment and work schedule.

Since her patient-care experience had focused on geriatric patients, the use of artwork alongside a concise summary of her qualifications conveys at a glance her value, while making a somewhat serious resume much warmer.

Nursing Experience

Licensed Practical Nurse / Shift Leader, Seniors Center, Hauppauge, NY 3/95 – present
Licensed Practical Nurse, St. Mary's, Jamaica, NY 10/93 – 2/95

- Responsible for the care of 40 geriatric patients through the direct supervision of four CNAs encompassing:
  - dispensing of medications
  - vital signs
  - intravenous therapy
  - oxygen therapy
  - wound care
  - application of dressings
  - tracheostomy care
  - nasopharengeal and oral suctioning
  - insertion of catheters
- Administer controlled narcotics, assuming full responsibility for possession of keys.
- Enforce regulations requiring the witnessed counting and charting of medications during change of shift.
- Develop and implement nursing care plans, and supervise patients' admission and discharge procedures.
- Oversee direct patient-care requirements within the Nursing Rehabilitation and Ambulatory Program.
- Liaison between patients, nurses, and physicians to address problems needing immediate/long-term resolution.
- Educate CNAs on the prevention of transmitted infectious bacteria.
- Ensure open lines of communication and accurate charting procedures; update computerized patient data.
- Facilitate a creative and safe environment with diversified and stimulating recreational activities.
- Assist staff nurses with initial assessment of patients upon admission.
- Report changes in patients' conditions and other matters of concern requiring prompt attention.
- Train and supervise personnel in all areas of patient care and daily procedures.

Education

Currently enrolled in Registered Nurse Program
Certificate of Completion, Licensed Practical Nursing Program, 1993
ST. JOHN'S COLLEGE, Jamaica, NY

Licenses & Certifications

New York State Licensed Practical Nurse, 1993

Mechanical Engineers

This person's goal was to demonstrate that he had abilities above and beyond his technical skills, which themselves are very strong. His resume shows that he also has significant background in sales and product development.

WESTON M. UBALLE

41 North 1st Avenue
Los Angeles, CA 90266

310-549-7162
westengineer@aol.com

PROFILE

Innovative technical expert offering extensive experience in product design, development, launch, and sale. Unique understanding of market demands, with proven ability to invent products that suit those requirements, thus driving and sustaining long-term revenue and profits. Knowledgeable regarding effective market research. Skilled at calling on customers directly and partnering with them.

EDUCATION

B.S., MECHANICAL ENGINEERING, California Institute of Technology, 1996

SOFTWARE

Unigraphics (UGS), Ansys, Adobe PhotoShop, Microsoft Word/Excel/PowerPoint, Internet, Email

EXPERIENCE

PACIFIC INTERNATIONAL CO., LOS ANGELES, CA
Major international manufacturer of industrial injection-molded products.

5/00 to Present — **Design Manager** (Los Angeles, CA)

- ☑ Lead project development from cycle inception, to completion, to initial sale. Specialize in products for the food and beverage industries, as well as environmental, material handling, and medical fields. Direct project managers and design engineers.

12/99 to 5/00 — **Project Manager, Design Department** (Los Angeles, CA)

- ☑ Directed the company's first product launch into Asia (The Philippines), the largest project internationally. Designed a collapsible RPC (reusable produce container) that reduced produce-packaging costs while simultaneously positioning national produce suppliers to receive special environmental tax considerations. Projections: $10 million-$12 million revenue; $4.5 million profit.
- ☑ Launched a packaging product custom-tailored for an international produce leader that features cube efficiency, wash capability, and complete compatibility with European wash systems. This project is currently performing well in trials in Costa Rica, and projections are $30 million in the initial year.

12/98 to 12/99 — **Sales Engineer** (Orlando, FL)

- ☑ Fueled sales growth of a self-designed and patented collapsible agriculture container from $0 to $13 million first year, then $30 million sales in year two, with projections for $45 million in year three.
- ☑ Identified and called upon key, high-volume produce growers for whom this product would yield lower packaging costs, thus positioning them to not only purchase this product, but, in turn, package and market their goods for the first time to major national retailers.

3/98 to 12/98 — **Senior Design Engineer** (Atlanta, GA)

- ☑ Designed a point-of-purchase merchandising display for 2-liter soft drink bottles to be placed on the top shelf and automatically advance bottles forward to maintain facings. *Patent No. US D5514406: "Facing Maintainer for Beverage Containers," filed 2/99.*
- ☑ Developed an interface for soft drink manufacturers that increased warehousing space 50% by allowing pallet stacking three pallets high instead of two, still with sufficient stability. *Patent pending: "Three-Tier Pallet Stacking Mechanism."*

7/97 to 3/98 — **Design Engineer** (Los Angeles, CA)

- ☑ Created a collapsible agriculture container for the produce industry that simultaneously met the requirements and specifications of produce growers and retailers and smoothly flowed through a closed loop supply chain network. *Patent No. US 6945567: "Collapsible Carrier," filed 12/97.*
- ☑ Managed the six-month tooling development and implementation process for the 12 initial molds for the collapsible container. Subsequent to extensive sales and product use by customers in the field, re-designed tooling for a second-generation version that was lighter, stronger, and more cube efficient, thus better suited to serve the market.

Paralegals and Legal Assistants

EVELYN R. JENKINS

98A Forrest Avenue, Melville, New York 11750 ♦ (631) 555-3029

PARALEGAL

Experienced, Certified Paralegal with excellent office management and client relations skills seeking a position within a corporate legal department where a working knowledge of legal terminology, general law, and legal proceedings pertaining to the following case types will be fully utilized and expanded:

Civil Litigation…Corporate Law…Wills & Estates…Negligence…Matrimonial and Mediation…
Personal Injury…Real Estate…Malpractice…Family Court…Bankruptcy…Criminal

Detail-oriented with excellent research, investigative, and reporting skills.
Exercise independent judgment and decision-making abilities and a high level of confidentiality.
Uphold the ethical standards of the legal profession.

♦ ♦ ♦

Windows 98, Word Perfect 8.0, LEXIS/NEXIS, MATLAW, McKinley's, DAKCS
Time Slips / Time Reporter, Forms of Bankruptcy (FOB)

This comprehensive introduction showcases Evelyn's solid knowledge, experience, and educational qualifications.

PROFESSIONAL EXPERIENCE

Paralegal, Nevins & Associates, Melville, New York — 8/99 – present

- Report directly to four attorneys with broad-ranged responsibilities that encompass the timely and complex preparation of cases from discovery to trial phase.
- Coordinate multifaceted office functions encompassing court calendar management, retainment of court reporters, and scheduling of conference rooms for deposition proceedings.
- Liaison between attorneys, clients, healthcare providers, insurance carriers, law firms, and government agencies.
- Ensure open lines of communication and satisfaction of deadlines through execution of dated correspondence.
- Perform computerized and law library research to obtain and gather case-relevant data and materials.
- Prepare content-specific case files for attorneys reflecting supporting forms, documentation, and photographs to use during client presentations, and index/cross reference network database information.

Legal Assistant, Funds Recovery, Inc., Levittown, New York — 2/94 – 8/99

- Collaborated with Collections, Medical Billing, and Finance departments to obtain documentation pertaining to the status of more than 50 weekly referred collections cases forwarded to the Legal Department.
- Carefully sourced and selected nationally based bonded attorneys utilizing the American Lawyers Quarterly, Commercial Bar Directory, National Directory List, and Columbia Directory List; determined the appropriate choice upon obtainment and review of résumés, copies of insurance policies, and court filing fees.
- Integrated traditional investigative methods and DAKCS database system to gather account histories and case-sensitive documentation for attorneys including credit bureau reports, court affidavits, judgments, skip-tracing records, bankruptcy notices, banking statements, proof of assets, and trial letters.
- Maintained ongoing communications with attorneys and clients from point of referral/discovery to trial phase, facilitating and expediting case settlements that awarded clients a minimum of 80% in recovered funds.

EDUCATION

Certificate of Completion, Paralegal Studies Program, 1998
LONG ISLAND UNIVERSITY at C.W. POST, Brookville, New York
Approved by the American Bar Association

Evelyn used this resume only once and landed a position at double her former salary!

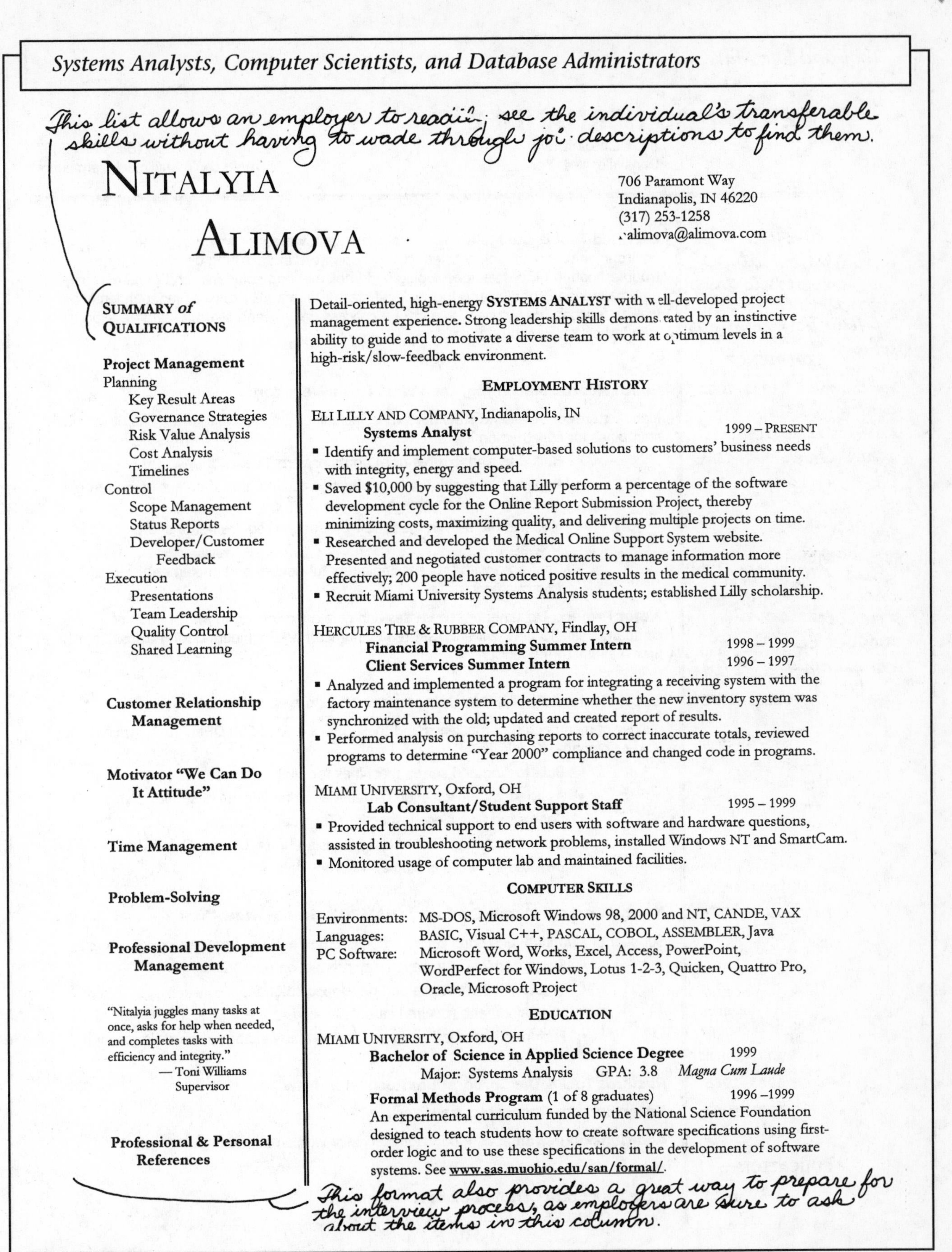

Systems Analysts, Computer Scientists, and Database Administrators

This list allows an employer to readily see the individual's transferable skills without having to wade through job descriptions to find them.

NITALYIA ALIMOVA

706 Paramont Way
Indianapolis, IN 46220
(317) 253-1258
·alimova@alimova.com

SUMMARY *of* QUALIFICATIONS

Project Management
Planning
- Key Result Areas
- Governance Strategies
- Risk Value Analysis
- Cost Analysis
- Timelines

Control
- Scope Management
- Status Reports
- Developer/Customer Feedback

Execution
- Presentations
- Team Leadership
- Quality Control
- Shared Learning

Customer Relationship Management

Motivator "We Can Do It Attitude"

Time Management

Problem-Solving

Professional Development Management

"Nitalyia juggles many tasks at once, asks for help when needed, and completes tasks with efficiency and integrity."
— Toni Williams
Supervisor

Professional & Personal References

Detail-oriented, high-energy **SYSTEMS ANALYST** with well-developed project management experience. Strong leadership skills demonstrated by an instinctive ability to guide and to motivate a diverse team to work at optimum levels in a high-risk/slow-feedback environment.

EMPLOYMENT HISTORY

ELI LILLY AND COMPANY, Indianapolis, IN
Systems Analyst 1999 – PRESENT
- Identify and implement computer-based solutions to customers' business needs with integrity, energy and speed.
- Saved $10,000 by suggesting that Lilly perform a percentage of the software development cycle for the Online Report Submission Project, thereby minimizing costs, maximizing quality, and delivering multiple projects on time.
- Researched and developed the Medical Online Support System website. Presented and negotiated customer contracts to manage information more effectively; 200 people have noticed positive results in the medical community.
- Recruit Miami University Systems Analysis students; established Lilly scholarship.

HERCULES TIRE & RUBBER COMPANY, Findlay, OH
Financial Programming Summer Intern 1998 – 1999
Client Services Summer Intern 1996 – 1997
- Analyzed and implemented a program for integrating a receiving system with the factory maintenance system to determine whether the new inventory system was synchronized with the old; updated and created report of results.
- Performed analysis on purchasing reports to correct inaccurate totals, reviewed programs to determine "Year 2000" compliance and changed code in programs.

MIAMI UNIVERSITY, Oxford, OH
Lab Consultant/Student Support Staff 1995 – 1999
- Provided technical support to end users with software and hardware questions, assisted in troubleshooting network problems, installed Windows NT and SmartCam.
- Monitored usage of computer lab and maintained facilities.

COMPUTER SKILLS

Environments: MS-DOS, Microsoft Windows 98, 2000 and NT, CANDE, VAX
Languages: BASIC, Visual C++, PASCAL, COBOL, ASSEMBLER, Java
PC Software: Microsoft Word, Works, Excel, Access, PowerPoint, WordPerfect for Windows, Lotus 1-2-3, Quicken, Quattro Pro, Oracle, Microsoft Project

EDUCATION

MIAMI UNIVERSITY, Oxford, OH
Bachelor of Science in Applied Science Degree 1999
Major: Systems Analysis GPA: 3.8 *Magna Cum Laude*
Formal Methods Program (1 of 8 graduates) 1996 –1999
An experimental curriculum funded by the National Science Foundation designed to teach students how to create software specifications using first-order logic and to use these specifications in the development of software systems. See **www.sas.muohio.edu/san/formal/**.

This format also provides a great way to prepare for the interview process, as employers are sure to ask about the items in this column.

Tool and Die Makers

FERNANDO HERNANDEZ

8961 Lincoln Drive
Dansville, New York 14435

(585) 278-5067
E-mail: toolman@machinemail.net

SUMMARY

This information is included to draw attention to his State of New York credentials, which is uncommon among candidates at this level.

Certified Tool & Die Maker with experience in a variety of manufacturing environments. Proven ability to effectively support manufacturing operations through troubleshooting processes, developing and implementing solutions, and general repair and maintenance of dies and tooling. Self-starter with ability to work independently as well as function productively as part of a process engineering team. Extensive experience with lathes, mills, and grinders.

EXPERIENCE

1992–2002 **South Monroe Machining, Inc.; West Rush, New York**

Manufacturer of brass and steel hardware, including hinges, chain locks, and other metal stampings for construction industry.

- Built dies and fixtures from prints or verbal descriptions.
- Repaired and maintained production tooling, including set-up and trouble-shooting of punch presses.
- Modified old dies to meet new production requirements and process changes.
- Conferred with Process Engineers to identify and implement process improvements.

This heading was included to put the best spin possible on the company's closing.

Major Project: Documented processes and cataloged tooling in anticipation of company's sale to new owners. Assisted Manufacturing Engineers during transition to new manufacturing facility.

1987–1992 **Royal Machine Tool & Die Corporation; Mendon, New York**

Builder of manufacturing machines and tooling for Fortune 500 OEM firms, including Ford Motor Company, Kodak, and Bausch & Lomb.

- Built tooling and set up machines for pilot production.
- Assisted with installation of manufacturing lines in customer plants.
- Conducted first-piece inspections for Quality Control.
- Served as Acting Foreman, supervising up to 15 Tool Makers.
- Directed activities of Apprentice Tool Makers.

1985–1987 **North American Metals Company (NAMCO); Greece, New York**

Supplier of aluminum and steel cans to the beverage and food processing industries.

- Supported 24-hour, seven-day production operation.
- Troubleshot processes and developed solutions.
- Maintained and repaired carbide tooling.
- Performed internal/external grinding and welding to maintain tooling.

1983–1985 **Richards Tool & Die Corporation; Rochester, New York**

1981–1983 **Ritter-Sybron; Rochester, New York**

Additional experience as Production Machinist with small manufacturing firms.

EDUCATION

1980 **New York State Certified Class A Toolmaker**
Mechanics Institute of Rochester; Chili, New York

Section Three

Important Trends in Jobs and Industries

In putting this section together, my objective was to give you a quick review of major labor market trends. To accomplish that, I chose two excellent articles that originally appeared in U.S. Department of Labor publications.

The first article is "Tomorrow's Jobs." It provides a superb—and short—review of the major trends that *will* affect your career in the years to come. Read it for ideas on selecting a career path for the long term.

The second article is "Employment Trends in Major Industries." While you may not have thought much about it, the industry you work in is just as important as your occupational choice. This great article will help you learn about the major trends affecting various industries.

Tomorrow's Jobs

Comments

This article, with minor changes, comes from the *Occupational Outlook Handbook* and was written by the U.S. Department of Labor staff. The material provides a good review of major labor market trends, including trends for broad types of occupations and industries.

Much of this article uses 2000 data, the most recent available at press time. By the time it is carefully collected and analyzed, the data used by the Department of Labor is typically several years old. Since labor market trends tend to be gradual, this delay does not affect the material's usefulness.

You may notice that some job titles in this article differ from those used elsewhere in *America's Top Computer and Technical Jobs*. This is not an error. The material that follows uses a different set of occupations than I used in choosing this book's described jobs.

Making informed career decisions requires reliable information about opportunities in the future. Opportunities result from the relationships between the populations, labor force, and the demand for goods and services.

Population ultimately limits the size of the labor force—individuals working or looking for work—which constrains how much can be produced. Demand for various goods and services determines employment in the industries providing them. Occupational employment opportunities, in turn, result from skills needed within specific industries. For example, opportunities for computer engineers and other computer-related occupations have surged in response to rapid growth in demand for computer services.

Examining the past and projecting changes in these relationships is the foundation of the Occupational Outlook Program. The following article presents highlights of Bureau of Labor Statistics projections of the labor force and occupational and industry employment that can help guide your career plans.

Population

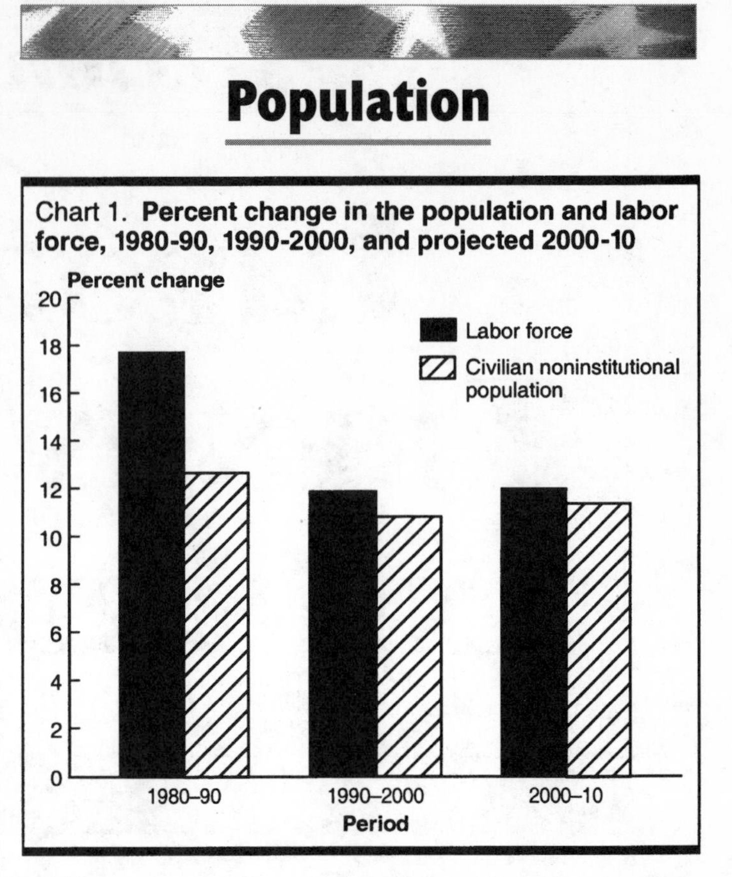

Population trends affect employment opportunities in a number of ways. Changes in population influence the demand for goods and services. For example, a growing and aging population has increased the demand for health services. Equally important, population changes produce corresponding changes in the size and demographic composition of the labor force.

The U.S. population is expected to increase by 24 million over the 2000–2010 period, at a slightly faster rate of growth than during the 1990–2000 period but slower than over the 1980–1990 period (chart 1). Continued growth will mean more consumers of goods and services, spurring demand for workers in a wide range of occupations and industries. The effects of population growth on various occupations will differ. The differences are partially accounted for by the age distribution of the future population.

The youth population, aged 16 to 24, will grow more rapidly than the overall population, a turnaround that began in the mid-1990s. As the baby boomers continue to age, the group aged 55 to 64 will increase by 11 million persons over the 2000–2010 period—more than any other group. Those aged 35 to 44 will be the only group to decrease in size, reflecting the birth dearth following the baby boom.

Minorities and immigrants will constitute a larger share of the U.S. population in 2010 than they do today. Minority groups that have grown the fastest in the recent past—Hispanics and Asians and others—are projected to continue to grow much faster than white, non-Hispanics.

Labor Force

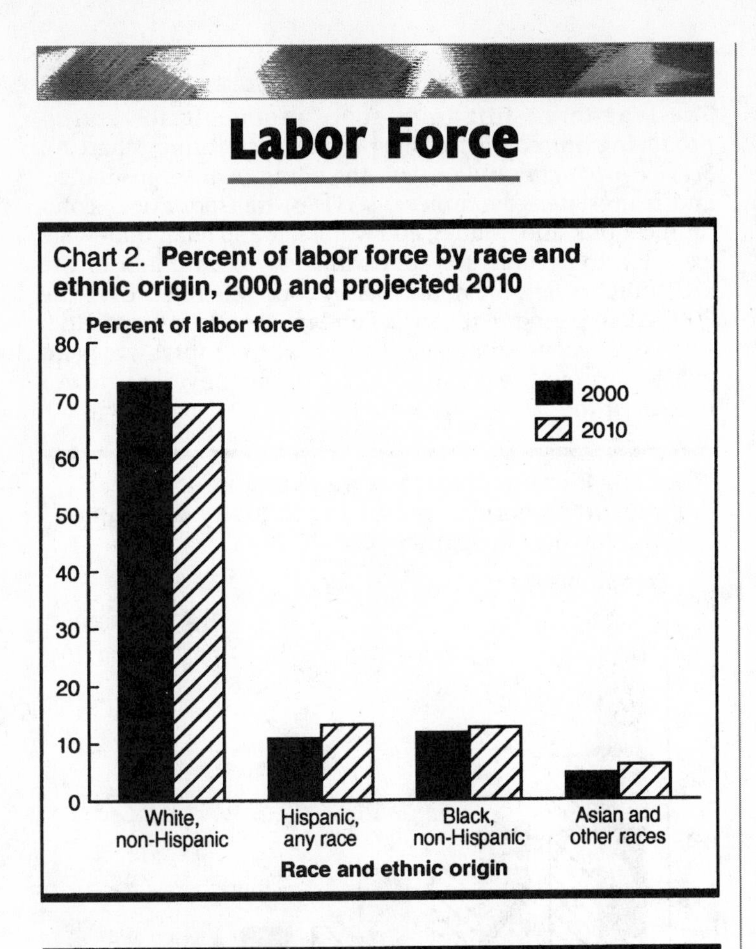

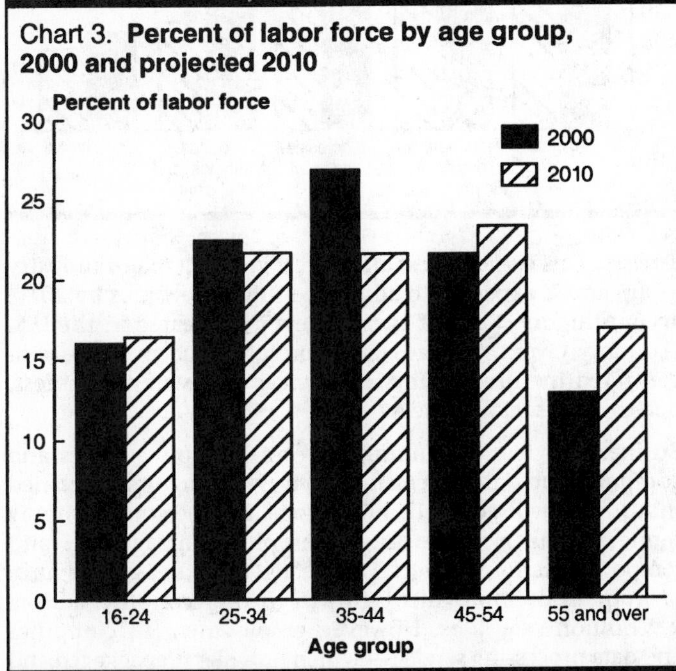

Population is the single most important factor in determining the size and composition of the labor force—comprising people who are either working or looking for work. The civilian labor force is projected to increase by 17 million, or 12 percent, to 158 million over the 2000–2010 period.

The U.S. workforce will become more diverse by 2010. White, non-Hispanic persons will continue to make up a decreasing share of the labor force, falling from 73.1 percent in 2000 to 69.2 percent in 2010 (chart 2). However, despite relatively slow growth, white, non-Hispanics will have the largest numerical growth in the labor force between 2000 and 2010, reflecting the large size of this group. Hispanics, non-Hispanic blacks, and Asian and other ethnic groups are projected to account for an increasing share of the labor force by 2010, growing from 10.9 to 13.3 percent, 11.8 to 12.7 percent, and 4.7 to 6.1 percent, respectively. By 2010, for the first time, Hispanics will constitute a greater share of the labor force than will blacks. Asians and others continue to have the fastest growth rates, but still are expected to remain the smallest of the four labor force groups.

The numbers of men and women in the labor force will grow, but the number of men will grow at a slower rate than the number of women. The male labor force is projected to grow by 9.3 percent from 2000 to 2010, compared with 15.1 percent for women. As a result, men's share of the labor force is expected to decrease from 53.4 to 52.1 percent, while women's share is expected to increase from 46.6 to 47.9 percent.

The youth labor force, aged 16 to 24, is expected to increase its share of the labor force to 16.5 percent by 2010, growing more rapidly than the overall labor force. The large group 25 to 54 years old, who made up 71 percent of the labor force in 2000, is projected to decline to 66.6 percent of the labor force by 2010. Workers 55 and older, on the other hand, are projected to increase from 12.9 percent to 16.9 percent of the labor force between 2000 and 2010, due to the aging of the baby-boom generation (chart 3).

Education and Training

Projected job growth varies widely by education and training requirements. All seven of the education and training categories projected to have faster-than-average employment growth require a postsecondary vocational or academic award (chart 4). These seven categories will account for two-fifths of all employment growth over the 2000–2010 period.

Employment in occupations requiring at least a bachelor's degree is expected to grow 21.6 percent and account for five out of the six fastest growing education or training categories. Two categories—jobs requiring an associate degree (projected to grow 32 percent over the 2000–2010 period, faster than any other category) and jobs requiring a postsecondary vocational award—together will grow 24.1 percent. The four categories of occupations requiring work-related training are projected to increase 12.4 percent, compared with 15.2 percent for all occupations combined.

Education is essential in getting a high-paying job. In fact, all but two of the 50 highest paying occupations require a college degree. Air traffic controllers and nuclear power reactor operators are the only occupations of the 50 highest paying that do not require a college degree.

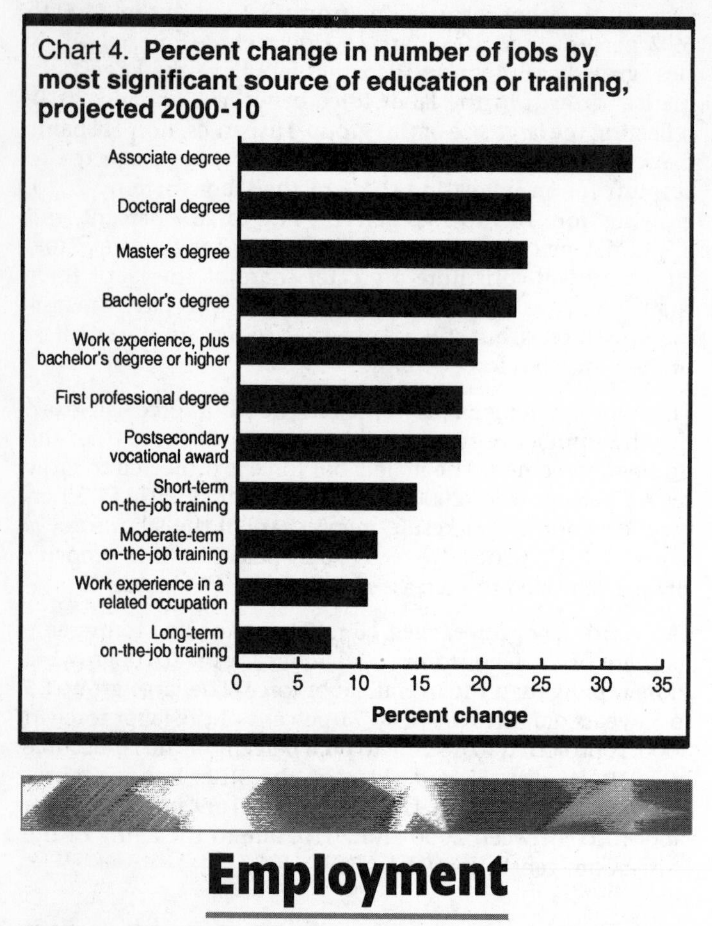

Employment

Total employment is expected to increase from 146 million in 2000 to 168 million in 2010, or by 15.2 percent. The 22 million jobs that will be added by 2010 will not be evenly distributed across major industrial and occupational groups. Changes in consumer demand, technology, and many other factors will contribute to the continually changing employment structure in the U.S. economy.

The following two sections examine projected employment change from both industrial and occupational perspectives. The industrial profile is discussed in terms of primary wage and salary employment. Primary employment excludes secondary jobs for those who hold multiple jobs. The exception is employment in agriculture, which includes self-employed and unpaid family workers in addition to wage and salary workers.

The occupational profile is viewed in terms of total employment—including primary and secondary jobs for wage and salary, self-employed, and unpaid family workers. Of the nearly 146 million jobs in the U.S. economy in 2000, wage and salary workers accounted for 134 million; self-employed workers accounted for 11.5 million; and unpaid family workers accounted for about 169,000. Secondary employment accounted for 1.8 million of all jobs. Self-employed workers held 9 out of 10 secondary jobs; wage and salary workers held most of the remainder.

Industry

The long-term shift from goods-producing to service-producing employment is expected to continue (chart 5). ***Service-producing industries***—including finance, insurance, and real estate; government; services; transportation, communications, and utilities; and wholesale and retail trade—are expected to account for approximately 20.2 million of the 22.0 million new wage and salary jobs generated over the 2000–2010 period. The services and retail trade industry divisions will account for nearly three-fourths of total wage and salary job growth, a continuation of the employment growth pattern of the 1990–2000 period.

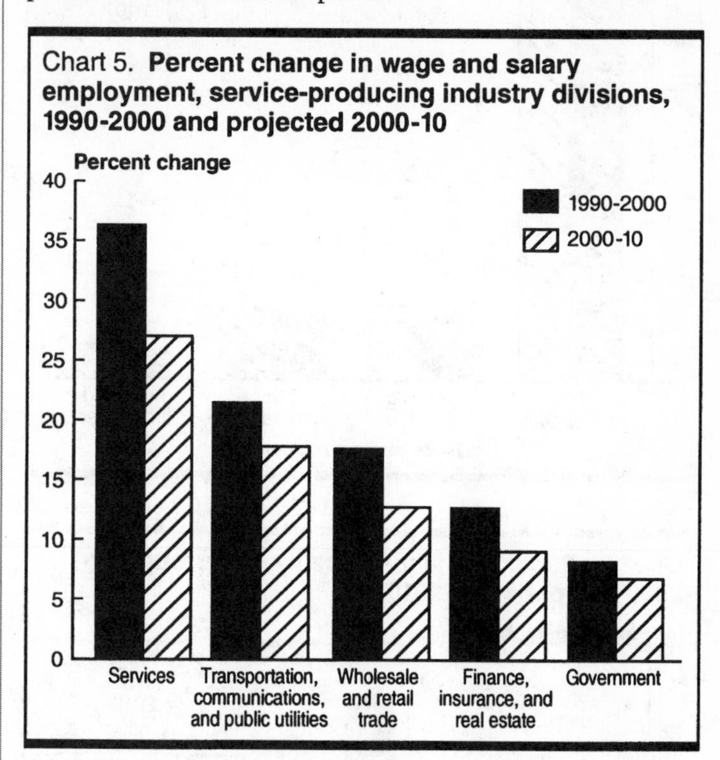

Services. This is the largest and fastest growing major industry group and is expected to add 13.7 million new jobs by 2010, accounting for 3 out of every 5 new jobs created in the U.S. economy. Over two-thirds of this projected job growth is concentrated in three sectors of services industries—business, health, and social services.

Business services—including personnel supply services and computer and data processing services, among other detailed industries—will add 5.1 million jobs. The personnel supply services industry, consisting of employment agencies and temporary-staffing services, is projected to be the largest source of numerical employment growth in the economy, adding 1.9 million new jobs. However, employment in computer and data processing services—which provides prepackaged and specialized software, data and computer systems design and management, and computer-related consulting services—is projected to grow by 86 percent between 2000 and 2010, ranking as the fastest growing industry in the economy.

Health services—including home healthcare services, hospitals, and offices of health practitioners—will add 2.8 million new jobs as demand for healthcare increases because of an aging population and longer life expectancies.

Social services—including child daycare and residential care services—will add 1.2 million jobs. As more women enter the labor force, demand for childcare services is expected to grow, leading to the creation of 300,000 jobs. An elderly population seeking alternatives to nursing homes and hospital care will boost employment in residential care services, which is projected to grow 63.5 percent and add 512,000 jobs by 2010.

Transportation, communications, and utilities. Overall employment is expected to increase by 1.3 million jobs, or by 17.9 percent. Employment in the transportation sector is expected to increase by 20.7 percent, from 4.5 million to 5.5 million jobs. Trucking and warehousing will provide the most new jobs in the transportation sector, adding 407,000 jobs by 2010. Due to population growth and urban sprawl, local and interurban passenger transit is expected to increase 31 percent over the 2000–2010 period, the fastest growth among all the transportation sectors.

Employment in the communications sector is expected to increase by 16.9 percent, adding 277,000 jobs by 2010. Half of these new jobs—139,000—will be in the telephone communications industry; however, cable and other pay television will be the fastest growing segment of the sector over the next decade, with employment expanding by 50.6 percent. Increased demand for residential and business wireline and wireless services, cable service, and high-speed Internet connections will fuel the growth in communications industries.

Employment in the utilities sector is projected to increase by only 4.9 percent through 2010. Despite increased output, employment in electric services, gas production and distribution, and combination utility services is expected to decline through 2010 due to improved technology that increases worker productivity. The growth in the utilities sector will be driven by water supply and sanitary services, in which employment is expected to increase 45.1 percent by 2010. Jobs are not easily eliminated by technological gains in this industry because water treatment and waste disposal are very labor-intensive activities.

Wholesale and retail trade. Employment is expected to increase by 11.1 percent and 13.3 percent, respectively, growing from 7 million to 7.8 million in wholesale trade and from 23.3 million to 26.4 million in retail trade. Increases in population, personal income, and leisure time will contribute to employment growth in these industries as consumers demand more goods. With the addition of 1.5 million jobs, the eating and drinking places segment of the retail trade industry is projected to have the largest numerical increase in employment within the trade industry group.

Finance, insurance, and real estate. Overall employment is expected to increase by 687,000 jobs, or 9.1 percent, by 2010. The finance sector of the industry—including depository and nondepository institutions and securities and commodity brokers—will account for one-third of these jobs. Security and commodity brokers and dealers are expected to grow the fastest among the finance segments; the projected 20.3 percent employment increase by 2010 reflects the increased number of baby boomers in their peak savings years, the growth of tax-favorable retirement plans, and the globalization of the securities markets. However, employment in depository institutions should continue to decline due to an increase in the use of Internet banking, ATM machines, and debit cards.

The insurance sector—including insurance carriers and insurance agents and brokers—is expected to add 152,000 new jobs by 2010. The majority of job growth in the insurance carriers segment will be attributable to medical service and health insurance, in which employment is projected to increase by 16 percent. The number of jobs with insurance agents and brokers is expected to grow about 14.3 percent by 2010, as many insurance carriers downsize their sales staffs and as agents set up their own businesses.

The real estate sector is expected to add the most jobs out of the three sectors—272,000 by 2010. As the population grows, demand for housing also will grow.

Government. Between 2000 and 2010, government employment, excluding public education and hospitals, is expected to increase by 6.9 percent, from 10.2 million to 10.9 million jobs. Growth in government employment will be fueled by growth at the state and local levels, in which the number of jobs will increase by 12.2 and 11.2 percent, respectively, through 2010. Growth at these levels is due mainly to an increased demand for services and the shift of responsibilities from the federal government to the state and local governments. Federal government employment is expected to decline by 7.6 percent as the federal government continues to contract out many government jobs to private companies.

Employment in the ***goods-producing industries*** has been relatively stagnant since the early 1980s. Overall, this sector is expected to grow 6.3 percent over the 2000–2010 period. Although employment is expected to increase more slowly than

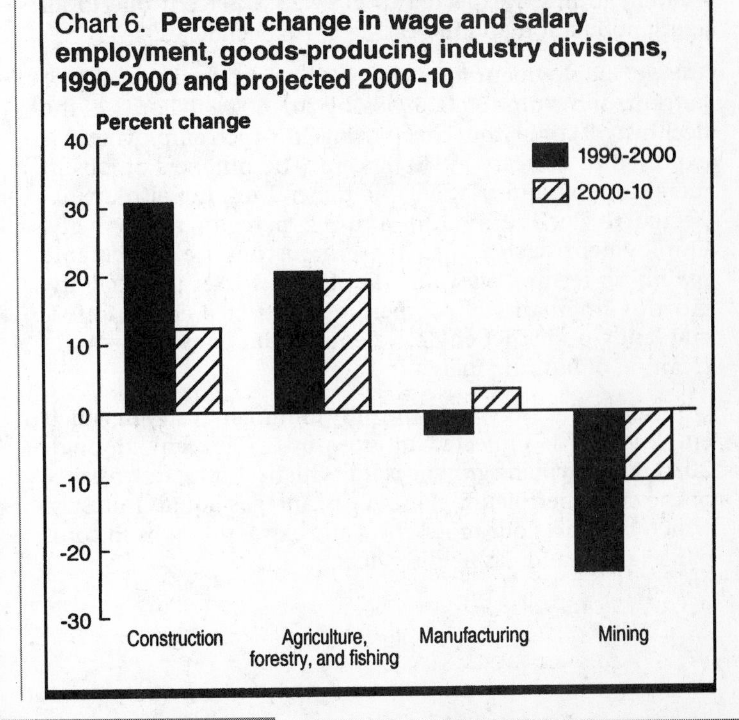

in the service-producing industries, projected growth within the goods-producing sector varies considerably (chart 6).

Construction. Employment in construction is expected to increase by 12.3 percent, from 6.7 million to 7.5 million. Demand for new housing and an increase in road, bridge, and tunnel construction will account for the bulk of job growth in this industry.

Agriculture, forestry, and fishing. Overall employment in agriculture, forestry, and fishing is expected to increase by 19.3 percent, from 2.2 million to 2.6 million. Three-fourths of this growth will come from veterinary services and landscape and horticultural services, which will add 96,000 and 229,000 jobs, respectively. Employment in crops, livestock, and livestock products is expected to continue to decline due to advancements in technology. The numbers of jobs in forestry and in fishing, hunting, and trapping are expected to grow only 1.9 percent by 2010.

Manufacturing. Rebounding from the 1990–2000 decline of 607,000 manufacturing jobs, employment in this sector is expected to grow modestly, by 3.1 percent, by 2010, adding 577,000 jobs. The projected employment growth is attributable mainly to the industries that manufacture durable goods. Durable goods manufacturing is expected to grow 5.7 percent, to 11.8 million jobs, over the next decade. Despite gains in productivity, the growing demand for computers, electronic components, motor vehicles, and communications equipment will contribute to this employment growth.

Nondurable manufacturing, on the other hand, is expected to decline by less than 1 percent, shedding 64,000 jobs overall. The majority of employment declines are expected to be in apparel and other textile products and leather and leather products industries, which together are expected to shed 131,000 jobs by 2010 because of increased job automation and international competition. On the other hand, drug manufacturing is expected to grow 23.8 percent due to an aging population and increasing life expectancies.

Mining. Employment in mining is expected to decrease 10.1 percent, or by some 55,000 jobs, by 2010. The majority of the decline will come from coal mining, in which employment is expected to decrease by 30 percent. The numbers of jobs in metal mining and nonmetallic mineral mining also are expected to decline by 13.8 and 3.2 percent, respectively. Employment decreases in these industries are attributable mainly to technology gains that boost worker productivity, growing international competition, restricted access to federal lands, and strict environmental regulations that require cleaning of burning fuels.

Oil and gas field services is the only mining industry in which employment is projected to grow, by 3.7 percent, through 2010. Employment growth is due chiefly to the downsizing of the crude petroleum, natural gas, and gas liquids industry, which contracts out production and extraction jobs to companies in oil and gas field services.

Occupation

Expansion of the service-producing sector is expected to continue, creating demand for many occupations. However, projected job growth varies among major occupational groups (chart 7).

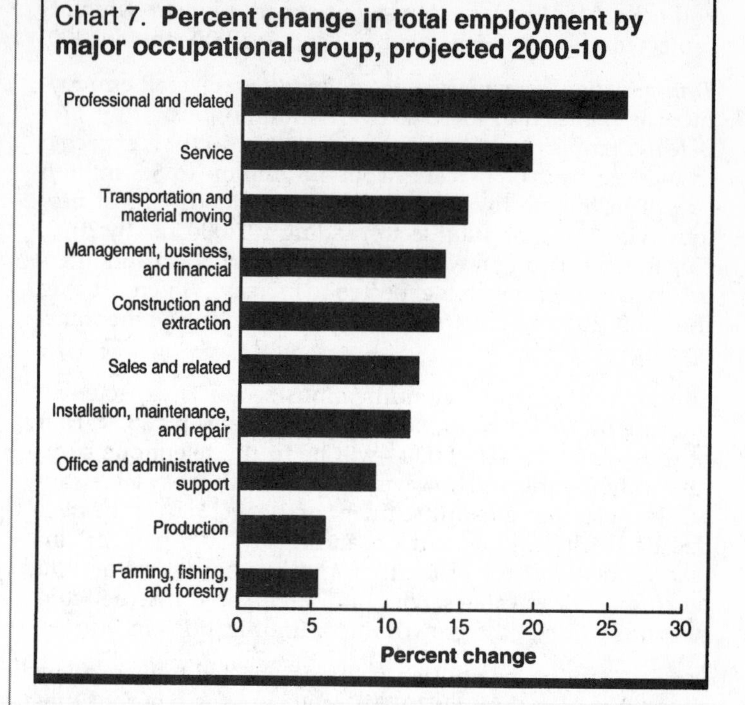

Professional and related occupations. Professional and related occupations will grow the fastest and add more new jobs than any other major occupational group. Over the 2000–2010 period, a 26-percent increase in the number of professional and related jobs is projected, a gain of 6.9 million. Professional and related workers perform a wide variety of duties and are employed throughout private industry and government. Nearly three-quarters of the job growth will come from three groups of professional occupations—computer and mathematical occupations; healthcare practitioners and technical occupations; and education, training, and library occupations—which will add 5.2 million jobs combined.

Service occupations. Service workers perform services for the public. Employment in service occupations is projected to increase by 5.1 million, or 19.5 percent, the second-largest numerical gain and second-highest rate of growth among the major occupational groups. Food preparation and serving-related occupations are expected to add the most jobs among the service occupations, 1.6 million by 2010. However, healthcare support occupations are expected to grow the fastest, 33.4 percent, adding 1.1 million new jobs.

Transportation and material moving occupations. Transportation and material moving workers transport and transfer people and materials by land, sea, or air. These occupations should grow 15.2 percent and add 1.5 million jobs by 2010. Among transportation occupations, motor vehicle operators will add the most jobs—745,000. Rail transportation occupations are the only group in which employment is projected to decline, by 18.6 percent, through 2010. Material moving occupations will grow 14 percent and will add 681,000 jobs.

Management, business, and financial occupations. Workers in management, business, and financial occupations plan and direct the activities of business, government, and other organizations. Employment is expected to increase by 2.1 million, or 13.6 percent, by 2010. Among managers, the numbers of computer and information systems managers and of public relations managers will grow the fastest, by 47.9 and 36.3 percent, respectively. General and operations managers will add the most new jobs—363,000 by 2010. Agricultural managers and purchasing managers are the only workers in this group whose numbers are expected to decline, losing 325,000 jobs combined. Among business and financial occupations, accountants and auditors and management analysts will add the most jobs—326,000 combined. Management analysts also will be one of the fastest growing occupations in this group, along with personal financial advisors, with job increases of 28.9 and 34 percent, respectively.

Construction and extraction occupations. Construction and extraction workers construct new residential and commercial buildings and also work in mines, quarries, and oil and gas fields. Employment of these workers is expected to grow 13.3 percent, adding 989,000 new jobs. Construction trades and related workers will account for the majority of these new jobs—862,000 by 2010. Most extraction jobs will decline, reflecting overall employment losses in the mining and oil and gas extraction industries.

Sales and related occupations. Sales and related workers transfer goods and services among businesses and consumers. Sales and related occupations are expected to add 1.9 million new jobs by 2010, growing by 11.9 percent. The majority of these jobs will be among retail salespersons and cashiers, occupations that will add almost 1 million jobs combined.

Installation, maintenance, and repair occupations. Workers in installation, maintenance, and repair occupations install new equipment and maintain and repair older equipment. These occupations will add 662,000 jobs by 2010, growing by 11.4 percent. Automotive service technicians and general maintenance and repair workers will account for 3 in 10 new installation, maintenance, and repair jobs. The fastest growth rate will be among telecommunications line installers and repairers, an occupation that is expected to grow 27.6 percent over the 2000–2010 period.

Office and administrative support occupations. Office and administrative support workers perform the day-to-day activities of the office, such as preparing and filing documents, dealing with the public, and distributing information. Employment in these occupations is expected to grow by 9.1 percent, adding 2.2 million new jobs by 2010. Customer service representatives will add the most new jobs—631,000. Desktop publishers will be among the fastest growing occupations, growing 66.7 percent over the decade. Order clerks, tellers, and insurance claims and policy processing clerks will be among the jobs with the largest employment losses.

Production occupations. Production workers are employed mainly in manufacturing, assembling goods, and operating plants. Production occupations will grow 5.8 percent and add 750,000 jobs by 2010. Metal and plastics workers and assemblers and fabricators will add the most production jobs—249,000 and 171,000, respectively. Textile, apparel, and furnishings occupations will account for much of the job losses among production occupations.

Farming, fishing, and forestry occupations. Farming, fishing, and forestry workers cultivate plants, breed and raise livestock, and catch animals. These occupations will have the slowest job growth among the major occupational groups—5.3 percent, adding 74,000 new jobs by 2010. Farm workers account for nearly 3 out of 4 new jobs in this group. The numbers of both fishing and logging workers are expected to decline, by 12.2 and 3.5 percent, respectively.

Computer occupations are expected to grow the fastest over the projection period (chart 8). In fact, these jobs account for 8 out of the 20 fastest growing occupations in the economy. In addition to high growth rates, these eight occupations combined will add more than 1.9 million new jobs to the economy.

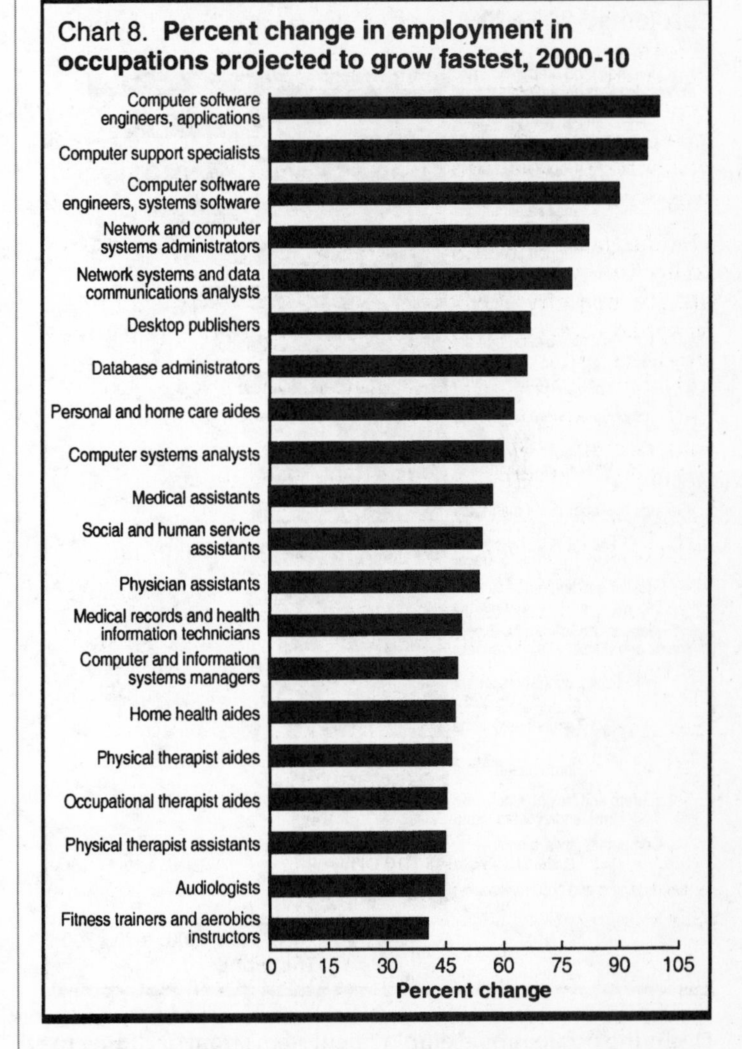

Health occupations comprise most of the remaining fastest growing occupations. High growth rates among computer and health occupations reflect projected faster-than-average growth in the computer and data processing and health services industries.

The 20 occupations listed in chart 9 will account for more than one-third of all new jobs, 8 million combined, over the 2000–2010 period. The occupations with the largest numerical increases cover a wider range of occupational categories than those occupations with the fastest growth rates. Computer and health occupations will account for some of these increases in employment, as will occupations in education, sales, transportation, office and administrative support, and food service. Many of these occupations are very large and will create more new jobs than those with high growth rates. Only 4 out of the 20 fastest growing occupations—computer software engineers, applications; computer software engineers, systems software; computer support specialists; and home health aides—also are projected to be among the 20 occupations with the largest numerical increases in employment.

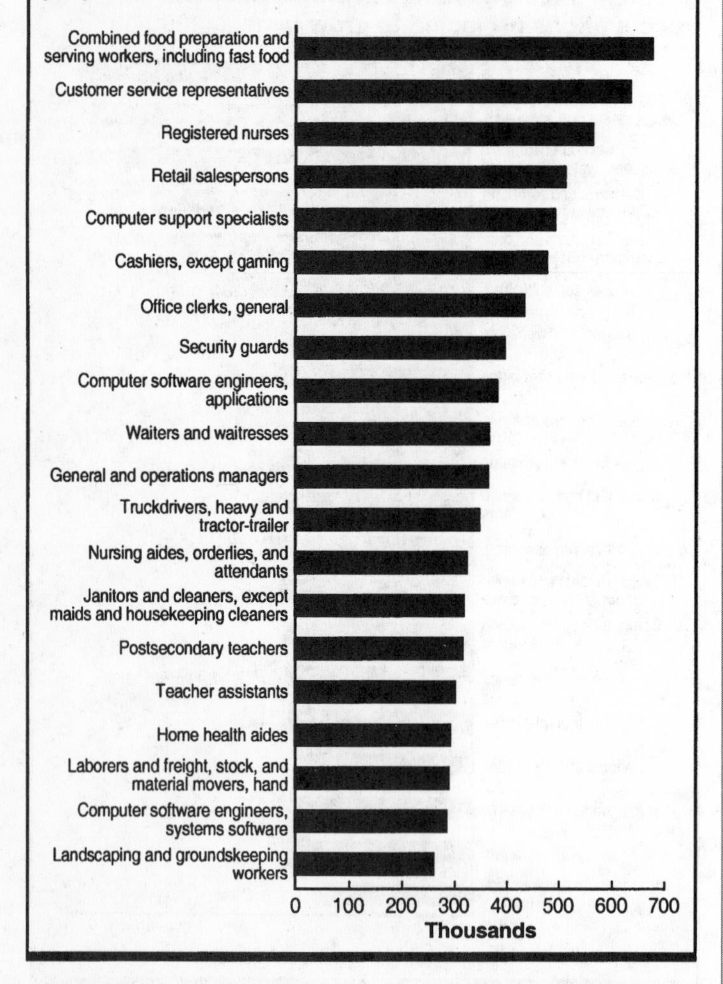

Declining occupational employment stems from declining industry employment, technological advancements, changes in business practices, and other factors. For example, increased productivity and farm consolidations are expected to result in a decline of 328,000 farmers over the 2000–2010 period (chart 10). The majority of the 20 occupations with the largest numerical decreases are office and administrative support and production occupations, which are affected by increasing automation and the implementation of office technology that reduces the need for these workers. For example, the increased use of ATM machines and Internet banking will reduce the number of tellers.

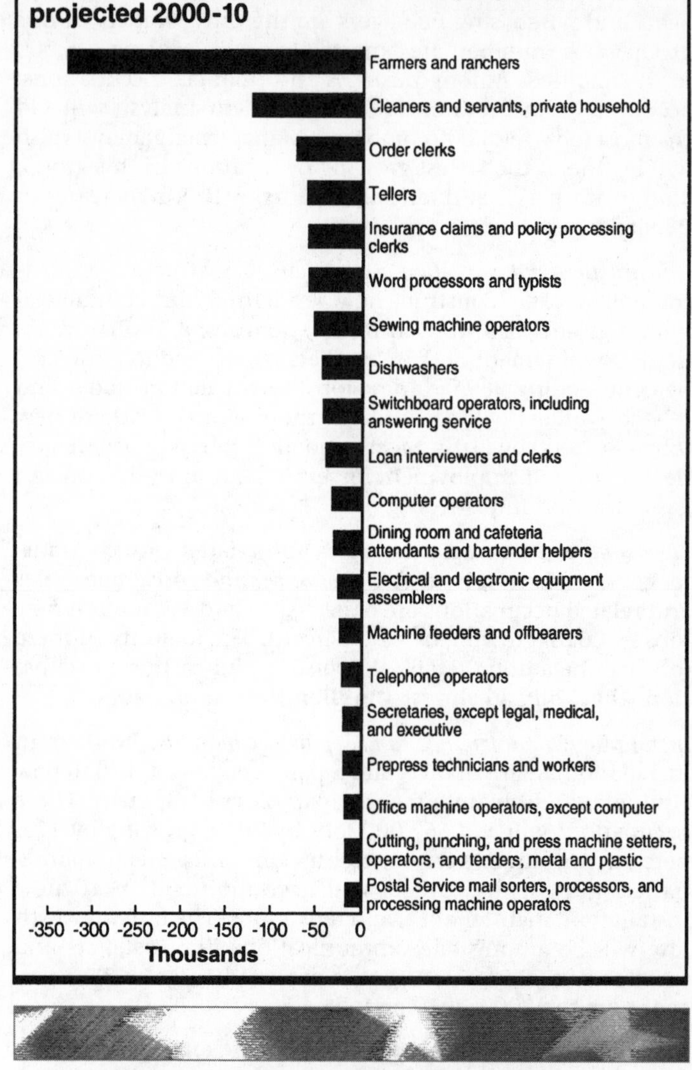

Total Job Openings

Job openings stem from both employment growth and replacement needs (chart 11). Replacement needs arise as workers leave occupations. Some transfer to other occupations while others retire, return to school, or quit to assume household responsibilities. Replacement needs are projected to account for 60 percent of the approximately 58 million job openings between 2000 and 2010. Thus, even occupations with little or no change in employment still may offer many job openings.

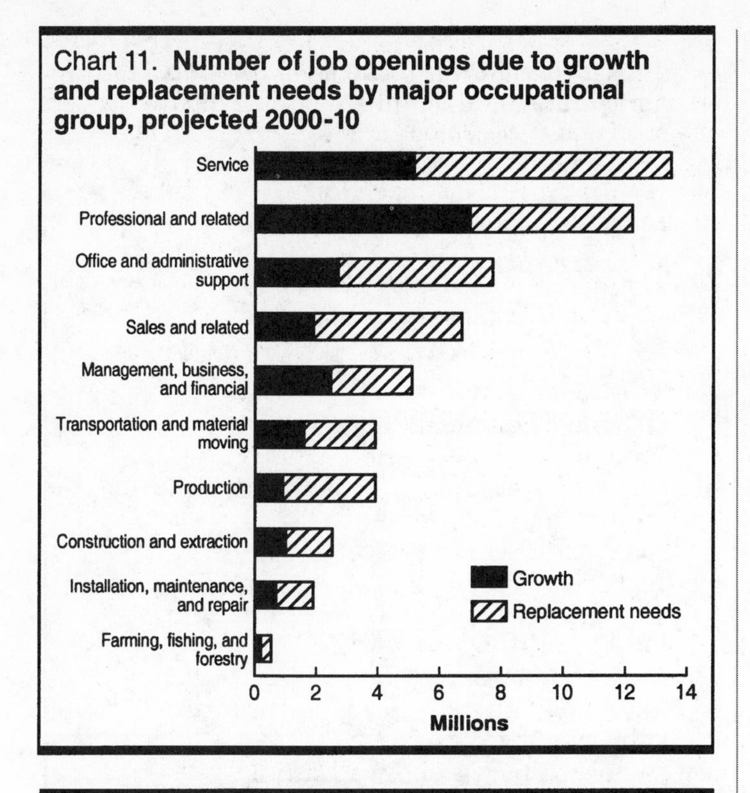

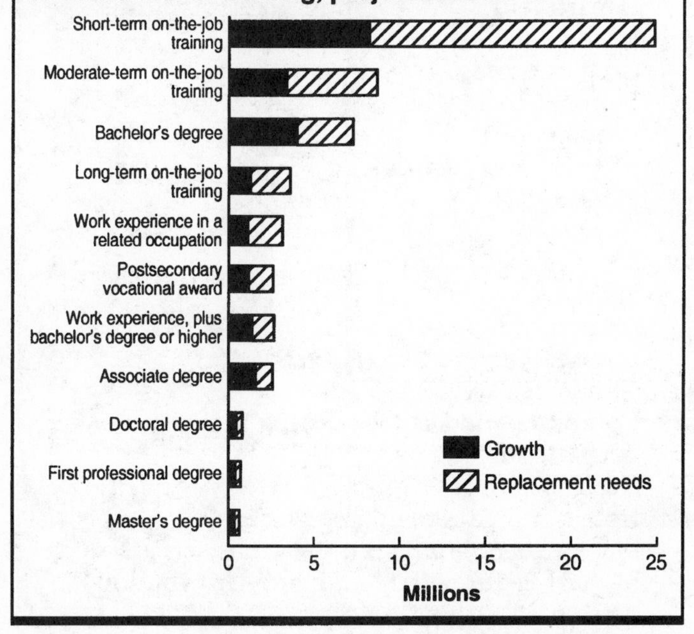

Professional and related occupations are projected to grow faster and add more jobs than any other major occupational group, with 7 million new jobs by 2010. Three-fourths of this job growth is expected among computer and mathematical occupations; healthcare practitioners and technical occupations; and education, training, and library occupations. With 5.2 million job openings due to replacement needs, professional and related occupations are the only major group projected to generate more openings from job growth than from replacement needs.

Due to high replacement needs, service occupations are projected to have the largest number of total job openings, 13.5 million. A large number of replacements are expected to arise as young workers leave food preparation and service occupations. Replacement needs generally are greatest in the largest occupations and in those with relatively low pay or limited training requirements.

Office automation will significantly affect many individual office and administrative support occupations. Overall, these occupations are projected to grow more slowly than the average, while some are projected to decline. Office and administrative support occupations are projected to create 7.7 million job openings over the 2000–2010 period, ranking third behind service and professional and related occupations.

Agriculture, forestry, and fishing occupations are projected to have the fewest job openings, approximately 500,000. Because job growth is expected to be slow, and levels of retirement and job turnover high, more than 80 percent of these projected job openings are due to replacement needs.

Employment in occupations requiring an associate degree is projected to increase 32 percent, faster than any other occupational group categorized by education or training. However, this category ranks only eighth among the 11 education and training categories in terms of job openings. The largest number of job openings will be among occupations requiring short-term on-the-job training (chart 12).

Almost two-thirds of the projected job openings over the 2000–2010 period will be in occupations that require on-the-job training and arise mostly from replacement needs. These jobs will account for 37.3 million of the projected 57.9 million total job openings through 2010. However, many of these jobs typically offer low pay and benefits; this is more true of jobs requiring only short-term on-the-job training, which will account for 24.8 million openings, than of the occupations in any other education or training category.

Jobs requiring a bachelor's degree, and which usually offer higher pay and benefits, will account for about 7.3 million job openings through 2010. Most of these openings will result from job growth.

Employment Trends in Major Industries

Comments

While hundreds of specialized industries exist, about 75 percent of all workers are employed in just 42 major ones. Space limitations do not allow us to provide you with detailed information on all industries; however, the following article presents a good overview of trends within industry types. It gives you important facts to consider in making your career plans. The article comes, with minor changes, from a U.S. Department of Labor publication titled *Career Guide to Industries*.

While you may not have thought much about it, the industry you work in will often be as important as the career you choose. For example, some industries pay significantly higher wages than others. So, if you found your way back to this article, read it over carefully and consider the possibilities.

For more information on a specific industry, look for the *Career Guide to Industries* in your library. A more widely available version of the same book is published by JIST under the title *Career Guide to America's Top Industries*. It is available through bookstores and many libraries. Here are the industries covered in the *Career Guide to Industries:*

Agriculture, Mining, and Construction
Agricultural Production
Agricultural Services
Construction
Mining and Quarrying
Oil and Gas Extraction
Manufacturing
Aerospace Manufacturing
Apparel and Other Textile Products
Chemical Manufacturing, Except Drugs
Drug Manufacturing
Electronic Equipment Manufacturing
Food Processing
Motor Vehicle and Equipment Manufacturing
Printing and Publishing
Steel Manufacturing
Textile Mill Products
Transportation, Communications, and Public Utilities
Air Transportation
Cable and Other Pay Television Services
Public Utilities
Radio and Television Broadcasting
Telecommunications
Trucking and Warehousing
Wholesale and Retail Trade
Department, Clothing, and Accessory Stores
Eating and Drinking Places
Grocery Stores
Motor Vehicle Dealers
Wholesale Trade
Finance and Insurance
Banking
Insurance
Securities and Commodities
Services
Advertising
Amusement and Recreation Services
Childcare Services
Computer and Data Processing Services
Educational Services
Health Services
Hotels and Other Lodging Places
Management and Public Relations Services
Motion Picture Production and Distribution
Personnel Supply Services
Social Services, Except Childcare
Government
Federal Government, Excluding the Postal Service
State and Local Government, Excluding Education and Hospitals

Overview

The U.S. economy is made up of industries with diverse characteristics. For each industry covered in the Department of Labor's publication *Career Guide to Industries*, detailed information is provided about specific characteristics: the nature of the industry, working conditions, employment, occupations in the industry, training and advancement, earnings, and outlook. This article provides an overview of these characteristics for the economy as a whole.

Nature of the Industry

Industries are defined by the goods and services they provide. Because workers in the United States produce such a wide variety of products and services, the types of industries in the U.S. economy range widely, from aerospace manufacturing to motion picture production. Although many of these industries are related, each industry has a unique combination of occupations, production techniques, inputs and outputs, and business characteristics. Understanding the nature of the industry is important, because it is this unique combination that determines working conditions, educational requirements, and the job outlook for the industry.

Industries are comprised of many different places of work, called *establishments*, which range from large factories and office complexes employing thousands of workers to small businesses employing only a few workers. Not to be confused with companies, which are legal entities, establishments are physical locations where people work, such as the branch office of a bank. Establishments that produce similar goods or services are grouped together into *industries*. Industries that produce related types of goods or services are, in turn, grouped together into *major industry divisions*. These are further grouped into the *goods-producing sector* (agriculture, mining, and construction; and manufacturing) or the *service-producing sector* (transportation, communications, and public utilities; wholesale and retail trade; finance and insurance; services; and government).

Distinctions within industries are also varied. Each industry is comprised of a number of subdivisions, which are determined largely by differences in production processes. An easily recognized example of these distinctions is in the food processing industry, which is made up of subdivisions that produce meat products, preserved fruits and vegetables, bakery items, beverages, and dairy products, among others. Each of these subdivisions requires workers with varying skills and employs unique production techniques. Another example of these distinctions is in public utilities, which employs workers in establishments that provide electricity, sanitary services, water, and natural gas. Working conditions and establishment characteristics often differ widely in each of these smaller subdivisions.

There were more than 7 million business establishments in the United States in 1999. The average size of these establishments varies widely across industries. Among industry divisions, manufacturing included many industries having among the highest employment per establishment in 1999. For example, the aerospace and steel manufacturing industries each averaged 200 or more employees per establishment.

Most establishments in the wholesale trade, retail trade, finance, and services industries are small, averaging fewer than 20 employees per establishment. Exceptions are the scheduled air transportation industry with 166.8 employees, and colleges, universities, and professional schools with 433.4. In addition, wide differences within industries can exist. Hospitals, for example, employ an average of 717.5 workers, while doctors' offices employ an average of 8.7. Similarly, although there is an average of 13 employees per establishment for all of retail trade, department stores employ an average of 164.3 people.

Establishments in the United States are predominantly small; 54.2 percent of all establishments employed fewer than 5 workers in 1999. The medium-sized to large establishments employ a greater proportion of all workers. For example, establishments that employed 50 or more workers accounted for only 5.4 percent of all establishments, yet employed 59 percent of all workers. The large establishments—those with more than 500 workers—accounted for only 0.3 percent of all establishments, but employed 20.3 percent of all workers. Table 1 presents the percent distribution of employment according to establishment size.

Establishment size can play a role in the characteristics of each job. Large establishments generally offer workers greater occupational mobility and advancement potential, whereas small establishments may provide their employees with broader experience by requiring them to assume a wider range of responsibilities. Also, small establishments are distributed throughout the nation; every locality has a few small businesses. Large establishments, in contrast, employ more workers and are less common, but they play a much more prominent role in the economies of the areas in which they are located.

TABLE 1

Percent Distribution of Establishments and Employment in All Industries by Establishment Size, 1999

Establishment Size (Number of Workers)	Establishments	Employment
Total	100.0	100.0
1–4	54.2	5.9
5–9	19.4	8.1
10–19	12.5	10.7
20–49	8.5	16.3
50–99	2.9	12.8
100–249	1.7	16.4
250–499	0.4	9.6
500–999	0.2	7.2
1,000 or more	0.1	13.2

SOURCE: Department of Commerce, County Business Patterns, 1999

Working Conditions

Just as the goods and services of each industry are different, working conditions in industries can vary significantly. In some industries, the work setting is quiet, temperature-controlled, and virtually hazard free. Other industries are characterized by noisy, uncomfortable, and sometimes dangerous work environments. Some industries require long workweeks and shift work; in many industries, standard 35- to 40-hour workweeks are common. Still other industries can be seasonal, requiring long hours during busy periods and abbreviated schedules during slower months. These varying conditions usually are determined by production processes, establishment size, and the physical location of work.

One of the most telling indicators of working conditions is an industry's injury and illness rate. Among the most common incidents causing work-related injury or illness are overexertion, being struck by an object, and falling on the same level. In 1999, approximately 5.7 million nonfatal injuries and illnesses were reported throughout private industry. Among major industry divisions, manufacturing had the highest rate of injury and illness—9.2 cases for every 100 full-time workers. Finance, insurance, and real estate had the lowest rate—1.8 cases. About 5,900 work-related fatalities were reported in 2000. Transportation incidents, contact with objects and equipment, assaults and violent acts, and falls were the most common events resulting in fatal injuries. Table 2 presents industries with the highest and lowest rates of nonfatal injury and illness.

Table 2

Nonfatal Injury and Illness Rates of Selected Industries, 2000

Industry	Cases Per 100 Full-Time Employees
All industries	6.3
High rates	
Transportation equipment	13.7
Transportation by air	13.3
Lumber and wood products	13.0
Primary metal	12.9
Food and kindred products	12.7
Fabricated metal products	12.6
Low rates	
Insurance carriers	1.9
Engineering and management	1.7
Depository institutions	1.5
Legal services	1.0
Insurance agents and brokers	0.9
Securities and commodities	0.6

Work schedules are another important reflection of working conditions, and the operational requirements of each industry lead to large differences in hours worked and in part-time versus full-time status. In retail trade, 28.9 percent of employees worked part time in 2000, compared with only 4.4 percent in manufacturing. Table 3 presents industries having relatively high and low percentages of part-time workers.

Table 3

Percent of Part-Time Workers in Selected Industries, 2000

Industry	Percent Part-Time
All industries	15.3
Many part-time workers	
Eating and drinking places	37.8
Department, clothing, and accessory stores	31.3
Grocery stores	30.4
Amusement and recreation services	29.3
Child-care services	28.2
Motion picture production and distribution	23.0
Educational services	21.9
Social services	20.9
Few part-time workers	
Public utilities	3.2
Chemical manufacturing, except drugs	3.0
Drug manufacturing	2.6
Electronic equipment manufacturing	2.4
Mining and quarrying	1.9
Aerospace manufacturing	1.8
Steel manufacturing	1.5
Motor vehicle and equipment manufacturing	1.5

The low proportion of part-time workers in some manufacturing industries often reflects the continuous nature of the production processes that makes it difficult to adapt the volume of production to short-term fluctuations in product demand. Once begun, it is costly to halt these processes; machinery must be tended and materials must be moved continuously. For example, the chemical manufacturing industry produces many different chemical products through controlled chemical reactions. These processes require chemical operators to monitor and adjust the flow of materials into and out of the line of production. Because production may continue 24 hours a day, 7 days a week, under the watchful eyes of chemical operators who work in shifts, full-time workers are more likely to be employed.

Retail trade and service industries, on the other hand, have seasonal cycles marked by various events, such as school openings or important holidays, that affect the hours worked. During busy times of the year, longer hours are common, whereas slack periods lead to cutbacks and shorter workweeks. Jobs in these industries are generally appealing to students and others who desire flexible, part-time schedules.

Employment

The total number of jobs in the United States in 2000 was 145.6 million. This included 11.5 million self-employed workers, 169,000 unpaid workers in family businesses, and more than 133.9 million wage and salary workers—including primary and secondary job holders. The total number of jobs is projected to increase to 167.8 million by 2010, and wage and salary jobs are projected to account for more than 155.9 million of them.

As shown in Table 4, although wage and salary jobs are the vast majority of all jobs, they are not evenly divided among the various industries. The services major industry division is the largest source of employment, with more than 50 million workers, followed by the wholesale and retail trade and manufacturing major industry divisions. Wage and salary employment ranged from only 216,000 in cable and other pay television services to 11.8 million in educational services. Three industries—educational services, health services, and eating and drinking places—together accounted for almost 31 million jobs, or nearly a quarter of the nation's employment.

Table 4

Wage and Salary Employment in Selected Industries, 2000, and Projected Change, 2000 to 2010

Industry	2000		2010		2000–2010	
	Employment (in 1000s)	Percent Distribution	Employment (in 1000s)	Percent Distribution	Percent Change	Employment Change
All industries	133,896	100.0	155,872	100.0	16.4	21,977
Goods-producing industries	27,984	20.9	29,728	19.1	6.2	1,745
Agriculture, mining, and construction	9,514	7.1	10,682	0.0	12.3	1,167
Agricultural production	1,120	0.8	1,092	0.7	–2.5	–28
Agricultural services	1,099	0.8	1,524	1.0	38.6	425
Construction	6,698	5.0	7,522	4.8	12.3	825
Mining and quarrying	231	0.2	199	0.1	–14.0	–32
Oil and gas extraction	311	0.2	289	0.2	–7.3	–23
Manufacturing	18,469	13.8	19,047	12.2	3.1	577
Aerospace manufacturing	551	0.4	655	0.4	18.9	104
Apparel and other textile products	633	0.5	530	0.3	–16.3	–103
Chemical manufacturing, except drugs	723	0.5	691	0.4	–4.5	–32
Drug manufacturing	315	0.2	390	0.3	23.8	75
Electronic equipment manufacturing	1,554	1.2	1,657	1.1	6.6	103
Food processing	1,684	1.3	1,634	1.0	–3.0	–50
Motor vehicle and equipment manufacturing	1,013	0.8	1,100	0.7	8.6	87
Printing and publishing	1,547	1.2	1,545	1.0	–0.2	–3
Steel manufacturing	225	0.2	176	0.1	–21.6	–49
Textile mill products	529	0.4	500	0.3	–5.4	–29
Service-producing industries	105,912	79.1	126,144	80.9	19.1	20,232
Transportation, communications, and public utilities	7,019	5.2	8,274	5.3	17.9	1,255
Air transportation	1,281	1.0	1,600	1.0	24.9	319
Cable and other pay television services	216	0.2	325	0.2	50.6	109
Public utilities	851	0.6	893	0.6	4.9	42
Radio and television broadcasting	255	0.2	280	0.2	9.7	25
Telecommunications	1,168	0.9	1,311	0.8	12.2	143
Trucking and warehousing	1,856	1.4	2,262	1.5	21.9	407
Wholesale and retail trade	30,331	22.7	34,200	21.9	12.8	3,869
Department, clothing, and accessory stores	4,030	3.0	4,198	2.7	4.2	168
Eating and drinking places	8,114	6.1	9,600	6.2	18.3	1,486
Grocery stores	3,107	2.3	3,281	2.1	5.6	174
Motor vehicle dealers	1,221	0.9	1,366	0.9	11.9	145
Wholesale trade	7,024	5.2	7,800	5.0	11.1	776
Finance, insurance, and real estate	7,560	5.6	8,247	5.3	9.1	687
Banking	2,029	1.5	1,999	1.3	–1.5	–31
Insurance	2,346	1.8	2,497	1.6	6.4	151
Securities and commodities	748	0.6	900	0.6	20.3	152
Services	50,764	37.9	64,483	41.4	27.0	13,719
Advertising	302	0.2	400	0.3	32.5	98
Amusement and recreation services	1,728	1.3	2,325	1.5	34.5	597
Childcare services	712	0.5	1,010	0.6	41.9	298
Computer and data processing services	2,095	1.6	3,900	2.5	86.2	1,805
Educational services	11,797	8.8	13,400	8.6	13.6	1,603
Health services	11,065	8.3	13,882	8.9	25.5	2,817
Hotels and other lodging places	1,912	1.4	2,167	1.4	13.3	255
Management and public relations services	1,090	0.8	1,550	1.0	42.2	460
Motion picture production and distribution	287	0.2	369	0.2	28.7	82
Personnel supply services	3,887	2.9	5,800	3.7	49.2	1,913
Social services, except childcare	2,191	1.6	3,118	2.0	42.3	927
Government	10,238	7.6	10,940	7.0	6.9	702
Federal government	1,917	1.4	1,772	1.1	–7.6	–145
State and local government	7,461	5.6	8,318	5.3	11.5	856

Although workers of all ages are employed in each industry, certain industries tend to employ workers of distinct age groups. For the reasons mentioned previously, retail trade employs a relatively high proportion of younger workers to fill part-time and temporary positions. The manufacturing sector, on the other hand, has a relatively high median age because many jobs in the sector require a number of years to learn and perfect skills that do not easily transfer to other firms. Also, manufacturing employment has been declining, providing fewer opportunities for younger workers to get jobs. As a result, almost one-third of the workers in retail trade were 24 years of age or younger in 2000, compared with only 10 percent of workers in manufacturing. Table 5 contrasts the age distribution of workers in all industries with the distributions in five very different industries.

TABLE 5

Percent Distribution of Wage and Salary Workers by Age Group, Selected Industries, 2000

Industry	Age Group			
	16 to 24	25 to 44	45 to 64	65 to 90
All industries	15	50	32	3
Textile mill products manufacturing	9	47	41	3
Public utilities	5	51	42	1
Eating and drinking places	45	39	15	2
Computer and data processing services	11	67	22	1
Educational services	11	43	43	3

Employment in some industries is concentrated in one region of the country, and job opportunities in these industries should be best in the states in which their establishments are located. Such industries are often located near a source of raw materials upon which the industries rely. For example, oil and gas extraction jobs are concentrated in Texas, Louisiana, and Oklahoma; many textile mill products manufacturing jobs are found in North Carolina, Georgia, and South Carolina; and a significant proportion of motor vehicle and equipment manufacturing jobs are located in Michigan. On the other hand, some industries such as grocery stores and educational services have jobs distributed throughout the nation, reflecting population density in different areas.

Occupations in the Industry

As mentioned above, the occupations found in each industry depend on the types of services provided or goods produced. For example, construction companies require skilled trades workers to build and renovate buildings, so these companies employ a large number of carpenters, electricians, plumbers, painters, and sheet metal workers. Other occupations common to the construction sector include construction equipment operators and mechanics, installers, and repairers. Retail trade, on the other hand, displays and sells manufactured goods to consumers, so this sector hires numerous sales clerks and other workers, including nearly 5 out of 6 cashiers. Table 6 shows the major industry divisions and the occupational groups which predominate in the division.

The nation's occupational distribution clearly is influenced by its industrial structure, yet there are many occupations, such as general manager or secretary, which are found in all industries. In fact, some of the largest occupations in the U.S. economy are dispersed across many industries. Because nearly every industry relies on administrative support, for example, this occupational group is the largest in the nation (see Table 7). Other large occupational groups include service occupations, professional specialty workers, and operators, fabricators, and laborers.

TABLE 6

Industry Divisions and Largest Occupational Concentration, 2000

Industry Division	Largest Occupational Group	Percent of Wage and Salary Jobs
Agriculture, forestry, and fishing	Farming, forestry, and fishing	45.5
Mining	Construction and extraction	33.6
Construction	Construction and extraction	67.6
Manufacturing	Production	51.7
Transportation, communications, and public utilities	Transportation and material moving	36.0
Wholesale and retail trade	Sales and related	32.4
Finance, insurance, and real estate	Office and administrative support	45.3
Services	Professional and related	36.9
Government	Office and administrative support	26.8

TABLE 7

Total Employment in Broad Occupational Groups, 2000, and Projected Change, 2000–2010 (Employment in Thousands)

Occupational Group	Employment, 2000	Percent Change, 2000–2010
Total, all occupations	145,571	15.2
Management, business, and financial	15,519	13.6
Professional and related	26,758	26.0
Service	26,075	19.5
Sales and related	15,513	11.9
Office and administrative support	23,882	9.1
Farming, fishing, and forestry	1,406	5.3
Construction and extraction	7,451	13.3
Installation, maintenance, and repair	5,820	11.4
Production	13,060	5.8
Transportation and material moving	10,088	15.2

Training and Advancement

Workers prepare for employment in many ways, but the most fundamental form of job training in the United States is a high school education. Fully 87.5 percent of the nation's workforce possessed a high school diploma or its equivalent in 2000. As the premium placed on education in today's economy increases, workers are responding by pursuing additional training. In 2000, 28.8 percent of the nation's workforce had some college or an associate degree, while an additional 27.5 percent continued in their studies and attained a bachelor's degree or higher. In addition to these types of formal education, other sources of qualifying training include formal company training, informal on-the-job training, correspondence courses, the Armed Forces, and friends, relatives, and other nonwork-related training.

The unique combination of training required to succeed in each industry is determined largely by the industry's occupational composition. For example, machine operators in manufacturing generally need little formal education after high school, but sometimes complete considerable on-the-job training. These requirements by major industry division are clearly demonstrated in Table 8.

Persons with no more than a high school diploma accounted for about 63.8 percent of all workers in agriculture, forestry, and fishing; 64.7 percent in construction; 53.4 percent in manufacturing; 44.8 percent in wholesale trade; and 57.4 in retail trade. On the other hand, workers who had acquired at least some training at the college level accounted for 72.4 percent of all workers in government; 71.4 percent in finance, insurance, and real estate; and 75.1 percent in professional and related services.

Education and training are also important factors in the variety of advancement paths found in different industries. In general, workers who complete additional on-the-job training or education increase their chances of being promoted. In much of the manufacturing sector, for example, production workers who receive training in management and computer skills increase their likelihood of being promoted to supervisors. Other factors which affect advancement opportunities include the size of the establishment or company, institutionalized career tracks, and the skills and aptitude of each worker. Each industry has some unique advancement paths, so persons who seek jobs in particular industries should be aware of how these paths may later shape their careers.

TABLE 8

Percent Distribution of Highest Grade Completed or Degree Received by Industry Division, 2000

Industry Division	Bachelor's Degree or Higher	Some College or Associate Degree	High School Graduate or Equivalent	Less than 12 Years or No Diploma
Agriculture, forestry, and fishing	14	22	34	30
Mining	17	26	45	12
Construction	10	25	44	21
Manufacturing	21	26	39	15
Transportation, communications, and public utilities	21	35	37	8
Wholesale trade	25	31	34	11
Retail trade	14	30	36	21
Finance, insurance, and real estate	37	33	25	4
Business and repair services	29	29	29	12
Personal services	13	28	38	22
Entertainment and recreation services	25	30	28	17
Professional and related services	48	27	20	5
Government	36	36	25	3

Earnings

As with other characteristics, earnings differ from industry to industry, the result of a highly complicated process that relies on a number of factors. For example, earnings may vary due to the occupations in the industry, average hours worked, geographical location, industry profits, union affiliation, and educational requirements. In general, wages are highest in metropolitan areas to compensate for the higher cost of living. And, as would be expected, industries that employ relatively few unskilled minimum-wage or part-time workers tend to have higher earnings.

A good illustration of these differences is shown by the earnings of all wage and salary workers in petroleum refining, which averaged $1,099 a week in 2000, and those in eating and drinking places, where the weekly average was $177. These differences are so large because petroleum-refining establishments employ more highly skilled, full-time workers, while eating and drinking places employ many lower skilled, part-time workers. In addition, many workers in eating and drinking places are able to supplement their low wages with money they receive as tips, which is not included in the industry wages data. Table 9 highlights the industries with the highest and lowest average weekly earnings.

Table 9

Average Weekly Earnings of Production or Nonsupervisory Workers on Private Nonfarm Payrolls in Selected Industries, 2000

Industry	Earnings
All industries	$474
Industries with high earnings	
Petroleum refining	1,099
Pipelines, except natural gas	956
Aircraft and parts	901
Computer and data processing services	897
Electric, gas, and sanitary services	895
Coal mining	871
Blast furnaces and basic steel products	870
Securities and commodities brokers	841
Engineering and architectural services	821
Motion picture production and services	811
Industries with low earnings	
Agricultural services	379
Nursing and personal care facilities	349
Apparel and other textile products manufacturing	338
Drug stores and proprietary stores	323
Hotels and motels	298
Food stores	281
Department stores	278
Amusement and recreation services	262
Child daycare services	258
Eating and drinking places	177

Employee benefits, once a minor addition to wages and salaries, continue to grow in diversity and cost. In addition to traditional benefits—including paid vacations, life and health insurance, and pensions—many employers now offer various benefits to accommodate the needs of a changing labor force. Such benefits are child care; employee assistance programs that provide counseling for personal problems; and wellness programs that encourage exercise, stress management, and self-improvement. Benefits vary among occupational groups, full- and part-time workers, public and private sector workers, regions, unionized and nonunionized workers, and small and large establishments. Data indicate that full-time workers and those in medium-size and large establishments—those with 100 or more workers—receive better benefits than part-time workers and those in smaller establishments.

Union penetration of the workforce varies widely by industry, and it also may play a role in earnings and benefits. In 2000, about 15 percent of workers throughout the nation were union members or were covered by union contracts. As Table 10 demonstrates, union affiliation of workers varies widely by industry. Approximately one-third of the workers in government and transportation, communications, and public utilities are union members or are covered by union contracts, compared with fewer than 4 percent in business and repair services; agriculture, forestry, and fishing; and finance, insurance, and real estate.

Table 10

Union Members and Other Workers Covered by Union Contracts as a Percent of Total Employment in Major Industry Divisions, 2000

Industry Division	Union Members or Covered by Union Contracts
Total, all industries	14.9
Government	35.7
Transportation, communications, and public utilities	31.9
Construction	20.4
Manufacturing	15.7
Mining	11.6
Entertainment and recreation services	10.5
Personal services	9.8
Wholesale trade	5.6
Retail trade	5.2
Professional and related services	5.2
Business and repair services	3.8
Agriculture, forestry, and fishing	3.3
Finance, insurance, and real estate	2.8

Outlook

Total employment in the United States is projected to increase by about 15 percent over the 2000–2010 period. Employment growth, however, is only one source of job openings; the total number of openings in any industry also depends on the industry's current employment level and its need to replace workers who leave their jobs. Throughout the economy, in fact, replacement needs will create more job openings than will employment growth. Employment size is a major determinant of job openings—larger industries generally provide more openings. The occupational composition of an industry is another factor. Industries with high concentrations of professional, technical, and other jobs that require more formal education—occupations in which workers tend to leave their jobs less frequently—generally have fewer openings resulting from replacement needs. On the other hand, more replacement openings generally occur in industries with high concentrations of service, labor, and other jobs that require little formal education and have lower wages, because workers in these jobs are more likely to leave their occupations.

Employment growth is determined largely by changes in the demand for the goods and services produced by an industry, worker productivity, and foreign competition. Each industry is affected by a different set of variables that impacts the number and composition of jobs that will be available. Even within an industry, employment in different occupations may grow at different rates. For example, changes in technology, production methods, and business practices in an industry might eliminate some jobs while creating others. Some industries may be growing rapidly overall, yet opportunities for workers in occupations that are adversely affected by technological change could be stagnant. Similarly, employment of some occupations may be declining in the economy as a whole yet may be increasing in a rapidly growing industry.

As shown in Table 4, employment growth rates over the next decade will vary widely among industries. Employment in goods-producing industries is expected to increase as growth in agricultural services, construction, and manufacturing is partially offset by declining employment in agricultural production and mining and quarrying. Rapid growth in agricultural services will be driven by its landscaping and veterinary services components. Growth in construction employment will stem from new factory construction as existing facilities are modernized; from new school construction, reflecting growth in the school-age population; and from infrastructure improvements such as road and bridge construction. Employment in agricultural production and mining and quarrying is expected to decline due to laborsaving technology. Reliance on foreign sources of energy also is expected to play a role in employment declines in mining and in oil and gas extraction.

Manufacturing employment will increase slightly, as strong demand continues for high-technology electrical goods and pharmaceuticals despite improvements in production technology and rising imports. Apparel manufacturing is projected to lose about 103,000 jobs over the 2000–2010 period—more than any other manufacturing industry—due primarily to increasing imports. But other manufacturing industries with strong domestic markets and export potential are expected to experience increases in employment. The drug manufacturing and aerospace manufacturing industries are two examples. Sales of drugs are expected to increase with growth in the population, particularly among the elderly, and the introduction of new drugs to the market. An increase in air traffic, coupled with the need to replace aging aircraft will generate strong sales for commercial aircraft. Both drug and aerospace manufacturing also have large export markets.

Growth in overall employment will result primarily from growth in service-producing industries over the 2000–2010 period, almost all of which are expected to have increasing employment. Rising employment in these industries will be driven by services industries, the largest and fastest-growing major industry group, which are projected to provide almost 2 out of 3 new jobs across the nation. Health, education, and personnel supply services will account for 6.3 million of these new jobs. In addition, employment in the nation's fastest-growing industry—computer and data processing services—is expected to nearly double, adding another 1.8 million jobs. Job growth in the services sector will result from overall population growth, the rise in the elderly and school-age populations, and the trend toward contracting out for computer, personnel, and other business services.

Wholesale and retail trade is expected to add 3.9 million jobs over the coming decade. More than 776,000 of these jobs will arise in wholesale trade, reflecting growth both in trade and in the overall economy. Retail trade is expected to add 3.1 million jobs over the 2000–2010 period, the result of increases in both population and personal income. Although most retail stores are expected to add employees, nonstore retailers will experience the fastest growth rate—35 percent—as electronic commerce and mail-order sales account for an increasing portion of retail sales. Eating and drinking places will have the largest number of new jobs, nearly 1.5 million.

Employment in transportation, communications, and public utilities is projected to increase by nearly 1.26 million new jobs. The trucking and warehousing industry will have the biggest increase—407,000 jobs. Trucking industry growth will be fueled by growth in the volume of goods that need to be shipped as the economy expands. Air transportation is expected to generate nearly 319,000 jobs. Air transportation will expand as consumer and business demand increases, reflecting a growing population and increased business activity. Demand for new telecommunications services, such as Internet and wireless communications, will lead to an expansion of the telecommunications infrastructure. Employment growth is projected to add 143,000 jobs. While radio and television broadcasting will show average growth due to consolidations in the industry, employment of cable and other pay-television companies will increase by 51 percent as they upgrade their systems to deliver a wider array of communication and programming services.

Overall employment growth in finance, insurance, and real estate is expected to be around 9 percent, with close to 687,000 jobs added by 2010. Securities and commodities will be the fastest-growing industry in this group, adding more than 152,000 jobs. A growing interest in investing and the popularity of 401(k) and other pension plans are fueling increases in this industry. In contrast, employment in the largest industry in this group, banking, will decrease by 1.5 percent, or –31,000 jobs, as technological advances and the increasing use of electronic banking reduce the need for large administrative support staffs. Nondepository institutions—including personal and business credit institutions, as well as mortgage banks—are expected to grow as fast as the average, adding 111,800 jobs. Insurance carriers will grow more slowly than average, increasing by only 42,600 jobs.

All 702,000 new government jobs are expected to arise in state and local government, reflecting growth in the population and its demand for public services. In contrast, the federal government is expected to lose more than 145,000 jobs over the 2000–2010 period, as efforts continue to cut costs by contracting out services and giving states more responsibility for administering federally funded programs.

In sum, recent changes in the economy are having far-reaching and complex effects on employment in all industries. Jobseekers should be aware of these changes, keeping alert for developments that affect job opportunities in each industry and the variety of occupations in each industry.